Embedded Systems

Architecture, Programming and Design

Series Foreword

The McGraw-Hill Companies has been a leader in providing trusted information and analysis for well over a century. From the Industrial Revolution to the Internet Revolution, The McGraw-Hill Companies has filled a critical need for information and insight by helping individuals and businesses in the field of Engineering.

As early as 1910, the McGraw-Hill book company was making a difference on college campuses with the publication of its first series, Electrical Engineering Texts, outlined and edited by Professor Harry E. Clifford of Harvard University. McGraw-Hill's Electrical Engineering textbooks have shaped engineering curricula worldwide. I am pleased that I have been invited to be the Global Series Editor in Electrical Engineering, helping to shape how the next generation of electrical engineering students around the globe will learn.

As advances in networking and communications bring the global academic community even closer together, it is essential that textbooks recognize and respond to this shift. It is in this spirit that we will publish textbooks in the McGraw-Hill Core Concepts in Electrical Engineering Series. The series will offer textbooks for the global electrical engineering curriculum that are reasonably priced, innovative, dynamic, and will cover fundamental subject areas studied by Electrical and Computer Engineering students. Written with a global perspective and presenting the latest in technological advances, these books will give students of all backgrounds a solid foundation in key engineering subjects.

Embedded Systems by Raj Kamal of Devi Ahilya University, India, is a textbook for undergraduate engineering students. This book describes an embedded system as one with embedded hardware and software and describes the fundamentals of the architecture, design, and applications for these systems. A full treatment of software and hardware engineering is provided and applications are described. Case studies are provided as well as exercises and a review of questions.

This book has been reviewed and assessed for use in engineering classrooms at all levels. Like Embedded Systems, each book in the Core Concepts series presents a comprehensive, straightforward, and accurate treatment of an important subject in Electrical & Computer Engineering. With their clear approach, contemporary technology, and international perspective, Core Concepts books are an unmistakable choice for professors wanting understandable, concise engineering textbooks that adhere to the standards of The McGraw-Hill Companies.

–**Richard C. Dorf,**
University of California, Davis
Series Editor, The Core Concepts Series in Electrical and Computer Engineering

Embedded Systems

Architecture, Programming and Design

Raj Kamal

Institute of Computer Sciences and Electronics
Devi Ahilya University
Indore
and
Arulmigu Kalasalingam College of Engineering
Krishnankoil

Higher Education

Boston   Burr Ridge, IL   Dubuque, IA   New York   San Francisco   St. Louis
Bangkok Bogotá   Caracas   Kuala Lumpur   Lisbon   London   Madrid   Mexico City
Milan   Montreal   New Delhi   Santiago   Seoul   Singapore   Sydney   Taipei   Toronto

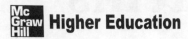
ISBN 978–0–07–340456–1
MHID 0–07–340456–X

Senior Sponsoring Editor: *Michael S. Hackett*
Developmental Editor: *Rebecca Olson*
Executive Marketing Manager: *Michael Weitz*
Project Coordinator: *Melissa M. Leick*
Senior Production Supervisor: *Kara Kudronowicz*
Associate Media Producer: *Christina Nelson*
Senior Designer: *David W. Hash*
Cover Designer: *Rokusek Design*
(USE) Cover Image: *©PhotoDisc*
Compositor: *Techbooks*
Printer: *Book-mart Press*

This book was previously published by Tata McGraw-Hill Publishing Company Limited, New Delhi, India, copyright © 2003.

Library of Congress Cataloging-in-Publication Data

Kamal, Raj.
    Embedded systems / Raj Kamal.—1st ed.
        p.   cm.
    Includes index.
    ISBN 978–0–07–340456–1 — ISBN 0–07–340456–X (alk. paper)
    1. Embedded computer systems. I. Title.

TK7895.E42K35   2008
004.16-dc22   2006041934
                                         CIP

www.mhhe.com

About the Author

Dr. Raj Kamal (b. 1949) did his M.Sc. at the age of 17. He published his first research paper in a UK journal at the age of 18, and his first program, written in FORTRAN that ran at ICT 1904, also at the age of 18 and completed his Ph.D. (1972) from Indian Institute of Technology. He is a strongly motivated ex-IITian with strong academic and research credentials. He has teaching and programming skills in Java, C, C++ and FORTRAN and assembly languages for 8086, 8051, 68HC11, and 8096. He has developed sound knowledge of web protocols, Internet and web technologies. He has also developed expertise in Embedded Processors and Systems, Multiprocessors, Microcontrollers and OOPs, Computer Architecture, Networking and Data Structure. He has about 30 years of teaching experience, guided 9 Ph.D.s, and published 43 research papers in International and 39 in National journals. Due to his constant drive for understanding emerging technologies and his passion for acquiring the latest knowledge and disseminating it, he has also authored several high technology area books. Dr Raj Kamal is a Senior Professor in Computer Science and Electronics, and teaches at Devi Ahilya University at Indore. Currently, he is associated with the TIFAC Center of Relevance and Excellence in Network Engineering, A K College of Engineering, Krishnankoil, Tamil Nadu.

Dedicated to
my mother

Ms. Rajeshwari Devi

on her seventieth birthday,
who taught me the ABCs in childhood that
enable me to write a few words today

Contents

Preface

People need embedded systems to play **video games** and to operate **vending machines**! They also need embedded systems for smart home appliances, such as microwaves, televisions, stereos, and so on. Drivers need embedded systems when using the cruise control in the car. Organisations require embedded systems for network systems and products. There are uncountable examples of the applications of embedded systems.

Three **classics** that not only influenced this author greatly but also provided a deep insight into the subject and fueled his interest in microprocessor- and microcontroller-based embedded systems are as follows: Foremost is the classic *Design with Microcontrollers and Microcomputers* by J.B. Peatman, McGraw-Hill (1988). The second is *Microcontrollers Architecture, Implementation and Programming - HD44795, MC68HC11, MCS-51, 80960CA* by Kenneth Hintz and Daniel Tabak, a McGraw-Hill International Edition (1992). The third is *Advanced Microprocessors* by Daniel Tabak, a McGraw-Hill International Edition (1995).

In the early days, embedded systems were designed using microprocessors like 8085. Applications were simple: for example, temperature-monitoring systems, data acquisition systems using ADC and DAC, music systems using appropriate interfaces and simple robotic systems using stepper motor interfaces. These are now not even referred to as embedded systems.

Since the early eighties, small-scale embedded systems have used microcontrollers from General Instruments Corporation, using their late seventies microcontrollers PIC 16xxx, Motorola microcontrollers 68HC05 and 08, and Intel microcontrollers of 8031 families. The remote of a television, watches, washing machines, ovens, calculators, and video games are all examples of commonly used small scale embedded systems. The advent of microcontrollers of Intel 8051/52, Motorola 68HC11/12, Intel 80196 and 80960 families in the late eighties gave a new dimension to the use of embedded systems hardware.

In the last few years, the advent of technology that embeds low level and high level processing hardware elements and application-specific processors into a chip has given an added dimension to embedded systems that are multiprocessor systems, on a single VLSI chip (called System-on-Chip) and are smart as well as highly sophisticated. A simple example is a smart card, and the latest, typical example of a sophisticated system is "Smart Cameras as Embedded Systems" developed at the Princeton University Embedded System Group and reported by Wayne Wolf and his team in *IEEE Micro*, IEEE Computer Society, September 2002 issue.

Embedded systems have definitions that vary with time as well as perception. *An embedded system can be defined as one that has computer hardware with software embedded in it as one of its most important components.* An embedded system is a dedicated, computer-based system for an application or product. *It addresses the issue of the response time constraints of various tasks of the system.* An embedded system may either be an independent system or a part of a larger system. Its software usually embeds in ROM(s) (Read Only Memory). Thus, it does not need secondary memories as in a computer.

While writing this book, the author had in mind the graduate engineering students, readers and keen learners who are the budding embedded system engineers of today and perhaps the master designers of these systems tomorrow. Also in mind were young software engineers interested in working on embedded software and real-time programming projects. **This book is intended as a textbook for students and as a reference for engineers to explain the concepts necessary for designing high-performance response-time constrained sophisticated systems.**

The author expects readers of this book to first learn about the embedded system architecture, its basic hardware and software elements, programming models and software engineering practices that are used during the system development process, and then learn the software techniques to embed codes into the systems. Readers will develop systems that make optimum use of the available system resources: *processor*, *memory*, *ports*, *devices* and *power*. This book has been written to fulfill these expectations.

What is the range of topics, innovative technologies and tools for designing an embedded system, simple or complex? This book will help the readers to understand these with ease. Next, students will develop useful projects on applications in the area of their choice, such as networking, communication, automobile electronics, data acquisition and storing, serving, processing and securing information, smart robots, real-time control and tracking systems, biomedical systems and sounds, images and video real-time processing, filtering, compressing and encrypting systems.

The organization of the chapters is as follows:

Chapter 1 gives a detailed introduction to embedded systems. Embedded systems hardware consists of a processor, memory devices, I/O devices and basic hardware units—power supply, clock and reset circuit, I/O ports to access peripheral and other on-chip or off-chip units. Examples of physical devices are UART, modem, transceiver, timer-counter, keypad, keyboard, LED display unit, LCD display unit, DAC and ADC and pulse dialer. This chapter introduces these hardware units, embedded software, state-of-the-art embedded systems and RTOS. It also provides a number of exemplary applications.

Chapter 2 explains embedded system architecture by its processor and memory organization. The readers will learn about the structural units in a processor that provides processing power in an embedded system. They will also learn about memory devices. This chapter explains processor and memory selection methods for a given embedded system. The basis on which the memory blocks and segments are allocated to the data structures is also explained. Memory map concepts and DMA concepts are described. The question of how the memories, devices, IO devices and processors interface is also answered.

Chapter 3 describes the devices—parallel and serial port devices, timing devices, devices for synchronous, iso-synchronous and asynchronous communications and important buses for networking these. Also described are the sophisticated interfacing features in device ports. This chapter also gives a description of I^2C, CAN, USB, advanced serial high speed buses, ISA, PCI, PCIX, advanced parallel high-speed buses.

Chapter 4 concentrates on device drivers. These are important service routines in an embedded system. Use of Linux internals as device drivers and for network functions is also described. Device drivers are explained with examples. An understanding of interrupt servicing and handling mechanisms is essential for an embedded system designer. This chapter fulfills that need. It explains thoroughly the concept of interrupt latencies and deadlines. This concept is helpful for real-time programming for an embedded system.

Chapter 5 explains the programming concepts and source code engineering tools for embedded system programming in embedded C/C++/Java. It thoroughly explains the use of pointers and data structures in embedded software. Important concepts explained include the use of multiple function calls in cyclic order and use of the function pointers, function queues and queues of interrupt service routines (thus also device drivers), and data structures; queues, stacks and lists. Object oriented programming concepts in C++ and Java are also described in this chapter. Memory optimization is critical in an embedded system. How is it done? This chapter has the answer.

Chapter 6 teaches program modeling concepts during the single and multi-processor system software development process. The uses of data flow and control data flow graphs are explained. Program models during real-time programming and uses of FSM and Petri Nets are described. This chapter answers the important question: How do we model multiprocessors, and schedule and synchronize the processing of instructions on them?

Chapter 7 is for learning software engineering practices and approaches in system development processes. Explained are the concepts of uses of linear sequential model, RAD (Rapid Development Phase) model and other important models, including the use of component-based (object-oriented) software development process models. Software requirement analysis, design, implementation, testing, debugging and validating strategies are described in this chapter. Also described is the UML language, which has emerged as an important design language.

Chapter 8 covers the most important aspect of real-time programming, namely, inter-process communications. It first explains the concept of the processes, tasks and threads. Next, it describes the use of semaphores. This chapter also explains thoroughly the use of the signals, mutex, message queues, mailboxes, pipes, virtual (logical) sockets and remote procedure calls.

Chapter 9 describes RTOS concepts. It first describes the OS structure and kernel functions. It then explains the process, memory, devices, files and IO subsystem management functions. It describes schedule management for multiple tasks by an RTOS. It further explains how the scheduling is done for multiple tasks in real time by periodic, cyclic, preemptive, time slicing and other scheduling models. IEEE standards are described for the RTOS functions. A highlight of this chapter is the description of a fifteen-point strategy for synchronization between the processes.

Chapter 10 describes the two most important RTOS tools, μC/OS-II and VxWorks, thoroughly with examples of the use of functions in these RTOSs.

Chapter 11 describes four case studies of programming with RTOS. These are a *vending machine* system, *TCP/IP network system*, *adaptive cruise control system* in a car and a *smart card*.

Chapter 12 details hardware and software designing and integration methods and tools. It explains the embedded system development process action plan. Uses of target system, emulator, ICE, use of device programmer for downloading the finalized codes into ROM, uses of code generation tools (assembler, compiler, loader and linker), simulator, exemplary prototype development tools and IDE are explained. Use of hardware testing tools is also described in this chapter.

Appendices give in brief the gist of CISC and RISC processor architectures, addressing modes and instruction sets, embedded high performance processors, ARM7, ARM9, ARM11 and IBM PowerPC 750. Also given is an overview of microcontroller architectures. DSPs are used in systems for imaging, video and convergent technology products. TriMedia (data, voice and video) processors are new innovative processors for real-time video, streaming networks and data networks. An appendix covers these too. Serial and parallel buses for interconnecting distributed devices and device hardware units in the embedded systems are also covered in brief. Embedded system topics are being introduced in both graduate and undergraduate programs. To guide the course designers and teachers, the suggested units in these courses are given in an appendix.

Also given are over one hundred references to published books, websites and journal papers. This will enable readers to carry out further in-depth study of the topics related to Embedded Systems.

The seven *Salient Features* of the book are:

1. Well structured, systematic coverage, and logical sequencing of topics.
2. Thorough explanation of embedded system programming concepts, OS, RTOS functions and inter-process synchronization.
3. Special coverage of modeling of programs and use of software engineering practices during the software development process for single as well as multiprocessor systems.
4. Thorough explanation of ports, devices, buses for networking devices and device drivers
5. Innovative case studies of RTOS programming in consumer electronics, communications, automobile electronics and secure transaction systems-on-chip.
6. Simultaneous coverage of two RTOSs, µC/OS-II and VxWorks, together to focus on finer points in applications of functions in the RTOSs.
7. Lucid presentation with strong emphasis on examples, well-designed figures and tables, list of keywords and their definitions followed by chapter-end review questions and practice exercises.

Every effort has been made to give correct information and exemplary codes and tools in the book. However, errors might have crept in despite the utmost care. The author will be grateful to readers who point these out to him.

Learned teachers, scholars, software and system engineers will oblige the author if they send their valuable suggestions for further improvement of this textbook. Suggestions and student queries will be heartily welcomed at the author's e-mail, `professor@rajkamal.org` and website `http://www.rajkamal.org`.

RAJ KAMAL

Acknowledgements

The author is grateful to his teachers at the Indian Institute of Technology, New Delhi, (1966–1972), and to the teachers at the University of Uppsala, Uppsala, Sweden, (1978–79, 1984). From these teachers, he learned about the great role of self-learning and the constant drive for understanding emerging technologies, and he developed a passion for knowledge. He also learned to keep abreast of the latest high technology areas.

The author is grateful to Professor M. S. Sodha, F.N.A., an eminent scientist, educationist and administrator. Prof. Sodha gave the author the opportunity to plan and set up from scratch the Electronics and Computer Laboratories at Devi Ahilya University. He has blessed the author all through his academic life since 1966.

The encouragement of Dr. Bharat Chapperwal, Vice Chancellor, and colleagues and friends at Devi Ahilya University in the first phase is unforgettable and the author is thankful to them. The author is especially thankful to Ms. Preeti Saxena (for checking and minutely tracing the errors in the C/C++ codes in the examples given in the book) and for the support of Dr. Sanjiv Tokekar, Ms. Vrinda Tokekar, Dr. Sumant Katyal, Mr. Sanjeev Shrivastava, of the University faculty and Ms. Vasanti G. Parulkar and Ms. Aparna Dev, his laboratory colleagues, during this phase.

The blessings of Kalvivallal T. Kalasalingam, chairman, A K College of Engineering and college Principal Dr. Chelliah Thangaraj's continuous support in the second but no-less important phase of this book are also unforgettable and the author is grateful to them for this encouragement. The help of the author's college colleagues, particularly Dr. S. Radhakrishnan and Dr. M. Shanmugasundram, at various stages is gratefully acknowledged.

The author is thankful to Mr. Sandeep Bansal (Indore) and Mr. Annathurai (AKCE) for making the AutoCad drawings. The author is grateful to the editorial team at McGraw-Hill for reviews and suggestions. Finally, he is grateful to his understanding spouse (Sushil), daughter (Shilpi) and son (Shalin).

RAJ KAMAL

Introduction to Embedded Systems

*What We Will Learn*_____

In this chapter, we will learn the following:

Section 1.1

1. Definitions of *system* and *embedded system*.
2. Classification of embedded systems into three types.
3. Skills needed to design an embedded system.

Section 1.2   The processing unit(s) of the embedded system

1. *Processor in an Embedded System.* A processor is an important unit in the embedded system hardware. A ***microcontroller*** is an integrated chip that has the processor, memory and several other hardware units in it; these form the microcomputer part of the embedded system. An ***embedded processor*** is a processor with special features that allow it to be embedded into a system. A ***digital signal processor (DSP)*** is a processor meant for applications that process digital signals. [For example, filtering, noise cancellation, echo elimination, compression and encryption].
2. Commonly used microprocessors, microcontrollers and DSPs in the small-, medium- and large-scale embedded systems.
3. A recently introduced technology that additionally incorporates the ***application-specific system processors (ASSPs)*** in the embedded systems.
4. Multiple processors in a system.

Section 1.3

1. Embedded system *power source(s)* and the need for controlled power-dissipation.
2. Embedded system *clock* oscillator circuit and clocking unit. It enables a processor to execute and process instructions.
3. *Real time clock (RTC),* timers and various timing needs of the system.
4. *Reset circuit* and *watchdog timer*.
5. *System memories*. [In the second part of Chapter 2, we will learn the system memories in detail].
6. System Input Output (*IO*) ports, serial Universal Asynchronous Receiver and Transmitter *(UART)* and other ports, multiplexers and demultiplexers and interfacing buses. [These will be dealt with in detail in Chapter 3].
7. Interrupt Handler [The latter part of Chapter 4 has the details].
8. Interfacing units—DAC (Digital to Analog Converter) using PWM (Pulse Width Modulation), ADC (Analog to Digital Converter), LED and LCD display units, keypad and keyboard, pulse dialer, modem and transceiver.
9. Hardware required for exemplary embedded systems.

Section 1.4   Different levels of languages that are used to develop the *embedded software* for a system [Chapter 5 details the high-level programming aspects, 'C' and 'C++' language structures and the application of these for coding embedded software].

1. System device drivers, device management and multitasking using an operating system (OS) and *real time operating system* (RTOS). [RTOSs are dealt with in detail in Chapters 9 and 10, and case studies using RTOSs in Chapter 11].
2. Software tools in system designing.
3. Software tools required in six exemplary cases.
4. Programming models for software designing. [Software designing models are detailed in Chapter 6].

Section 1.5   *Exemplary applications* of each type of embedded system.

Section 1.6   Designing an embedded system on a VLSI chip.

1. Embedded *SoC* (System on Chip) and *ASIC* (Application Specific Integrated Circuit) and examples of their applications. These use (i) Application Specific Instruction Processor (ASIP), (ii) Intellectual Property (IP) core, and (iii) Field Programmable Gate Arrays (FPGA) core with single or multiple processor units on an ASIC chip.
2. *Smart card,* an example of the units of an embedded system on a chip (SoC).

1.1 AN EMBEDDED SYSTEM

1.1.1 System

A system is a way of working, organizing or doing one or many tasks according to a fixed plan, program, or set of rules. A system is also an arrangement in which all its units assemble and work together according to the plan or program. Let us examine the following two examples.

Consider a watch. It is a **time-display system**. Its parts are its hardware, needles and battery with the beautiful dial, chassis and strap. *These parts organize to show* the real time every second and continuously update the time every second. The system program updates the display using three needles after each second. *It follows a set of rules.* Some of these rules are as follows: (i) All needles move clockwise only. (ii) A thin and long needle rotates every second such that it returns to same position after a minute. (iii) A long needle rotates every minute such that it returns to same position after an hour. (iv) A short needle rotates every hour such that it returns to same position after twelve hours. (v) All three needles return to the same inclinations after twelve hours each day.

Consider a washing machine. It is an **automatic clothes-washing system**. The important hardware parts include its status display panel, the switches and dials for user-defined programming, a motor to rotate or spin, its power supply and control unit, an inner water-level sensor, a solenoid valve for letting water in and another valve for letting water drain out. *These parts organize* to wash clothes automatically according to a program preset by a user. *The system program* is to wash the dirty clothes placed in a tank, which rotates or spins in pre-programmed steps and stages. *It follows a set of rules.* Some of these rules are as follows: (i) Follow the steps strictly in the following sequence. Step I: Wash by spinning the motor according to a programmed period. Step II: Rinse in fresh water after draining out the dirty water, and rinse a second time if the system is not programmed in water-saving mode. Step III: After draining out the water completely, spin the clothes drum fast for a programmed period of time to dry the clothes. Step IV: Show the wash status by a blinking display. Sound the alarm for a minute to signal that the wash cycle is complete. (ii) At each step, display the process stage of the system. (iii) In case of an interruption, execute only the remaining part of the program, starting from the position when the process was interrupted. There can be no repetition from Step I unless the user resets the system by inserting another set of clothes and resets the program.

1.1.2 Embedded System

A computer is a system that has the following or more components.
1. A microprocessor
2. A large memory comprising the following two kinds:
   (a) Primary memory (*semiconductor* memories—RAM, ROM and fast accessible caches)
   (b) Secondary memory (*magnetic* memory located in hard disks, diskettes and cartridge tapes and *optical* memory in CD-ROM)
3. Input units like keyboard, mouse, digitizer, scanner, etc.
4. Output units like video monitor, printer, etc.
5. Networking units like Ethernet card, front-end processor-based drivers, etc.

6. I/O units like a modem, fax, modem, etc.

An embedded system has computer hardware with embedded software as one of its most important components. It is a dedicated computer-based system for an application or product. It may be either an independent system or a part of a larger system. As its software usually embeds in ROM (Read Only Memory) it does not need secondary memories as in a computer. An embedded system has three main components:

1. It has hardware. Figure 1.1 shows the units in the hardware of an embedded system.
2. It has main application software. The application software may concurrently perform the series of tasks or multiple tasks.
3. It has a real time operating system (RTOS) that supervises the application software and provides a mechanism to let the processor run a process as per scheduling and do the context-switch between the various processes (tasks). The RTOS defines the way the system works. It organizes access to a resource in sequence of the series of tasks of the system. It schedules their working and execution by following a plan to control the latencies and to meet the deadlines. [Latency refers to the waiting period between running the codes of a task and the instance at which the need for the task arises.] It sets the rules during the execution of the application software. A small-scale embedded system may not need an RTOS.

An embedded system has software designed to keep in view three constraints: (i) available system-memory, (ii) available processor speed and (iii) the need to limit power dissipation when running the system continuously in cycles of wait for events, run, stop and wake-up.

There are several definitions of embedded systems given in recent books from others in the field:

Wayne Wolf, author of *Computers as Components—Principles of Embedded Computing System Design*: "What is an *embedded computing system*? Loosely defined, it is any device that includes a programmable computer but is not itself intended to be a general-purpose computer" and "a fax machine or a clock built from a microprocessor is an embedded computing system".

Todd D. Morton, author of *Embedded Microcontrollers*: "*Embedded Systems* are electronic systems that contain a microprocessor or microcontroller, but we do not think of them as computers—the computer is hidden or embedded in the system."

David E. Simon, author of *An Embedded Software Primer*: "People use the term *embedded system* to mean any computer system hidden in any of these products."

Tim Wilmshurst, author of *An Introduction to the Design of Small-Scale Embedded Systems with examples from PIC, 80C51 and 68HC05/08 Microcontrollers*: (1) "An *embedded system* is a system whose principal function is not computational, but which is controlled by a computer embedded within it. The computer is likely to be a microprocessor or microcontroller. The word *embedded* implies that it lies inside the overall system, hidden from view, forming an integral part of greater whole". (2) "An embedded system is a microcontroller-based, software-driven, reliable, real time control system, autonomous, or human- or network-interactive, operating on diverse physical variables and in diverse environments, and sold into a competitive and cost-conscious market".

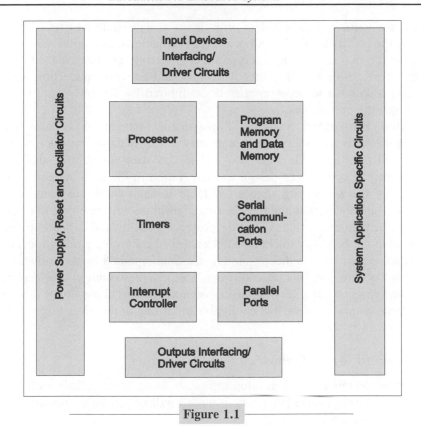

Figure 1.1

The components of an embedded system hardware.

1.1.3 Classification of Embedded Systems

We can classify embedded systems into three types as follows [Section 1.5 will give examples of each type later]:

1. **Small-Scale Embedded Systems:** These systems are designed with a single 8- or 16-bit microcontroller; they have little hardware and software complexities and involve board-level design. They may even be battery operated. When developing embedded software for these, an editor, assembler and cross assembler, specific to the microcontroller or processor used, are the main programming tools. Usually, 'C' is used for developing these systems. 'C' program compilation is done into the assembly, and executable codes are then appropriately located in the system memory. The software has to fit within the available memory and keep in view the need to limit power dissipation when the system is running continuously.

2. **Medium-Scale Embedded Systems:** These systems are usually designed with a single or few 16- or 32-bit microcontrollers or DSPs or Reduced Instruction Set Computers (RISCs). These have both hardware and software complexities. For complex software design, there are the following programming tools: RTOS, Source code engineering tool, Simulator, Debugger and Integrated Development Environment (IDE). Software tools also provide the solutions to the

hardware complexities. An assembler is of little use as a programming tool. These systems may also employ the readily available ASSPs and IPs (explained later) for the various functions—for example, for the bus interfacing, encrypting, deciphering, discrete cosine transformation and inverse transformation, TCP/IP protocol stacking and network connecting functions. [ASSPs and IPs may also have to be appropriately configured by the system software before being integrated into the system-bus.]

3. **Sophisticated Embedded Systems**: Sophisticated embedded systems have enormous hardware and software complexities and may need scalable processors or configurable processors and programmable logic arrays. They are used for cutting-edge applications that need hardware and software co-design and integration in the final system; however, they are constrained by the processing speeds available in their hardware units. Certain software functions such as encryption and deciphering algorithms, discrete cosine transformation and inverse transformation algorithms, TCP/IP protocol stacking and network driver functions are implemented in the hardware to obtain additional speeds by saving time. Some of the functions of the hardware resources in the system are also implemented by the software. Development tools for these systems may not be readily available at a reasonable cost or may not be available at all. In some cases, a compiler or retargetable compiler might have to be developed for these. [A retargetable compiler is one that configures according to the given target configuration in a system.]

1.1.4 Skills Required for an Embedded System Designer

An embedded system designer has to develop a product using the available tools within the given specifications, cost and time frame. [Chapters 7 and 12 will cover the design aspects of embedded systems. See also Section 1.5.]

1. *Skills for Small-Scale Embedded System Designer:* Author Tim Wilmshurst, in the book noted above (see page 4), has said that the following skills are needed in the individual or team that is developing a small-scale system: "Full understanding of microcontrollers with a basic knowledge of computer architecture, digital electronic design, software engineering, data communication, control engineering, motors and actuators, sensors and measurements, analog electronic design and IC design and manufacture". Specific skills will be needed in specific situations. For example, control engineering knowledge will be needed for design of control systems and analog electronic design knowledge will be needed when designing the system interfaces. Basic aspects of the following topics will be described in this book to prepare the designer who already has a good knowledge of the microprocessor or microcontroller to be used: (i) Computer architecture and organization. (ii) Memories. (iii) Memory allocation. (iv) Interfacing the memories. (v) Burning (a term used for porting) the executable machine codes in PROM or ROM (Section 2.3.1). (vi) Use of decoders and demultiplexers. (vii) Direct memory accesses. (viii) Ports. (ix) Device drivers in assembly. (x) Simple and sophisticated buses. (xi) Timers. (xii) Interrupt servicing mechanism. (xiii) C programming elements. (xiv) Memory optimization. (xv) Selection of hardware and microcontroller. (xvi) Use of ICE (In-Circuit-Emulators), cross-assemblers and testing equipment. (xvii) Debugging the software and hardware by using test vectors. Basic knowledge in the other areas—data communication, control engineering, motors and actuators, sensors and measurements, analog electronic design and IC design and manufacture—can be obtained from the standard textbooks available.

A designer interested in small-scale embedded systems may not need at all concepts of interrupt latencies and deadlines and their handling, the RTOS programming tools described in Chapters 9 and 10 and program designing models given in Chapter 6.

2. *Skills for a Medium-Scale Embedded System Designer:* 'C' programming and RTOS programming and program modeling skills are a must to design a medium-scale embedded system. Knowledge of the following becomes critical: (i) Tasks and their scheduling by RTOS. (ii) Cooperative and preemptive scheduling. (iii) Inter-processor communication functions. (iv) Use of shared data, and programming the critical sections and re-entrant functions. (v) Use of semaphores, mailboxes, queues, sockets and pipes. (vi) Handling of interrupt-latencies and meeting task deadlines. (vii) Use of various RTOS functions. (viii) Use of physical and virtual device drivers. [Refer to Chapters 8 to 10 for detailed descriptions of these seven skills along with examples and to Chapter 11 to learn their use with the help of case studies.] A designer must have access to an RTOS programming tool with Application Programming Interfaces (APIs) for the specific microcontroller to be used. Solutions to various functions like memory-allocation, timers, device drivers and interrupt handing mechanisms are readily available as the APIs of the RTOS. The designer needs to know only the hardware organization and use of these APIs. The microcontroller or processor then represents a small system element for the designer and a little knowledge may suffice.

3. *Skills for a Sophisticated Embedded System Designer:* A team is needed to co-design and solve the high level complexities of the hardware and software design. An embedded system hardware engineer should have expertise in hardware units and basic knowledge of 'C', RTOS and other programming tools. Software engineers should have basic knowledge in hardware and a thorough knowledge of 'C', RTOS and other programming tools. A final optimum design solution is then obtained by system integration.

1.2　PROCESSOR IN THE SYSTEM

A *processor* is the heart of the embedded system. For an embedded system designer, knowledge of microprocessors and microcontrollers is a prerequisite. In the following explanations, too, it has been presumed that the reader has a thorough understanding of microprocessors or microcontrollers. [The reader may refer to a standard text or the texts listed in the References at the end of this book for an in-depth understanding of microprocessors, microprocessors and DSPs that are incorporated in embedded system design.]

1.2.1　Processor in a System

A processor has two essential units: Program Flow Control Unit (CU) and Execution Unit (EU). The CU includes a fetch unit for fetching instructions from the memory. The EU has circuits that implement the instructions pertaining to data transfer operations and data conversion from one form to another. The EU includes the Arithmetic and Logical Unit (ALU) and also the circuits that execute instructions for a program control task, say, halt, interrupt, or jump to another set of instructions. It can also execute instructions for a call or branch to another program or for a call to a function.

A processor runs the cycles of fetch and execute. The instructions, defined in the processor instruction set, are executed in the sequence that they are fetched from the memory. A processor is mostly in the form of an IC chip; alternatively, it could be in core form in an ASIC or at a SoC. Core means a part of the functional circuit on the VLSI chip.

An embedded system processor chip or core can be one of the following.

1. General Purpose Processor (GPP):
   a. Microprocessor. [Refer to Section 1.2.2.]
   b. Microcontroller. [Refer to Section 1.2.3.]
   c. Embedded Processor. [Refer to Section 1.2.4.]
   d. Digital Signal Processor (DSP). [Refer to Section 1.2.5.]
   e. Media Processor. [Refer to Appendix E Section E.1.]
2. Application Specific System Processor (ASSP) as additional processor [Refer to Section 1.2.6.]
3. Multiprocessor system using General Purpose processors (GPPs) and Application Specific Instruction Processors (ASIPs) [Refer to Section 1.2.7.]
4. GPP core(s) or ASIP core(s) integrated into either an Application Specific Integrated Circuit (ASIC), or a *Very Large Scale Integrated Circuit* (VLSI) circuit or an FPGA core integrated with processor unit(s) in a VLSI (ASIC) chip. [Refer to Section 1.6.]

For a system designer, the following are important considerations when selecting a processor:

1. Instruction set.
2. Maximum bits in an operand (8 or 16 or 32) within a single arithmetic or logical operation.
3. Clock frequency in MHz and processing speed in Million Instructions Per Second (***MIPS***). [Refer to Appendix B for an alternate metric, ***Dhyrystone***, for processing performance.]
4. Processor ability to solve the complex algorithms used in meeting the deadlines for their processing.

A general-purpose processor is used because of the following: (i) Processing by the known instructions available in the predefined general-purpose instruction set results in fast system development. (ii) Once the board and I/O interfaces are designed for a GPP, these can be used for a new system by just changing the embedded software in the board ROM. (iii) Readiness of a compiler facilitates embedded software development in high-level language. (iv) Readiness of well-tested and debugged processor-specific APIs and the codes previously designed for other applications results in fast development of a new system.

1.2.2   Microprocessor

The CPU is a unit that centrally fetches and processes a set of general-purpose instructions. The CPU instruction set (Section 2.4) includes instructions for *data transfer* operations, *ALU* operations, *stack* operations, *input and output* (I/O) operations and *program control*, *sequencing* and *supervising* operations. The general-purpose instruction set (refer to Appendix A, Section A.1) is always specific to a specific CPU. Any CPU must possess the following basic functional units.

1. A control unit to fetch and control the sequential processing of a given command or instruction and for communicating with the rest of the system.
2. An ALU for the arithmetic and logical operations on the bytes or words. It may be capable of processing 8, 16, 32 or 64-bit words at an instant.

A microprocessor is a single VLSI chip that has a CPU and may also have some other units (for examples, *caches, floating point processing arithmetic unit, pipelining* and *super-scaling units*) that are additionally present and that result in faster processing of instructions. [Refer to Section 2.1.]

The earlier generation microprocessor's fetch-and-execute cycle was guided by clock frequency of the order of ~1 MHz. Processors now operate at clock frequency of 2 GHz. [Intel released a 2 GHz processor on August 25, 2001. This also marked the twentieth anniversary of the introduction of the IBM PC. Intel released 3 GHz Pentium 4 on April 14, 2003.] Since early 2002, a few highly sophisticated embedded systems (for examples, Gbps transceiver and encryption engine) have incorporated the GHZ processor. [Gbps means gigabit per second. Transceiver means a transmitting via receiving circuit with appropriate processing and controls for, example, bus-collisions.]

One example of an older generation microprocessor is Intel 8085. It is an 8-bit processor. Others are Intel 8086 or 8088, which are 16-bit processors, *Intel 80x86 (also referred as x86) processors are the 32-bit successors of 8086.* [The *x* here means extended 8086 for 32 bits.] Examples of 32-bit processors in 80x86 series are Intel 80386 and 80486. Mostly, the IBM PCs use 80x86 series of processors and the embedded systems incorporated inside the PC for specific tasks (like graphic accelerator, disk controllers, network interface card) use these microprocessors.

An example of the new generation 32- and 64-bit microprocessor is the classic **Pentium** series of processors from Intel. These have superscalar architecture [Section 2.1]. They also possess powerful ALUs and Floating Point Processing Units (FLPUs) [Table 2.1]. An example of the use of Pentium III operating at 1 GHz clock frequency in an embedded system is the 'Encryption Engine'. This gives encrypted data at the rate of 0.464 Gbps.

Table 1.1. lists the important microprocessors used in the embedded systems and indicates various streams of families.

Table 1.1
Important Microprocessors Used in the Embedded Systems

Stream	Microprocessor Family	Source	CISC or RISC or Both features
Stream 1	68HCxxx	Motorola	CISC
Stream 2	(a) 80x86	Intel	CISC
	(b) i860	Intel	CISC with RISC
Stream 3	SPARC	Sun	RISC
Stream 4	(a) PowerPC 601, 604	IBM	RISC
	(b) MPC 620	Motorola	RISC

The microprocessors from Streams 1 and 2 have Complicated Instruction Set Computer (CISC) architecture [Section A.1]. Microprocessors form Streams 3 and 4 have Reduced Instruction Set Computer (RISC) architecture [Section A.1.4]. A RISC processor provides speedy processing of the instructions, each in a single clock-cycle. Further, besides the greatly enhanced capabilities mentioned above, there is great enhancement of the speed by which an instruction from a set is processed. The Thumb® instruction set is a new industry standard that also gives a reduced code density in a RISC processor. [The concepts of architecture features of the processor in an embedded system, *CISC and*

RISC processors and processor instruction-set will be explained later in Appendices A and B.] *RISCs are used when the system needs to perform intensive computation, for example, in a speech processing system.*

How does a system designer select a microprocessor? This will be explained in Section 2.2.

 A microprocessor is used when large embedded software is to be located in the external memory chips. A RISC core microprocessor is used when intensive computations are to be performed.

1.2.3   Microcontroller

Just as a microprocessor is the most essential part of a computing system, a microcontroller is the most essential component of a control or communication circuit. *A microcontroller is a single-chip VLSI unit (also called 'microcomputer') which, though having limited computational capabilities, possesses enhanced input-output capabilities and a number of on-chip functional units*. [Refer to Section 1.3 for various functional units.] Microcontrollers are particularly suited for use in embedded systems for real-time control applications with *on-chip* program memory and devices.

Figure 1.2 shows the functional circuits present (in solid boundary boxes) in a microcontroller. It also shows the application-specific units (in dashed boundary boxes) in a specific version of a given microcontroller family. A few of the latest microcontrollers also have high computational and superscalar processing capabilities. [For the meaning of superscalar architecture, refer to Section 2.1.] Appendix C gives the comparative functionalities of select microcontroller representatives from these families.

Important microcontroller chips for embedded systems are usually among the following five streams of families given in Table 1.2.

Table 1.2
Important Microcontrollers@ Used in the Embedded Systems

Stream	Microcontroller Family	Source	CISC or RISC or Both features
Stream 1	68HC11xx, HC12xx, HC16xx	Motorola	CISC
Stream 2	8051, 80251	Intel	CISC
Stream 3	80x86$	Intel	CISC
Stream 4	PIC 16F84 or 16C76, 16F876 and PIC18	Microchip	CISC
Stream 5*	Enhancements of ARM9, ARM7	ARM, Texas, etc.	CISC with RISC Core

@ Other popular microcontrollers are: (i) Hitachi H8x family and SuperH 7xxx. (ii) Mitsubishi 740, 7700, M16C and M32C families. (iii) National Semiconductor COP8 and CR16 /16C. (iv) Toshiba TLCS 900S (v) Texas Instruments MSP 430 for low-voltage battery-based system. (vi) Samsung SAM8. (vii) Ziglog Z80 and eZ80.
$ 80x86 Microcontroller versions (typically 80188 8-bit processor or 80386 16-bit processor), with each there are the 64 kb memory, 3 timers and 2 DMA channels.
* Refer to Sections 1.2.4 and B.1.

Figure 1.3 shows commonly used microcontrollers in the small-, medium- and large-scale embedded systems. Section C.1 (refer to Tables C.1.1 to C.1.3 therein) describes those features that have to be considered by a system designer before choosing a microcontroller as a processing unit.

A microcontroller is used when a small part of the embedded software has to be located in internal memory and when the on-chip functional units like interrupt-handler, port, timer, ADC and PWM are needed.

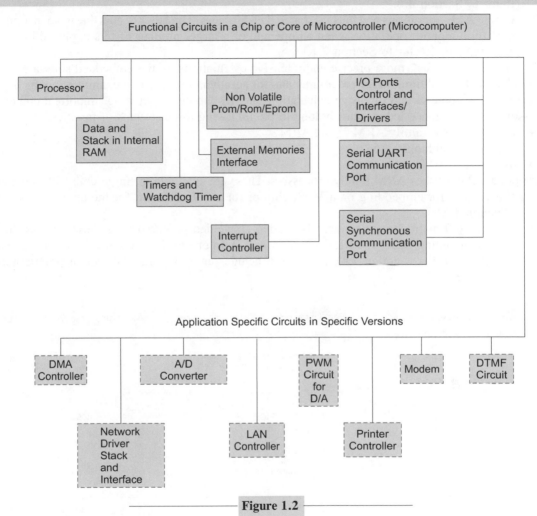

Figure 1.2

Various functional circuits (solid boundary boxes) in a microcontroller chip or core in an embedded system. Also shown are the application-specific units (dashed boundary boxes) in a specific version of a microcontroller.

1.2.4 Embedded Processor for a Complex System

For fast, precise and intensive calculations and for complex real time applications, the microcontrollers and microprocessors mentioned above do not suffice. An electronics warfare system, for example, an Advanced Warning and Control System (AWACS), which also associates tracking radar, is an example of a complex real time system. Special microprocessors and microcontrollers, often called embedded processors, are required. When a microcontroller or microprocessor is specially designed such that it has the following capabilities, then the term *embedded processor* is preferred instead of *microcontroller or microprocessor*.

1. Fast context switching and thus lower latencies of the tasks in complex real time applications. [Refer to Section 4.6.]
2. Atomic ALU operations and thus no shared data problem. The latter occurs due to an incomplete ALU (non-atomic) operation when an operand of a larger number of bits is placed in two or four registers. [Refer to Section 2.1.]
3. RISC core for fast, more precise and intensive calculations by the embedded software.

Calculations for real time image processing and for aerodynamics are two examples in which there is a need for fast, precise and intensive calculations and fast context-switching. Important embedded processor chips for embedded systems belong to the following three streams of families.

- Stream 1: ARM family ARM 7[*] and ARM 9[*]
- Stream 2: Intel family i960.
- Stream 3: AMD family 29050.

[*] Appendix B describes ARM family processors. These are available in single-chip CPU version as well in file version for embedding on a VLSI chip or for a SoC solution for the embedded system. Refer to Section 1.6.

Intel family i960 microcontrollers are also called embedded processors, as these possess the required features including CISC and RISC (Section A.1.4). In one of the versions, these also have a 4-channel DMA controller (Section 2.6). An 80960 includes an 8-channel, 248-vector programmable interrupt controller.

An embedded processor is used when fast processing, fast context-switching and atomic ALU operations are all needed.

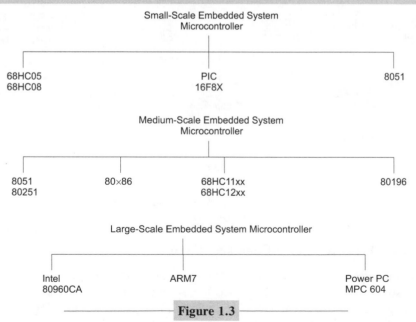

Figure 1.3

Commonly used microcontrollers in small-, medium- and large-scale embedded systems.

1.2.5 Digital Signal Processor (DSP)

Just as a microprocessor is the most essential unit of a computing system, a digital signal processor (DSP) is an essential unit of an embedded system for a large number of applications needing processing of signals. Exemplary applications are in image processing, multimedia, audio, video, HDTV, DSP modem and telecommunications processing systems. DSPs are also used in systems for recognizing an image pattern or a DNA sequence fast. Appendix D describes in detail the embedded system DSPs.

The DSP as a GPP is a single chip VLSI unit. It possesses the computational capabilities of a microprocessor and also has a Multiply and Accumulate (MAC) unit. Nowadays, a typical DSP has a 16 x 32 MAC unit.

A DSP provides fast, discrete-time, signal-processing instructions. It has Very Large Instruction Word (VLIW) processing capabilities; it processes Single Instruction Multiple Data (SIMD) instructions fast; and it processes Discrete Cosine Transformations (DCT) and inverse DCT (IDCT) functions fast. The latter are a must for fast execution of the algorithms for signal analysing, coding, filtering, noise cancellation, echo-elimination, compressing and decompressing and so on.

Important DSPs for the embedded systems are from three streams as given in Table 1.3.

Table 1.3
Important Digital Signal Processor[@] Used in the Embedded Systems

Stream	DSP Family	Source
Stream 1	TMS320Cxx[+]	Texas
Stream 2	SHARC	Analog Device
Stream 3	5600xx	Motorola

[+] For example, TMS320C62XX for fixed-point DSP at clock speed of 200 MHz. Refer to Section D.4 for a detailed description.

A DSP is used when signal-processing functions need to be processed fast.

1.2.6 Application Specific System Processors (ASSPs) in Embedded Systems

Lately a new class of embedded systems has emerged. Systems additionally incorporate the Application Specific System Processor (ASSP) chip(s) or core(s) in their design. These ASSPs have recently become available.

Assume that there is an embedded system for real time video processing. Real time processing need in embedded systems arises for digital television, high definition TV decoders, set-top boxes, DVD (Digital Video Disc) players, Web phones, video conferencing and other systems. The processing needs a video compression and decompression system, which incorporates an MPEG 2 or MPEG 4 standard. [MPEG stands for Motion Picture Expert Group.] MPEG 2 or MPEG 4 compression of signals is done before storing or transmitting; decompression is done before retrieving or receiving these signals. For MPEG compression algorithms, if a GPP embedded software is run, separate DSPs are required to achieve real time processing. An ASSP that is dedicated to these specific tasks alone provides a faster solution. The ASSP is configured and interfaced with the rest of the embedded system.

Assume that there is an embedded system that interconnects using a specific protocol from the system units through a specific bus architecture to another system. Also, assume that there is a need for suitable encryption and decryption. [The output bit stream encryption protects the messages or design from passing to an unknown external entity.] For these tasks, besides embedding the software, it may also be necessary to embed some RTOS features [Section 1.4.6]. If the software alone is used for the above tasks, it may take a longer time than a hardwired solution for application-specific processing. An ASSP chip provides such a solution. For example, an ASSP chip [from i2Chip (http://www.i2Chip.com)] has a TCP, UDP, IP, ARP, and Ethernet 10/100 MAC (Media Access Control) hardwired logic included into it. The chip from i2Chip, W3100A, is a unique hardwired Internet connectivity solution. Much needed TCP/IP stack processing software for networking tasks is thus available as a hardwired solution. This gives an output that is five times faster than a software solution using the system's GPP. It is also an RTOS-less solution. Using the same microcontroller in the embedded system to which this ASSP chip interfaces, Ethernet connectivity can be added. [For terms TCP, UDP, IP, ARP, Ethernet 10/100 and MAC, refer to a suitable reference book on computer networking. See also *Internet and Web Technologies* by RajKamal, Tata McGraw-Hill, 2002, to understand the meaning of each bit in these protocols.]

Another ASSP example is 'Serial-to-Ethernet Converter' (IIM7100). It does real time data processing by a hardware protocol stack. It needs no change in the application software or hardware and provides the economical and smallest RTOS solution.

> An ASSP is used as an additional processing unit for running the application-specific tasks in place of processing using embedded software.

1.2.7 Multiprocessor Systems using General Purpose Processors (GPP)

In an embedded system, several processors may be needed to execute an algorithm fast and within a strict deadline. For example, in real-time video processing, the number of MAC operations needed per second may be more than is possible from one DSP unit. An embedded system then may have to incorporate two or more processors running in synchronization.

In a cell phone, a number of tasks have to be performed: (a) Speech signal-compression and coding. (b) Dialing. (c) Modulating and Transmitting. (d) Demodulating and Receiving. (e) Signal decoding and decompression. (f) Keypad interface and display interface handling. (g) Short Message Service (SMS) protocol-based messaging. (h) SMS message display. For all these tasks, a single processor does not suffice. Suitably synchronized multiple processors are required.

Consider a video conferencing system. In this system, a quarter common intermediate format (Quarter-CIF) is used. Image pixel size is just 144 x 176 as against 525 x 625 pixels in a video picture on TV. Even then, samples of the image have to be taken at a rate of 144 x 176 x 30 = 760320 pixels per second and have to be processed by compression before transmission on a telecommunication device or Virtual Private Network (VPN). [Note: The number of frames should be 25 or 30 per second (as per the standard adopted) for real-time displays and in motion pictures and between 15 and 10 for video conferencing.] A single DSP-based embedded system does not suffice to get real-time images. *Real-time video processing and multimedia applications most often need a multiprocessor unit in the embedded system.* [A media processor, described in Appendix E, is an alternate solution in place of use of multiprocessors for real time video processing.]

Multiple processors are used when a single processor does not meet the needs of the different tasks that have to be performed concurrently. The operations of all the processors are synchronized to obtain an optimum performance.

1.3 OTHER HARDWARE UNITS

1.3.1 Power Source and Managing the Power Dissipation and Consumption

Most systems have a *power supply* of their own. The supply has a specific operation range or a range of voltages. Various units in an embedded system operate in one of the following four operation ranges:

(i) $5.0V \pm 0.25V$

(ii) $3.3V \pm 0.3V$

(iii) $2.0 \pm 0.2V$

(iv) $1.5V \pm 0.2V$

Additionally, a $12V \pm 0.2V$ supply is needed for a flash (a memory form used in systems like data sticks and digital cameras) or Electrically Erasable and Programmable Read Only memory (EEPROM) when present in the microcontroller of an embedded system and for RS232C serial Interfaces (Section 2). [Lately, flash memories require supply voltages of 5V or less.]

Voltage is applied on the chips of an embedded system as follows. The flow of voltage and the connections depend on the number of supply pins provided within the processor, plus the pins in the associated chips and circuits. The pins are in pairs, consisting of the supply in and the ground line. The following points have to be taken care of while connecting the supply rails (lines):

1. A processor may have more than two pins of V_{DD} and V_{SS}. This distributes the power in all the sections and reduces interference between sections. There should be a separate radio frequency interference bypassing capacitor as close as possible to each pair of V_{DD} and V_{SS} pins in the system processor as well as in other units.

2. Supply should separately power the (a) external I/O driving ports, (b) timers and (c) clock and reset circuits. Clock and reset circuits (Sections 1.3.2 and 1.3.3) need to be specially designed to be free from any radio frequency interference. An I/O device may dissipate more power than the other internal units of the processor. A timer may dissipate a constant power even in *wait* state. Hence, these three circuits are powered separately.

3. From the supply, there should be separate interconnections for pairs of V_{DD} and V_{SS} pins, analog ground, analog reference and analog input voltage lines, the ADC unit digital ground and other analog parts in the system. An ADC needs stringent noise-free supply inputs.

Certain systems do not have a power source of their own: they connect to an external *power supply* or are powered by the use of *charge pumps*. (1) Network Interface Card (NIC) and Graphic Accelerator are examples of embedded systems that do not have their own power supply and connect to PC power-supply lines. (2) A charge pump consists of a diode in the series followed by a charging capacitor. The diode gets forward bias input from an external signal; for example, from an RTS signal in the case of the mouse used with a computer. Charge pumps bring the power from a non-supply line. [Ninepins COM port has a signal called Request To Send (RTS). It is an active low signal. Most of the time it is in inactive state logic '1' (~5V). The charge pump inside the mouse uses it to store the charge when the mouse is in an idle state; the pump dissipates the power when the mouse is used]. A regulator circuit getting input from this capacitor gives the required voltage supply. A charge pump in a contact-less smart card uses the radiations from a host machine when inserted into that [Section 1.6.6].

Low voltage systems are built using LVCMOS (Low Voltage CMOS) gates and LVTTL (Low Voltage TTL). Use of 3.3V, 2.5V, 1.8V and 1.5 Volt systems and I/O (Input-Output) Interfaces other than the conventional 5V systems results in significantly reduced power-consumption and can be advantageously used in the following cases: (a) In portable or hand-held devices such as a cellular phone [Compared to 5V, a CMOS circuit power dissipation reduces by half, $\sim(3.3/5)^2$, in 3.3V operation. This also increases the time intervals needed for recharging the battery by a factor of two]. (b) In a system with smaller overall geometry, the low voltage system processors and I/O circuits generate lesser heat and thus can be packed into a smaller space.

There is generally an inverse relationship between the propagation delay in the gates and operational voltage. Therefore, the 5V system processor and units are also used in most systems.

An embedded system may need to be run continuously, without being switched off; the system design, therefore, is constrained by the need to limit power dissipation while it is running. Total power consumption by the system in running, waiting and idle states should also be limited. The *current* needed at any instant in the processor of an embedded system depends on the state and mode of the processor. The following are the typical values in six states of the processor:

(i) 50 mA when only the processor is running; that is, the processor is executing instructions.

(ii) 75 mA when the processor plus the external memories and chips are in running state; that is, fetching and execution are both in progress.

(iii) 15mA when only the processor is in stop state; that is, fetching and execution have both stopped and the clock has been disabled from all structural units of the processor.

(iv) 15 mA when the processor plus the external memories and chips are in stop state; that is, fetching and execution have both stopped and the clock disabled from all system units.

(v) 5 mA when only the processor is in waiting state; that is, fetching and execution have both stopped but the clock has not been disabled from the structural units of the processor, such as timers.

(vi) 10 mA when the processor, the external memories and the chips are in waiting state. Waiting state now means that fetching and execution have both stopped; but the clock has not been disabled from the structural units of the processor and the external I/O units and dynamic RAMs refreshing also has not stopped.

An embedded system has to perform tasks continuously from power-up and may also be left in power-on state; therefore, power saving during execution is important. A microcontroller used in the embedded system must provide for executing *wait* and *stop* instructions and operation in power-down mode. One way to do this is to cleverly incorporate into the software the *wait* and *stop* instructions. Another is to operate the system at the lowest voltage levels in the idle state by selecting power-down mode in that state. Yet another method is to disable use of certain structural units of the processor— for example, caches—when not necessary and to keep in a disconnected state those structure units that are not needed during a particular software-portion execution, for example timers or I/O units. In a CMOS circuit, power dissipates only at the instance of change in input. Therefore, unnecessary glitches and frequent input changes increase power dissipation. VLSI circuit designs have a unique way of avoiding power dissipation. A circuit design is made such that it eliminates all removable glitches, thereby eliminating any frequent input changes.

Note 1 The processor goes into a stop state when it receives a *stop* instruction. The stop state also occurs in the following conditions: (1) On disabling the clock inputs to the processor. (2) On stopping the external clock circuit functions. (3) On the processor operating in auto-shutdown mode. When in stop state, the processor disconnects with the buses. [Buses become in tri-state.] The stop state can change to a running state. The transition to the running state is either because of a user interrupt or because of the periodically occurring wake-up interrupts.

Note 2 The processor goes into a waiting state either on receiving (i) an instruction for *wait*, which slows or disables the clock inputs to some of the processor units including ALU, or (ii) when an external clock-circuit becomes non-functional. The timers are still operating in the waiting state. The waiting state changes to the running state when either (i) an interrupt occurs or (ii) a reset signals.

Note 3 Power dissipation occurs typically by 2.5 mW per 100 kHz reduced clock rate. So a decrease from 8000 kHz to 100 kHz reduces power dissipation by about 200 mW, which is nearly similar to when the clock is non-functional. [Remember, total power dissipated (energy required) may not reduce. This is because on reducing the clock rate the computations will take a longer time at the lower clock rate and the total energy required equals the power dissipation per second multiplied by the time.] The power 25 mW is typically the residual dissipation needed to operate the timers and few other units. By operating the clock at lower frequency or during the power-down mode of the processor, the two advantages are: (i) Heat generation reduces and (ii) Radio frequency interference also then reduces due to the reduced power dissipation within the gates. [Radiated RF (Radio Frequency) power depends on the RF current inside a gate, which reduces due to increase in ON state resistance between the drain and channel when there is reduced heat generation.]

Lately, a new technology is the use of clock manager circuits in conjunction with oscillator circuits. It is used in sophisticated embedded systems on chips (SoCs). Two to sixteen synchronous clocks are created by the combination of clock doublers and clock dividers (by 2). Further, incoming clock signals at the bus may be divided first and then multiplied before being applied to a fast operation circuit. This reduces the power consumption between gates. The clock manager circuit is configured for the smart delivery of the appropriate frequency clock to each section of the circuit being managed during real-time processing. [Note: A sophisticated technology—*phased delay locked loops*—has to be used. When using the common logic gates of counters, there are continuously varying delays at the gates (say, for example, 10 ns plus or minus 2 ns). The synchronous clocks cannot be designed by using the counters alone.]

An internal power source or a charge pump is essential in every system. An embedded system has to perform tasks continuously from power-up to power-off and may even be kept ON continuously. Clever real-time programming by using Wait and Stop instructions and disabling certain units when not needed is one method of saving power during program execution. Operations can also be performed at reduced clock rate when needed in order to control power dissipation; yet all the tasks must execute within the set deadlines and all tasks needing full speed processing must process fast. [For a definition of 'deadline', refer to Section 4.6.] For embedded system software, a performance analysis during its design phase must also include the analysis of power dissipation during program execution and during standby. Good design must optimize the conflicting needs of low power dissipation and fast and efficient program execution.

1.3.2   Clock Oscillator Circuit and Clocking Units

After the power supply, the clock is the next important unit of a system. A processor needs a **clock oscillator** circuit. The clock controls the various clocking requirements of the CPU, the system timers and the CPU machine cycles. The machine cycles are for

(i) fetching the codes and data from memory and then decoding and executing at the processor, and

(ii) transferring the results to memory.

The *clock* controls the time for executing an instruction. The clock circuit uses either a crystal resonator (external to the processor), a ceramic resonator (internally associated with the processor) or an external oscillator IC attached to the processor. (a) The crystal resonator gives the highest stability in frequency with temperature and drift in the circuit. The crystal, in association with an appropriate resistance in parallel and a pair of series capacitance at both pins, resonates at the frequency, which is either double or single times the crystal-frequency. Further, the crystal is kept as near as feasible to two pins of the processor. (b) The internal ceramic resonator, if available in a processor, saves the use of the external crystal and gives a reasonable though not very highly stable frequency. [A typical drift of the ceramic resonator is about ten minutes per month compared to the typical drift of 1 or 5 minutes per month of a crystal]. (c) The external IC-based clock oscillator has a significantly higher power dissipation compared to the internal processor-resonator. However, it provides a higher driving capability, which might be needed when the various circuits of embedded system are concurrently driven. For example, a multiprocessor system needs the clock circuit, which should give a high driving capability and enables control of all the processors concurrently.

For the processing unit(s), a highly stable oscillator is required and the processor clock-out signal provides the clock for synchronizing all the system units.

1.3.3   Real Time Clock (RTC) and Timers for Various Timing and Counting Needs of the System

A timer circuit suitably configured is the *system clock*, also called real time clock (RTC). An RTC is used by the schedulers and for real time programming. An RTC is designed as follows: Assume a processor generates a clock output every 0.5 μs. When a system timer is configured by a software instruction to issue timeout after 200 inputs from the processor clock outputs, then there are 10,000 interrupts (ticks) each second. The RTC ticking rate is then 10 kHz and it interrupts every 100 μs. The RTC is also used to obtain software-controlled delays and time-outs.

More than one timer using the system clock (RTC) may be needed for the various timing and counting needs in a system. Refer to Section 3.2 for a description of timers and counters.

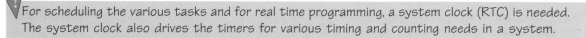

For scheduling the various tasks and for real time programming, a system clock (RTC) is needed. The system clock also drives the timers for various timing and counting needs in a system.

1.3.4 Reset Circuit, Power-up Reset and Watchdog-Timer Reset

Reset means that the processor starts the processing of instructions from a starting address. That address is one that is set by default in the processor program counter (or instruction pointer and code segment registers in x86 processors) on a power-up. From that address in memory, the fetching of program-instructions starts following the *reset* of the processor. [In certain processors, for example, 68HC11 and HC12, there are two start-up addresses. One is as per power-up reset vector and other is as per reset vector after the *reset* instruction or after a time-out (for example from a watchdog timer)].

The reset circuit activates for a fixed period (a few clock cycles) and then deactivates. The processor circuit keeps the reset pin active and then deactivates to let the program proceed from a default beginning address. The reset pin or the internal reset signal, if connected to the other units (for example, I/O interface or Serial Interface) in the system, is activated again by the processor; it becomes an outgoing pin to enforce reset state in other sister units of the system. On deactivation of the *reset* that succeeds the processor activation, a program executes from start-up address.

Reset can be activated by one of the following:

1. An external reset circuit that activates on the power-up, on the switching-on reset of the system or on the detection of a low voltage (for example < 4.5V when what is required is 5V on the system supply rails). This circuit output connects to a pin called the reset pin of the processor. This circuit may be a simple RC circuit, an external IC circuit or a custom-built IC. The examples of the ICs are MAX 6314 and Motorola MC 34064.
2. By (a) software instruction, (b) time-out by a programmed timer known as watchdog-timer (or on an internal signal called COP in 68HC11 and 68HC12 families) or (c) a clock monitor detecting a slowdown below certain threshold frequencies due to a fault.

The watchdog-timer is a timing device that resets the system after a predefined timeout. This time is usually configured and the watchdog-timer is activated within the first few clock cycles after power-up. It has a number of applications. In many embedded systems, reset by a watchdog-timer is very essential because it helps in rescuing the system if a fault develops and the program gets stuck. On restart, the system can function normally. Most microcontrollers have on-chip watchdog timers.

Consider a system controlling the temperature. Assume that when the program starts executing, the sensor inputs work all right. However, before the desired temperature is achieved, the sensor circuit develops some fault. The controller will continue delivering the current nonstop if the system is not reset. Consider another example of a system for controlling a robot. Assume that the interfacing motor control circuit in the robot arm develops a fault during the run. In such cases, the robot arm may continue to move unless there is a watchdog-timer control. Otherwise, the robot will break its own arm!

An important circuit that associates a system is its reset circuit. A program that is reset and runs on a power-up can be one of the following: (i) A system program that executes from the beginning. (ii) A system boot-up program. (iii) A system initialization program.
The watchdog-timer reset is a very useful feature in control applications.

1.3.5 Memories

In a system, there are various types of memories. Figure 1.4 shows a chart for the various forms of memories that are present in systems. These are as follows: (i) Internal RAM of 256 or 512 bytes in a microcontroller for registers—***temporary data and stack.*** (ii) Internal ROM/PROM/EPROM for about 4 kb to 16 kb of program (in the case of microcontrollers). (iii) External RAM for the ***temporary data and stack*** (in most systems). (iv) Internal caches (in the case of certain microprocessors). (v) EEPROM or flash (in many systems saving the results of processing in nonvolatile memory: for example, system status periodically and digital camera images, songs, or speeches after a suitable format compression). (vi) External ROM or PROM for embedding software (in almost all non-microcontroller-based systems). (vii) RAM memory buffers at the ports. (viii) Caches (in superscaler microprocessors). [Refer to Sections 2.1 and 2.3 for further details of these.]

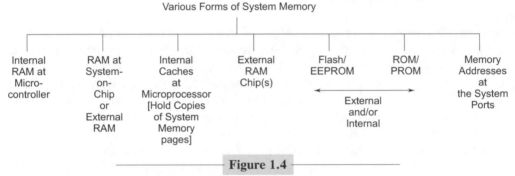

Figure 1.4

The various forms of memories in the system.

Table 1.4 gives the functions assigned in the embedded systems to the memories. ROM, PROM or EPROM embeds the embedded software specific to the system.

Table 1.4

Functions Assigned to the Memories in a System

Memory Needed	Functions
ROM or EPROM	Storing application programs from where the processor fetches the instruction codes. Storing codes for system booting, initializing, initial input data and strings. Codes for RTOS. Pointers (addresses) of various service routines.
RAM (Internal and External) and RAM for buffer	Storing the variables during program run and storing the stack. Storing input or output buffers, for example, for speech or image.
EEPROM or Flash	Storing non-volatile results of processing.
Caches	Storing copies of instructions and data in advance from external memories and storing temporarily the results during fast processing.

A system embeds (locates) the following either in the microcontroller's internal ROM, PROM or in an external ROM or PROM: boot-up programs, initialization data, strings for an initial screen-display or initial state of the system, the programs for various tasks, ISRs and kernel. The system has RAMs for saving temporary data, stack and buffers that are needed during a program run. The system also has flash for storing non-volatile results.

1.3.6　Input/Output and I/O Ports, I/O Buses and I/O Interfaces

The system gets inputs from physical devices (such as, for example, the key-buttons, sensors and transducer circuits) through the input ports. A controller circuit in a system gets the inputs from the sensor and transducer circuits. A receiver of signals or a network card gets the input from a communication system. [A communication system could be a fax, modem, or broadcasting service]. Signals from a network are also received at the ports. Consider the system in a vending machine. It gets inputs from a port that collects the coins that a child inserts—as if only a child would use this wonderful machine! Consider the system in a mobile phone. The user inputs the mobile number through the buttons, directly or indirectly (through recall of the number from its memory). A panel of buttons connects to the system through the input port or ports. The processor identifies each input port by its memory buffer address(es), called port address(es). Just as a memory location holding a byte or word is identified by an address, each input port is also identified by the address. *The system gets the inputs by the read operations at the port addresses*.

The system has output ports through which it sends output bytes to the real world. An output may be to an LED (Light Emitting Diode) or LCD (Liquid Crystal Display) panel. For example, a calculator or mobile phone system sends the output-numbers or an SMS message to the LCD display. A system may send output to a printer. An output may be to a communication system or network. A control system sends the outputs to alarms, actuators, furnaces or boilers. A robot is sent output for its various motors. Each output port is identified by its memory buffer address or port address. *The system sends the output by a write operation to the port address.*

There are also general-purpose ports for both the input and output (I/O) operations. For example, a mobile phone system sends output as well as gets input through a wireless communication channel. Each I/O port is also identified by an address to which the *read* and *write* operations both take place.

Refer to Section 3.1 for more details regarding ports. Basically, there are *two types of I/O ports: Parallel and Serial*. From a serial port, a system gets a serial stream of bits at an input or sends a stream at an output. For example, through a serial port, the system gets and sends the signals as the bits through a modem. A serial port also facilitates long distance communication and interconnections. A serial port may be a *Serial UART port,* a *Serial Synchronous port* or some other *Serial Interfacing port*. [UART stands for Universal Asynchronous Receiver and Transmitter].

A system port may get inputs from multiple channels or may have to send output to multiple channels. A *demultiplexer* takes the inputs from various channels and transfers the input from a select channel to the system. A *multiplexer* takes the output from the system and sends it to another system.

A system might have to be connected to a number of other devices and systems. For networking the systems, there are different types of buses: for example, I^2C, CAN, USB, ISA, EISA and PCI. [Refer to Sections 3.3, 3.4 and Appendix F for buses in detail.]

A system connects to external physical devices and systems through parallel or serial I/O ports. Demultiplexers and multiplexers facilitate communication of signals from multiple channels through a common path. A system often networks to the other devices and systems through an I/O bus: for example, I^2C, CAN, USB, ISA, EISA and PCI bus.

1.3.7 Interrupts Handler

A system may possess a number of devices and the system processor has to control and handle the requirements of each device by running an appropriate Interrupt Service Routine (ISR) for each. *An interrupts-handling mechanism must exist in each system to handle interrupts from various processes in the system:* for example, to transfer data from a keyboard or a printer. [Refer to Chapter 4 for a detailed description of the interrupts and their control (handling) mechanism in a system]. Important points regarding the interrupts and their handling by programming are as follows:

1. There can be a number of interrupt sources and groups of interrupt sources in a processor. [Refer to Section 4.5.] An interrupt may be a hardware signal that indicates the occurrence of an event. [For example, a real-time clock continuously updates a value at a specified memory address; the transition of that value is an event that causes an interrupt.] An interrupt may also occur through timers, through an interrupting instruction of the processor program or through an error during processing. The error may arise due to an illegal op-code fetch, a division by zero result or an overflow or underflow during an ALU operation. An interrupt can also arise through a software timer. A software interrupt may arise in an exceptional condition that may have developed while running a program.
2. The system may prioritize the sources and service them accordingly [Section 4.6.5].
3. Certain sources are not maskable and cannot be disabled. Some are defined to the highest priority during processing.
4. The processor's current program has to divert to a service routine to complete that task on the occurrence of the interrupt. For example, if a key is pressed, then an ISR reads the key and stores the key value in the processor memory address. If a sequence of keys is pressed, for instance in a mobile phone, then an ISR reads the keys and also calls a task to dial the mobile number.
5. There is a programmable unit on-chip for the interrupt handling mechanism in a microcontroller.
6. The application program or scheduler is expected to schedule and control the running of routines for the interrupts in a particular application.

The scheduler always gives priority to the ISRs over the tasks of an application.

> A system must have an interrupt handling mechanism for executing the ISRs in case of the interrupts from physical devices, systems and software exceptions.

1.3.8 DAC (Using a PWM) and ADC

Suppose a system needs to give an analog output of a control circuit for automation. The analog output may be to a power system for a DC motor or furnace. A *Pulse Width Modulator* (PWM) unit in the microcontroller operates as follows: Pulse width is made proportional to the analog-output needed. PWM inputs are from 00000000 to 11111111 for an 8-bit DAC operation. The PWM unit outputs to an external integrator and then provides the desired analog output.

Suppose an integrator circuit (external to the microcontroller) gives an output of 1.024 Volt when the pulse width is 50% of the total pulse time period, and 2.047V when the width is 100%. When the width is made 25% by reducing by half the value in PWM output control-register, the integrator output will become 0.512 Volt.

Now assume that the integrator operates with a dual (plus-minus) supply. Also assume that when an integrator circuit gives an output of 1.023 Volt, the pulse width is 100% of total pulse time period and −1.024 Volt when the width is 0%. When the width is made 25% by reducing by half the value in an output control register, the integrator output will be 0.512 Volt; at 50% the output will be 0.0 Volt.

From this information, finding the formulas to obtain converted bits for a given PWM register bits ranging from 00000000 to 11111111 in both the situations is left as an exercise for the reader.

The ADC in the system microcontroller can be used in many applications such as Data Acquisition System (DAS), analog control system and voice digitizing system. Suppose a system needs to read an analog input from a sensor or transducer circuit. If converted to bits by the ADC unit in the system, then these bits, after processing, can also give an output. This provides a control for automation by a combined use of ADC and DAC features.

The converted bits can be given to the port meant for digital display. The bits may be transferred to a memory address, a serial port or a parallel port.

A processor may process the converted bits and generate a Pulse Code Modulated (PCM) output. PCM signals are used digitizing the voice in the digital format].

Important points about the ADC are as follows:

1. Either a single or dual analog reference voltage source is required in the ADC. It sets either only the analog input's upper limit or both the lower and upper limits. For a single reference source, the lower limit is set to 0V (ground potential). When the analog input equals the lower limit the ADC generates all bits as 0s and when it equals the upper limit it generates all bits as 1s. [As an example, suppose in an ADC the upper limit or reference voltage is set as 2.255 Volt. Let the lower limit reference Voltage be 0.255V. Difference in the limits is 2 Volt. Therefore, the resolution will be (2/256) Volt. If the 8-bit ADC analog-input is 0.255V, the converted 8 bits will be 00000000. When the input is (0.255V + 1.000V) = 1.255V, the bits will be 10000000. When the analog input is (0.255V + 0.50V), the converted bits will be 01000000. [From this information, finding a formula to obtain converted bits for a given analog input = v Volt is left as an exercise for the reader].

2. An ADC may be of 8, 10, 12 or 16 bits depending upon the resolution needed for conversion.

3. The start of the conversion signal (STC) signal or input initiates the conversion to 8 bits. In a system, an instruction or a timer signals the STC.

4. There is an end of conversion (EOC) signal. In a system, a flag in a register is set to indicate the end of conversion and generate an interrupt.

5. There is a conversion time limit in which the conversion is definite.

6. A Sample and Hold (S/H) unit is used to sample the input for a fixed time and hold until conversion is finished.

An ADC unit in the embedded system microcontroller may have multi-channels. It can then take the inputs in succession from the various pins interconnected to different analog sources.

For automatic control and signal processing applications, a system must provide necessary interfacing circuit and software for the Digital to Analog Conversion (DAC) unit and Analog to Digital Conversion (ADC) unit. A DAC operation is done with the help of a combination of PWM unit in the microcontroller and an external integrator chip. ADC operations are needed in systems for voice processing, instrumentation, data acquisition systems and automatic control.

1.3.9   LCD and LED Displays

A system requires an interfacing circuit and software to display the status or message for a line, multi-line displays, or flashing displays. An LCD screen may show a multi-line display of characters or show a small graph or icon (called a pictogram). One innovation in the mobile phone system turns the screen blue to indicate an incoming call. Third-generation system phones have both image and graphic displays. An LCD needs little power, which it gets from a supply or battery (or a solar panel in the calculator). An LCD is a diode that absorbs or emits light on application of 3 V to 4 V and 50 or 60 Hz voltage-pulses with currents less than ~50 μA. The pulses are applied with the same polarity on the crystal's front and back plane for no light, or with opposite polarity for light. Here polarity at an instance means logic '1' or '0'. An LSI (Lower Scale Integrated Circuit) display-controller is often used in the case of matrix displays.

For indicating ON status of the system there may be an LED, which glows when it is ON. A flashing LED may indicate that a specific task is under completion or is running. It may indicate a wait status for a message. The LED is a diode that emits yellow, green, red (or infrared light in a remote controller) on application of a forward voltage of 1.6 to 2 V. An LED needs current up to 12 mA above 5 mA (less in flashing display mode) and is much brighter than the LCD. Therefore, for flashing display and for display limited to few digits, LEDs are used in a system.

> For displaying and messaging, the LCD matrix displays and LED arrays are used in a system. The system must provide necessary interfacing circuit and software for the output to LCD display controller and LED interfacing ports.

1.3.10   Keypad /Keyboard

The keypad or keyboard is an important device for getting user inputs. The system must provide the necessary interfacing and key-debouncing circuit as well as the software for the system to receive input from a set of keys or from a keyboard or keypad. A keypad has up to a maximum of 32 keys. A keyboard may have 104 or more keys. The keypad or keyboard may interface serially or as parallel to the processor directly through a parallel port, a serial port or a controller.

> For inputs, a keypad or board may interface to a system. The system must provide necessary interfacing circuit and software to receive inputs directly from the keys or through a controller.

1.3.11   Pulse Dialer, Modem and Transceiver

For user connectivity through the telephone line or a network, a system provides the necessary interfacing circuit. It also provides the software for pulse dialing through the telephone line, for modem interconnection for fax, for Internet packets routing, and for transmitting and connecting a WAG (Wireless Gateway) or cellular system. *A transceiver is a circuit that can transmit as well as receive byte streams.*

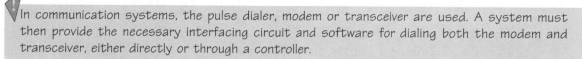

In communication systems, the pulse dialer, modem or transceiver are used. A system must then provide the necessary interfacing circuit and software for dialing both the modem and transceiver, either directly or through a controller.

1.3.12   GPIB (IEEE 488) Link

A system may need linking to another instrument or system. The IEEE 488 GPIB (General Purpose Interface Bus) link is a standard bus originally developed by HP [Hewlett Packard] that links the measuring and instrumentation systems. The embedded system used in the instrumentation systems uses this interfacing standard.

1.3.13   Linking and Interfacing Buses and Units of the Embedded System Hardware

The buses and units in the embedded system hardware need to be linked and interfaced. One way to do this is to incorporate a **glue logic circuit.** [Instead of using individual gates, buffers and decoders, we use the glue logic circuit.] A glue circuit is a circuit that is placed (glued) for all the bus logic actions between circuits and between all chips and main chips (processors and memories). The glue logic circuit of an embedded system may be a circuit for interconnecting the processor to external memories so that the appropriate chip-select signals, according to the system memory, map each of the memory chips [Section 2.5]. The glue logic circuit also includes a circuit to interconnect the parallel and serial ports to the peripherals. [Refer to Chapter 3 for more information on ports.] The glue circuit simplifies the overall embedded system circuit greatly. An example of the use of the glue circuit is to connect the processor, memories and the ports interfacing the LCD display matrix and the keypad.

Programming and configuring one of the followings gives a glue circuit. (i) PAL (Programmable Array Logic). (ii) GAL (Generic Array Logic). (iii) PLD (Programmable Logic Device). (iv) CPLD (Combined PLD). (v) FPGA. These devices are configurable and programmable by a system called a *device programmer*.

PAL has the AND–OR logic arrays. PAL implements only the combinational logic circuit. GAL is another array logic, an advanced version of PAL, which provides sequential circuit implementation. It thus also provides the latches, counters and register circuits. PLD is another logic device that is programmable. A ROM is also a PLD. CPLD is a combination circuit integrated with a PLD. It is a logic device for implementing mixed functions, analog and digital. A CPLD also helps in the control functions and designing a PLC (Programmable Logic Controller). FPGA has a macro cell, which is a combination of gates and flip-flops. An array has many macro cells. The links within the array or in between macro cells are fusable by a device programmer in these devices.

A glue-circuit designed by configuring and programming PAL, GAL, CPLD or FPGA links provides the interfaces for the buses and all the units of the embedded system hardware in a single chip.

1.3.14   Hardware Units Required in Sample Cases

Table 1.5 lists the hardware units that must be present in the embedded systems. Six examples have been chosen to represent systems of varying sophistication. These are:

- Vending machine (a case study for its programming can be found in Section 11.1),
- Data Acquisition System,
- Robot,
- Mobile Phone,
- Adaptive Cruise Control (ACC) system with car string stability (a case study for its programming is in Section 11.3) and
- Voice Processor and Storage System (input, compression, store, decompression, recording and replay).

Remember, RTCs, Timers, Idle-mode, Power-down mode, Watchdog timer and Serial I/O Port, UART port and glue-logic circuit are needed practically in all the applications and have, therefore, not listed in the Table. The values given here refer to a typical system only.

Table 1.5

Hardware Required in Six Sample Embedded Systems with Typical Values

Hardware Required	Vending Machine[&]	Data Acquisition System	Robot	Mobile Phone	Adaptive Cruise Control System with String Stability[#]	Voice Processor
Processor	Micro-controller	Micro-controller	Micro-controller	Multi-processor SoC	Micro-processor	Micro-processor +DSP
Processor Internal Bus Width in Bits	8	8	8	32	32	32
CISC or RISC Processor Architecture	CISC	CISC	CISC	RISC	RISC	RISC
Caches and MMU	No	No	No	Yes	No	Yes
PROM or ROM Memory	4 kB	8 kB	8 kB	1 MB	64 kB	1 MB
EEPROM + Flash	No	512 B	256 B	32 kB	4 kB	4 MB [+]
RAM Interrupts Handler	256 B On-chip	256 B On-chip	256 B On-chip	1 MB[+] On SoC	4 kB Off-chip	1 MB[@] Off-chip
Input–output Ports	Multiple ports: Input for coin sorter port, delivery port and display port	Multiple ports for sensors and actuators	Multiple ports for motors and for angle encoders	Keypad and display ports	Switch buttons and display ports	Input port for speech and output port for replay

Hardware Required	Vending Machine[&]	Data Acquisition System	Robot	Mobile Phone	Adaptive Cruise Control System with String Stability[#]	Voice Processor
Transceiver	No	No	No	Yes for connection to cell service	Yes for tracking radar	No
GPIB Interface	No	Yes	No	No	No	No
Real time detection of an event or signal (Capture and Compare time on an event)	No	Yes	No	Yes	Yes	Yes
Pulse Width Modulation for DAC	No	Yes	Yes	Yes	Yes	Yes
Analog to Digital conversion (bits)	No	Yes	Yes	Yes	Yes	Yes
Modulation Demodulation	No	No	No	Yes	No	No
Digital Signal Processing Instructions	No	No	No	Yes	No	Yes
Non linear controller Instructions	No	No	No	No	Yes	No

[&] Refer to case study in Section 11.1.

[#] Refer to case study 11.2. String stability means maintaining a constant distance between the cars. A radar and transceiver pair is used to measure in-front car distance. EEPROM in needed to store adaptive algorithm parameters.

[@] Excessive need of RAM is due to buffer memory RAM for voice inputs being processed.

[+] Excessive need of RAM is due to buffer memory RAM for voice inputs and outputs being processed.

[&] Buffer-memory for the speech signals.

[+] For storing the voice.

! Embedded systems for a wide spectrum of applications may need different processing hardware platforms. However, it is also true that by changing the embedded software, the same hardware platform can be used for entirely different applications or for new upgrades of the same system.

1.4 SOFTWARE EMBEDDED INTO A SYSTEM

The software is the most important aspect, the brain, of the embedded system.

1.4.1 Final Machine Implementable Software for a Product

An embedded system processor and the system need software that is specific to a given application of that system. The processor of the system processes the instruction codes and data. In the final stage, these are placed in the memory (ROM) for all the tasks that have to be executed. The final stage software is also called ROM image. Why? Just as an image is a unique sequence and arrangement of pixels, embedded software is also a unique placement and arrangement of bytes for instructions and data.

Each code or datum is available only in bits and bytes format. The system requires bytes at each ROM address, according to the tasks being executed. *A machine implementable software file is therefore like a table of address and bytes at each address of the system memory.* The table has to be readied as a ROM image for the targeted hardware. Figure 1.5 shows the ROM image in a system memory. The image consists of the boot up program, stack address pointer(s), program counter address pointer(s), application tasks, ISRs (Section 4.12), RTOS, input data and vector addresses. [Refer to Section 2.5 for the details.]

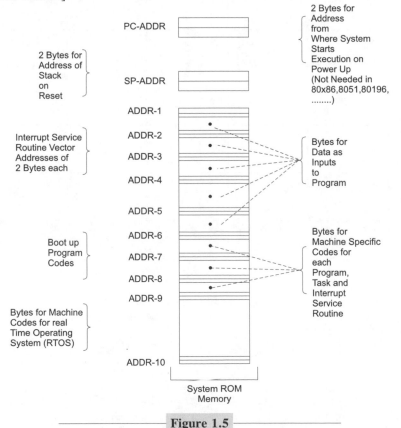

Figure 1.5

System ROM memory embedding the software, RTOS, data, and vector addresses.

> Final stage software is also called the ROM image. The final machine implementable software for a product embeds in the ROM (or PROM) as an image at a frame. Bytes at each address must be defined for creating the ROM image. By changing this image, the same hardware platform will work differently and can be used for entirely different applications or for new upgrades of the same system.

1.4.2　Coding of Software in Machine Codes

During coding in this format, the programmer defines the addresses and the corresponding bytes or bits at each address. In configuring some specific physical device or subsystem, machine code–based coding is used. For example, in a transceiver, placing certain machine code and bits can configure it to transmit at specific Mbps or Gbps, using a specific bus protocol and networking protocol. Another example is using certain codes for configuring a control register with the processor. During a specific code-section processing, the register can be configured to enable or disable use of its internal cache. However, coding in ***machine implementable codes*** is done only in specific situations: ***it is time-consuming because the programmer must first understand*** the processor instruction set and then memorize the instructions and their machine codes.

1.4.3　Software in Processor Specific Assembly Language

When a programmer understands the processor and its instruction set thoroughly, a program or a small specific part can be coded in the ***assembly language***. An exemplary assembly language program in ARM processor instruction set will be shown in an example given in Section A.2.

Coding in assembly language is easy to learn for a designer who has gone through a microprocessor or microcontroller course. Coding is extremely useful for configuring physical devices like ports, a line-display interface, ADC and DAC and reading into or transmitting from a buffer. These codes can also be device driver codes. [Section 4.1]. They are useful to run the processor or device specific features and provide an optimal coding solution. *Lack of knowledge of writing device driver codes or codes that utilize the processor-specific, features-invoking codes in an embedded system design team can cost a lot. Vendors may not only charge for the API, but also charge intellectual property fees for each system shipped out of the company.*

To do all the coding in *assembly language* may, however, be very time-consuming. Full coding in assembly may be done only for a few simple, small-scale systems, such as toys, vending machines, robots or data acquisition systems.

Figure 1.6 shows the process of converting an *assembly language program* into the machine implementable software file and then finally obtaining a ROM image file.

1. An ***assembler*** translates the assembly software into the machine codes using a step called *assembling*.

2. In the next step, called *linking,* a ***linker*** links these codes with the other required assembled codes. Linking is necessary because of the number of codes to be linked for the final binary file. For example, there are the standard codes to program a delay task for which there is a reference in the assembly language program. The codes for the delay must link with the assembled codes. The delay code is sequential from a certain beginning address. The assembly software code is also sequential from a certain beginning address. Both of the codes have to be at the distinct addresses and available addresses in the system. Linker links these. The linked file in binary for

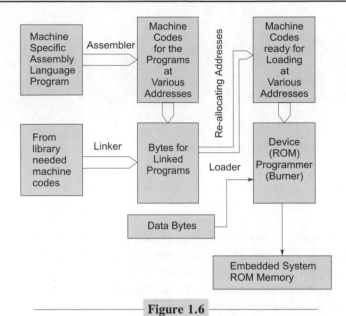

Figure 1.6

*The process of converting an assembly language program
into the machine codes and finally obtaining the ROM image*

run on a computer is commonly known as an executable file or an '.exe' file. After linking, there has to be reallocation of the sequences of placing the codes before the actual placement of the codes in the memory.

3. In the next step, the ***loader*** program performs the task of *reallocating* the codes after finding the physical RAM addresses available at a given instant. The loader is a part of the operating system and places codes into the memory after reading the '.exe' file. This step is necessary because the available memory addresses may not start from 0x0000, and binary codes have to be loaded at the different addresses during the run. The loader finds the appropriate start address, and in a computer, it loads into a section of RAM the program that is ready to run.

4. The final step of the system design process is *locating* the codes as a ROM image and permanently placing them at the actually available addresses in the ROM. In embedded systems, there is no separate program to keep track of the available addresses at different times during the running, as in a computer. The designer has to define the available addresses to load and create files for permanently locating the codes. A program called ***locator*** reallocates the linked file and creates a file for permanent location of codes in a standard format. This format may be Intel Hex file format or Motorola S-record format. [Refer to Appendix G for details.]

The locator locates the I/O tasks and hardware device driver codes at the unchanged addresses. This is because the port addresses for these are fixed for a given system.

5. Lastly, either (i) a laboratory system, called ***device programmer,*** takes as input the ROM image file and finally *burns* the image into the PROM or EPROM or (ii) at a foundry, a mask is created for the ROM of the embedded system from the image file. [The process of placing the codes in

PROM or EPROM is also called burning.] **The mask** created from the image gives the ROM in IC chip form.

For configuring some specific physical device or subsystem like transceiver, the machine codes can be immediately used. For physical device driver codes or codes that utilize the processor-specific features-invoking codes, 'processor-specific' assembly language is used. A file is then created in three steps using 'Assembler', 'Linker' and 'Locator'. The file has the ROM image in a standard format. A device programmer finally burns the image in PROM or EPROM. A mask created from the image gives the ROM in IC chip form.

1.4.4 Software in High Level Language

To do all the coding in *assembly language* may be very time consuming in most cases. Software is therefore developed in a high-level language, 'C' or 'C++' or 'Java'. Most of the times, 'C' is the preferred language. [Refer to Sections 5.1, 5.8 and 5.9 to understand the advantages available in each and to Section 5.11 for the use of 'C' source-code programming tools.] For coding, there is little need to understand assembly language instructions and the programmer does not have to know the machine code for any instruction at all. The programmer needs to understand only the hardware organization. As an example, consider the following problem:

Add 127, 29 and 40 and print the square root.

An exemplary C language program for all the processors is as follows: (i) *# include <stdio.h>* (ii) *# include <math.h>* (iii) *void main (void) {* (iv) *int i1, i2, i3, a;* float result; (v) i1 = 127; i2 = 29; i3 = 40; *a = i1 + i2 + i3; result = sqrt (a);* (vi) *printf (result);}*

It is evident, then, that coding for square root will need many lines of code and can be done only by an expert assembly language programmer. To write the program in a high-level language is very simple compared to writing it in the assembly language. 'C' programs have a feature that adds the assembly instructions when using certain processor-specific features and coding for the specific section, for example, a port device driver. Figure 1.7 shows the different programming layers in a typical

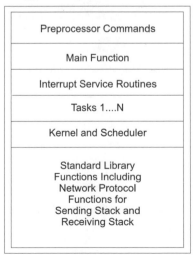

Preprocessor Commands
Main Function
Interrupt Service Routines
Tasks 1....N
Kernel and Scheduler
Standard Library Functions Including Network Protocol Functions for Sending Stack and Receiving Stack

Figure 1.7

The different program layers in the embedded software

embedded 'C' software. [Refer to appropriate sections in Chapters 3 to 5.] These layers are as follows. (i) Processor Commands. (ii) Main Function. (iii) Interrupt Service Routine. (iv) Multiple tasks, say, 1 to N. (v) Kernel and Scheduler. (vi) Standard library functions, protocol functions and stack allocation functions.

Figure 1.8 shows the process of converting a C program into the ROM image file. A *compiler* generates the object codes. The compiler assembles the codes according to the processor instruction set and other specifications. The 'C' compiler for embedded systems must, as a final step of compilation, use a *code-optimizer*. It optimizes the codes before linking. After compilation, the linker links the object codes with other needed codes. For example, the linker includes the codes for the functions, printf and sqrt codes. Codes for device management and driver (device control codes) also link at this stage: for example, printer device management and driver codes. After linking, the other steps for creating a file for ROM image are the same as shown earlier in Figure 1.6.

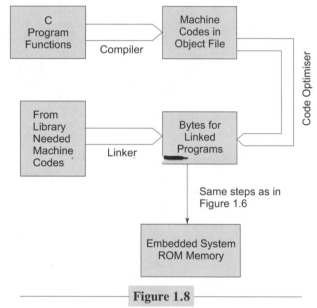

Figure 1.8

The process of converting a C program into the file for ROM image

For most systems, software codes are developed in 'C'. A 'C' program has various layers: processor commands, main function, task functions and library functions, interrupt service routines and kernel (scheduler). The compiler generates an object file. Using linker and locator, the file for ROM image is created for the targeted hardware. C++ and Java are other languages used for software coding.

1.4.5 Software for the Device Drivers and Device Management Using an Operating System

In an embedded system, there are a number of *physical devices*. Types of physical devices are keyboard, display, disk, parallel port and network card.

An innovative concept is the use of virtual devices during programming. A virtual device example is a *file* (for reading and writing the stream of bytes or words) *or a pipe* (for buffering a stream of bytes). The term *virtual device* follows from the analogy that just as a keyboard gives an input to the processor on a *read*, a file also gives an input to the processor. The processor gives an output to a printer on a *write*. Similarly, the processor writes an output to the file. Most often, an embedded system is designed to perform multiple functions and has to control multiple physical and virtual devices.

A **device** for the purpose of control, handling, reading and writing actions can be taken as consisting of three components. (i) Control Register or Word—It stores the bits that, on setting or resetting by a device driver, control the device actions. (ii) Status Register or Word—It provides the flags (bits) to show the device status. (iii) Device Mechanism that controls the device actions. There may be input data buffers and output data buffers on a device. Device action may be to get input into or send output from the buffer. (*The control registers, input data buffers, output data buffers and status registers form part of the device hardware.*)

A **device driver** is software for controlling, receiving and sending a byte or a stream of bytes from or to a device. In case of physical devices, a driver uses the hardware status flags and control register bits that are in set and reset states. In case of virtual devices also, a driver uses the status and control words and the bits that exist in set and reset states.

A *driver* controls three functions: (i) Initialization that is activated by placing appropriate bits at the control register or word. (ii) It calls an ISR on interrupt or on setting a status flag in the status register and runs (drives) the ISR (also called Interrupt Handler Routine). (iii) It resets the status flag after interrupt service. A driver may be designed for asynchronous operations (multiple times use by tasks one after another) or synchronous operations (concurrent use by the tasks). This is because a device may get activated when an interrupt arises and the device driver routine runs.

Using Operating System (OS) functions, a device driver code can be made such that the underlying hardware is hidden as much as possible. An API then defines the hardware separately. This makes the driver usable when the device hardware changes in a system.

A device driver accesses a parallel port or serial port, keyboard, mouse, disk, network, display, file, pipe and socket at specific addresses. An OS may also provide device driver codes for the system-port addresses and for the access mechanism (read, save, write) for the device hardware.

Device management software modules provide codes for detecting the presence of devices, for initializing these and for testing the devices that are present. The modules may also include software for allocating and registering port addresses (in fact, it may be a register or memory) for the various devices at distinctly different addresses, including codes for detecting any collision between these, if any. It ensures that any device has access to one task at any given instant. It takes into account that virtual devices may have addresses that can be relocated by a *locator* (for PROM). [The actual physical or hardware devices have predefined fixed addresses (the addresses are not relocated by the locator)].

An OS also provides and executes modules for managing devices that associate with an embedded system. The underlying principle is that at an instant, only one physical device should get access to or from one task only. The OS also provides and manages the virtual devices like pipes and sockets [Section 8.3].

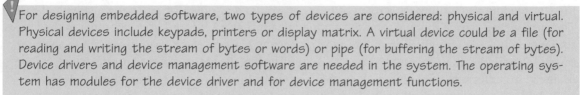

For designing embedded software, two types of devices are considered: physical and virtual. Physical devices include keypads, printers or display matrix. A virtual device could be a file (for reading and writing the stream of bytes or words) or pipe (for buffering the stream of bytes). Device drivers and device management software are needed in the system. The operating system has modules for the device driver and for device management functions.

1.4.6 Software Design for Scheduling Multiple Tasks and Devices Using an RTOS

Most often, an embedded system is designed to perform scheduling of multiple functions while controlling multiple devices. An embedded system program is therefore designed as a multitasking system program. [Refer to Section 8.1 for definitions of the *tasks* (functions) and *task states*.]

In a multitasking OS, each process (task) has a distinct memory allocation of its own and a task has one or more functions or procedures for a specific job. A task may share the memory (data) with other tasks. A processor may process multiple tasks separately or concurrently. The OS software includes scheduling features for the processes (tasks, ISRs and device drivers). An OS or RTOS has a kernel. [Refer to Section 9.2 for understanding kernel functions in detail.] The kernel's important function is to schedule the transition of a task from a ready state to a running state. It also schedules the transition of a task from a blocked state to the running state. The kernel may block a task to let a higher priority task be in running state. [It is called preemptive scheduling]. The *kernel* coordinates the use of the processor for the multiple tasks that are in ready state at any instant, such that only one task among many is in the running state. This is because there is only one processor in the system. The kernel schedules and dispatches a task to a different state than the present. [For multiprocessor systems, scheduling and synchronization of various processors are also necessary]. The kernel controls the interprocess (task) messaging and sharing of variables, queues and pipes.

RTOS functions can thus be highly complex. Chapters 9 to 11 will describe the RTOS functions in an embedded system. In an embedded system, RTOS has to be scalable. *Scaleable OS* is one in which memory is optimized by having only a part of the needed features associate with the final system software.

There are a number of popular and readily available RTOSs. Chapters 9 and 10 will describe these. Case studies employing these RTOSs will be discussed in Chapter 11.

Embedded software is most often designed for performing multiple actions and controlling multiple devices and their ISRs. Multitasking software is therefore essential. For scheduling multiple tasks, popular, readily available RTOS kernels with functions (like device drivers) are most often used.

1.4.7 Software Tools for Designing an Embedded System

Table 1.6 lists the applications of software tools for assembly language programming, high-level language programming, RTOS, debugging and system integration tools.

Table 1.6
Software Modules and Tools for the Detailed Designing of an Embedded System

Software Tool	Application
Editor	For writing C codes or assembly mnemonics using the keyboard of the PC for entering the program. Allows the entry, addition, deletion and insert, appending previously written lines or files, and merging record and files at the specific positions. Creates a source file that stores the edited file. It also has an appropriate name [provided by the programmer].
Interpreter	For expression-by-expression (line-by-line) translation to the machine executable codes.
Compiler	Uses the complete sets of the codes. It may also include the codes, functions and expressions from the library routines. It creates a file called object file.
Assembler	For translating the assembly mnemonics into binary opcodes (instructions), i.e., into an executable file called a binary file. It also creates a list file that can be printed. The list file has address, source code (assembly language mnemonic) and hexadecimal object codes. The file has addresses that adjust during the actual run of the assembly language program.
Cross-Assembler	For converting object codes or executable codes for a processor to other codes for another processor and vice versa. The cross-assembler assembles the assembly codes of the target processor as the assembly codes of the processor of the PC used in the system development. Later, it provides the object codes for the target processor. These codes will be the ones actually needed in the final developed system.
Simulator	To simulate all functions of an embedded system circuit including additional memory and peripherals. It is independent of a particular target system. It also simulates the processes that will execute when the codes execute on the targeted particular processor.
Source-code Engineering Software	For source code comprehension, navigation and browsing, editing, debugging, configuring (disabling and enabling the C++ features) and compiling.
RTOS	Refer to Chapters 9 and 10.
Stethoscope	For dynamically tracking the changes in any program variable. It tracks the changes in any parameter. It demonstrates the sequences of multiple processes (tasks, threads, service routines) that execute. It also records the entire time history.
Trace Scope	To help in tracing the changes in the modules and tasks with time on the X-axis. A list of actions also produces the desired time scales and the expected times for different tasks.
Integrated Development Environment	Software and hardware environment that consists of simulators with editors, compilers, assemblers, RTOS, debuggers, stethoscope, tracer, emulators, logic analyzers, EPROM EEPROM application codes' burners for the integrated development of a system.
Prototyper~	For simulating source code engineering including compiling, debugging and, on a browser, summarizing the complete status of the final target system during the development phase.
Locator#	Uses cross-assembler output and a memory allocation map and provides the locator program output. It is the final step of software design process for the embedded system.

For locator refer to Section 2.5. Locator program output is in the Intel hex file or Motorola S-record format.
~ An Example is Tornado Prototyper from WindRiver® for integrated cross-development environment with a set of tools.

The assembler codes in an assembly language. For high-level language programming, a special source code engineering tool may be needed while designing sophisticated systems. RTOS is necessary in most embedded systems, as a system, in most cases, has to schedule multiple tasks to meet their deadline, drive a number of physical and virtual devices and handle many ISRs. Debugging tools like stethoscope and trace scope are needed for debugging. It is an important step in the system development. A sophisticated tool—such as Integrated Development Environment or Prototype development tools—is needed for integrated development of system software and hardware.

1.4.8 Needed Software Tools in the Exemplary Cases

Table 1.7 gives the various tools needed to design exemplary systems.

Table 1.7
Software Tools Required in Exemplary Systems

Software Tools	Vending Machine	Data Acquisition System	Robot	Mobile Phone	Adaptive Cruise Control System with String Stability[#]	Voice Processor
Editor	Yes	Yes	Yes	NR	NR	NR
Interpreter	Yes	NR	Yes	NR	NR	NR
Compiler	NR	Yes	No	Yes	Yes	Yes
Assembler	Yes	Yes	Yes	No	No	No
Cross-Assembler	NR	Yes	No	No	No	No
Locator[#]	Yes	Yes	Yes	Yes	Yes	Yes
Simulator	NR	Yes	Yes	Yes	Yes	Yes
Source-code Engineering Software	NR	NR	NR	Yes	Yes	Yes
RTOS	MR	MR	MR	Yes	Yes	Yes
Stethoscope	NR	NR	NR	Yes	Yes	Yes
Trace Scope	NR	NR	NR	Yes	Yes	Yes
Integrated Development Environment	NR	Yes	Yes	Yes	Yes	Yes
Prototyper~	NR	No	No	Yes	Yes	Yes

~ An Example is Tornado prototyper WindRiverâ for integrated cross-development environment with a set of tools.
Note: NR means not required. MR means may be required in specific complex system but not mandatory.
[#] For locator, refer to Section 2.5.

RTOS is needed in most embedded systems, as a system may have to schedule multiple tasks, drive a number of physical and virtual devices and handle many ISRs. Embedded systems for medium-scale and sophisticated applications may need a number of sophisticated software tools.

1.4.9 Models for Software Designing

In complex or multiprocessor systems, there are different models that are employed during the design processes of the embedded software and its RTOS, including: (i) Finite State Machine (FSM). (ii) Petri Net model. (iii) Control and Data flow graph. (iv) Activity diagrams based UML Model. For multiprocessor systems, the following additional models are needed: (i) Synchronous Data Flow (SDF) Graph. (ii) Timed Petri Nets and Extended Predicate/Transition Net. (iii) Multi-Thread Graph (MTG) System. These models are explained in Chapter 6.

1.5 EXEMPLARY EMBEDDED SYSTEMS

1.5.1 Exemplary Applications of Each Type of Embedded System

Embedded systems have very diversified applications. A few select application areas of embedded systems are telecom, smart cards, missiles and satellites, computer networking, digital consumer electronics, and automotive. Figure 1.9 shows the applications of embedded systems in these areas.

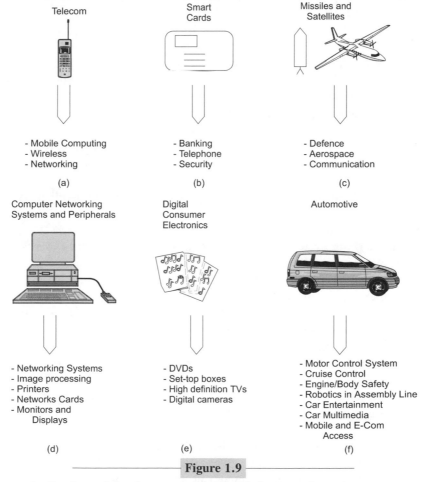

Figure 1.9

Applications of the three types of embedded systems in various areas.

A few examples of *small-scale embedded system* applications are:

- Vending Machine
- Stepper motor controllers for a robotics system
- Washing or cooking system
- Multitasking toys
- Microcontroller-based single or multi-display digital panel meter for voltage, current, resistance and frequency
- Keyboard controller
- Serial port cards
- Computer mouse
- CD drive or Hard Disk drive controller
- The peripheral controllers of a computer, for example, a CRT display controller, a keyboard controller, a DRAM controller, a DMA controller, a printer-controller, a laser printer-controller, a LAN controller, a disk drive controller
- Fax or photocopy or printer or scanner machine
- Digital diary
- Remote control of TV
- Telephone with memory, display and other sophisticated features
- Motor controls systems—for example, an accurate control of speed and position of DC motor, robot, and CNC machine; automotive applications such as a closed-loop engine control, a dynamic ride control and an anti-lock braking system monitor
- Electronic data acquisition and supervisory control system
- Electronic instruments, such as an industrial process controller
- Electronic smart weight display system and an industrial moisture recorder-controller
- Digital storage system for a signal wave form or electricity or water meter readings
- Spectrum analyzer
- Biomedical systems such as an ECG LCD display-recorder, a blood-cell analyzer and a patient monitor system

Some examples of *medium-scale embedded systems* are:

- Computer networking systems, for example, a router, a front-end processor in a server, a switch, a bridge, a hub and a gateway
- For Internet appliances, there are numerous application systems: (i) An intelligent operation, administration and maintenance router (IOAMR) in a distributed network and (ii) mail client card to store e-mail and personal addresses and to smartly connect to a modem or server
- Entertainment systems, such as a video game and a music system
- Banking systems, for example, bank ATM and credit card transactions
- Signal Tracking Systems, for example, an automatic signal tracker and a target tracker
- Communication systems, such as a mobile-communication SIM card, a numeric pager, a cellular phone, a cable TV terminal, and a FAX transceiver with or without a graphic accelerator
- Image Filtering, Image Processing, Pattern Recognizer, Speech Processing and Video Processing
- A system that connects a pocket PC or PDA (Personal Digital Assistant) to the automobile driver mobile phone and a wireless receiver. The system then connects to a remote server for Internet or e-mail or to a remote computer at an ASP (Application Service Provider). This system forms the backbone of m-commerce (mobile e-commerce) and mobile computing.

- A personal information manager using frame buffers in handheld devices
- Thin Client [A Thin Client provides diskless nodes with remote boot capability]. Application of thin clients accesses a data center from a number of nodes; in an Internet laboratory it accesses the Internet through a remote server.
- Embedded Firewall/ Router using ARM7/ i386 multiprocessor and 32 MB of Flash ROM. The load balancing and two Ethernet interfaces are its other important functions. These interfaces support PPP, TCP/IP and UDP protocols.
- DNA sequence and pattern storage card and DNA pattern recognizer

Examples of *sophisticated embedded systems* are:

- Embedded systems for wireless LAN and for convergent technology devices
- Embedded systems for real time video and speech or multimedia processing systems
- Embedded interface and networking systems using high speed (400 MHz plus), ultra high speed (10 Gbps) and large bandwidth: routers, LANs, switches and gateways, SANs (Storage Area Networks), WANs (Wide Area Networks), video, interactive video and broadband IPv6 (Internet Protocol version 6) Internet and other products
- Security products, high-speed network security and Gigabit rate encryption rate products
- Embedded sophisticated system for space lifeboat (NASA's X-38 project) under development that will be used in the future with the ISS (International Space Station). In an emergency, it will bring the astronauts and crew members back to Earth from the ISS. With a press of a button this lifeboat will detach from ISS and travel back to Earth resisting all the climatic/ atmospheric conditions and meeting exact timing constraints. This will also be a fault-tolerant system.

1.6　EMBEDDED SYSTEM-ON-CHIP (SOC) AND IN VLSI CIRCUIT

Lately, embedded systems are being designed on a single silicon chip, called *System-on-chip (SoC)*. *SoC is a new design innovation for embedded systems*. An embedded processor is a part of the SoC VLSI circuit. A SoC may be embedded with the following components: multiple processors, memories, multiple standard source solutions, called IP (Intellectual Property) cores and other logic and analog units. A SoC may also have a network protocol embedded into it. It may also embed an encryption function unit. It can embed discrete cosine transforms for signal processing applications. It may embed FPGA (Field Programmable Gate Array) cores [Section 1.6.5].

For a number of applications, the GPP (microcontrollers, microprocessors or DSPs) cores may not suffice. For security applications, killer applications, smart card, video game, PDA computer, cellphone, mobile Internet, handheld embedded systems, Gbps transceivers, Gbps LAN systems and satellite or missile systems, we need special processing units in a VLSI designed circuit to function as a processor. These special units are called Application Specific Instruction Processors (ASIP). For an application, both the configurable processors (called FPGA cum ASIP processors) and non-configurable processors (DSP or microprocessor or microcontrollers) might be needed on a chip. One example of a killer application using multiple ASIPs is high-definition television signals processing. [High definition means that the signals are processed for a noise-free, echo-canceled transmission, and for obtaining a flat high-resolution image (1920 x 1020 pixels) on the television screen.] A cell phone is another killer application. [A killer application is one that is useful to millions of users.]

Recently, embedded SoCs have been designed for functioning as *DNA chips*. Consider an FPGA with a large number of gate arrays. Now, using VLSI design techniques, we can configure these arrays to process the specific tasks on an SoC. This gives an SoC as a DNA chip. Each set of arrays has a specific and distinct DNA complex structure. These structures as well as the processor embeds on the DNA chip.

1.6.1 Exemplary SoC for Cell Phone

Figure 1.10 shows an SoC that integrates two internal ASICs, two internal processors (ASIPs), shared memories and peripheral interfaces on a common bus. Besides a processor and memories and digital circuits with embedded software for specific applications, the SoC may possess analog circuits as well]. An exemplary application of such an ASIC embedded SoC is the cell phone. One ASIP in it is configured to process encoding and deciphering and another does the voice compression. One ASIC dials, modulates, demodulates, interfaces the keyboard and multiple line LCD matrix displays, stores data input and recalls data from memory. ASICs are designed using the VLSI design tools with processor GPP or ASIP and analog circuits embedded into the design. The designing is done using the Electronic Design Automation (EDA) tool. [For design of ASIC digital circuits, a 'High Level Design Language (HDL)' is used].

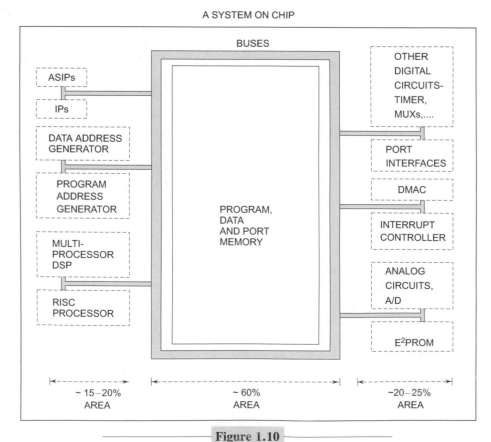

Figure 1.10

A SoC embedded system and its common bus with two internal ASICs, two internal processors, shared memories and peripheral interfaces.

1.6.2 ASIP

Using VLSI tools, a processor itself can be designed. A system specific processor (ASIP) is the one that does not use the GPP (standard available CISC or RISC microprocessor or microcontroller or signal processor). The processor on chip incorporates a section of the CISC or RISC instruction set. This specific processor may have an especially configurable instruction set for an application. An ASIP can also be configurable. Using the appropriate tools, an ASIP can be designed and configured for the instructions needed in the following exemplary functions: DSP functions, controller signals processing function, adaptive filtering functions and communication protocol–implementing functions. On a VLSI chip, an embedded ASIP in a special system can be a unit within an ASIC or SoC.

1.6.3 IP Core

On a VLSI chip, there may be high-level components, ones that possess gate-level sophistication in circuits above that of the counter, register, multiplier, floating point operation unit and ALU. A standard source solution for synthesizing a higher-level component by configuring FPGA core or a core of VLSI chip may be available as an Intellectual Property (IP). The copyright for the synthesized design of a higher-level component for gate-level implementation of an IP is held by the designer or designing company. One has to pay a royalty for every chip shipped. An embedded system may incorporate IPs.
- An IP may provide a hardwired implementable design of a *transform*, or of an *encryption algorithm* or a *deciphering algorithm*.
- An IP may provide a design for *adaptive filtering* of a signal.
- An IP may provide full design for implementing Hyper Text Transfer Protocol (HTTP) or File Transfer Protocol (FTP) to transmit a web page or a file on the Internet.
- An IP may be designed for the PCI or USB bus controller. [Sections 3.3 and 3.4.]

1.6.4 Embedding a GPP

A General Purpose Processor (GPP) can be embedded on a VSLI chip. Recently, GPPs called ARM 7 and ARM 9, which embed onto a VLSI chip, have been developed by ARM and have been enhanced by Texas Instruments. [Refer to http:/www.ti.com/sc/docs/asic/modules/arm7.htm and arm9.htm]. An ARM-processor VLSI-architecture is available either as a CPU chip or for integrating it into VLSI or SoC. [The instruction set features of the ARM is given in Section A.1.] ARM provides CISC functionality with RISC architecture at the core. An application of ARM-embedded circuits is ICE [Section 12.3.2]. ICE is used for debugging an embedded system. Exemplary ARM 9 applications are setup boxes, cable modems, and wireless devices such as mobile handsets.

ARM9 [Section 2.2.5] has a single cycle 16 x 32 multiply accumulate unit. It operates at 200 MHz. It uses 0.15 µm GS30 CMOSs. It has a five-stage pipeline. It incorporates RISC. It integrates with a DSP when designing an ASIC solution having multiprocessors' architecture [Section 6.3]. An example is its integration with DSP with TMS320C55x. A lower capability but very popular version of ARM9 is ARM7. It operates at 80 MHz clock. It uses 0.18 µm based GS20 CMOSs. Using ARM7, a large number of embedded systems have recently become available. One recent application is in integrating the operating system Linux Kernel 2.2 and the device drivers into an ASIC.

1.6.5 FPGA (Field Programmable Gate Arrays) Core with a Single or Multiple Processor

A recent innovation is the Field Programmable Gate Arrays (FPGA) core with a single or multiple processor units on chip. One example is Xilinx Virtex-II Pro FPGA XC2VP125. Other example is 90 nm Spartan-3 FPGA released in April 2003 by Xilinx. [An FPGA consists of a large number of programmable gates on a VLSI chip. There is a set of gates in each FPGA cell, called 'macro cell'. Each cell has several inputs and outputs. All cells interconnect like an array (matrix). Each interconnection is fusible (detachable) using a FPGA programming tool.] The interested reader can refer to articles in *Xcell Journal* 2002 and *Xcell Journal* 2003.

Consider the algorithms for the following: an SIMD instruction, Fourier transform and its inverse, DFT or Laplace transform and its inverse, compression or decompression, encrypting or deciphering, a specific pattern-recognition (for recognizing a signature or fingerprint or DNA sequence). We can configure (fuse) an algorithm into the logic gates of the FPGA. It gives us hardwired implementation for a processing unit. It is specific to the needs of the embedded system. The embedded software can implement an algorithm in one of the FPGA sections and another algorithm in another section.

An exemplary latest SoC design is on the XC2VP125 system. It has 125136 logic cells in the FPGA core with four IBM PowerPCs. It has been very recently used for developing embedded systems integrated with programmable logic. For example, there is a solution reported for data security with an encryption engine and data rate of 1.5 gigabits per second. Other types of embedded systems integrated with logic arrays are DSP-enabled, real-time video processing systems and line echo eliminators for the Public Switched Telecommunication Networks (PSTN) and packet switched networks. [A packet is a unit of a message or a flowing data such that it can follow a programmable route among the number of optional open routes available at an instance].

1.6.6 Components in an Exemplary SoC–Smart Card

Figure 1.11 shows embedded system hardware components on an SoC for a contactless smart card. Its components are as follows: [Section 11.4 will describe a case study of embedded software design.]

- ASIP (Application Specific Instruction Processor)
- RAM for temporary variables and stack
- ROM for application codes and RTOS codes for scheduling the tasks
- EEPROM for storing user data, user address, user identification codes, card number and expiration date
- Timer and interrupt controller
- A carrier frequency ~16 MHz generating circuit and Amplitude Shifted Key (ASK) Modulator. ASK modulator gives 10% excess amplitude of carrier pulses for bit '1' and 10% less for bit '0'. A load modulation sub-carrier has one-sixteenth of this frequency and modulates the 1s and 0s by Binary Phase Shifted Keying (BPSK).
- Interfacing circuit for the I/Os
- Charge pump for delivering power to the antenna (of ~5 mm range) for transmission and for system circuits. The charge pump stores charge from received RF (radio frequency) at the card antenna that it hears from the host in the vicinity. [The charge pump is a simple circuit that consists of a diode and a high value ferroelectrics material-based capacitor.]

An Embedded System
Contactless Smart Card Components

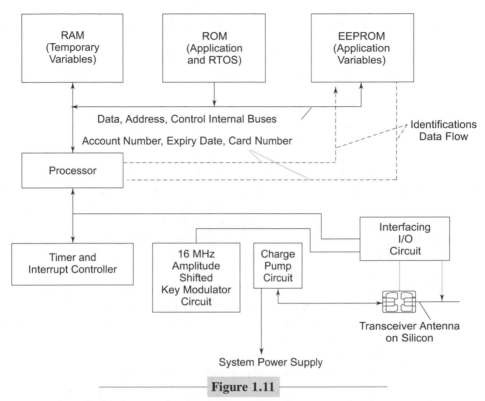

Figure 1.11

An embedded system hardware components in a contactless smart card.

■ SUMMARY ■

- The embedded system is a sophisticated system consisting of several hardware and software components, and its design may be several times more complex than that of a PC and the programs running on a PC.
- The embedded system processor can be a general-purpose processor chosen from number of families of processors, microcontrollers, embedded processors and digital signal processors (DSPs). Alternatively, an application specific instruction processor (ASIP) may be designed for specific application on a VLSI chip. An ASSP may be additionally used for fast hardwired implementation of a certain part of the embedded software. A sophisticated embedded system may use a multiprocessor unit also.
- Embedded system embeds (locates) a software image in the ROM. The image mostly consists of the following: (i) Boot up program. (ii) Initialization data. (iii) Strings for an initial screen-display or system state. (iv) Programs for the multiple tasks that the system performs. (v) RTOS kernel.

- The embedded system needs a power source and controlled and optimized power-dissipation from the total energy requirement for given hardware and software. The charge pump provides a power-supply-less system in certain embedded systems.
- The embedded system needs clock and reset circuits. Use of the clock manager is a recent innovation.
- The embedded system needs interfaces: Input Output (I/O) ports, serial *UART* and other ports to accept inputs and to send outputs by interacting with the peripherals, display units, keypad or keyboard.
- The embedded system may need bus controllers for networking its buses with other systems.
- The embedded system needs timers and a watchdog timer for the system clock and for real-time program scheduling and control.
- The embedded system needs an *interrupt controlling unit.*
- The embedded system may need ADC for taking analog input from one or multiple sources. It needs DAC using PWM for sending analog output to motors, speakers, sound systems, etc.
- The embedded system may need an LED or LCD display units, keypad and keyboard, pulse dialer, modem, transmitter, multiplexers and demultiplexers.
- Embedded software is usually made in the high-level languages C or C++ or Java with certain features added, enabled or disabled for programming. C and C++ also facilitates the incorporation of assembly language codes.
- The embedded system most often needs a real-time operating system for real-time programming and scheduling, device drivers, device management and multitasking.
- There are a number of software tools needed in the development and design phase of an embedded system. [Refer to Table 1.7.]
- There are a large number of applications and products that employ embedded systems.
- A VLSI chip can embed IPs for the specific applications, besides the ASIP or a GPP core. A system-on-chip is the latest concept in embedded systems, such as a mobile phone. A contactless smart card is one such application, the details of which are shown in Figure 1.11.

■ LIST OF KEYWORDS AND THEIR DEFINITIONS ■

- *System*: A way of working, organizing or doing some task or series of tasks by following the fixed plan, program and set of rules.
- *Embedded system*: A sophisticated system that has a computer (hardware with application software and RTOS embedded in it) as one of its components. An embedded system is a dedicated computer-based system for an application or product.
- *Processor*: A processor implements a process or processes as per the command (instruction) given to it.
- *Process*: A program or task or thread that has a distinct memory allocation of its own and has one or more functions or procedures for specific job. The process may share the memory (data) with other tasks. A processor may run multiple processes separately or concurrently.
- *Microcontroller*: A unit with a processor. Memory, timers, watchdog-timer, interrupt controller, ADC or PWM, etc., are provided as required by the application.

- *GPP (General-purpose processor)*: A processor from a number of families of processors, microcontrollers, embedded processors and digital signal processors (DSPs) having a general-purpose instruction set and readily available compilers to enable programming in a high-level language.
- *ASSP (Application specific system processor)*: A processing unit for specific tasks, for example, image compression, which is integrated through the buses with the main processor in the embedded system.
- *ASIP (Application specific instruction processor)*: A processor designed for specific application on a VLSI chip.
- *FPGA*: These are Field Programmable Gate Arrays on a chip. The chip has a large number of arrays with each element having fusable links. Each element of the array consists of several XOR, AND, OR, multiplexer, demultiplexer and tristate gates. By appropriate programming of the fusable links, a design of a complex digital circuit is created on the chip.
- *Registers*: These are associated with the processor and temporarily store the variable values from the memory and the execution unit during instruction processing.
- *Clock*: Fixed-frequency pulses that an oscillator circuit generates and that controls all operations during processing and all timing references of the system. Frequency depends on the needs of the processor circuit. A processor, if it needs a 100 MHz clock, then its minimum instruction processing time is a reciprocal of it, which is 10 ns.
- *Reset*: A processor state in which the processor registers acquire initial values and from which an initial program starts; this program is usually the one that also runs on power-up.
- *Reset circuit*: A circuit to force reset state and that gets activated for a short period on power-up. When reset is activated, the processor generates a reset signal for the other system units needing reset.
- *Memory*: This stores all the programs, input data and output data. The processor fetches instructions from it to execute and gives the processed results back to it as per the instruction.
- *ROM*: A read-only memory that locates the following in its ROM-embedded software: initial data and strings and operating system or RTOS.
- *RAM*: This is a random-access read and write memory that the processor uses to store programs and data that are volatile and which disappear on power down or off.
- *Cache*: A fast *read and write* on-chip unit for the processor execution unit. It stores a copy of a page of instructions and data. It has these fetched in advance from the ROM and RAM so that the processor does not have to wait for instruction and data from external buses.
- *Timer*: A unit to provide the time for the system clock and real-time operations and scheduling.
- *Watchdog-timer*: The timeout from this timer resets the processor in case the program gets stuck for an unexpected length of time.
- *Interrupt controller*: A unit that controls the processor operations arising out of an interrupt from a source.
- *ADC*: A unit that converts, as required, the analog input between + and − pins with respect to the reference voltage to digital 8 or 10 or 12 bits.
- *PWM*: Pulse width modulator to provide a pulse of width scaled to the analog output desired. On integrating PWM output, the DAC operation is achieved.
- *DAC*: Digital bits (8 or 10 or 12) converted to analog signal scaled to a reference voltage.

- *Input Output (I/O) ports*: The system gets the inputs and outputs from these. Through these, the keypad or LCD units attach to the system.
- *UART*: Universal Asynchronous Receiver and Transmitter.
- *LED*: Light emitting diode—a diode that emits red, green, yellow or infrared light on forward biasing between 1.6V to 2 V and currents between 8–15mA. Multi-segment and multi-line LED units are used for bright displays of digits, characters, charts and short messages.
- *LCD*: Liquid crystal diode—a diode that absorbs or emits light on application of 3 to 4 V 50 or 60 Hz voltage pulses with currents ~ 50 mA. Multi-segment and multi-line LCD units are used for a display of digits, characters, charts and short messages with very low power dissipation.
- *Modem*: A circuit to modulate the outgoing bits into pulses usually used on the telephone line and to demodulate the incoming pulses into bits for incoming messages.
- *Multiplexer*: A digital circuit that has digital inputs from multiple channels. It sends only one channel output at a time. The channel at the output has the same address as the channel address bits in its input.
- *Demultiplexer*: A digital circuit that has digital outputs at any instance in multiple channels. The channel that is connected is the one that has the same address as the channel address bits in its input.
- *Compiler*: A program that, according to the processor specification, generates machine codes from the high-level language. The codes are called object codes.
- *Assembler*: A program that translates assembly language software into the machine codes placed in a file called an '.exe' (executable) file.
- *Linker*: A program that links the compiled codes with the other codes and provides the input for a loader or locator.
- *Loader*: It is a program that reallocates the physical memory addresses for loading into the system RAM memory. Reallocation is necessary, as available memory may not start from 0x0000 at a given instant of processing in a computer. The loader is a part of the OS in a computer.
- *Locator*: It is a program to reallocate the linked files of the program application and the RTOS codes at the actual addresses of the ROM memory. It creates a file in a standard format, a ROM image.
- *Device Programmer*: It takes the inputs from a file generated by the locator and burns the fusable link to actually store the data and codes at the ROM.
- *Mask and ROM mask*: Created at a foundry for fabrication of a chip. The ROM mask is created from the ROM image file.
- *Physical Device*: A device like a printer or keypad connected to the system port.
- *Virtual Device*: A file or pipe that is programmed for opening and closing and for reading and writing, such as a program for attaching and detaching a physical device and for input and output.
- *Pipe*: A data structure (*or virtual device*) through which is sent a byte stream from a data source (for example, a program structure) and which delivers the byte stream to the data sink (for example, a printer).
- *File*: A data structure (*or virtual device*) which sends the records (characters or words) to a data sink (for example, a program structure) and which stores the data from the data source (for example, a program structure). A file in computer may also be stored at the hard disk.

- *Device Driver*: Interrupt service routine software, which runs after the programming of the control register (or word) of a peripheral device (or virtual device) and to let the device get the inputs or outputs. It executes on an interrupt to or from the device.
- *Device manager*: Software to manage multiple devices and drivers.
- *Multitasking*: Processing codes for the different tasks as directed by the scheduler.
- *Kernel*: A program with functions for memory allocation and deallocation, task scheduling, inter-process communication, effective management of shared memory access by using the signals, exception (error) handling signals, semaphores, queues, mailboxes, pipes and sockets [See Section 8.3], I/O management, interrupts control (handler), device drivers and device management.
- *Real-time operating system*: Operating System software for real-time programming and scheduling, process and memory manager, device drivers, device management and multitasking.
- *VLSI chip*: A very large scale integrated circuit made on silicon with ~ 1M transistors.
- *System on Chip*: A system on a VLSI chip that has all of needed analog as wells as digital circuits, for example, in a mobile phone.

▪ REVIEW QUESTIONS ▪

1. Define a system. Now define embedded system.
2. What are the essential structural units in (a) microprocessor, (b) Embedded processor, (c) microcontroller, (d) DSP, (e) ASIP and (f) ΛSIP? List each of thcsc.
3. How does a DSP differ from a general-purpose processor (GPP)? Refer to Sections 1.2.5 and D.2.
4. What are the advantages and disadvantages of (a) a processor with only fixed-point arithmetic unit and (b) a processor with additional floating-point arithmetic processing unit?
5. A new innovation is media processors [Section E.1.] Refer to Sections 1.2.5, D.2 and E.1. How does a media processor differ from a DSP?
6. Explain the media processor use in a convergence-technology embedded system like a mobile phone with mail client, Internet connectivity and image-frame downloads.
7. Compare features in an exemplary family chip (or core) of each of the following: Microprocessor, Microcontroller, RISC Processor, Digital Signal Processor, ASSP, Video processor and Media processor. Refer to Section 1.2 and Appendices A to E.
8. Why do late generation systems operate a processor at low voltages (<2 V) and I/O at (~3.3V)?
9. What are the techniques of power and energy management in a system?
10. What is the advantage of running a processor at reduced clock speed in certain sections of instructions and at full speed in other sections of instructions?
11. Name the advantages of the following: (a) Stop instruction, (b) Wait instruction, (c) Processor idle mode operation, (d) Cache-use disable instruction and (e) Cache with multi-ways and blocks in an embedded system.
12. What do we mean by charge pump? How does a charge pump supply power in an embedded system without using the power supply lines?
13. What do you mean by 'real time' and 'real time clock'?

14. What is the role of *processor reset* and *system reset*?
15. Explain the need of watchdog timer and reset after the watched time.
16. What is the role of RAM in an embedded system?
17. Why do we need multiple actions and multiple controlling tasks for the devices in an embedded system? Explain it with an example of the embedded system—a TV remote control.
18. When do we need multitasking OS?
19. When do we need an RTOS?
20. Why should be embedded system RTOS be scalable?
21. Explain the terms IP core, FPGA, CPLD, PLA and PAL.
22. What do you mean by System-on-Chip (SoC)? How will the definition of embedded system change with System-on-Chip?
23. What are the advantages offered by an FPGA for designing an embedded system?
24. What are the advantages offered by an ASIC for designing an embedded system?
25. What are the advantages offered by an ASIP for designing an embedded system?
26. Real time video processing needs sophisticated embedded systems with specific real time constraints. Why? Explain it.
27. Why does a processor system always need an 'Interrupts Handler (Interrupt Controller)'?
28. What role does a linker play?
29. Why do we use a loader in a computer system and a locator in an embedded system?
30. Why does a program reside in ROM in the embedded system?
31. Define ROM image and explain each section of an ROM image in an exemplary system.
32. When will you use the compressed program and data in ROM? Give five examples of embedded systems having these in their ROM images.
33. When will you use SRAM and DRAM? Explain your replies.
34. What do we mean by the following: physical device, virtual device, plug and play device, bus self-powered device, device management and device specific processor.

■ PRACTICE EXERCISES ■

35. Locate definitions of an embedded system from books listed in References and list these definitions in column 1 with their references in column 2.
36. Classify the embedded systems into small-scale, medium-scale and sophisticated systems. Now, reclassify these embedded systems with and without real-time (response time constrained) systems and give 10 examples of each.
37. An automobile cruise control system is to be designed in a project. What skills will be needed in the team of hardware and software engineers?
38. Take a value, $x = 1.7320508075688$. It is squared once again by a floating-point arithmetic processor unit. Now x is squared by a 16-bit integer fixed point arithmetic processing unit. How does the result differ? [Note: Fixed-point unit will multiply only 17320 with 17320, divide the result by 10000 and then again divide the result by 10000.]
39. Design four columns table two examples of embedded systems in each row's columns 2 and 3. Column 1: the type of processor needed among the followings: Microprocessor, Microcontroller, Embedded Processor, Digital Signal Processor, ASSP, Video-processor and Media processor. Give reasoning in column 4.

40. How does a CMOS I/O circuit power dissipation reduction compare to 5V, factor of half, $\sim(3.3/5)^2$, in I/O 3.3V operation?

41. How much reduction in power dissipation for a processor CMOS circuit occurs when V reduces from 5V to 1.8V operation?

42. Refer to Sections 1.3.5 and G.1 to G.3. List various types of memories and application of each in the following: Robot, Electronic smart weight display system, ECG LCD display-cum-recorder, Router, Digital Camera, Speech Processing, Smart Card, Embedded Firewall/ Router, Mail Client card, and Transceiver system with a collision control and jabber control [Collision control means transmission and reception when no other system on the network is using the network. Jabber control means control of continuous streams of random data flowing on a network, which eventually chokes a network.]

43. Tabulate hardware units needed in each of the systems mentioned in Question 42 above.

44. Give two examples of embedded systems, which need one or more of following units. (a) DAC (Using a PWM) (b) ADC (c) LCD display (d) LED Display (e) Keypad (f) Pulse Dialer (g) Modem (h) Transceiver (i) GPIB (IEEE 488) Link.

45. An ADC is a 10-bit ADC? It has reference voltages, $V_{ref-} = 0.0V$ and $V_{ref+} = 1.023V$. What will be the ADC outputs when inputs are (a) -0.512 V (b) $+0.512$ V and (c) $+2047V$? What will be the ADC outputs in three situations when (i) $V_{ref-} = 0.512$ V and $V_{ref+} = 1.023V$ (ii) $V_{ref-} = 1.024$ V and $V_{ref+} = 2.047V$ and (iii) $V_{ref-} = -1.024$ V and $V_{ref+} = +2.047V$.

46. Refer to Sections 1.4, 5.1, 5.8 and 5.9. Tabulate the advantages and disadvantages of using coding language as following: (a) Final Machine Implementable, (b) ALP (Assembly Language Programming, (c) 'C', (d) C++ and (e) Java.

47. List the software tools needed in designing each of the embedded system examples in Question 42.

48. Justify the importance of device drivers in an embedded system. Refer to Sections 1.4.5 and 4.1.3.

49. The cost of designing an embedded system may be thousands of times the cost of its processor and hardware units. Explain this statement.

50. FPGA (Field Programmable Gate Arrays) core integrated with a single or multiple processor units on chip and FPSLIC (Field Programmable System Level Integrated Circuits) are recent novel innovations. How do these help in the design of sophisticated embedded systems for real time video processing?

# Chapter 2

Processor and Memory Organisation

*What We Have Learnt*_____

The previous chapter dealt with the following:

1. An embedded system is a dedicated computer-based system or a part of a large system of an application. Its processor may be a microprocessor with CISC or RISC architecture. It can be a microcontroller or a Digital Signal Processor (DSP).
2. Embedded systems, which have many applications, have software dedicated to the system and have a number of hardware units.
3. The processor and the memory are two important elements in the system. The processor runs software using various memory units: ROM, RAM and caches.
4. The system software embeds in the ROM. It is called the ROM image. The image consists of the boot-up program, initialisation data, strings for an initial screen-display or system state, programs for multiple tasks that the system performs and the RTOS kernel.
5. While the ROM stores the ROM image consisting of the finally designed codes, the RAM stores the temporary values of the variables and the stack during program-execution. Caches store copies of instructions and data in advance from external memories and temporarily store the results during fast processing.

An important question for the systems hardware designer is how to select the appropriate micro-processor, microcontroller, or DSP, which will give an optimum performance with an appropriate set of memory devices. The designer must also know how to organise (i) the chosen processor and memories and (ii) how to design an appropriate interfacing circuit, which interconnects the processor, memories, I/O devices and other units of the system. The designer of a system must have, therefore, a thorough understanding of the basic structural unit, memories, memory allocation maps, bus signals and the sequence and speed of action of all these in a processor. This chapter addresses these questions.

An organisation of the processor and memories in the system provides the basic platform for good performance. The design goal is to get the best processor performance with optimum use of memory for the targeted embedded system and with the least dissipation of energy.

What We Will Learn

In this chapter we will learn the following:

1. Structural units of a processor
   (a) The internal buses and external buses that interconnect the processor with the system memories, I/O devices and all other system units
   (b) Superscalar, processing, pipelining and cache units for improved computational performance of the processor by faster program execution
   (c) Arithmetic and logic operation on the data to process the integers using the ALU and floating point operations using a processing unit for the floating point numbers
   (d) Atomic operations processing unit in certain embedded-processors to solve the share data problems that arise during multitasking operations, and
   (e) Processor registers, register-windows and register-files
2. Various types of memory devices and their uses
3. Memory blocks for the data-structures and data set elements
4. Concept of memory map
5. Device registers and addresses of I/O devices
6. Location of ROM image as per memory map by designing a locator program
7. Direct memory access feature that enables the devices of the system to access the system memories directly
8. Memories and I/O devices interface circuit with the processor

2.1 STRUCTURAL UNITS IN A PROCESSOR

Table 2.1 lists the structural units in general-purpose processors and the functions of each. Figure 2.1 shows interconnections between twenty-five structural units of a processor by a block diagram.

Table 2.1
Structural Units and Functions of a Processor

Structural Unit[&]	Functions
MAR Memory Address Register	It holds the address of the byte or word to be fetched from external memories. Processor issues the address of instruction or data to MAR before it initiates a fetch cycle.
MDR Memory Data Register	It holds the byte or word fetched (or to be sent) from (to) external memory or I/O address.
Internal Bus	It internally connects all the structural units inside the processor. Its width can be of 8, 18, 32, or 64 bits.
Address Bus	It is an external bus that carries the address from the MAR to the memory as well as to the I/O devices and the other units of the system.

Structural Unit	Functions
Data Bus	It is an external bus that carries, during a read operation or write operation, the bytes of an instruction or data from or to an address. The address is determined by the MAR.
Control Bus	It is an external bus to carry control signals to or between the processor and memory (or devices).
BIU Bus Interface Unit	It is the interface unit between the processor's internal units and the external buses.
IR Instruction Register	It takes sequentially the instruction codes (opcodes) to the execution unit of the processor.
ID Instruction Decoder	It decodes the instruction opcode received at the IR and passes it to the processor CU.
CU Control Unit	It controls all the bus activities and unit functions needed for processing.
ARS Application Register set	It is a set of on-chip registers used during processing the instructions of the *application* program of the user. A register window[#] consists a subset of registers with each subset storing static variables of a software routine. A register file is a file that associates a unit like ALU or FLPU.
ALU Arithmetic Logical Unit	It is a unit to execute the arithmetic or logical instructions according to the current instruction present at the IR.
PC Program Counter	It generates the instruction cycle to fetch the address from the memory through MAR. It auto-increments as the instructions are fetched regularly and sequentially. It is called the instruction pointer in 80x86 processors.
SRS System Register Set	It is a set of registers used while processing the instructions of the supervisory system program.
SP Stack Pointer	It is a pointer for an address, which corresponds to a stack-top in the memory.
IQ Instruction Queue	It is a queue of instructions so that the IR does not have to wait for the next instruction, after one has been carried out.
PFCU Prefetch Control Unit	It is a unit that controls the *fetching* of data into the I-cache and D-cache in advance from the memory units. The instructions and data are delivered when needed to the processor's execution unit(s). The processor does not have to fetch data just before executing the instruction.
I-Cache Instruction Cache	It sequentially stores, like an instruction queue, the instructions in FIFO mode. It lets the processor execute instructions at great speed while, through PFCU; the processor accesses external system-memories at relatively much slower speeds.
BT Cache Branch Target Cache	It facilitates ready availability of the next instruction set when a branch instruction like *jump, loop or call* is encountered. Its fetch unit foresees a branching instruction at the I-cache.

Structural Unit	Functions
D-Cache Data Cache	It stores the prefetched data from the external memory. A data cache generally holds both the key (address) and value (word) together at a location. It also stores write-through data when so configured. Write-through data is the resulting data from the execution unit that transfer through the cache to the external memory addresses.
MMU Memory-Management Unit	It manages the memories* such that the instructions and data are readily available for processing. [See note 2 also.]
FLPU Floating Point Processing Unit	A unit separate from ALU for floating point processing, which is essential in processing mathematical functions fast in a microprocessor or DSP.
FRS Floating Point Register Set	A register set dedicated for storing floating point numbers in a standard format and used by FLPU for its data and stack.
Advanced Processing Units	These are units used for multistage pipeline processing, multiline superscalar processing to obtain processing speeds higher than one instruction per clock cycle[$]. There is also a MAC[£] unit(s) for multiplying coefficients of a series and accumulating these during computations.
AOU Atomic Operation Unit	It lets a user (compiler) instruction, when broken into a number of processor instructions called atomic operations, finish before an interrupt of a process occurs. This prevents problems from arising out of shared data between various routines and tasks.

Notes: 1. [&]Units in the high performance-giving processor are listed in italics. 2. [$]Instruction cycle time becomes many times less than the clock cycle time due to the advanced processing unit 3. *MMU (Memory Management Unit) manages the pages at the RAM memory as well as the copies at the internal and external caches. Managing has to be done in such a way that when the instructions execute, there are a minimum number of page and cache faults (misses) as possible. 4. [#]This helps in fast context switching from one task to another when the processor is concurrently running multiple tasks. [Refer to Section 4.6 for an explanation of context switching and its function.] 5.[£] MAC units are invariably needed in DSPs. [Refer to the main text.]

Almost all processors possess the structural units shown by blocks with the full-line boundary in Fig. 2.1. A high performance processor possesses the structural units that are shown by blocks with dashed-line boundary in Fig. 2.1. [Generally, performance is measured in term of MIPS (Million Instructions Per Second) or MFLOPS (Million Floating Point Operations Per Second) or Dhrystones. Refer to Appendix B.]

Advanced processing units include the Instruction Pipelining unit, which improves performance by processing instructions in multiple stages, and the superscalar execution unit, which improves performance on execution of two or more instructions in parallel. [Refer to *Advanced Microprocessors* by Daniel Tabak, McGraw-Hill, International Edition, 1995.]

The prefetching unit improves performance by fetching instructions and data in advance for processing. Caches and MMU improve performance by giving fast the instructions and data to the processor execution unit.

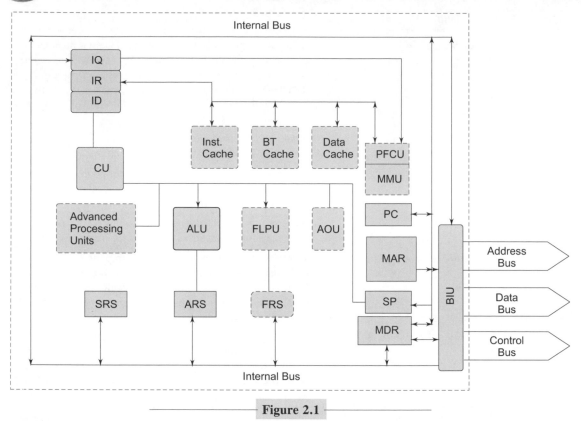

Figure 2.1

Block diagram of structural units in a processor in the embedded system.
[The units shown with the dashed boundary are present in the high performance processors only.]

RISC architecture improves performance by executing instructions in a single clock cycle (by hardwired implementation of instructions), by using multiple register sets, windows and files and by greatly reducing dependency on the external memory accesses for data due to the reduced number of addressing modes.

The floating point unit and FRS *process* the mathematical functions faster than when employing integers-processing ALU.

The MAC (Multiply and Accumulate) unit at a DSP provides fast multiplication of two operands and accumulating results at a single address. It quickly computes an expression such as the following summation: $y_n = \Sigma (a_i.x_{n-i})$ where the sum is made for $i = 0, 1, 2, …, N$-1. Here i, n and N are the integers, a is a coefficient, x is independent variable or an input element and y is the dependent variable or an output element.

How do pipeline and superscalar units give such higher performance? Let us look at Example 2.1.

Example 2.1 Pipeline and Superscalar Execution

Step 1 Let us assume that the processor instruction cycle time is 0.02 ms (at 50 MHz operation) and that the processor executes an instruction in one clock cycle. The processor performance expected without advanced processing units will be 50 MIPS.

Step 2 Assume there is a three-stage pipeline as in ARM 7. Let us, for the moment, ignore the effect of branching (called *branch penalty*). Three instructions will process in one clock cycle. The maximum expected performance of the processor without superscalar but with pipeline will be 150 MIPS.

Step 3 Assume there is a two-line superscalar. Let us ignore the effects of unaligned data (*data dependency penalty*). Six instructions can process in single clock cycle with the three-stage pipeline and two superscalar units. The maximum performance will now be six times the processor cycle time, 300 MIPS.

1. A term used in this example is *branch penalty*. If a branching instruction is encountered at a multistage pipeline, then the instructions executed in part at the preceding stages become redundant. These instructions have to be executed in full again later on after completion of the loop or return from a routine. The time required for reprocessing these is the branch penalty.
2. Another term used in this example is *data dependency penalty*. Assume that there are two instructions in two execution lines during a superscalar operation. Further, that one instruction depends on the data output of another. This is known as improper alignment. Thus, the two instructions are not aligned before putting them in separate lines. One instruction will now have to wait and can't proceed further till the other instruction is executed. The waiting time is the data dependency penalty.

The following description is a list of the essential characteristics of a processor structure. Every system designer needs to consider these during the processor selection and hardware design.

   1. *Instruction Cycle Time*: This is the time taken by a processor to execute a simple instruction, which is ~1 μs for the 8051 processor operating at ~12 MHz, and 1.6 ns for the PowerPC MPC 604 processor. The system designer uses the Instruction Cycle Time as an indicator to match the processor speed with the application. For applications that need fast processing, the PowerPC MPC 604 processor would be considered suitable; for other applications, in which a slower processing will suffice, the 8051, 68HC11 or 80196 processors can be chosen.
   2. *Internal Bus Width*: The ALU gets inputs through the buses. Bits in a single operand to ALU (during a single arithmetic or logical operation) are equal to the bus width. A 32-bit bus will facilitate the availability of arithmetic operations on 32-bit operands in a single cycle. The 32-bit bus becomes a necessity for signal processing and control system instructions. When the bus-width is 32 bits, it reads or writes an integer of 32 bits and will process about four times faster than when the width is 8. An internal bus of 64 bits can be found in the PowerPC MPC 604 and in Pentium processors.
   3. *CISC or RISC Architecture*: CISC or RISC architecture may affect the system design. CISC has the ability to process complex instructions and complex data sets with fewer registers. RISC executes simpler instructions and in a single cycle per instruction. [Refer to Appendix A for an understanding of CISC and RISC architecture and for an exemplary RISC ARM 7 and its instruction set]. *CISC with RISC implementation* means that for most instructions, there are hardwired implementations that process in a single instruction cycle. [Examples are ARM, 80960 and Pentium processors].
   4. *Program-Counter* (PC) *bits and its reset value*: The number of PC bits decides the maximum possible size of the physical memory that can be accessed by the processor. The reset value

tells the designer where the initial program that runs on system reset or power up should be stored. The processor will start execution from there. [The initial instruction pointer and code segment register bits decide the initial program memory address in 80x86 processors.]

5. *Stack-Pointer bits and initial reset value*: Stack pointer values must point to addresses of the words stored at the *stack*. These addresses must be within the ones allocated for the stack in the system. The software designer defines the initial reset value and sets the stack pointer accordingly for a process in the initial program. The hardware designer chooses between two processors, one using external memories for the stack and other with internal stacks.

6. *Pipelined and Superscalar Units*: High processor performance is required in many cases. For example, real-time signal processing. Pipelining and superscalar operations have now become essential [Example 2.1]. The hardware designer must select the processor as per the required MIPS or MFLOPS performance. [A multi-processor system (Section 6.3) will be needed for very high performance requirement.]

7. *On-chip Memories as RAM and/or Register Files*, *Windows*, *Caches* and ROM: These are the integral parts of the memory-organisation within a system. The software designer should enable the use of caches in a processor by an appropriate instruction, to obtain greater performance during run of a section of a program, while simultaneously disabling the remaining sections in order to reduce the power dissipation and minimise the system energy requirement. Hardware designers should select a processor with multi-way cache units so that only that part of a cache unit gets activated that has the data necessary to execute a subset of instructions. This also reduces power dissipation.

8. *External Interrupts*: There are a number of pins in the processor where the external circuits can send the interrupt signals. Section 4.1 will explain its use later.

9. *Interrupt Controller*: A processor may possess an internal interrupt controller to program the service routine priorities and to allocate vector addresses. An internal interrupt controller is of great help in most applications.

10. *Bit Manipulation Instructions*: These instructions help in easy manipulation of bits at the ports and memory addresses.

11. *Floating Point Processor*: A processor possessing the FLPU and FRS units performs the floating point operations fast. These permit higher computational capabilities in the processor; they are essential for signal processing and sophisticated control applications.

12. *Direct Memory Access (DMA) Controller with multiple channels*: When there are number of I/O devices and an I/O device needs to access a multibyte data set fast, the system memory on-chip DMA controller is of great help. Section 2.6 will explain it later.

13. *On-Chip MMU*: It is needed when using caches.

Table 2.2 lists the features for the CISC family microcontrollers and microprocessors. Table 2.3 lists features for the exemplary RISC processors from ARM, Intel and IBM. Tables show the availability or unavailability of these and characteristics in the various families of processors that are used in the embedded system hardware.

Table 2.2
Features in Four CISC Microcontroller and Processor Families

Capability	Intel 8051 and Intel 8751	Motorola M68HC11 E2	Intel 80196KC	Intel Pentium
Processor instruction cycle in μs (typical)	1	0.5	0.5	0.01[@]
Internal Bus Width in Bits	8	8	16	64
CISC or RISC Architecture	CISC	CISC	CISC	CISC with RISC feature[$]
Program Counter bits with reset value	16 (0x0000)	16 [(0xFFFE)]	16 (0x2080)	32⁻ (0xFFFF FFFF)
Stack Pointer bits with initial reset value in case a processor defines these.	8 (0x07)	16	16	32[l]
Atomic Operations Unit	No	No	No	No
Pipeline and Superscalar Architecture	No	No	No	Yes
On-Chip RAM and/or Register file Bytes	128 & 128 RAM	512 RAM	256 & 232	......
Instruction Cache	No	No	No	8 kB^
Data Cache	No	No	No	8 kB^
Program memory EPROM/EEPROM	4 k	8k	8 k	No
Program memory capacity in Bytes	64 k[1]	64k	64 k	4 GB
Data/Stack Memory Capacity in Bytes	64 k[1]	64k	64 k	4 GB
Main Memory Harvard or Princeton Architecture	Harvard	Princeton	Princeton	Princeton
External Interrupts	2	2	2	1[#]
Bit Manipulation Instructions	Yes	Yes	Yes	Yes
Floating Point Processor	No	No	No	Yes
Interrupt Controller	No	No	No	Yes
DMA Controller Channels	No	No	1 (PTS)[*]	4
On-Chip MMU	No	No	No	Yes

[@] It is Maximum time in a typical Pentium 100 MHz version [Latest Pentium 4 operates at 3 GHz].

[$] Single clock-cycle hardwired implementation for simple instructions like a RISC.

[l] Stack Pointer ESP 32 bits together with the Stack Segment ES 16 bits point to physical stack address at the memory. It equals ES * 0x10000+ ESP.

[^] This is in a standard version.

[1] Program and data memory spaces are separate in Intel 8051 family members. It is common in others.

[#] Using one INTR pin and an external programmable interrupt controller, up to 256 interrupts can be handled

[*]PTS means there is a Peripheral Transactions Server providing a DMA-like feature.

Memory addresses are in hexadecimal. Let us remember 0x10000 means the hexadecimal memory address 10000. 0x100FF means the hexadecimal memory address 100FF. This is the convention in C language, which helps us to distinguish a decimal number from a hexadecimal number.

Table 2.3
Features in Three RISC Processor Families

Capability	ARM 7~	Intel 80960CA	PowerPC MPC 604
Processor instruction cycle in μs (typical)	0.012~	0.03	0.0016
Internal Bus Width in Bits	32	32	64
CISC or RISC Architecture	Both[b]	Both[b]	RISC
Program Counter bits	32	32⁻	32[+]
Stack Pointer bits	32	32[l]	32[++]
Atomic Operations Unit	No	Yes	No
Superscalar Architecture	Yes	Yes	Yes
On-Chip RAM and/or Register file Bytes	16 kB	1536B RAM	32 kB[@]
Instruction Cache	16 kB[$]	1 kB	16 kB
Data Cache	16 kB[$]	No	16 kB
Program memory EPROM/EEPROM	-	—	...
Program Physical memory capacity in Bytes	4 GB	4 GB	4 GB
Data/Stack Physical Memory Capacity in Bytes	4 GB	4 GB	4 GB
Harvard or Princeton Main memory Architecture	Princeton~	Princeton	Princeton
External Interrupts	32	8	2
Bit Manipulation Instructions	Yes	Yes	Yes
Floating Point Processor	Yes	Yes#	Yes
Interrupt Controller	Yes	Yes	No
DMA Controller Channels	No	4	No
On-Chip MMU	Yes	Yes#	Yes

Notes:

~ARM9 processor operates at 200 MHz and has 16 x 32 MAC unit.

[b] Both RISC and CISC means the instruction set provides for both types of instructions, though internally it has RISC based core to provide fast single clock cycle processing.

32[l] means it has three 32-bit pointers at 80960, 32-means that a 32-bit Instruction Pointer replaces the program counter. There is also a register in it that stores the address of next instruction that follows a branch and link instruction.

[@] 32 kB means 16 kB instruction and 16 kB data in same cache. It has a four-way set associative 32 bytes per line in the cache in PowerPC 601.

[$] Instruction and data cache are common. [An interesting example is use of 16 kB, 32-way I-cache with 32 byte block for data and 16 kB, 32 way D-cache with 32 byte instruction block in StrongARM (Intel).]

~~ Its successor, the ARM 9 processor, has Harvard architecture.

means in 80960MC version only.

[+]PowerPC has a CR (Count Register), which replaces the program counter.

[++]LR (link register) replaces stack pointer in PowerPC.

Registers organise onto a common internal bus of the processor. A register is of 32, 16 or 8 bits depending on whether the ALU performs at an instance a 32- or 16- or 8-bit operation.

A hardware designer must also keep in mind the following processor-specific features.

1. It must be remembered when coding in assembly language or designing a compiler or locator program that data may store in *big-endian* mode in a processor. Lower order bytes store at the higher address. Examples are *Motorola processors*. A data may also store in *little-endian* mode in a processor. Lower order bytes store at the lower addresses and vice versa. Examples are

Intel processors. A processor may also be *configurable* at the initial program stage as *big-endian or little-endian* in a program. Examples are *ARM processors.*

2. It must be remembered when considering processor performance in certain high performance processors, the external data bus may access in a mode called burst mode from a memory. An exemplary processor is a 32-bit processor, 80960 that uses the burst mode bus. A memory access from the memory chip 27960 is through this burst mode. Four words transfer from the memory in four successive clock cycles, after the processor issues an address on the address bus once every four sequential accesses to the words. For example, if 80960 issues an address, 0x00000000. The bytes from 0x00000000, 0x00000001, 0x00000002 and 0x00000003 are accessed in bursts mode in four successive processor cycles. Fetching of instructions, as well as integers and floating-point numbers, becomes fast.

3. It must be remembered when coding in assembly and when organizing the main memories of certain processors, that their memory organisation may be Harvard architecture instead of Princeton architecture. Program memory and data memory have separate set of addresses and have separate instructions for instruction and data memory area accesses. A processor having Harvard architecture is needed for access to streams of data. Examples are (i) single instruction multiple data type instructions and (ii) DSP instructions.

    For example, consider a DSP computation of the following expression in a 'Finite Impulse Response (FIR) filter'. An n-th filtered output sequence, $y_n = \Sigma(a_i.x_{n-i})$ where the sum is made for i = 0, 1, 2, …, N-1. Here *i, n* and *N* and the integers. If N = 10, then for each value of *y*, first multiplication of one of the 10 coefficients, a_i and one of the 10 input sequences, *x,* is done and then the summation is done. The total computations for all 10 values of *n* will need 100 multiplications and 100 summations. Storing and accessing the coefficients from a separate set of memory-addresses in a separate memory make this access fast by using a separate set of buses.

    80x86 processors and ARM 7 have Princeton architecture for main memory. 8051-family microcontrollers have Harvard architecture. The ARM 9 processor has Harvard architecture. DSP processors invariably have Harvard architecture. Caches are now mostly organised in Harvard architecture (I-cache and D-cache).

4. A processor may have separate I/O address space for the memory and also have separate input-output and memory load-store instructions. This is so when the processor does not have to perform arithmetic and logical operations directly on I/O data and when a limited number of external I/O units need to be considered. These are the requirements in a computer. Processors of Intel 80x86 family process and access the I/O units and I/O devices by separate instructions. These processors also have a separate set of addresses for accessing inputs and outputs. It simplifies the I/O units interfacing circuit that connects the processor. When the arithmetic, logical and bit manipulation instructions that are available for data in memory are also to be made available for the I/O data, a processor should have the common I/O and memory address space and common instructions to read and write. Almost all microcontrollers, therefore, have no separate instructions for the I/O processing. Motorola processors also have no separate instructions for the I/O processing.

5. It must be remembered when coding that an embedded system has to perform multiple tasks and some variables have to be shared between the tasks. Atomic operations are necessary when processing multiple tasks. [Refer to Section 8.2.1 for shared data problem.] An embedded

processor such as 80960 provides hardware implementation. Also, special programming skills and efforts are needed in case the atomic operations are not provided in a processor. [For example, use of IPCs like semaphores. Refer to Section 8.2.2].

Let the number of ticks generated by a real-time clock (or a system clock) store in a 32-bit variable, RTC_Count. RTC_Count increases by one every time there is an interrupt from this clock. Now, consider an instruction, RTC_Count ++. It means that there is a memory address that allocates to RTC_Count. An instruction at 16-bit or 8-bit processor may implement in the following steps.

1. Step 1: Send the addresses twice or four times on the address bus. This is because the processor has to fetch the RTC_Count through twice or four times use of the data bus.
2. Step 2: Incrementing using the ALU. Incrementing will be in two or four cycles as ALU is for 16- or 8-bit operands and instruction is for 32-bit increment.
3. Stage 3: Send back at the same addresses, the incremented value of RTC_Count.
4. There is a single machine instruction in 80x86 for incrementing at a memory address. Therefore, the compiler generates that instruction. If there is an interrupt, then this instruction must first get completed, and then only should the processor permit a diversion of the current program to a service routine. There should be no diversion on partial completion of an instruction. All steps 1 to 3 must complete. If an interrupt occurs after completing the machine instructions for all three steps, this operation will be atomic because the finer details of actual implementation have been fully taken care of.
5. There is no single instruction to implement the above at 8051. The RTC_Count bits must transfer in four machine instructions to the four 8-bit registers. Then, in another instruction, the word in these increments by the successive register-increment instructions. The bytes of that register transfer again in the third step to the RTC_Count addresses. If an interrupt occurs after a machine instruction but before the transfer of the incremented bits is completed, the operation is non-atomic. Any other program that uses the memory address of RTC_Count will get the old value. On the atomic scale (processor execution level), the instruction may not complete when a processor does not have an AOU (Table 2.1). Further, on return to this program, if the RTC_Count incremented by some another function or routine, the processor will use the old (unchanged) value to complete the remaining instructions. [This is because the register bits saved earlier are retrieved on return from the interrupt service routine]. The number of counts reverts to the wrong value in spite of its change by another function.

    Consider an add operation on two memory locations (m1) + (m2) and to store the result in a location (m3). If there is an interrupt by a system unit during execution of this instruction using the ALU and if the interrupting unit service routine also uses these memory locations in between, the result from ALU will be wrong because of the shared memory between the interrupted process and interrupting process. An atomic operation ensures that the specified operation on a shared memory address will first be completed before that address location is modified as a result of a system-bus access by another unit of the embedded system. An example of another unit is a DMA Controller (Section 2.6) or a co-processor. An AOU is present at 80960. (i) Consider its Atomic ADD instruction, atadd (m1), (m2), (m3) means complete (m3)← (m1) + (m2) before that (m1) or (m2) modifies during a system bus access by another unit. (ii) Consider its Atomic Modify instruction, atmod (m1), (m2), (m3) means complete (m3)← ((m3) AND (m2)) OR ((m1) AND (m2)) before that (m1) or (m2) or (m3) modifies in the system bus access.

6. It must be remembered when considering a processor that an embedded system has to be energy efficient. A processor must have *auto-shutdown features in its various structural units* when these are not employed during a particular time interval. A processor must also have *programmability such that the use of advanced processing units in it can be disabled when processing normal instructions* and enabled when processing time-critical instruction sets. Time-critical instruction sets are needed in real-time voice or video processing systems. PowerPC has both the required features and also has special cache design for power efficiency. [Processor thus has high computing power at a lower power dissipation]. Recent StrongArm family processors have complex high power-efficiency-providing features.

(1) A processor which can operate at a higher clock speed processes more instructions per second. (2) A processor gives high computing performance when there exist (a) Pipeline(s) and superscalar architectures, (b) prefetch cache unit, caches, and register-files and MMU and (c) RISC architecture. (3) A processor with register-windows provides fast context switching in a multitasking system. (4) A power-efficient embedded system requires a processor that has auto-shutdown feature for its units and programmability for disabling these when the processing need for a function or instruction set does not have execution time constraint. It is also required to have Stop, Sleep and Wait instructions. It may also require special cache design. (5) A processor with burst mode accesses external memories fast, reads fast and writes fast. (6) A processor having Harvard main-memory architecture is needed when streams of data are required to be accessed in cases of single instruction multiple data type instructions and DSP instructions. Separate data buses ensure simultaneous accesses for instructions and data. (7) When the arithmetic, logical and bit manipulation instructions that are available for data in memory are also to be made available for the I/O operations, a processor with memory mapped inputs and outputs must be used. (8) A processor with an atomic operation unit provides hardware solutions to shared data problems when designing embedded software, or else special programming skills and efforts should be made when sharing the variables among the multiple tasks.

2.2 PROCESSOR SELECTION FOR AN EMBEDDED SYSTEM

The processor selection process can be understood by considering four representative cases. First a design-table similar to Table 2.4 is built. A processor having the required structural units and capable of giving the desired processor performance in a system is then chosen.

1. Case 1: Systems in which processor instruction cycle time ~ 1 μs and on-chip devices and memory can suffice. Examples are Vending Machine, 56 kbps Modem, Robots, Data Acquisition System (like an ECG recorder or weather recorder or multipoint temperature and pressure recorder) and Real Time Robotic Controller.
2. Case 2: Systems in which processor instruction cycle times ~ 10 to 40 ns suffice, on-chip devices and memory do not suffice and medium processor performance is required. Examples are 2 Mbps router, image processing, Voice data acquisition, Voice compression, Video decompression, Adaptive Cruise Control System with String Stability and Network Gateway.
3. Case 3: Systems in which instruction cycle times of 5 ns to 10 ns are required and high MIPS or MFLOPS performance is needed. Examples are Multi-port 100 Mbps Network Transceiver, Fast 100 Mbps Switches, Routers, Multichannel Fast Encryption and Decryption System.

4. Case 4: These are systems in which an instruction cycle time of even 1 ns does not suffice and very high processor performance is required along with use of the floating point and MAC units. Examples are Voice Processor, Video Processing, Real Time Audio or Video Processing and Mobile Phone Systems.

Table 2.4
Needed Processor Capabilities in Four Exemplary Set of Systems

Processor capability Required	Case 1: Vending Machine, Data Acquisition System, Real Time Robotic Control	Case 2: Voice data acquisition, Voice-data Compression, Video Compression, Adaptive Cruise Control System with String Stability, Network Gateway	Case 3: Multi-port Network Transceiver, Fast Switches, Routers, Multi-channel Fast Encryptions and decryptions	Case 4: Voice Processor, Video Processing and Mobile Phone Systems
Required Processor	Microcontroller	Microprocessor	Multiprocessor System	Microprocessor +DSP based Multi-processor System
Processor instruction cycle in μs (typical)	~ 0.5 to 1	0.01 – 0.04	0.005 – 0.01	0.001 – 0.005
Processor Performance	Low suffices	Medium to High	High	Very High
Internal Bus Width in Bits	8	32	32	64
CISC or RISC Architecture	CISC	CISC or RISC	RISC	RISC
Program Counter and Stack Pointers	16	32	32	32
Stack at External Memory or Internal	External	External or Internal	Internal	Internal
On-chip Atomic Operations Unit	-	-	Yes[&]	-
Pipelined and Superscalar and Pipelined Architecture	No	Yes	Yes	Yes
Off-Chip RAM in view of excessive RAM needs	No, on-chip Suffices	Yes	Yes	Yes
On-chip Register Windows and files due to fast context switching needs	No	Yes	Yes	Yes
On-Chip Interrupts Handler	Yes	Yes	Yes	Yes
Instruction and Data Caches and MMU	No	Yes	Yes	Yes
On-chip memory EPROM	Yes, on chip Suffices	No, on chip does not Suffice	No, on chip does not Suffice	No, on chip does not Suffice

External Interrupts	1 to 16	1	128-256	16
Bit Manipulation Instructions	Used	Heavily used	Heavily used	Heavily used
Floating Point Processor	No	Yes	No	Yes
Streams of Data requiring Harvard main Memory Architecture	No	Mostly Not necessary	May be Yes	Invariably Yes
DMA Controller Channels	No	No	Yes	May be Yes
Exemplary Processor Family	8051, 68HC11 or 12 or 16, 80196, PIC16F84	80x86, 80860, 80960	ARM 7, PowerPC	ARM 9, TMS family DSPs, PowerPC

& Needed when multiple ports and multichannel operations need data sharing.

Example 2.2 Case Study of a Real Time Robot Control System

1. A robotic system motor needs signaling at the rate above 50 to 100 ms. Hence there is enough time available for signaling and real time control of multiple motors at the robot when we use a processor with instruction cycle time ~ 1 μs.
2. The processor speed need not be very high and performance needed is much below 1 MIPS. So no caches and advanced processing units like pipeline and superscalar processing are required.
3. Four-coil stepper motor needs only 4-bit input and a DC motor needs a 1-bit pulse width modulated output. Therefore, 8-bit processor suffices.
4. Frequent accesses and bit manipulations at I/O ports are needed. CISC architecture therefore suffices.
5. Program can fit in 4 kB or 8 kB of internal ROM on-chip and stack sizes needed in the program are small that can be stacked in on-chip 256 or 512 Byte RAM. Microcontroller is thus needed. No floating-point unit is needed.

   Microcontrollers that are appropriate for this case are 8051, 68HC11 or 68HC12 or 68HC16 or 80196. Microcontroller 68HC12 or 68HC16 are better choices due to a large number of ports available. [68HC12 instruction cycle and clock cycle time = 0.125 ms. Number of ports = 12 in 68HC12. Therefore, 6 or more degree of freedom robot with 6 or more motors can be driven directly through these ports. STOP and WAIT instructions exist in the processor to save power when robot is at rest!]

Example 2.3 Case Study of Voice Data Compression System

1. Voice signals are pulse code modulated. The rate at which bits are generated is 64 kbps. A suitable algorithm can process the data compression of these bits with instruction cycle time ~ 0.01 to 0.04 μs (100 to 25 MHz) when the processor uses advanced processing units and caches.
2. Let us assume that the processor instruction cycle time is 0.02 μs (50 MHz). With a three-stage pipeline and two lines superscalar architecture, the highest performance will be 300 MIPS. [Refer to Example 2.1 for an understanding of the computations of MIPS.] It suffices not only for voice but also for video compression.

3. Frequent accesses and complex instructions may not be needed. Either a CISC or RISC can be used.
4. Program can't fit in 4 kB or 8 kB of internal ROM on-chip and stack sizes needed in the program are big. Instead large ROM and RAM as well as caches are needed.
5. No floating point is needed as mostly the bit manipulation instructions are processed during compression.
6. Exemplary processors that are appropriate for the above case are 80x86, 80860 and 80960.

Example 2.4 Case Study of Fast Network Switching System

1. Transfer rates of 100 MHz plus are needed in fast switches on a network. Assuming 10 instructions per switching and transceiver action, instruction cycle time should be 0.001 plus. A multiprocessor system is needed for GHz transfer rates,
2. Let us assume that the processor instruction cycle time is 0.01 μs (100 MHz). With five-stage pipeline and two lines superscalar architecture, the highest performance will be 1000 MIPS. [Refer to Example 2.1 for understanding the computations of MIPS]. The multiprocessor system is thus needed for 100 MHz plus switches.
3. The processor should have RISC architecture for single cycle instruction processing.
4. ROM and RAM as well as caches are needed.
5. No floating-point is needed as mostly the bits are processed for input and output.
6. Exemplary processors that are appropriate for the above case are PowerPC and ARM 7.

Example 2.5 Real-Time Video Processing

1. Real-time video processing requires fast compression of an image. The use of DSP is also essential. A number of real-time functions have to be processed: for instance, scaling and rotation of images, corrections for shadow, colour and hue, image sharpening and filter functions. In such cases, a multiprocessor system with DSP(s) that has the best processing performance is required.
2. All advanced processing units, which are listed in Section 2.1, are needed.
3. Exemplary processors that are appropriate for multiprocessor system are ARM 9 integrated with TMS family DSP(s) or ASIC solution using Xilinx FPGA with multiple processors Virtex–II Pro^{TM} (Section 1.6.5).

Different systems require different processor performance in terms of processing speed; they also require different processor features. A hardware designer takes these into view and selects an optimum performance–giving processor.

───┤ 2.3 MEMORY DEVICES ├──────────────────

Section 1.3.5 explained the uses of memory in a system. We learnt the functions that are assigned to the memories in embedded systems (Table 1.4). We saw six examples (Table 1.5) and these pointed to the fact that the memory requirement varies from system to system. A simple credit-debit transacting card may require just 2 kB of memory. On the other hand, a smart card for secure transactions when embedding a Java program for cryptographic functions may require 32 kB (typical value) memory. A complex embedded system may need huge memory amounts in MB.

From a memory device, a data byte, a word, a double word, or a quad word may be accessed from or at all addressable locations. A similar process would be used to access from all locations and there would be equal access time for a read or write that is independent of a memory address location. This mode is called random access mode to differentiate from a mode called serial access mode. The following subsections explain and look at certain important aspects of the memory devices. The various memory devices are described from the point of view of an embedded system hardware or software designer.

2.3.1 ROM: Its Uses, Forms and Variants

ROM non-volatility is a most important asset and it is extremely useful to embed codes and data in a system. ROM is a loosely used term. It may mean masked ROM, PROM, OTP-ROM, EPROM and EEPROM for a hardware designer. In a strict sense, ROM means a masked ROM made at a foundry from the programmer's ROM image file (Section 1.4.2). ROM that embeds the software or an application logic circuit is in one of the three forms: masked ROM, PROM and EPROM. When ROM is to be programmed during run-time and is to hold the processed result, it is used as either an EEPROM or a flash memory device.

(i) Masked ROM

A **masked ROM** is built from a circuit, which has r inputs (A_0 to A_{r-1}) and 8 outputs (D_0 to D_7). [Byte storing at an address is most common.] The circuit for masked ROM is one of a set of 2^r combinational circuits. Appropriate masking gives a desired set of outputs at each combinational circuit. Certain links fuse and others that are masked do not fuse. [A combination circuit is a circuit made up of logic gates with a distinct set of output logic states during distinct input logic states. It has a distinct truth table for r inputs x 8 outputs. As soon as the inputs change (or withdraw), the output also changes in this circuit.].

The embedded software designer (after thorough testing and debugging) provides to a manufacturing foundry a file having a table of desired output bits for the various combinations of the input address bits. [Locator program (Section 2.5) creates this table.] The manufacturer prepares the programming masks and then programs the ROM at a foundry. This ROM is returned to the system manufacturer. [Normally, a one-time masking charge could be about Rupees 100,000 (~US $ 2,000).] Generally, therefore, a system manufacturer will place the order and the manufacturing foundry will accept the order for a minimum of 1,000 pieces. The ROM is a cost effective solution to a bulk user of ROMs for the manufacture of embedded systems. An embedded system manufacturer using a masked ROM does not have to use a device programmer (ROM burner) each time a system ROM is made using EPROM or PROM. *Masked ROMs are never written at the system manufacturer level.*

(ii) An EPROM, an E^{2}PROM and an OTP ROM

Special versions of ROM can be programmed at the designer or manufacturer site for an embedded system with the help of a *device programmer*. [Refer to Appendix G.2 for device programmer working.] One version is EPROM. It is an ultraviolet ray *Erasable* and device programmer *Programmable Read Only Memory* device. Erasing the device means restoring 1 at each bit in the cell arrays at each ROM address. Another version is E^{2}PROM (EEPROM). It is an *Electrically Erasable and Programmable Read Only Memory* device. Examples of EPROM and EEPROM are 2732, which is a 4 kB EPROM; 28F256, which is a 32 kB EEPROM and 28F001 is a 512 K x 16-bit EEPROM.

EEPROM erasing during an application-program run is done by sending all eight data bus bits as 1s for the write in the presence of a high voltage (+5V or 12V) and a short duration write pulse. A 12V EEPROM is used when EEPROM is to be used as EPROM. The 5V form is used when erase and write is done during the system run. Erase operation is as follows: Write a byte having all 1s at an address. Programming operation is as follows: Write a byte having 1s and 0s at the erased address. The processor within the system can do the erasing and programming. The write operation is similar to write in RAM. What then, is the difference between the EEPROM and RAM? The difference is that in RAM, the read and write timing cycles are identical. Here, the write cycle has to be longer than in the case of RAM, and it must succeed the erase of the byte by writing 0xFF. Further, an addition a voltage V_{pp} signal is needed when erase and write occurs at the EEPROM. The number of times an EEPROM can be written is 1 million times plus. There is no limit for RAM, and a practically infinite number of writes is possible without first writing 1s.

Flash memory is a recent 5V form of EEPROM in which a sector of bytes can be erased in a flash (very short duration corresponding to a single clock cycle). [Lately flashes of even ~ 3V form have become available.] A sector can be from 256 kB to 16 kB. The advantage over EEPROM is that the erasing of many bytes simultaneously saves time in each erase cycle that precedes a write cycle. The disadvantage is that once a sector is erased, each byte writes into it again one by one, and that takes too long a time. A new version of flash is **boot block flash**. A sector is reserved to store once only at the time of first boot. Later on it is protected from any further erase. In other words, it has an OTP sector also that can be used to store ROM image like in a ROM.

PROM (an OTP ROM, a one time device programmer programmable) is another form. A PROM, once written, is not erasable.

(iii) Uses of ROM or EPROM

Figure 1.5 showed what a ROM embeds: *Program codes for various tasks, interrupt service routines, operating system kernel, initialisation (bootstrap program and data) and the standard data or table or constant strings*.

The ROM is not only used for program and data storage, but also used for obtaining the preprogrammed logic outputs and output sequences for the given sets and sequences of inputs. [Inputs are given analogous to an address signal by the processor and outputs are obtained analogous to those obtained during a processor read cycle.] Assume that there are 8 inputs ($r = 8$). The truth table for it will have 256 combinations. 8 x 8 ROM can be programmed to generate 256 sets of 8 bit-outputs for each combination. Examples of applications of preprogrammed logic outputs are as follows.

1. A ROM is used to hold the language specific bits for the fonts corresponding to each character in a printer.

2. A ROM is used to hold the images bits for a display. A pictogram generates from these bits. A ROM is used in a display circuit. It stores the bytes for the full bit-image corresponding to the pixels for a pictogram. Sequential changes at the inputs repeatedly generate the full pictogram.

There is a use of the ROM in a CISC as a control ROM at a micro-programmed unit [Section A.1.4.]. It stores a set of microinstructions for each processor instruction. A set of microinstructions is stored in a sequence such that it specifies a set of signals for the various fetch and executing units for an instruction.

(iv) Uses of EEPROM, Flash and OTP

An EEPROM is usable by erase over 1 million times. It can be erased during run-time itself. Flash memory is usable about 10,000 times for repeated erasing followed by programming during the run-time. The PROM is written only once by a device programmer or first time system run.

Three examples of EEPROM memory device applications are as follows: **(i)** Storing current date and time in a machine. **(ii)** Storing port statuses. **(iii)** Storing driving, malfunctions and failure history in an automobile for use by mechanics later on.

Three examples of flash memory device applications are as follows: **(i)** Storing the photographs in a digital camera. **(ii)** Storing voice compressed form in a voice recorder. [Recall of prerecorded message from a telephone exchange.] **(iii)** Storing messages in a mobile phone.

Examples of uses of an OTP ROM are as follows: **(i)** Smart card identity number and user's personal information. **(ii)** Storing boot programs and initial data like a pictogram displaying a seal or monogram. **(iii)** ATM card or credit card or identity card. Once the various details are written at the bank and handed over to the account holder, there is no modification possible in the embedded PROM of the card. Just as a paper holds information permanently once written or printed, so does a PROM.

2.3.2 RAM Devices

A system designer considers RAM devices of eight forms. These forms are SRAM, DRAM, NVRAM, EDO RAM, SDRAM, RDRAM, Parameterised Distributed RAM and Parameterised Block RAM. They are explained later in Appendix G (Section G.2).

(i) Uses of RAM

As explained in Section 1.3.5, the RAM is for storing the variables during program run and storing the stack. It stores input and output buffers, for example, of speech or image. It can also store the application program and data when the ROM image is stored in a compressed format in an embedded system and decompression is done before the actual run of the system.

SRAM is used most commonly.

DRAM is used mostly in computers or high memory density systems.

1. EDO RAM is used in systems with buses to the devices when operating with clock rates up to 100 MHz, and when a zero-wait state is needed between two fetches and there is single cycle read or write.

2. SDRAM synchronises the read operations and keeps the next word ready while the previous one is being fetched. This device is used when buses can fetch to the processor up speed of to 1 GHz.

3. RDRAM accesses in burst and gets four successive words in a single fetch and thus gives greater than 1 GHz performance of the total system.

4. In Parameterised Distributed RAM, the RAM distributes in various system subunits. I/O buffers and transceiver subunits can have a slice of RAM each and a system stack can be at another slice. Distribution provides buffering of memory at the subunits before they are fetched and processed by the processor. It facilitates faster inputs at the I/O devices than the processor system buses access the memory devices.

5. Parameterised Block RAM is used when a specific block of the RAM is dedicated for use by a subunit only, for example, a MAC unit. A parameterised block RAM is used when an access by the bus is slow compared with the processing speed of the subunit.

Different types of memory devices in varying sizes are available for use as per requirement. (1) Masked ROM or EPROM or flash stores the embedded software (ROM image). Masked ROM is for bulk manufacturing. (2) EPROM or EEPROM is used for testing and design stages. (3) EEPROM (5V form) is used to store the results during the system program run time. It is erased byte by byte and written during the system-run. It is useful to store modifiable bytes, for example, the run-time system status, time and date and telephone number. (4) Flash stores the results byte by byte during a system run after a full sector erase. (5) Flash is thus very useful when a processed image or voice is to be stored or a data set or system configuration data is to be stored, which can be upgraded as and when required. For example, in flash, a new image (after compressing and processing) can be stored and the old one erased from a sector in single instruction cycle. (6) Boot block flash has an OPT sector(s) also to store boot program and initial data or permanent system configuration data. It serves by storing ROM image or its part in OTP sector(s) and at the same time serves by storing as a flash in other sectors. (7) RAM is mostly used in SRAM form in a system. (8) Sophisticated systems use RAM in the form of a DRAM, EDO RAM, SDRAM or RDRAM. (9) Parameterised distributed RAM is used when the I/O devices and subunits require a memory buffer and a fast write by another system. (10) Subunits like MAC use separate blocks of RAM when operating at fast speeds.

2.4 MEMORY SELECTION FOR AN EMBEDDED SYSTEM

Once software designer coding is over and the ROM image file is ready, a hardware designer of a system is faced with the questions of what type of memory devices and what size of each should be used. First a design table like Table 2.5 is built. The memory device with the needed feature and size is chosen. With the following case studies as the examples, let us understand how these questions are answered. The actual memory requirement can be known only after coding as per the given specification. ROM image size and RAM allocations for various segments, data sets and structures will be available from the software designer. However, the following case studies suggest a method to get a prior estimate of the memory type and size requirements. [Remember, the memory devices are available in steps. Examples are 1 kB, 4 kB, 16 kB, 32 kB, 64 kB, 128 kB, 256 kB, 512 kB and 1 MB. Therefore, if 92 kB memory is needed then a device of 128 kB is selected.]

Table 2.5
Required Memory Devices in Four Exemplary Sets of Systems

Memory Required	Case 1: Vending Machine, System, Real time Robotic Control	Case 2: Data Acquisition System	Case 3: Multi-port Network Transceiver, Fast Switches, Routers, Multi-channel Fast Encryptions and decryptions	Case 4: Voice Processor, Video processing and Mobile Phone Systems	Case 5: Digital Camera or Voice Recorder System
Processor Used	Micro-controller	Micro-controller	Multi-processor System	Micro-processor +DSP based Multi-processor System	Micro-processor
Internal ROM or EPROM	4 to 8 kB	8 kB	-	-	-
Internal EEPROM	256 B to 512 B	256 B to 512 B	-	-	-
Internal RAM	256 B to 512 B	256 B to 512 B	-	-	-
ROM or EPROM device	No	No	64 kB	64 kB	64 kB
EEPROM or Flash$ device	No to 126 kB	64 kB	512 B	32 kB	256 kB to 16 MB
RAM Device	No	4 kB to 8 kB	64 kB	1 MB	1 MB
Parameterised Distributed RAM	No	No		Yes for I/O buffers. 4 kB per channel	-
Parameterised Block RAM	No	No		Yes for MAC unit, Dialing I/O unit	-

Note: $ Flash with a boot block can be used to store the protected part of the boot program in its OTP sector (s).

Examples 2.6a and b—Case Studies of an clothes washing machine and chocolate vending machine systems

1. Consider an automatic washing machine system. (a) An EEPROM first byte is required to store the state (wash, rinse cycle 1, rinse cycle 2 and drying), which has been finished at any instant of machine run. An EEPROM second byte is required to store the time in minutes already spent at the current stage. The third byte is needed to store the status of the user set buttons. Thus 128 bytes EEPROM at best should suffice in microcontrollers. (b) Embedded software can be within 4 kB ROM at the microcontroller. (c) RAM is needed only for a few variables and stacks. An internal RAM of 128 B should suffice. (d) There-fore, no external memory device is required with the system when using a microcontroller.

2. Consider a *vending machine* system. (a) EEPROM bytes are required to store the time and date. EEPROM bytes are needed to store the machine status and number of each type of coins in the cash collection channel and in the coin-refunding channel. Thus 128 bytes EEPROM in microcontroller at best shall suffice. (b) Embedded software can be within 4 kB ROM at the microcontroller. RAM is needed just for the few variables and stacks. An internal RAM of 128 B should suffice. (c) Therefore, no external memory device is required with the system when using a microcontroller.

3. Consider a robotic system. (a) EEPROM bytes are required to store the rest status of each degree of freedom. Thus 512 bytes EEPROM in microcontroller at best shall suffice. (b) Embedded software can be within 8 kB ROM at the microcontroller. (c) RAM is needed only for the variables and only one stack is needed for the return address of the subroutine calls. Internal RAM of 512 B shall suffice. Therefore, no external memory device is required with the system when using a microcontroller.

Examples 2.7a and b—Case Studies of the Data Acquisition Systems for the sixteen-parameter channels and for the ECG waveforms

1. Consider a data acquisition system. Assume that there are sixteen channels and at each channel the 4 B data store every minute. (a) Bytes are to be stored at a flash. Assume that the results are stored at a flash for a day before it is printed or transferred to a computer. Thus 92 kB is the data acquired per day. 128 kB flash will thus suffice. (b) Embedded software can be within 8 kB ROM at the microcontroller. (c) RAM is needed only for the variables and only one stack is needed for the return address of the subroutine calls. An internal RAM of 512 B will suffice. (d) Intermediate calculations are needed for storing ADC results in proper format. Unit conversion functions need to be calculated, which may necessitate a RAM of about 4 kB to 8 kB. (e) Therefore, a microcontroller with 8 kB EPROM and 512 B RAM is required and the external flash (or 5V EEPROM) of 128 kB and external RAM of 4 kB to 8 kB are required with the system.

2. Consider another data acquisition system, which is used for recording the ECG waveforms. Let each waveform be recorded at 256 points. A 64 kB flash will be required for 256 patient records.

Example 2.8—Case Study of a Multichannel Fast Encryption-cum-decryption Transceiver System

1. Consider a system with multiple channels. There are encrypted inputs at each channel. These are decrypted for retransmission to other systems.

2. EEPROM is required for configuring ports and storing their statuses. Assume 16 channels. The 512 kB shall suffice for a 16 B need per channel.

3. Encryption and decryption algorithms can be in 64 kB ROM.

4. Multichannel data buffers are required before the caches process the algorithms. Therefore, 1 MB RAM may be required.

5. I/O buffer storage of 4 kB per channel is needed. If a parameterised distributed RAM is employed at each channel, the system performance will be increased.

6. The system will thus be expected to need memory devices, 64 kB ROM, 512 B EEPROM, 1 MB RAM and 4 kB per channel distributed parameterised RAM.

Example 2.9—Case Study of a Mobile Phone system

1. Due to the need for processing voice compression-decompression and encryption-decryption algorithms and for the algorithms for DSP processing, the ROM image will be large. Assume it is can be taken as 1 MB. Now if the ROM image is stored in a compressed format, a boot-up program first runs a decompression program also. The decompressed program and data first load at RAM and then the application program runs from there. The RAM is obviously of bigger size in these systems. ROM can be reduced as per compression factor.
2. A large RAM is also needed. It can be taken as 1 MB for storing the decompressed program and data and the data buffers.
3. The phone memory for entering important telephone numbers can be at 16 kB EEPROM. The EEPROM is taken due to a byte-by-byte need when the numbers change. A flash of 16 kB can be taken for recording of the messages.
4. Parameterised block RAM at MAC subunit and other subunits will improve the system performance.
5. The system will need many memory devices, 1 MB ROM, 16 kB EEPROM, 16 kB Flash, 1 MB RAM and block RAM at the subunits as per requirement.

Example 2.10—Case Study of Digital Camera and voice recorder

1. Assume a low resolution uncoloured digital camera system. The gif (graphic image format) compressed format images are to be recorded. (a) Assume that an image has Quarter-CIF (Common Intermediate Format) of 144 x 176 pixels. Then, 25,344 pixels are to be stored per image. Assume that compression reduces the image by a factor of 8 and then the 3 kB per image will be needed. Flash required will be 0.2 M for 64 camera images. Therefore, a 256 kB flash will be required. (b) A 64 digital uncoloured images camera system will thus be estimated to need memory devices, 64 kB ROM, 256 kB flash and 1 MB RAM.
2. Assume a voice recording system. 64 kbps are required, assuming 8-bit pulse code modulation of the voice signals. [Average frequency is taken as 8 kHz.] Assume voice data compression factor of 8, 1 kB flash is require per second. 4 MB flash shall be required for each hour record.
   Since voice compression–decompression algorithms have to be processed, the ROM image will be large. It can be taken as 1 MB. Using compression techniques, a 64 kB ROM can be store the ROM image.
4. The RAM needed is large for storing the decompressed program. It can therefore be estimated as 1 MB.
5. A one-hour voice recorder system is estimated to require the memory devices, 64 kB ROM, 4 MB flash and 1 MB RAM

Simple systems like vending machines or robots needs no external memory devices. The designer selects a microcontroller, which has the on-chip memory required by the system. The data acquisition system needs EEPROM or flash. A mobile phone system needs 1 MB plus RAM device and 32 kB plus EEPROM or flash device. Image or voice or video recording systems require a large flash memory.

2.5   ALLOCATION OF MEMORY TO PROGRAM SEGMENTS, BLOCKS AND MEMORY MAP OF A SYSTEM

2.5.1   Functions, Processes, Data and Stacks at the Various Segments of Memory

Program routines and processes can have different segments. For example, a program code can be segmented and each segment stored at a different memory block. A pointer address points to the start of the memory block storing a segment and an *offset* value is used to retrieve for a memory address within that segment. The data can also be segmented with each segment at different blocks. Similarly, strings can be segmented. Figure 2.2 (a) shows different segment types that may be required in a software design. A segment can have partitions of fixed sizes called pages. A processor, for example, an 80x86, may have segment registers and offset registers [Figure 2.2 (b)].

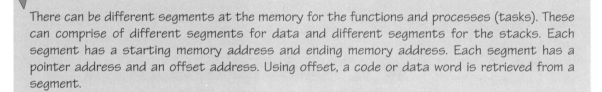

There can be different segments at the memory for the functions and processes (tasks). These can comprise of different segments for data and different segments for the stacks. Each segment has a starting memory address and ending memory address. Each segment has a pointer address and an offset address. Using offset, a code or data word is retrieved from a segment.

2.5.2   Memory Blocks for Elements of the Different Data Structures and Data-sets

The software design approach is to use data sets and data structures in a program. [Section 5.4.2.] It is therefore important for the system software designer to understand the following details.

There can be different sets and different structures of data at the memory. Following are the data structures and data sets that are commonly used during processing in a system and that are stored at the different memory blocks in a system.

A. A data structure, called a *stack*, is a special program element and an allotted memory block from which a data element is always read in a LIFO (*Last In First Out*) way by the processor. Various stack structures may be created during processing. Figure 2.3 shows the various stack structures that are created during execution of embedded software.

1. A call can be made for another routine during running of a routine. On completion of the called routine, in order for the processor to return only to the one calling, the instruction address for return must be saved on the stack. There can also be nesting, which means one routine calling another. The return from the called routine is always to the calling routine. Therefore, at the memory a block of memory address is allocated to the stack that saves the ***return addresses*** of the nested calls.
2. There may be at the beginning an input data saved as a stack at RAM in order to be retrieved later in LIFO mode. An application may create the run-time stack structures. There can be ***multiple data stacks*** at the different memory blocks, each having a separate pointer address. [Section 5.6 will describe uses of the *stack* structures by a software designer.]

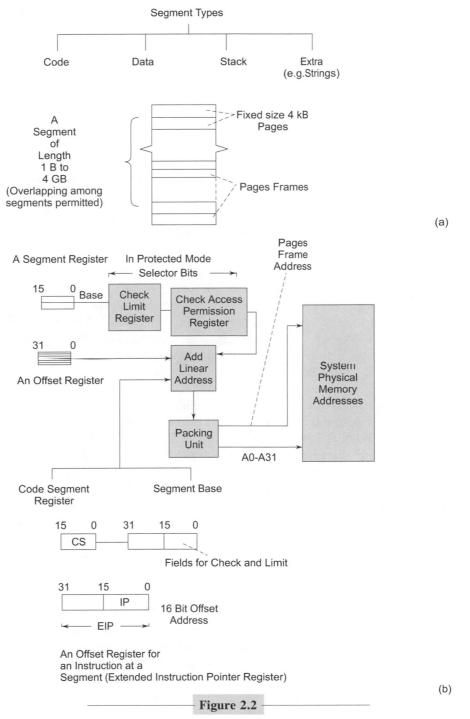

Segment Types

Code   Data   Stack   Extra
(e.g.Strings)

A
Segment
of
Length
1 B to
4 GB
(Overlapping among
segments permitted)

Fixed size 4 kB
Pages

Pages Frames

(a)

A Segment Register   In Protected Mode
Selector Bits

Pages
Frame
Address

15   0   Base   Check Limit Register   Check Access Permission Register

31   0

Add Linear Address

An Offset Register

Packing Unit

A0-A31

System Physical Memory Addresses

Code Segment Register   Segment Base

15   0   31   15   0

CS

Fields for Check and Limit

31   15   0

IP

EIP

16 Bit Offset Address

An Offset Register for
an Instruction at a
Segment (Extended Instruction Pointer Register)

(b)

Figure 2.2

*(a) The segment types and pages in an exemplary program (b) An exemplary use
of its registers, CS and EIP in an 80x86 for fetching or writing to a physical memory address.*

3. Each task or thread in a multitasking or multithreading software design (Section 8.1) should have its own stack where its *context* is saved. The context is saved on the processor switching to another task or thread. The context includes the return address for the program counter for retrieval on switching back to the task. There can be ***multiple stacks*** at the memory for the different contexts at the different memory blocks, each having a separate pointer address. Application programs and supervisory programs (OS) have separate stacks at separate memory blocks.

Each processor has at least one stack pointer so that the instruction stack can be pointed and calling of the routines can be facilitated. There is a stack pointer in 8051, 68HC11 and 80196. Some advanced processors have multiple stack pointers. There are four pointers, RIP, SP, FP and PFP in 80960. MC68010 provides USP (User Stack Pointer) and SSP (Supervisory Stack Pointer). MC68040 provides for USP (User Stack Pointer), SSP (Supervisory Stack Pointer), Memory Stack frames pointers, and Instruction Stack Pointer. [A block may also be called a frame.] When a processor has only one stack pointer, the OS allocates the memory addresses that are used as the pointers for the multiple instruction and data stacks.

> Stack is a special data structure at the memory. It has a pointer address that always points to the top of a stack. This pointer address is called a stack pointer. A processor has at least one stack pointer. A value from the stack is retrieved from the memory in LIFO mode, while a row of data or a data in a table or a data in the queue is accessed in a FIFO (First In First Out) mode. Each process should have a separate top-of-stack pointer and a separate block at its allocated memory. Since there are multiple processes in an embedded system, there are multiple stacks.

B. A data structure, *array,* is an important programming element. Let us look at the results of a test in a class with 30 students with roll numbers 1 to 30. Let i be an index used instead of roll number. Let marks in the test of roll number 1 be in the scalar integer variable, M [0]. Let M [0], M [1], ..., M [28] and M [29] be the variables for the marks of roll numbers 1, 2, ..., 29, 30, respectively. There is a pointer that points to the first scalar value M [0]. It may be pointed out by a register called index pointer. The index register could then be incremented from 0 to 29 by an instruction within a loop to point to the marks of students of succeeding roll numbers. Take another example. An expression, $y_n = \Sigma (a_i.x_{n-i})$ has the coefficient a_i. These are stored as an array. Input x_i also stores as another array and output y_k as yet another array. Here i, j and k are the integers each varying from - N to N-1, where N is per the limits. Figure 2.4 (a) shows an array in memory block with a pointer and index that jointly point to its element.

> One-dimensional array is a special data structure at the memory. It has a pointer address that always points to the first element of the array. From the first element pointer and index of that element, an address is constructed from which the processor can access one of the array elements. Index is an integer that stars from 0. Data word can be retrieved from any element address in the block that is allocated to the array. A processor register may also be used for storing the index and another register for array base pointer.

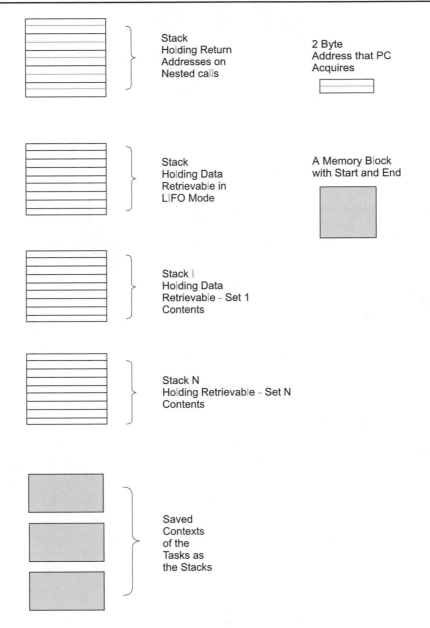

Stack
Holding Return
Addresses on
Nested calls

2 Byte
Address that PC
Acquires

Stack
Holding Data
Retrievable in
LIFO Mode

A Memory Block
with Start and End

Stack
Holding Data
Retrievable - Set 1
Contents

Stack N
Holding Retrievable - Set N
Contents

Saved
Contexts
of the
Tasks as
the Stacks

Figure 2.3

An example of different stack structures at the memory blocks. Each Stack Pointer points to the top of the stack to where the processor can read or write. A data word always retrieves in LIFO mode from a stack.

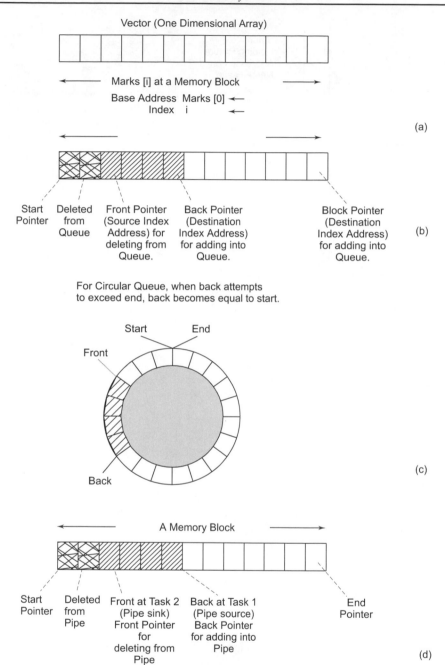

Figure 2.4

(a) An array at a memory block with one pointer for its base, first element with index = 0. Data word can retrieve from any element by defining the pointer and index. (b) A queue at a memory block with two pointers to point to its two elements at the front and back. A data word always retrieves in FIFO mode from a queue. (c) A circular queue at a memory block with two pointers to point front and back. (d) A memory block for a pipe with front and back points at two different tasks.

C. A data structure, called a *queue,* is another important programming element. [A *queue* is an allotted memory block from which a data element is always retrieved in FIFO mode.] Using the queues, the bytes are sent on a network or to a file or to a printer. There are two pointers needed. One is for pointing to an address in a memory block where an element can be inserted (add-on writing). This pointer should increment on each addition. It is called queue back or tail pointer. The other pointer is for pointing to an address in a memory block from where an element can be deleted (remove on reading). This pointer should increment on each deletion. Both pointers, at the beginning, point to starting memory address at the block. It is called queue front or head pointer. Insertions into a queue are usually faster than deletions. For example, in a queue at the printer where system inserts the values faster than the values are printed. The difference in addresses at the two pointers is as per queue current length. Section 4.5 will describe the uses of the queues during coding for the embedded software. Since there is a possibility that the tail pointer may increment beyond a limit set by the end address of the memory block, an exception (an error indication) is usually thrown whenever the pointer increments beyond the block end-boundary. Further increments beyond the limit may prompt intrusions into other blocks. Figure 2.4 (b) shows a memory block with the two pointers needed for insertions and deletions.

A queue is a data structure with an allotted memory block (buffer) from which a data element is always retrieved in FIFO mode. It has two pointers, one for its head and the other for its tail. Any deletion is made from the head address and any insertion is made at the tail address. An exception (an error indication) must be thrown whenever the pointer increments beyond the block end boundary so that appropriate action can be taken.

D. A queue is called a circular queue when a pointer on reaching a limit, returns to its starting value. [A circular *queue* is also a bounded memory block allotted to a queue such that its pointer on increasing never exceeds the set limit.] From a circular queue also, the data element is always retrieved in FIFO way mode and no exception is thrown on exceeding the limit of the memory block allocated. Figure 2.4 (c) shows a memory block with a circular queue with its two pointers needed for insertions and deletions.

A circular queue is a queue in which both pointers cannot increment beyond the memory block (buffer) and reset to starting value on insertion beyond the boundary.

E. A queue is called a *pipe,* generally when the source from where the insertions are made has an identity distinct from a destination (sink) entity where deletions are made. A *pipe* means a common memory block allotted for a queue to two distinct entities that are interconnected in some way. For example, one writes (inserting) into the queue and other reads (deleting). A pipe usually connects two tasks. Figure 2.4 (d) shows a memory block with a *pipe.*

A pipe is a queue with distinctly defined source and destination entities. Sometimes, programmers loosely refer to a 'pipe' as a 'queue' and vice versa.

F. A *table* is a two-dimensional array (matrix) and is an important data set that is allocated a memory block. There is always a base pointer for a table. It points to its first element—the first column first row. There are two indices, one for a column and the other for a row. Figure 2.5 (a) shows a memory block with the pointers for a table. Like an array, any element can be retrieved from three addresses; table base, column index and row index. Instead of a pointer, value is used in an instruction that value is called *displacement*. Displacement can be for a column or row.

> A table is a data set allocated with a memory block. Three pointers, table base, column index and destination index pointers can retrieve any element of the table.

G. A *hash table* is a data set that is a collection of pairs of a key and a corresponding value. A hash table has a key or name in one column. The corresponding value or object is in the second column. The keys may be at non-consecutive memory addresses. Look-up tables store like a hash. If the first column of a table is used as a key (pointer to the value) and the second column as a value, we call that table a look-up table. Figure 2.5 (b) shows a memory block with the pointers for a hash.

> A hash table is a data set allocated with a memory block, usually for the look-up table. Just as an index identifies an array element, a hash-key identifies a hash element.

H. A *list* is a data structure with a number of memory blocks, one for each element. A list has a top (head) pointer for the memory address—where it starts. Each list-element of the memory also stores the pointer to the next element. The last element points to *null*. A *list* is for non-consecutively located objects at the memory. Figure 2.5 (c) shows the memory blocks with the pointers for a list.

> A list is a data structure in which each element also stores a pointer to the next element at the list. It has one memory block allotted to each of its elements. The list-top pointer points to its first element and the last element points to null.

Example 2.11   A function in an application program has the following design. Allocation of the memory blocks for the function's data sets and structures are given below.

1. There are ten nested calls to the system or other functions. Memory allocation required for a stack structure for saving return instruction addresses = 40 B. [Assume that program counter address is of 4 B.]

2. On each call the following are also saved on a stack. (i) Four pointers (addresses each of 4 bytes). (ii) Four integers (each of 4 bytes) and (iii) Four floating point numbers (each of 4 bytes). Memory allocation required for a stack structure for saving function parameters = (4 x 4) + (4 x 4) + (4 x 4) = 48 B.

3. There are three arrays for calculating 20 filtered output sequences by the expression, $y_n = \Sigma (a_i . x_{n-i})$. [Section 2.1.] Memory allocations required for each array structure = 20 x 8 = 160 B. [Assume each element is a double precession floating pointer number of 8 B (64 bits IEEE 754 format).]

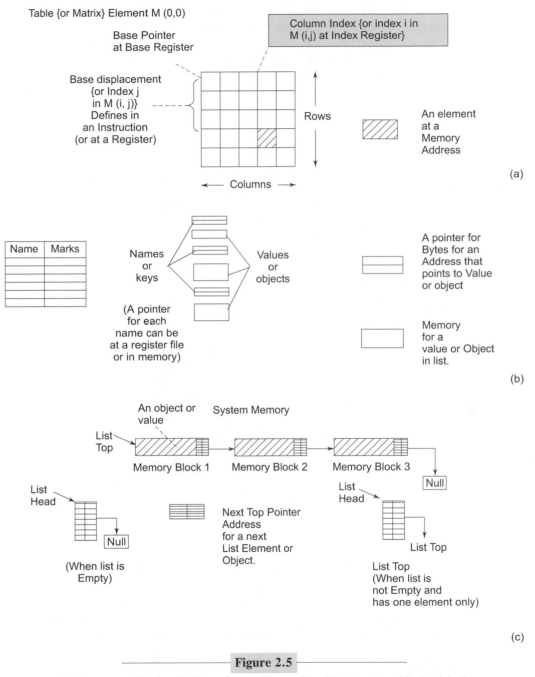

Figure 2.5

(a) A memory block with the pointers for a table (b) A memory block with the pointers for a hash or lookup table (c) The memory blocks with the pointers for a list.

2.5.3 The Memory Maps

Figure 2.6 (a) shows memory areas needed in the case of Princeton architecture in the system. Figure 2.6 (b) shows memory areas needed in the case of Harvard architecture.

1. Vectors and pointers, variables, program segments and memory blocks for data and stacks have different addresses in the program in Princeton memory architecture.
2. Program segments and memory blocks for data and stacks have separate sets of addresses in Harvard architecture. Control signals and read-write instructions are also separate. [Refer to Sections 2.5.1 and 2.5.2 for segments and blocks.]

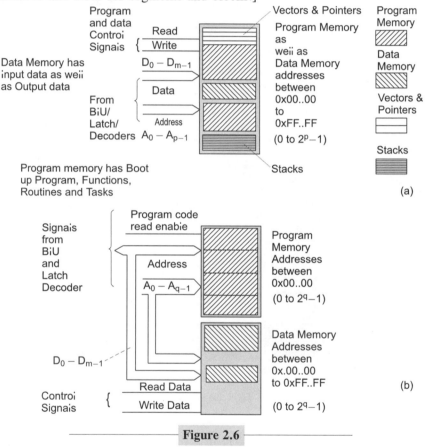

Figure 2.6

(a) Memory map (Princeton architecture) (b) Memory map (Harvard architecture).

Figure 2.7 shows external memory devices in 80960 and the memory block in it coexisting with the port addresses and the memory addresses in 80960. The *system memory allocation map* is not only a reflection of addresses available to the memory blocks—and the program segments and addresses available to I/O devices—but also reflects a description of the memory and I/O devices in the system hardware. It maps guides to the actual presence of the various memories at the various units, EPROM, PROM, ROM, EEPROM, Flash Memory, SRAM (static RAM), DRAM (Dynamic RAM) and I/O devices. It reflects memory allocation for the programs, and data and I/O operations by the locator program. It shows the memory blocks and ports (devices) at these addresses. Figures 2.8 (a) and (b) show memory and I/O devices memory allocation map for the 68HC11 (having memory mapped I/O architecture), and for the 80x86 PC (having I/O mapped I/O architecture), respectively.

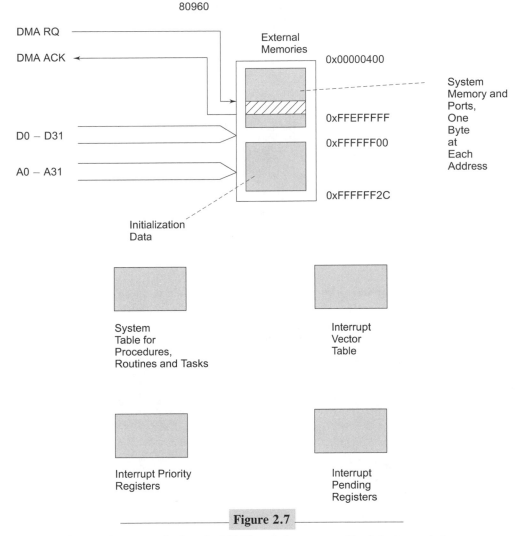

Figure 2.7

External memory devices in 80960 and the memory block in it coexisting with the port addresses. [Bus signals also shown.]

Four examples of memory allocation maps are given in Figures 2.9 (a) to (d). System I/O devices maps may be designed separately. This not only reflects the actual presence of the I/O devices, but also guides the available addresses of the various device registers and port-data. [An example of a device is a timer. I/O devices are the peripheral units of the system.]

The following is a description of memory allocation maps in the locator programs for four exemplary systems in Figure 2.9.

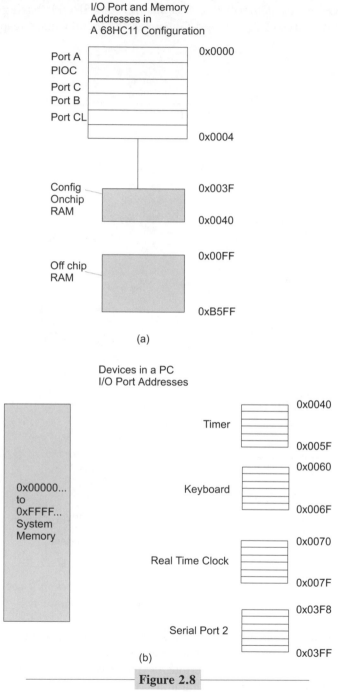

Figure 2.8

(a) I/O port, memory and devices address spaces in 68HC11 (b) Device addresses in 80x86-based PC.

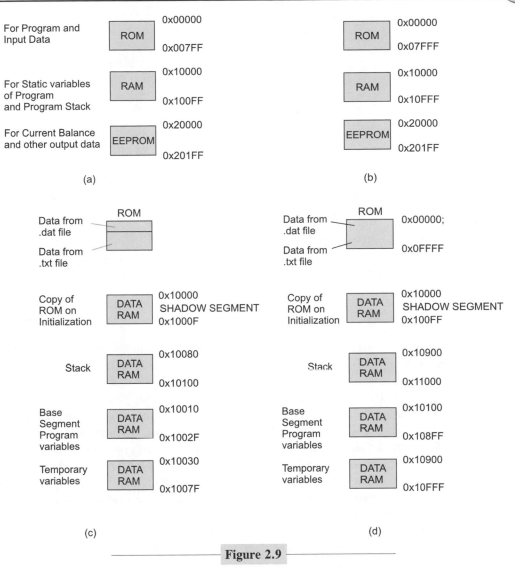

Figure 2.9

(a) to (d) Four memory allocation maps in four exemplary systems and their locator programs.

Example 2.12 The memory map for an exemplary embedded system, a debit-credit card, needs a 2 kB memory. It also needs a 256 B RAM mainly for stacks. It also needs EEPROM 512 B for storing the balance amount under credit or debit and the previous transaction records on the card. [This is an alternative to a magnetic strip on the card.] So the memory locator or linker script program for this system may define memory map as follows.

1. Memory
2. {ram : ORIGIN = 0x10000, LENGTH = 256

3. eeprom : ORIGIN = 0x20000, LENGTH = 512
4. rom : ORIGIN = 0x00000, LENGTH = 2K
5. }

Example 2.13 The memory map for an exemplary embedded system, a Java-embedded card with software for encrypting and deciphering transactions, needs a 32 kB ROM memory and a RAM 4 kB. It also needs EEPROM 512B for storing not only the balance amount under credit or debit but cryptographic keys and the previous transaction records on the card. [This is an alternative to a magnetic strip on the card.] So the memory locator or linker script program for this system may define the memory map as follows.

1. Memory
2. { ram : ORIGIN = 0x10000, LENGTH = 4K
3. eeprom : ORIGIN = 0x20000, LENGTH = 512
4. rom : ORIGIN = 0x00000, LENGTH = 32K
5. }

One can also make the following important observation from the above examples. There are memory space gaps between the origin of ROM, RAM and EEPROM in spite of the very small lengths of available memory. This gap is due to a design feature of its hardware designer providing for expansion of these memories in the future hardware that will need no change in the interfacing decoder circuit [Section 2.7] between the memories and processor. Further, its software program has to make minimal changes. The changes will only be in length. This is because when there is no gap the origin will also change. This feature ensures that any future changes in the program code sizes and data sizes will not need the change in the locator codes. A feature of a locator program must also be that it does not relocate the addresses of the special purpose ports that are dedicated to a particular I/O task or dedicated to the device driver read and write operations. Section G.2 gives the formats of the locator output records that are inputs for a device program.

The final step of the design process in an embedded system is to locate the bytes at the ROM from: the image for the bootstrap (reset) program and data, initialisation data as well as the standard data or table or constant strings device driver data and programs, the program codes for various tasks, interrupt service routines and operating system kernel. [The bootstrap program consists of the instructions that are executed on the system reset. The bootstrap data example is for stack pointer initialisation. The initialisation data may be for defining initial state and system parameters. The constant strings may be for the initial screen display.] There is a shadow segment at ROM. The shadow segment has the initialisation data, constant strings, and the start-up codes that are copied into the RAM by a shadow segment copy program at the system boot up. When a start-up code (booting) program is executed, a copy of the shadow segment from the ROM is generated at the RAM. The RAM also holds the data (intermediate and output data) and stack. [*A compressed program format locates at the ROM in case of long ROM image of the system program.*] *This is because the decompression program plus compressed image will need less memory than the uncompressed ROM image. The start-up code task is to generate the decompressed program codes and store into the RAM. The processor executes these subsequently by fetches from the RAM.*

A processor may have predefined memory locations for the initialisation boot record. For example, in 80960 a record consists of twelve words. These are stored in between the ROM addresses, 0xFFFFFF00 to 0xFFFFFF2C (Figure 2.7).

Example 2.14   There are sections at the memory allocation map. Consider its description in a locator program. The sections for an exemplary embedded system, debit-credit card memory in example 2.12 may be defined as follows.

1. SECTIONS
2. {/* Stack Top Location for 256 B RAM*/
3. _TopOfStack = 0x10100;
4. /* Bottom of Heap */
5. _BottomOfHeap = 0x10080;
6. text rom :
7. {/* Debit-credit card program instructions are at the text file named here*/
8. * (————.txt)
9. }
10. data ram :
11. {
12. /* Shadow Segment for 16 byte of Initialised Data at RAM for a copy from ROM from */
13. _DataStart = 0x10000;
14. /* Debit-credit card shadow segment data at the data file named here*/
15. * (————.data)
16. _DataEnd = 0x1000F;
17. }
18. /* Command for copy into the RAM */
19. > rom
20. bss :
21. {
22. /* Base Segment for 32 byte of Program Variables Data at RAM */
23. _bssStart = 0x10010;
24. /* Debit-credit card base segment data at the base segment data file named here*/
25. * (————.bss)
26. _bssEnd = 0x1002F;
27. }
28. }

Example 2.15   Consider the sections at the memory allocation map for an exemplary embedded system, Java-embedded card in Example 2.13.

1. SECTIONS

2. {/* Stack Top Location in 4 kB RAM*/
3. _TopOfStack = 0x11000;
4. /* Bottom of Heap */
5. _BottomOfHeap = 0x10900;
6. text rom :
7. {/* Encrypting Java Card program instructions are at the text file named here*/
8. * (———.txt)
9. }
10. data ram :
11. {
12. /* Shadow Segment for 256 bytes of Initialised Data at RAM for a copy from ROM from */
13. _DataStart = 0x10000;
14. /* The card shadow segment data at the data file named here*/
15. * (———.data)
16. _DataEnd = 0x100FF;
17. }
18. /* Command for copy into the RAM */
19. > rom
20. bss :
21. {
22. /* Base Segment for 2 kB Program Variables Data at RAM */
23. _bssStart = 0x10100;
24. /* Java card base segment data at the base segment data file named here*/
25. * (———.bss)
26. _bssEnd = 0x108FF;
27. }
28. }

The memory map that includes the device I/O addresses is designed after appropriate address allocations of the pointers, vectors, data sets and data structures. From the map, the locator program can be easily designed. A designer must remember that if the main memory is of Harvard architecture, the program memory map will be separate. For example, 8051 reads from the program memory by a separate set of instructions (input–output instructions).

2.5.4 Addresses at Map for Internal Devices and I/O Devices

All I/O ports and devices have addresses. These are allocated to the devices according to the system processor and the system hardware configuration. The following examples clarify this and show systems with the three different processors.

1. Consider a system with an 80960 embedded processor. Figure 2.7 shows that device addresses are in between the addresses allocated to the system memory addresses and memory addresses for the system table for procedures and routines, interrupt vector table.
2. Consider another system with a 68HC11 microcontroller. Figure 2.8 (a) shows that the device addresses are within the RAM and are also distinct from the memory addresses.

3. Consider another system with an 80x86 processor. Figure 2.8 (b) shows the memory addresses on the left side. It shows the IBM PC system allocated port addresses for the timer, keyboard, real time clock and serial port (called COM2) on the right side. This figure shows that the device addresses can be the same as the memory addresses and they need not be distinct. Further, these depend on the system hardware configuration.

I/O device addresses are considered as part of the memory addresses by 80960 as well as 68HC11 [Figures 2.7 and 2.8 (b)]. Figure 2.8 (a) shows a specific case of 68HC11 configuration register bits. The configuration shown in the figure shows that Port A, I/O control register PIOC, Port C, B and port control registers have addresses between 0x0000 and 0x0004. On-chip RAM has been configured between 0x003F to 0x0040. The port addresses and on-chip RAM are configurable by the bits of the configuration register in 68HC11. For example, the above device addresses can also be configured and allocated between 0x0100 and 0x1040. On the other hand, the 8051, 80196 and 80196 microcontrollers have pre-assigned device addresses for its internal devices and these are unconfigurable addresses.

I/O device addresses are not the part of the memory addresses in 80x86 processors [Figure 2.8(b)].

Sections 3.1 and 3.2 will describe I/O and other devices in detail. Device addresses are used for processing by the *driver*. [Chapter 4.] A device has an address, which is usually according to the system hardware or may also be the processor assigned ones. These addresses allocate to:

1. Device Data Register(s) or RAM buffer(s).
2. Device Control Register(s). It saves control bits and may save configuration bits also.
3. Device Status Register(s). It saves flag bits as device status. A flag may indicate the need for servicing and show occurrence of a device-interrupt.

Each device, and thus each device register, must be allocated addresses at the memory map. *A very important point to remember is that in most cases, each set of I/O device addresses is often fixed by the system hardware. A locator or loader cannot reallocate these to any other set of addresses. Another point to remember depends on the device: at a device address there can be one or a number of device-registers.* A physical or virtual device can be configured to attach or detach from receiving input and sending output. A device address can also be just like a file, making it read only or write only or both read and write only.

Example 2.16 gives the details of addresses of the registers of an I/O device, *a serial line device*. [Also called UART device].

Example 2.16 A *serial line device* has the addresses of device registers as follows. [Refer to Section 3.1 and Figure 3.1 (b) for UART signals and serial I/O format.] These addresses are fixed by its hardware configuration of UART port interface circuit in a system employing a 80x86 processor. They are from 0x2F8 to 0x2FE at COM1 in a PC.

1. **(A)** *Two I/O data buffer registers (one for receiving and other for transmitting) are at a common address*, 0x2F8. Provided a control bit at address 0x2FB is 0, **(i)** during read from the address, the processor accesses from the RBR (Receiver Data Buffer Register) of the device and **(ii)** during write to the address, processor accesses the TRH (Transmitter Holding Register) of the device. Both RBR and TRH are at 0x2F8. **(B)** Provided a control bit at address 0x2FB is 1, data of two bytes of *Divisor Latch* are at the distinct addresses, 0x2F8 (LSB) and 0x2F9 (MSB). Divisor latch holds a 16-bit value for dividing the system clock. This then selects the rate of serial transmission of bits at the line. [While writing a device driver, remember that a bit in another register (control register) at 2FB changes the 0x2F8 from an I/O register (RBR or TRH) to lower byte of the divisor latch register. Refer to examples in Section 4.3.]

2. *Three Control Registers of the device are at three distinct addresses,* 0x2FA, 0x2FB and 0x2FC during a write operation. These are as follows. **(i)** IER (Interrupt Enabling Register). It enables the device interrupts. **(ii)** LCR (Line Control Register). It defines how and how many bits will be on the line. **(iii)** MCR (Modem Control Register). It defines how the modem handshakes and communicates.

3. *Three Status Registers of the device are at three distinct addresses,* 0x2FA, 0x2FD and 0x2FE during a read operation. These are as follows: **(i)** IIR (Interrupt Identification Register) at 0x2FA. It has the flags. A flag sets on a device-interrupt and resets at the system reset and at the servicing of the corresponding device-interrupt. **(ii)** LCR (Line Control Register) at 0x2FD. It defines how and how many bits will be on the line. **(iii)** MCR (Modem Control Register) at 0x2FE. It defines how modem handshakes and communicates.

Each I/O device is at a distinct address. Each device has three sets of registers—data (buffer) register(s), control register(s) and status register(s). There can be one or more device registers at a device address. The addresses of a device are according to the system processor and the system hardware configuration. Most processors process the memory devices and other devices with the same instructions. The designer must remember that 80x86 processors process these with a different set of instructions (input-output instructions).

2.6 DIRECT MEMORY ACCESS

I/O devices need to transfer the data of other systems to the memory addresses in a system. A system may also need to transfer data to the I/O devices to be transmitted to other systems. A direct memory access (DMA) is required when a multi-byte data set or a block of data is to be transferred between two systems without the CPU intervening, except at the start and at the end.

Three modes are usually supported in DMA operations. (a) Single transfer at a time and then release of the hold on the system bus. (b) Burst transfer and then release of the hold on the system bus. A burst may be the size of a few kB. (c) Bulk transfer and then release of the hold on the system bus after the transfer has completed.

DMA transfer is facilitated by the DMAC (DMA Controller). Data transfer occurs efficiently between I/O devices and system memory with the least processor intervention using DMAC. The system address and data buses become unavailable to the processor and available to the I/O device that connects DMAC. For example, in a computer, the transfer between hard disk system and memory uses a channel of DMAC. Besides processors, other devices also get the right to use the system memory by a direct transfer. Figure 2.10 shows a DMAC. It also shows the buses and control signals—between processor, memory, DMAC and the data-transferring I/O device.

A DMAC may provide memory access to multiple channels. There is a separate set of registers for programming each channel. There may be separate interrupt signals in the case of a multi-channel DMAC. A multi-channel DMAC provides DMA action from system memories and two (or more) I/O) devices.

The 80x86 processors does not have on-chip DMAC units. The 8051 family member 83C152JA (and its sister JB, JC and JD microcontrollers) have two DMA channels on-chip. The 80196KC has a PTS (Peripheral Transactions Server) that supports DMA functions. [Only single and bulk transfer modes are supported, not the burst transfer mode by PTS.] The MC68340 microcontroller has two

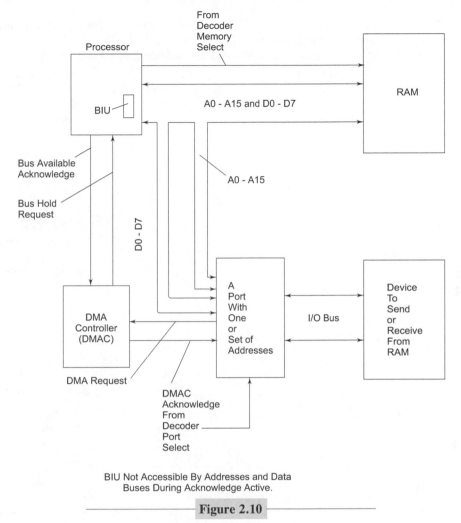

Figure 2.10

DMA-Controller with the buses and control signals in-between.

DMA channels on chip. The 80960CA has a four-channel DMAC on chip, with a mode called demand transfer mode also provided.

> On-chip or a separate DMAC facilitates direct byte transfers between memory and I/O devices. Designers can use DMAC in sophisticated systems so that the system performance improves by separate processing of the transfers from and to the peripherals.

2.7 INTERFACING PROCESSOR, MEMORIES AND I/O DEVICES

Section 1.3.13 describes the glue circuit for interfacing the system buses between the processor, memory-devices and I/O devices. Interconnections are through the data and address buses and control signals. Understanding of timing diagrams of bus signals is essential for the appropriate design of interfacing circuit and fusing (burning) that in the GAL or FPGA. Figures 2.11 (a) and (b) show the circuits for the interfacing memory devices and ports in 8051 and 68HC11, respectively. The 8051 microcontroller uses an additional signal $\overline{\text{PSEN}}$ (Program Store Enable by program memory read). [This is because of the use of Harvard architecture for main memories.]

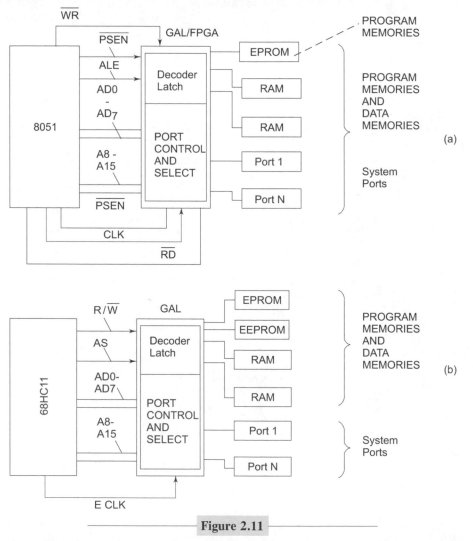

Figure 2.11

(a) Circuits for the interfacing memory devices and ports in 8051 using a GAL or an FPGA.

(b) Circuits for the interfacing memory devices and ports in 68HC11 using a GAL or an FPGA.

1. ***Demultiplexing of the buses for the address and data in the memories***: Time division multiplexing (TDM) means that in different time slots, there are different sets (channels) of signals. There may be a time division multiplexing of the buses for the address and data to (or

from) the memories. There are the address signals during one time slot and the data bus signals in another time slot. There is a control signal called Address Latch Enable (**ALE**) in 8051. The control signal is Address Strobe (AS) in 68HC11. It is *address valid* (ADV) in 80196. An ALE or AS or ADV demultiplexes the address and data buses to the devices.

2. ***Demultiplexing in the case of Harvard Architecture***: Demultiplexing of the common address bus for the program and data memory devices is done by $\overline{PSEN}$ in 8051. When it activates, the address signals to program memory activates. When $\overline{RD}$ activates the address signals to data memory activate. Demultiplexer and decoder circuits use $\overline{PSEN}$ and ALE in 8051.

3. ***Decoder***(s): Each chip of the memory-device or port that connects the processor has a separate chip select signal (*CS*). The decoder is a circuit, which has appropriate bits of the address bus at the input and generates corresponding CS signals for each device (memory and ports).

An interfacing circuit consists of decoders and demultiplexers and is designed as per available control signals and timing diagrams of the bus signals. This circuit connects all the units, processor, memory devices and the I/O device. It is a part of the glue circuit used in the system and is in a GAL or FPGA.

▪ SUMMARY ▪

A summary of the important points in this chapter is given below.

- An embedded system hardware designer must select an appropriate processor and an appropriate set of memories for the system and design an appropriate interfacing circuit between the processor, memories and I/O devices. This is done after taking into account the various available processors, structural units and architecture, memory types, sizes and speeds, bus signals and timing diagrams.

- Structural units of a processor that interconnects through a bus are memory address and data registers, system and arithmetic unit registers, control unit, instruction decoder, instruction register and arithmetic and logical unit. *Registers in processor* register sets are important and are meant for various functions during processing.

- Advanced processors have the following additional structural units: prefetch control unit, instruction queuing unit, caches for instruction, data and branch transfer, floating point registers and floating point arithmetic unit. Pipelining and superscalar features and caches in the processors are used in high performance systems. MIPS or MFLOPS or Dhrystone per second define the computing performance. The goal is to provide the optimal computing performance at the least cost and least power dissipation and total energy requirement.

- There is a set of instructions and various addressing modes for the arithmetic and logical instructions, data transfer instructions, I/O instructions and program-flow control instructions. The format of instructions and data and the addressing modes differ in the CISC and RISC processors. Each processor circuit supports a unique instruction set.

- The CISC processor supports a number of addressing modes in its instructions. It gives improved programming features. A programmer uses the multiple types of instructions for the multiple types of data structures. There are variable cycles of execution for an instruction.

- The RISC processor supports only a few addressing modes in its instructions and gives improved processor performance and single-cycle execution per instruction. RISC architecture, superscalar processing, pipelining, and cache units in advanced processors improve the processor performance and provide faster program execution.

- A data sharing problem arises when various functions or tasks share a common variable. Atomic operations solve the share data problem. Certain processors have an atomic operations control unit.
- Processor selection can be done using a design table.
- A system needs ROM and RAM memory of various types and sizes. Various forms of ROM are masked ROM, PROM, EPROM, EEPROM, flash and boot back flash. Basics details of the memories are the addresses available, speed for the read and write operations and modes of memory access.
- The memory has the blocks of addresses for the program segments, data, stacks, and array addresses. Memories also hold vector addresses (pointers), boot up program instruction set, instruction sets for the tasks, interrupt service routines and functions. Stacks are also at the memory. The designer designs a memory allocation map (and also an I/O allocation map in the 80x86 case). The size of the memory should be adequate to hold all the codes, data, data sets and structures and stacks.
- Memory access speed should match with processor structure and access needs.
- A DMA controller should be used to improve system performance by providing direct accesses to the I/O devices and peripherals.
- According to the memory map with I/O device addresses, a locator program is designed to locate the linked object code file and generate a ROM image.
- Memory selection can be done using a design table.
- Each I/O device has a distinct set of addresses. Each I/O device also has a distinct set of device registers—data registers, control registers and status registers. At a device address, there may be more than one register. The device addresses are according the system hardware.
- Bus signals interface the processor, memory and devices. The interface circuit takes into account the timing diagram in reference to processor clock output. The circuit uses the processor control, address and data bus signals and takes into account the timing diagram for the bus signals. A PAL-, GAL- or FPGA-based circuit provides a single-chip solution for the latches, decoders, multiplexers, demultiplexers and any other necessary interfacing circuit.

▪ LIST OF KEYWORDS AND THEIR DEFINITIONS ▪

- *Internal Bus:* A set of paths that carry in parallel the signals between various internal structural units of a processor. Its size is 64 bit in a 64-bit processor.
- *Memory Address Register:* A register that holds the address for a memory unit for placing it on the bus using a bus interface unit.
- *Memory Data Register:* A register that holds the data for or from a memory unit.
- *Bus Interface Unit:* A unit to interconnect the internal buses with the external buses for control, address and data bits.
- *System Registers:* Processor registers.
- *Arithmetic Unit Registers:* Registers hold the input and output operands and flags with the ALU.
- *ALU:* A unit to perform arithmetic and logic operations as per the instructions.

- *Control Unit:* To control and sequence all the processing actions during an instruction execution.
- *Instruction Register:* A register to hold the current instruction for execution.
- *Instruction Decoder:* The circuit to decode the opcode of the instruction and direct the control unit accordingly.
- *Prefetch Control Unit:* A unit to fetch instructions and data in advance from the memory units.
- *Instruction Queuing Unit:* A unit to hold a queue of instructions and place these into the cache.
- *Instruction Cache:* A place to hold sequentially the instructions that have been prefetched for superscalar- and pipeline-based parallel execution.
- *Data Cache:* Cache to hold the data in content addressable memory format.
- *Branch Transfer Cache:* A place to hold in advance the next set of instructions to be executed on the program branching to this set.
- *Share Data Problem:* A problem in which a variable, when shared between the tasks, becomes modified by another task instruction before a running task completes operation on that variable.
- *Atomic Operation:* It lets a running task instruction complete operations on a shared variable in a critical region of the codes.
- *Superscalar processor:* A processor with the capacity to fetch, decode and execute more than one instruction in parallel at an instant.
- *Pipelining:* There is pipelining also in the superscalar processor. It means than its ALU circuit divides into n substages. If in its last stage the processing of a pth instruction is taking place at an instant, then at the first stage processing of $(p + n)$th instruction is taking place. There may be multiple pipelines in a processor to process in parallel.
- *Memory Management Unit:* A unit to manage the prefetch, paging and segmentation of the memories.
- *Accumulator:* A register that provides input to an ALU and that accumulates a resulting operand from the ALU.
- *Program Counter:* A processor register to hold the current instruction address to be executed after a fetch cycle on the buses.
- *Stack Pointer:* A register that hold an address to define the available memory address to where the processor can push the registers and variables on a stack operation and from where they can be popped.
- *Stack:* A block to memory that holds the pushed values for "last in first out" data transfer.
- *Queue:* A block to memory that holds the pushed values for "first in first out" and "last in last out" data transfer.
- *List:* A data structure in which each object has a pointer to the next object. The first object in the list is pointed to by the list head and the last one points to a NULL pointer.
- *Hash Table:* A table in which keys are stored as a first column in the table and values are stored as a second column of a table.
- *Index Register:* A register holding a memory address of a variable in an array, queue, table or list.
- *Segment Register:* A register to point to the start of a segment for a program code or data set or string or stack.
- *Special Function Register:* A register in 8051 for special functions of accumulator, data pointer, timer control, timer mode, serial buffer serial control, power-down control, ports, etc.

- *Opcode:* First byte of an instruction for the instruction decoder of the processor. It defines the operation or process to be performed on the operand(s).
- *CISC:* A Complicated Instruction Set Computer that has one feature that provides a big instruction set for permitting multiple addressing modes for the source and destination operands in an instruction. The hardware executes the instructions in different number of cycles. It is done as per the addressing mode used in an instruction.
- *RISC:* A Reduced Instruction Set Computer that has one feature that provides a small instruction set and permits limited addressing modes for the source and destination operands in an instruction. The hardware executes each instruction in a one-cycle period.
- *Instruction Set:* A unique processor-specific set of instructions.
- *ARM 7 and ARM 9:* Two families of a RISC processor for a SoC from ARM and Texas Instruments, also available in single-chip CPU versions and in file versions for embedding at a VLSI chip. ARM 7 has Princeton architecture of main memory and ARM 9 has Harvard Architecture. [Refer to Section B.1.]
- *PROM or OTP:* A type of memory which is programmable only once by a device programmer. OTP is one time programmable memory.
- *EPROM:* A type of memory that is erasable many times by UV light exposure and programmable by a device programmer.
- *EEPROM:* A type of memory in which each byte is erasable many times and then programmable by the instructions of a program as well as by a device programmer.
- *Flash:* A type of memory that has a sector of bytes that is erasable many times (maximum ~ 10,000) in a flash at the same instance in a single cycle. Each erased byte is then programmable by the write instruction of a program as well as by a device programmer.
- *Boot Back Flash:* A flash with a few sectors similar to an OTP device, it can enable storage of boot up program and initial data.
- *Memory Map:* A memory-addresses-allocation table such that the map reflects the available memory addresses for various uses of the processor. A memory map defines the addresses of the ROMs and RAMs of the systems.
- *Device:* A physical or virtual unit that has three sets of registers—data registers, control registers and status registers—and the processor addresses these like a memory.
- *Device Address:* A device address used by processor to access its set of registers. At each address there may be one or more device registers.
- *Device Register:* A register in a device for byte, word of data, flags or control bits. Several device registers may have a common address.
- *Timing Diagram:* A diagram that reflects the relative time intervals of the signals on the external buses with respect to the processor clock pulses.
- *Interface circuit:* A circuit consisting of the latches, decoders, multiplexers and demultiplexers.
- *DMA:* A direct memory access by a controller internal or external. DMA operation facilitates the peripherals and devices of the system to obtain access to the system memories directly without a processor controlling the transfer of the bytes in a memory block.

■ REVIEW QUESTIONS ■

1. What are the common structure units in most processors?

2. What are the special structural units in processors for digital camera systems, real-time video processing systems, speech compression systems, voice compression systems and video games?

3. How do separate caches for instruction, data and branch transfer help?

4. What is the advantage of having multi-way cache units so that only that part of a cache unit that has the necessary data to execute a subset of instructions is activated? List four exemplary processors with multi-way caches.

5. When do you need MAC units at a processor in the system?

6. Explain three stage pipeline and superscalar processing and branch and data dependency penalties.

7. What are the advantages in Harvard architecture? Why is the ease of accessing stack and data-table at program memory less in Harvard memory architecture compared to Princeton memory architecture?

8. Explain three performance metrics of a processor: MIPS, MFLOPS and Dhrystone per Second.

9. Why should a program be divided into functions (routines or modules) and each placed in different memory blocks or segments?

10. Why should data be divided into data type and data structures and each placed in different memory blocks or segments? Explain how do the following data structures store at the memory: Stack, Vector, Array, Circular Queue, List and Look-up-table.

11. What do you mean by device registers and device address?

12. How does a boot block flash differ from a flash memory? How do flash, EEPROM and flash EEPROM differ? When do you use masked ROM for ROM image and when boot flash ROM in an embedded system?

13. Refer to Section B.1. How do the ARM 7, ARM 9, ARM 11 and StrongArm differ? When will you prefer to use ARM 7, ARM 9 or ARM 11?

14. How do the 68HC12 and 68HC16 differ? When will you prefer to use 68HC12 and 68HC16?

15. How does a memory map help in designing a locator program? What are the Intel and Motorola formats for the ROM image records?

16. What is meant by the terms: atomic operation, burst mode, PowerPC special power-saving modes, encryption key, Quarter-CIF, EDO RAM, RDRAM, peripheral transactions server, shadow segment, on-chip DMAC and time-division multiplexing.

■ PRACTICE EXERCISES ■

17. A two-by-three matrix multiplies by another three-by-two matrix. If data transfer from a register to another takes 2 ns, addition takes 20 ns and multiplication takes 50 ns, what will be the execution time? How will a MAC unit help, assuming that these times are the same in a DSP with a MAC unit?

18. An array has 10 integers, each of 32 bits. Let an integer be equal to its index in the array multiplied by 1024. Let the base address in memory be 0x4800. How will the bits be stored for the 0th, 4th and 9th element in (a) big-endian mode and (b) little-endian mode?

19. We can assume that the memory of an embedded system is also a device. List the reasons for it. [Hint: The use of pointers, like access control registers, and the concept of virtual file and RAM disk devices.]

20. What are the advantages of parameterised block RAM and parameterised distributed RAM in a fast transceiver?

21. Nowadays, high-performance embedded systems use either a RISC processor or processor with a RISC core with a code-optimised CISC instruction set. Why?

22. A circular queue has 100 characters at the memory addresses, each of 32 bits. What will be the total memory space required, including the space for both the queue pointers?

23. Estimate the memory requirement for a 500-images-capacity digital camera when the resolution is (a) 1024 x 768 pixels, (b) 640 x 480, (c) 320 x 240 and (d) 160 x 120 pixels and each image stores in a compressed jpg format. Assume each pixel colour is defined by 24 bits.

24. What are the special structural units in processors for the systems for digital camera, real-time video processing, speech compression and video game?

25. How does a decoder help in memory and I/O devices interfacing? Draw four exemplary circuits.

||||||||||||||||||||||||||||||||||||||||||||||||||||||||||||||||||||||||||||||||||||||

Devices and Buses for Device Networks

What We Have Learnt

Let us summarize the facts that have been presented in the previous chapters:

1. Embedded systems hardware consists of processor(s), memory devices, input/output (I/O) devices and basic hardware units such as power supply, clock circuits and reset circuits.
2. I/O devices consist of I/O ports to access peripheral and other on-chip or off-chip units—such as UART, modem, transceiver, timer-counter, keypad, keyboard, LED display unit, LCD unit, DAC, ADC and pulse dialer.
3. Processor caches, pipelines, superscalar and other advanced processing units are instrumental in the high-computing performance of a processor, and power dissipation during high-speed computation is managed by appropriate processor instructions.
   - We have also learnt the organization of a processor, memory-devices and I/O devices; Methods of selections of appropriate processor and memory devices for optimum system performance; Memory devices (ROM and RAM) of different types, based on size and speed of access; Addresses at the memory and I/O devices and their registers in a system; Interfacing of system buses with the memory and I/O devices; and the use of the DMA controller to improve system performance by enabling the I/O units a direct access to system memories.

Can we think of a computer without devices for video output, mouse, keyboard input and magnetic storage? *No*. So also in embedded systems. *System I/O devices and timing devices play the most significant role in any embedded system.* A device connects to and accesses the system processor either internally or through a port with each port having an assigned port address similar to a memory address. Advanced networking devices such as transceivers and encrypting and decrypting devices operate at MHz and GHz rates. Distributed devices are networked using the sophisticated I/O buses in a system. For example, take the case of an automobile. Most devices are distributed at different locations in any automobile. These are networked using a bus called Control Area Network (CAN) bus.

A hardware engineer designing an embedded system must, therefore, clearly understand the *features of new, sophisticated devices, interface circuits and their speed of operations, and the buses for networking the devices.*

What We Will Learn

This chapter will cover the following topics:

1. Devices and parallel and serial input, output and I/O ports;
2. Synchronous and asynchronous serial devices and their examples, High-level Data Link Control (HDLC) and UART, respectively;
3. Internal serial-communication devices of microcontrollers;
4. Features in parallel ports;
5. Sophisticated interfacing features in the systems for fast I/Os, fast transceivers, and real time voice and video I/Os;
6. Timing and counting devices and concept of software timers as the virtual timing devices;
7. Inter Integrated Circuit communication (I^2C) bus between multiple distributed ICs;
8. CAN bus as the control network between the distributed devices in the automobiles;
9. Universal Serial Bus (*USB*) for fast serial transmission and reception between the host embedded system and distributed serial devices like the keyboard, printer, scanner and ISDN system;
10. IBM Standard Architecture (ISA) and Peripheral Component Interconnect (PCI)/PCI-X (PCI Extended) interfacing buses between a host computer or system and the PC-based devices, systems or cards. [For example, PCI bus between the PC and Network Interface Card (NIC).]

3.1 I/O DEVICES

3.1.1 Types and Examples of I/O Devices

I/O devices can be classified into the following I/O types: (i) Synchronous Serial Input (ii) Synchronous Serial Output (iii) Asynchronous Serial UART input (iv) Asynchronous Serial UART output (v) Parallel one bit input (vi) Parallel one bit output (vii) Parallel Port Input (viii) Parallel Port Output. Some devices function both as input and as output. For example, *the modem*. Table 3.1 gives a classification of I/O devices into various types. It also gives examples of each type.

Table 3.1
Examples of Various Types of I/O Devices

I/O Device Type	Examples
Serial Input	Audio Input, Video Input, Dial Tone, Network Input, Transceiver Input, Scanner, Remote Input and Serial I/O Bus Input
Serial Output	Audio Output, Video Output, Dialing Number, Network Output, Remote TV Control, Transceiver Output, Multiprocessor Communication and Serial I/O Bus Output
Serial UART Input	Keypad, Mouse, Keyboard, Modem, Character Inputs on serial line
Serial UART Output	Modem, Printer, Character Outputs on serial line

I/O Device Type	Examples
Parallel Port Single Bit Input	(i) Completion of a revolution of a wheel, (ii) Achieving preset pressure in a boiler, (iii) Exceeding the upper limit of the permitted weight over the pan of an electronic balance, (iv) Presence of a magnetic piece in the vicinity of or within reach of a robot arm to its end point and (v) Filling a liquid up to a fixed level
Parallel Port Single Bit Output	(i) PWM output for DAC, which controls liquid level, temperature, pressure, speed or angular position of a rotating shaft or a linear displacement of an object or a DC motor control (ii) Pulses to an external circuit
Parallel Port Input	(i) ADC input from liquid-level measuring sensor or temperature sensor or pressure sensor or speed sensor or DC motor rpm sensor (ii) Encoder inputs for bits for angular position of a rotating shaft or a linear displacement of an object
Parallel Port Output	(i) Multilane LCD display matrix unit in a cellular phone to display on the screen the phone number, time, messages, character outputs or pictogram bit-images or e-mail or web page (ii) Printer or robot stepper motor coil driving output bits

Figure 3.1(a) shows the input serial port, output serial port, bi-directional half-duplex serial port, and bi-directional full duplex serial port signals. [Half-duplex means that, at an instant, communication can only be one way. An example of half-duplex mode is telephone communication. On one telephone line, we can talk only in the half-duplex mode.] Full-duplex means that, at an instant, the communication can be both ways.

An example of the full-duplex mode of communication is the communication between the modem and the computer. Figure 3.1 (a) also shows the serial I/O bit format and line states as a function of time. Figure 3.1(b) shows handshaking signals of port and UART serial bits at the serial line device. [Example 2.16 (modem)]. When sending or receiving a character (byte) the logic states during 10 or 11 periods as a function of time are also shown. A bit period, δT = Reciprocal of Baud rate, the rate at which the bits from an UART change. [Baud is taken from a German word for a raindrop.] UART bits consist of a start bit, 8 character bits, an option programmable bit (P bit) and a stop bit.

Serial line device (UART) communication is either in a 10-bit or 11-bit format. Communication can be full duplex, which is simultaneously both ways, or half-duplex, which is one way. Serial line device is an important communication mode.

3.1.2 Synchronous, Iso-synchronous and Asynchronous Communications from Serial Devices

When a byte (character) or a frame (a collection of bytes) of the data is received or transmitted at constant time intervals with uniform phase differences, the communication is called *synchronous*. Bits of a data frame are sent in a fixed maximum time interval. Iso-synchronous communication is a special case when the maximum time interval can be varied.

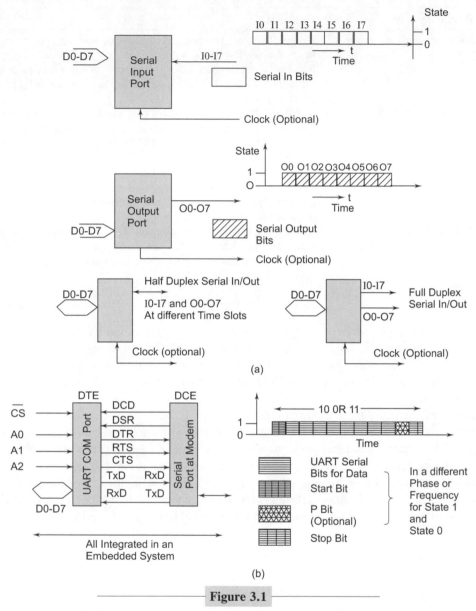

Figure 3.1

(a) Input serial port, Output Serial port, Bi-directional half-duplex serial port, and bi-directional full-duplex serial port (b) The handshaking signals used in UART serial port.

An example of synchronous serial communication is frames sent over a LAN. Frames of data communicate with the time interval between each frame remaining constant. Another example is the inter-processor communication in a multiprocessor system. Table 3.2 gives a *synchronous* port device bits. ***Two characteristics of synchronous communication are as follows.***

1. Bytes (or frames) maintain a constant phase difference. It means they are synchronous, i.e., in synchronization. Permission is not required for sending either the bytes or the frames at random

time intervals; this mode provides for no handshaking *during* the communication. The transmitter is the master and the receiver its slave.

2. A clock ticking at a certain rate must be present for transmitting serially the bits of all the bytes (or frames). The clock is *not always implicit* to the synchronous data receiver. The transmitter *generally* transmits the clock rate information in the synchronous communication of the data. Figure 3.2 gives ten methods by which synchronous signals, with the clocking information, are sent. (*i*) There are two separate lines for data bits and the clock and the parallel-in serial-out (PISO) and serial-in parallel-out (SIPO) are used for transmitting and receiving, respectively. (*ii*) There is a common line and the clock information is encoded by modulating the clock with the stream of bits. (*iii*) There are preceding and succeeding additions synchronizing and signaling bits. There are five common methods of encoding the clock information into a serial stream of the bits. (*a*) Frequency Modulation (FM). (*b*) Mid Frequency Modulation (MFM). (*c*) Manchester coding. (*d*) Quadrature amplitude modulation (QAM). (*e*) Bi-phase coding. The synchronous receiver separates the serial bits of the message and synchronizing clock.

When a byte (characters) or a frame (a collection of bytes) of data is received or transmitted at variable time intervals, communication is called *asynchronous*. An example of this is a telephone where the words are spoken at variable time intervals. Voice data on the line is asynchronously sent. Another example of asynchronous data is the characters between a keyboard and computer. Communication between the UART device (DTE) and a modem (DCE) is also asynchronous communication. [DTE stands for 'Data Terminal Equipment'. DCE stands for 'Data Communication Equipment'.] RS232C is an interfacing signal standard between DCE and DTE.

Two characteristics of asynchronous communication are as follows:

1. Bytes (or frames) need not maintain a constant phase difference and are asynchronous, i.e., not in synchronization. Bytes or frames can be sent at variable time intervals. This mode also facilitates in-between handshaking between the serial transmitter port and serial receiver port.

2. Though the clock must tick at a certain rate to transmit bits of a single byte (or frame) serially, it is *always implicit* to the asynchronous data receiver. The transmitter *does not* transmit (neither separately nor by encoding using modulation) along with the serial stream of bits any clock rate information in the asynchronous communication. The receiver clock thus does not maintain identical frequency and constant phase difference with transmitter clock.

Table 3.2
Synchronous Port Device Bits

S.No.	Bits at Port	Compulsory or Optional	Explanation
1.	Sync code bits or bi-sync code bits or frame start and end signaling bits	Optional	A few bits (each separated by interval ΔT) as sync code for frame synchronization or signaling precedes the data bits. There may be inversion of code bits after each frame. Flag bits at start and end are also used in certain protocols.
2.	Data bits	Compulsory	*m* frame bits or 8 bits transmit such that each bit is at the line for time ΔT or, each frame is at the line for time $m.\Delta T$.[@]
3.	Clock bits	Mostly not Optional	Either on a separate clock line or on a single line such that the clock information is also embedded with the data bits by an appropriate encoding or modulation.

[@]Reciprocal of ΔT is the bit per second (bps). Note: *m* may be a large number. It depends on the protocol.

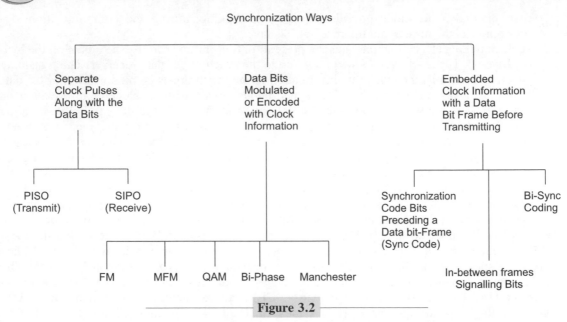

Figure 3.2

Ten ways by which the synchronous signals with the clocking information transmit from a master device to slave device.

When a device sends data using the serial communication frame, it may not be as simple as shown in Table 3.2 or as in an UART device (Figure 3.1). It can be complex and has to be as per the communication protocol followed by the transmitting and receiving device. Consider the following example.

When data is communicated using the physical devices on a network, most often a synchronous serial communication is used. **HDLC** (High-Level Data Link Control) is an international standard protocol for a data link network. It is used for linking data from point to point and between multiple points. It is used in telecommunication and computer networks. It is a bit-oriented protocol. The total number of bits is not necessarily an integer multiple of a byte or a 32-bit integer. Communication is full duplex.

Table 3.3 gives the synchronous network port device bits in HDLC protocol. The reader may refer to a standard textbook, for example, *Data Communications, Computer Networks and Open Systems* by Fred Halsall from Pearson Education (1996) for details of HDLC and its field bits.

Table 3.3
Format of Bits in a Synchronous HDLC Protocol-based Network Device

S.No.	Bits at Port	Compulsory or Optional	Explanation
1.	Frame start and end signalling flag bits	Compulsory	Flag bits at start and end are (01111110)
2.	Address bits for destination	Compulsory	8 bits in standard format and 16 bits in extended format
3a.	Control bits Case 1: Information Frame	Compulsory as per case 1 or 2 or 3	First bit 0, next 3-bits N(S), next bit [$] P/F and last 3-bits N(R) in standard format [&] N(R) and N(S) = 7-bits each in extended format

S.No.	Bits at Port	Compulsory or Optional	Explanation
3b.	Control bits Case 2: Supervisory Frame	-	First two bits (10), next 2-bits[#] RR or RNR or REJ or SREJ, next bit P/F and last 3-bits N(R) in standard format. N(R) and N(S) = 7 bits each in extended format
3c.	Control bits Case 3: Un-numbered Frame	-	First two bits (11), next 2-bits ^M, next bit P/F and last 3-bit remaining bits for M. [8-bits are immaterial after M bits in extended format]
4.	Data bits	Compulsory	m frame bits transmit such that each bit is at the line for time ΔT or, each frame is at the line for time $m.\Delta T$.[@]
5.	FCS (Frame Check Sequence) bits	Compulsory	16 bits in standard format and 32 in extended format
6.	Frame End flag bits	Compulsory	Flag bits at end are also (01111110)

Notes: 1. Bits are given in order of their transmission or reception.
2. [$]P/F = 1 and P means when a primary station (Command device) is polling the secondary station (receiving device). P/F = 1 and F means when receiving device has no data to transmit. Usually it is done in last frame.
3. [&]N(R) and N(S) means received (earlier) and sending (now) frame sequence numbers. These are modulo 8 or 128 in standard or extended format frame, respectively.
4. RR, RNR, REJ and SREJ are messages to convey 'Receiver ready,' 'Receiver not ready,' 'Reject,' and 'Selective reject'. REJ or SREJ is a negative acknowledgement (NACK). NACK is sent only when the frame is rejected. [*A parallel example is that a child cries out only when something is wrong.*] 'Reject' means that the receiver received a frame out-of-sequence; it is rejected and a repeat transmission of all the frames from the point of frame rejection is requested using REJ. 'Selective reject' means that a frame is received out-of-sequence; it is to be rejected and a selective repeat transmission is requested for this frame using SREJ.
5. ^M five bits are for a command (or response) from a transmitter. Examples of a command are *reset*, *disconnect* or *set a defined mode type*; examples of a response are a message from the receiver for a disconnect mode accepted, frame rejected, command rejected, and for an unnumbered acknowledgement.
6. [@] When five 1s transmit for the data, one 0 is stuffed additionally. This prevents misinterpretation by the receiver of the data bits as flag bits (01111110).

Besides HDLC, a communication system may use another protocol for synchronous or asynchronous transmission from a device port. X.25, Frame Relay, ATM, DSL and ADSL are other protocols for networking the physical devices in telecommunication and computer networks. Ethernet and token ring are protocols used in LAN networks. An *application* also uses an *application* protocol. Examples of the *application* protocols are HTTP, HTTPS, SMTP, POP3, ESMTP, TELNET, FTP, DNS, IMAP 4 and Bootp. Embedded Internet appliances use application protocols and Web protocols. WAP is a protocol used in wireless networks. Embedded system designers use the protocols in embedded network devices such as bridges and routers. The interested reader may refer to a standard textbook on data communication and computer networks for understanding these protocols.

Synchronous, iso-synchronous and asynchronous are three methods of communication for a device. Clock information is transmitted explicitly or implicitly in synchronous communication. The receiver clock continuously maintains constant phase difference with the transmitter clock. HDLC is an important data link protocol for computer networks and telecommunication devices. UART communication is asynchronous.

3.1.3   Examples of Internal Serial-Communication Devices

Most microcontrollers have internal serial communication devices. Table 3.4 gives the features of on-chip serial devices in select microcontrollers.

Table 3.4

Processor with On-chip Serial Ports (Devices) in Microcontrollers

Features	Intel 8051 and Intel 8751	Motorola M68HC11E2	Intel 80196
Synchronous Serial Port (Half or Full Duplex)	Half	Full	Half
Asynchronous UART Port (Half or Full Duplex)	Full	Full	Full
Programmability for 10 as well as 11 bits per byte from UART	Yes	Yes	Yes
Separate un-multiplexed Port Pins for Synchronous and UART Serial ports	No	Yes (Separate 4 Pins)	No
Synchronous Serial Port as a Master or Slave define by a Software or its Hardware	Software	Hardware and Software	Software
UART Serial Port as a Master or Slave define by Software programming for P bit	Yes	Yes	Yes
Synchronous Serial Port Registers	SCON, SBUF and TL-TH 0-1	SPCR, SPSR and SPDR	SPCON SPSTAT BAUD_ RATE and SBUF
UART Serial Port Registers	SCON, SBUF and TL-TH 0-1 (Timer 2 in 8052)	BAUD, SCC1, SCC2, SCSR, SCIRDR and SCITDR	SPCON SPSTAT BAUD_ RATE and SBUF
Uses Internal Timer or Uses Separate programmable BAUD rate generator	Timer	Separate	Separate as well as the Timer

Notes: 1. Intel 80960 and PowerPC 604 do not possess internal serial ports.
2. 68HC12 provides SPI (Serial Peripheral Interface) communication device operations at 4 Mbps. SPI device operates up to 2 Mbps in 68HC11. 68HC12 provides two SCI (Serial Communication Interface) communication devices that can operate at two different clock rates. Standard baud rates can be set up to 38.4 kbps. There is only one SCI and standard baud rates can be set up to 9.6 kbps only in 68HC11.

1. There is an on-chip common USART-like hardware device in 8051. USART means Universal Synchronous and Asynchronous Receiver and Transmitter. This device is called SI (Serial Interface) in 8051. Its features are as follows: SCON (Serial Control Register) is a special function register. It saves the control bits as well as status flags of SI. It sets its mode of communication. An SFR (Special Function Register), SBUF (Serial Buffer) is a serial buffer that is either a *read* or a *write* in an instruction in both synchronous and UART communication. There are no programmable instances of occurrence of negative edges and positive edges within an interval ΔT of the serial out data. SI is double buffered. It means that an intermediate

used. The great advantage of an built-in device for dynamically matched impedances is that when resistors are replaced with digitally dynamically controlled and matched impedances in the devices, there are no line reflections and therefore no missing bits or bus faults.

5. An I/O device may consist of multiple gigabit (622 Mbps to 3.125 Gbps) transceivers (MGTs). Special support circuitry is needed for this rate. Rocker I/O^{TM} serial 3.125 Gbps transreceivers are the examples of circuits that provide the support circuitry for this rate.

6. A device for I/O may integrate a SerDes (Serialization and De-serialization) subunit. SerDes is a standard subunit in a device where the bytes placed at 'transmit holding buffer' serialize on transmission, and once the bits are received these are de-serialized and are placed at the 'receiver buffer'. Once the device SerDes subunit is configured, serialization and de-serialization is done automatically without the use of the processor instructions. The great advantage of the SerDes unit is that these operations are fast when compared to operations without a SerDes. [A device for I/O may integrate a DAA or McBSP subunit when serializing. Refer to Sections D.4.4 and D.4.5.]

7. Recently, multiple I/O standards have been developed for I/O devices. A support to the multiple I/O standards may be needed in certain embedded systems. A technology, Flexible Select I/O Ultra^{TM} technology supports over 20 single-ended and differential I/O signaling standards. Advantages of multiple standard support devices are obvious.

8. An I/O device may integrate a digital Physical Coding Sub layer (PCS). Analog audio and video signals can then be pulse code modulated (PCM) at the sub layer. The PCS layer directly provides the codes from the analog inputs within the device itself. The codes are then saved in the device data buffers. The advantage of a built-in PCS at a port device is that there is no need of external PCM coding. Besides, these operations are in the background as well as fast. A PCS improves the system's performance for multimedia inputs at the devices.

9. A device for I/O may integrate an analog Physical Media Attachment (PMA) unit for connecting direct inputs and outputs of voice, music, video and images. The great advantage of a built-in PMA is that the device directly connects to the physical media. PMA is needed for real-time processing of video and audio inputs at the device.

Nowadays, I/O devices have sophisticated features. Schmitt trigger inputs are used for noise elimination. Devices with low-voltage gates and devices using power management by preventing unnecessary toggling at the inputs are used for sophisticated applications. Dynamically controlled impedance matching is new technology and it eliminates line reflections when interfacing the devices. A SerDes subunit serializes and de-serializes outputs and inputs in the devices. A port may have PCS and PMA subunits for analog inputs for video and audio I/O devices.

3.2 TIMER AND COUNTING DEVICES

Can we think of even a simple system like a TV remote controller or a washing machine without a timer device? *No*. The same answer applies to embedded systems. A timer device is fairly complex and it has a number of states (Table 3.5).

Table 3.5
States in a Timer

S. No.	States
1.	Reset State
2.	Initial Load (Idle) State
3.	Present State
4.	Overflow State
5.	Overrun State
6.	Running (Active) or Stop (Blocked) State
7.	Finished (Done) State
8.	Reset enabled/disabled State
9.	Load enabled/disabled State
10.	Auto Re-Load enabled/disabled State
11.	Service Routine Execution enable/disable State

Table 3.6 lists twelve uses of a timer device. It also explains the meaning of each use.

Table 3.6
Uses of a Timer Device

S. No.	Applications and Explanation
1.	Real Time Clock Ticks (System Heart Beats). [Real time clock is a clock, which, once the system starts, does not stop and can't be reset and its *count value* can't be reloaded. *Real time endlessly flows and never returns!*].
2.	Initiating an event after a preset delay time. Delay is as per *count value* loaded.
3.	Initiating an event (or a pair of events or a chain of events) after a comparison(s) between the pre-set time(s) with counted value(s). Preset time is loaded in a Compare Register. [It is similar to presetting an alarm].
4.	Capturing the *count value* at the timer on an event. The information of *time* (instance of the event) is thus stored at the *capture register.*
5.	Finding the time interval between two events. *Time* is captured at each event and the intervals are thus found out.
6.	Wait for a message from a queue or mailbox or semaphore for a preset time when using RTOS. There is a predefined waiting period before RTOS lets a task run.
7.	Watchdog timer. It resets the system after a defined time.
8.	Baud or Bit Rate Control for serial communication on a line or network. Timer timeout interrupts define the time δT of each baud or ΔT for each bit.
9.	Input pulse counting when using a timer, which is ticked by giving non-periodic inputs instead of the clock inputs. The timer acts as a counter if, in place of clock inputs, the inputs are given to the timer for each instance to be counted.
10.	Scheduling of various tasks. A chain of software-timers interrupt and RTOS uses these interrupts to schedule the tasks. [Refer Section 8.1.2 for definition of *task*.]
11.	Time slicing of various tasks. RTOS switches after preset time-delay from one running task to the next. Each task can therefore run in predefined slots of time.
12.	Time division multiplexing (TDM). Timer device is used for multiplexing the input from a number of channels. Each channel input is allotted a distinct and fixed-time slot to get a TDM output. [For example, multiple telephone calls are the inputs and TDM device generates the TDM output for launching it into the optical fiber.]

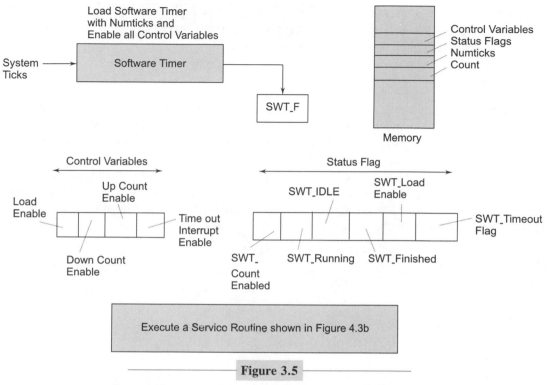

Figure 3.5

Control bits, status flags and variables of a software timer.

Timing devices are needed for a number of uses in a system. (i) There can only be a limited number of hardware timers present in the system. A system has at least one hardware timer, from which the system clock is configured. A microcontroller may have 2, 3 or 4 hardware timers. One of the hardware timer ticks form the inputs from the internal clock of the processor and generates the system clock. Using the system clock or internal clock, the number of hardware timers that are present can be driven. These timers are programmable by the device driver programs. (ii) A software timer is software that executes and increases or decreases a count-variable (count-value) on an interrupt on a timer output or on a real-time clock interrupt. A software timer can also generate interrupt on overflow of count-value or on finishing value of the count variable. Software timers are used as virtual timing devices. There are number of control bits and a time-out status flag in each timer device. A timer device when given count inputs, in place of clock-pulses, and performs as a counting device.

3.3 SERIAL COMMUNICATION USING THE I²C, CAN AND ADVANCED I/O BUSES BETWEEN THE NETWORKED MULTIPLE DEVICES

3.3.1 I²C Bus

Assume that there are a number of device circuits, some for measuring temperatures and some for measuring pressures in a number of processes in a plant. How can these ICs mutually network through a common bus? The I²C bus has become the standard bus for these circuits. [There are three standards: Industrial 100 kbps I²C, 100 kbps SM I²C, and 400 kbps I²C.]

The I²C Bus has two lines that carry its signals—one line is for the clock and one is for bi-directional data. There is a protocol for the I²C bus. Figure 3.6(a) shows the signals during a transfer of a byte when using the I²C bus; Figure 3.6(b) shows the format of the bits at the I²C bus.

From *master*, a data frame has fields beginning from start bit as per Table 3.9.

Table 3.9

Field and its length	Explanation
First field of 1 bit	It is start bit like in an UART.
Second field of 7 bits	This is the address field. It defines the slave address, which is being sent the data frame (of many bytes) by the master.
Third field of 1 control bit	It defines whether a read or write cycle is in progress.
Fourth field of 1 control bit	Bit defines whether the present data is an acknowledgement
Fifth field of 8 bits	It is for IC device data byte.
Sixth field of 1-bit	It is a bit NACK (negative acknowledgement). If active then an acknowledgement after a transfer is not needed from the slave; if not active, acknowledgement is expected from the slave.
Seventh field of 1 bit	It is a stop bit like in an UART.

The disadvantage of this bus is the time taken by the algorithm in the master hardware that analyzes the bits through I²C in case the slave hardware does not provide for the hardware that supports it. Some ICs support the protocol and some do not. Also, there are open collector drivers at the master. Therefore, a pull-up resistance of 2.2 K on each line is essential.

> I²C is a serial bus for interconnecting ICs. It has a start bit and a stop bit like in UART. It has seven fields for start, 7-bits address, defining a read or write, defining byte as acknowledging byte, data byte, NACK and end.

3.3.2 CAN Bus

A number of devices are located and are distributed in a car. These are networked and are controlled through a network bus. The 'CAN' bus is a standard bus in distributed network. It is mainly used in automotive electronics. [Refer to Section 11.3 for standards and an application of the embedded

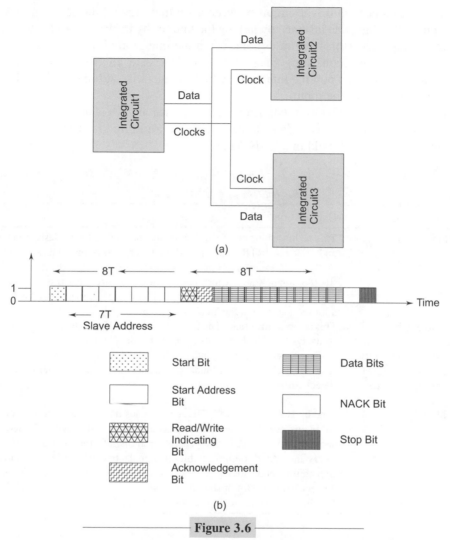

Figure 3.6

(a) Signals during a transfer of a byte when using the I^2C (Inter Integrated Circuit) bus
(b) Format of the bits at the I^2C bus.

system in automotive electronics.] It has a serial line, which is bi-directional. It receives or sends a bit at an instance by operating at the maximum rate of 1 Mbps. It employs a twisted pair connection to each node. The pair runs up to a maximum length of 40 m.

1. CAN serial line is pulled to logic level '1' by a resistor between the line and + 4.5V to +12V. Line is at logic '1' in its idle state, also called 'recessive state'.

2. Each node has a buffer gate between an input pin and a CAN serial line. A node gets the input at any instance from the line after sensing that instant when the line is pulled down to '0'. The latter is called 'dominant state'.

3. Each node has a current driver circuit between an output pin and the serial line. A node sends the output to the line at an instance by pulling the line '0' by its driver. An NPN transistor is the current-driving transistor. Its emitter connects to the line ground. Its collector connects to the line. Using a driver (consisting of a buffer inverter gate connected to base of the NPN transistor), the node can pull the line to '0', which is otherwise at logic '1' in its idle state. This lets other nodes sense the input.

4. A node sends the data bits as a data frame. Data frames always start with '1' and always end with seven '0's. Between two data frames, there are a minimum of three fields. Table 3.10 gives the details of each field in a CAN frame.

Table 3.10
Each Field in a CAN Frame

Field and its length	Explanation
First field of 12 bits	This is the arbitration field. It contains the packet 11-bit destination address and RTR bit. [RTR stands for 'Remote Transmission Request'.] When this bit is at '1', it means this packet is for the destination address. If this bit is at '0' (dominant state) it means this packet is a request for the data from a device. The device is 1 defined by an identifier. The device is at the destination address specified in this field.
Second field of 6 bits	This is the control field. The first bit is the identifier extension. The second bit is always '1'. The last 4 bits are code for data length.
Third field of 0 to 64 bits	Its length depends on the data length code in the control field.
Fourth field (third if data field has no bit present) is of 16 bits	It is the CRC (Cyclic Redundancy Check) word. The receiver node uses it to detect errors, if any, during the transmission.
Fifth field of 2 bits	First bit is the 'ACK slot'. The sender sends it as '1' and receiver sends back '0' in this slot when the receiver detects an error in the reception. Sender after sensing '0' in the ACK slot, generally retransmits the data frame. The second bit is the 'ACK delimiter' bit. It signals the end of ACK field. If the transmitting node does not receive any acknowledgement of data frame within a specified time slot, it should retransmit.
Sixth field of 7-bits	It is for end-of-the-frame specification and has seven '0's.

5. A CAN bus line usually interconnects to a CAN controller between the line and host at the node. It basically gives the input and gets the output between the physical and data link layers at the host node. The CAN controller has a BIU (bus interface unit consisting of buffer and driver), a protocol controller, status-cum-control registers, receiver-buffer and message objects. These units connect the host node through the host interface circuit.

6. There is an arbitration method called CSMA/AMP (Carrier Sense Multiple Access with Arbitration on Message Priority). A node stops transmitting on sensing a dominant bit, which indicates that another node is transmitting.

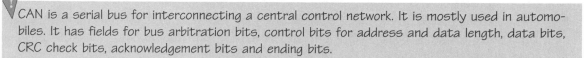

CAN is a serial bus for interconnecting a central control network. It is mostly used in automobiles. It has fields for bus arbitration bits, control bits for address and data length, data bits, CRC check bits, acknowledgement bits and ending bits.

3.3.3 USB Bus

The Universal Serial Bus (USB) is a bus between a host system and a number of interconnected peripherals. It provides a fast (up to 12 Mbps) and as well as a low-speed (up to 1.5 Mbps) serial transmission and reception between host and serial devices like scanner, keyboard, printer, scanner, ISDN, etc. [There are two standards: USB 1.1 (a low-speed, 1.5 Mbps 3-meter channel along with a high-speed, 12 Mbps 25-meter channel) and USB 2.0 (high-speed, 480 Mbps 25-meter channel).] *USB protocol has this feature—a device can be attached, configured and used, reset, reconfigured and used, share the bandwidth with other devices, detached (while others are in operation) and reattached.* The host schedules the sharing of bandwidth among the attached devices at an instance. A device can be either bus-powered or self-powered. In addition, there is a power management by the software at the host for the USB ports.

The host connects to the devices or nodes using USB port-driving software and host controller. The host computer or system has a host-controller, which connects to a root hub. A hub is one that connects to other nodes or hubs. A tree-like topology forms as follows. The root hub connects to the hub and node at level 1. A hub at level 1 connects to the hub and node at level 2 and so on. Only the nodes are present at the last level. Root hub and each hub at a level have a star topology with the next level.

A USB bus cable has four wires, one for +5V, two for twisted pairs and one for ground. There are termination impedances at each end as per the device speed. The Electromagnetic Interference (EMI)-shielded cable is for 15 Mbps devices.

Serial signals are Non Return to Zero (NRZI) and the clock is encoded by inserting synchronous code (SYNC) field before each packet. [Refer to Table 3.2.] The receiver synchronizes its bit recovery clock continuously. The data transfer is of four types: (a) Controlled data transfer. (b) Bulk data transfer. (c) Interrupt driven data transfer. (d) Iso-synchronous transfer.

USB is a polled bus. The host controller regularly polls the presence of a device as scheduled by the software. It sends a token packet. The token consists of fields for type, direction, USB device address and device end-point number. The device does the handshaking through a handshake packet, indicating successful or unsuccessful transmission. A CRC field in a data packet permits error detection.

USB supports three types of pipes. (a) 'Stream' with no USB-defined protocol. It is used when the connection is already established and the data flow starts. (b) 'Default Control' for providing access. (c) 'Message' for the control functions of the device. The host configures each pipe with the data bandwidth to be used, transfer service type and buffer sizes.

USB is a serial bus for interconnecting a system. It attaches and detaches a device from the network. It uses a root hub. Nodes containing the devices can be organized like a tree structure. It is mostly used in networking the I/O devices like scanner in a computer system.

3.3.4 Advanced Serial High-Speed Buses

An embedded system may need to connect multiple gigabits per second (Gbps) transceiver (transmit and receive) serial interface(s). Exemplary products are wireless LAN, Gigabit Ethernet, SONET (OC-48, OC-192, OC-768). The following are examples of the new advanced bus protocols.

1. IEEE 802.3-2000 [1 Gbps bandwidth Gigabit Ethernet MAC (Media Access Control)] for 125 MHz performance
2. IEE P802.3oe draft 4.1 [10 Gbps Ethernet MAC] for 156.25 MHz dual direction performance
3. IEE P802.3oe draft 4.1 [12.5 Gbps Ethernet MAC] for four channel 3.125 Gbps per channel transceiver performance
4. XAUI [10 Gigabit Attachment Unit]
5. XSBI [10 Gigabit Serial Bus Interchange]
6. SONET OC-48
7. SONET OC-192
8. SONET OC-768
9. ATM OC-12/46/192

Section E.1 in Appendix E gives a table of serial bus device standards and summarizes the features of emerging serial bus standards.

3.4 HOST SYSTEM OR COMPUTER PARALLEL COMMUNICATION BETWEEN THE NETWORKED I/O MULTIPLE DEVICES USING THE ISA, PCI, PCI-X AND ADVANCED BUSES

We need an interconnection bus within a PC or embedded system to a number of PC-based I/O cards, systems and devices. This bus needs to be separated from the system bus that connects the processor to the memories. The system bus and the interconnection bus have to operate at different levels of speeds. Exemplary devices are display monitor, printer, character devices, network subsystems, video card, modem card, hard disk controller, a thin client, digital video capture card, streaming displays, a 10/100 Base T card and a card using DEC 21040 PCI Ethernet LAN controller. Each of these devices, which perform a specific function, may contain a processor and software. Each device has the specific memory address-range, specific interrupt-vectors (pre-assigned or auto-configured) and the device I/O port addresses. *A bus of appropriate specifications and protocol interfaces these to the host system or computer.* Two old interconnection buses for communication between the host and a device are ISA and EISA (Extended ISA). A new interconnection for the bus is either PCI or PCI/X. [A variant of it is Compact PCI (cPCI).] The following subsections describe ISA and PCI buses.

3.4.1 ISA Bus

The ISA bus (used in IBM Standard Architecture) connects only to a card (or an embedded device) that has an 8086 or 80186 or 80286 processor, and in which the processor addressing and IBM PC architecture addressing limitations and interrupt vector address assignments are taken into account. There is no geographical addressing.

We can explain the limitation for memory access by a system using the ISA bus of the original IBM PC as follows. ISA bus memory accesses can be in two ranges, 640 kB to 1 MB and 15 MB to 16 MB. The former range also overlays with the range used by video boards and BIOS. [Note: Linux OS does not support the second range directly for accessing a device.]

We can explain the I/O port addresses limitations for the devices that use this bus as follows. The 8086 to 80286 processor has I/O mapped I/Os, not memory mapped I/Os. Though the instruction set provides for I/O instructions for 64 kB I/O addresses, the IBM PC configuration ignores the address lines A_{10} to A_{15} and these are not decoded. Therefore, only 1024 I/O port addresses are available. A hexadecimal addressing scheme with three nibble addresses between 000 to 3FF only can be used for a device. The A_{10} to A_{15} bits are thus immaterial. Following are the addresses allocated in IBM Standard Architecture (ISA).

1. Addresses allocated are hex 000–00F for DMA chip 8237. The addresses for other devices are as follows.
2. Hex 020–021 addresses allocated are for programmable interrupt controller 8255; it is hex 040–043 for timer 8253.
3. These are hex 060–063 for a parallel-port-programmable parallel interface.
4. The hex 080–083, 0A0–0AF, 0C0 to 0CF, 0E0 to 0EF allocated are for components on the motherboard.
5. Reserved addresses from peripherals are hex 220–24F, 278–27F, 2F0–2F7, 3C0–3CF and 3E0 to 3F0.
6. The addresses allocated are hex 2F8–2FF and 3F8–3FF for IBM COM ports.
7. Addresses are hex 320–32F and 3F0–3F7 for hard disk and floppy diskette, respectively.
8. Only thirty-two addresses between hex 300 to 31F are available for prototype card, for example, ADC card.
9. Addresses allocated are between hex 380–389 and 3A0–3A9 for synchronous communication,
10. Synchronous Data Link Control (SDLC) addresses allocated are between hex 380-38C.
11. Display monitor ports are within hex 380-38F (monochrome) and 3D0-3DF for (colour and graphics).

There is a limited availability of interrupt vectors in the IBM PC 80x86 family. Interrupt service functions are now shared at software level: for example, SWT interrupt. Original ISA specifications did not allow that.

EISA bus is a 32-bit data and address-lines version of ISA, and the ISA devices (system using this bus for I/Os) are also supported. An EISA device driver first checks the EISA bus availability on the hosting computer or system. It supports the sharing of the interrupt functions, SCI (Serial Communication Interface) controller and Ethernet devices. Unix and Linux support the EISA-bus-driven cards and devices.

ISA and EISA buses are compatible with IBM architecture. They are used for connecting devices following I/O addresses and interrupt vectors as per IBM PC architecture. EISA is 32-bit extension of ISA. It also supports software interrupt functions and Ethernet devices.

3.4.2 PCI and PCI/X Buses

Recently, the most used bus in the computer system for interfacing PC-based devices is PCI, which provides a superior throughput than EISA. It is almost platform-independent, unlike the ISA, which depends on the IBM platform, the IBM PC interrupt vectors and I/O and memory allocations. Its clock rate is nearest to the sub-multiple of system clock. PCI provides three types of synchronous parallel interfaces. Its two earlier versions are 32/33 MHz and 64/66 MHz. There is a recently introduced version, called PCI-X 64/100 MHz.

Lately, two super-speed versions of PCI have been introduced. These are PCI Super V2.3 264/528 MBps 3.3V (on a 64-bit bus), 132/264 (on a 32-bit bus) and PCI-X Super V1.01a, for 800MBps 64-bit bus 3.3 Volt.

A PCI bus has a 32-bit data bus extendible to 64 bits. In addition, it has 32-bit addresses extendible to 64 bits. Its protocol specifies the interaction between the different components of a computer. [A specification is version 2.1]. Its synchronous/asynchronous throughput is up to 132/528 MB/s [33M x 4/66M x 8 Byte/s]. It operates on 3.3V to 5V signals. A typical application is as follows. An exemplary PCI Card has a 16 MB Flash ROM with a router gateway for a LAN.

A PCI driver can access the hardware automatically as well as by the programmer-assigned addresses. The PCI feature of automatically detecting the interfacing systems for assigning new addresses is important for coding a device driver. The PCI bus therefore simplifies the addition and deletion (attachment and detachment) of the system peripherals. A manufacturer registers a global number. For example, 68HC11 or 80386 are globally registered numbers. A 16-bit register in a PCI device identifies this number to let that device auto-detect it. Another 16-bit register identifies a device ID number. These two numbers allow the device to carry out its auto-detection by its host computer. Each device may use a FIFO controller with a FIFO buffer for maximum throughput.

There are three identification numbers by which a device identifies its address space. (i) I/O port. (ii) Memory locations. (iii) Configuration registers of total 256 B with a 4-byte unique ID. Each PCI device has address space allocation of 256 bytes to access it by the host computer. The unique feature of PCI bus is its configuration address space. Interrupts are handled by the uniquely assigned interrupt type (a number). [Refer to Section 4.5.3 for understanding meaning of type and the types in IBM PC]. A configuration register number 60 stores the one byte for the interrupt type that defines this unique number. Figure 3.7 shows the 64-bytes standard-configuration registers in a PCI device. The following are the abbreviations used in the figure.

VID: Vendor ID. *DID*: Device ID. *RID*: Revision ID. *CR*: Common Register. *CC*: Class Code. *SR*: Status Register. *CL*: Cache Line. *LT*: Latency Timer. *BIST*: Base Input Tick. *HT*: Header Type. *BA*: Base Address. *CBCISP*: Card Base CIS Pointer. *SS*: Sub System. *ExpROM*: Expansion ROM. *MIN_GNT*: Minimum Guaranteed time. *MAX_GNT*: Maximum Guaranteed Time.

VID, *DID*, *RID*, *CR*, *SR*, and *HT* are compulsorily configured. Rests are optional.

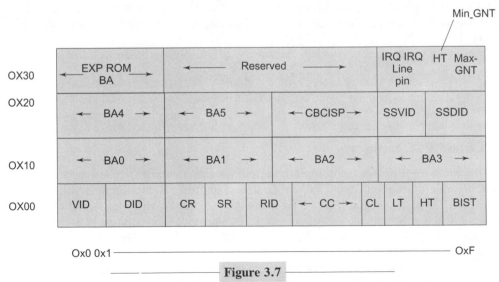

Figure 3.7

The 64 bytes at the standard device independent-configuration registers in a PCI device.

A PCI controller must access one device at a time. Thus, all the devices within the host computer can share I/O port addresses and memory locations but cannot share the configuration registers. That means that a device cannot modify other configuration registers but can access other device resources or share the work or assist the other device. If there are reasons for doing it so, a PCI driver can change the default boot-up assignments on configuration transactions.

A device can initialize at booting time. This helps to avoid any address collision. A PCI device on boot-up disables its interrupt and closes the door to its address space except to the configuration registers space. The PCI BIOS with the device performs the configuration transactions and then memory and address spaces automatically map to the address space in the host computer.

> PCI and PCI/X buses are the most commonly used buses and are independent from the IBM architecture. PCI/X is an extension of PCI and supports 64/100 MHz transfers. Lately, new versions have been introduced for the PCI bus architecture.

3.4.3 Advanced Parallel High-Speed Buses

Many telecommunication, computer, and embedded processor–based products need parallel buses for system I/Os. Three versions of PCI parallel synchronous/asynchronous buses provide the system-synchronous parallel interfaces. These three versions may not have sufficiently high speeds and ultra high speeds and large bandwidths that are needed in systems like I/Os, routers, LANs, switches and gateways, SANs (Storage Area Networks), WANs (Wide Area Networks) and other products. These do not suffice the source-synchronous parallel interfacing requirements. Bandwidth needs increase exponentially in the order of audio, graphics, video, interactive video and broadband IPv6 Internet. An embedded system may need to connect the I/O using gigabit parallel synchronous interfaces. The following are the new advanced bus standard protocols and a number of proprietary protocols developed recently.

1. GMII (Gigabit Ethernet MAC Interchange Interface).
2. XGMI (10 Gigabit Ethernet MAC Interchange Interface).
3. CSIX-1. 6.6 Gbps 32 bit HSTL with 200 MHz performance.
4. RapidIO^{TM} Interconnect Specification v1.1 at 8 Gbps with 500 MBps performance or 250 MHz dual direction registering performance using 8 bit LVDS (Low Voltage Data Bus).

Section E.2 in Appendix E gives a table of parallel bus device standards and summarizes the features of emerging parallel bus standards.

> PCI Parallel Buses are important for distributed devices. The latest high-speed sophisticated systems use new sophisticated buses.

■ SUMMARY ■

I/O Devices and timing devices are essential in any system. The following is a summary of the important points raised in this chapter.

- An embedded system connects to external devices like keypad, multi-line display unit, printer or modem through ports. A device connects and accesses from and to the system processor through either a parallel or serial I/O port. A device port may be full-duplex or half-duplex. Each port has an assigned port address. The processor accesses that address in a memory-mapped I/O, as if it accesses a memory address. A decoder takes the address bus signals as the input and generates a chip select signal, CS, for the port address selection. It is activated when the processor initiates a read or write cycle from the port. This device can use the handshaking signals before storing the bits at the port buffer or before accepting the bits from the port buffer.

- Bits are received at the receiver according to the clock phases of the transmitter at the synchronous serial input and output.
- Bits are received at the receiver independent of the clock phases at the UART (asynchronous serial input and output port) transmitter. UART (in microcontrollers) usually sends a byte in a 10-bit or 11-bit format. HDLC is a standard protocol for a synchronous communication data link network between the devices. Embedded networking devices use the networking protocols that are recommended for a given application.
- Special purpose ports exist at the microcontroller. These provide specific functions in specific cases. *On-chip devices* with the processor exist in the microcontrollers.
- Timer devices have a large number of uses in a system. There has to be at least one hardware timer in a system. A software timer is a virtual timing device. There can be a number of software timers in a system. A timer is essentially a counter getting the count-inputs (ticks) at regular time intervals. Internal programmable timing devices with a processor (microcontroller) unit can be used for many applications and to generate interrupts for the ticks for the software timers.
- I^2C bus is used between multiple ICs for inter-integrated circuit communication.
- CAN bus is used in centrally controlled networks in automobile electronics.
- USB (Universal Serial Bus) is a standard for serial bus communication between the system and devices like scanner, keyboard, printer and mouse. There is a root-hub and all nodes have a tree-like structure.
- If, for example, an I/O card interconnects to a PC or embedded system, then the ISA or PCI bus can be used. These buses are to enable a host computer to communicate with other systems, for example, to a network interface card (NIC). Lately, a number of systems use PCI or PCI/X bus.
- Devices and buses for sophisticated embedded systems have a high degree of sophistication.

■ LIST OF KEYWORDS AND THEIR DEFINITIONS ■

- *Parallel Port*: A port for read and write operations on multiple bits at an instance.
- *Serial Port*: A port for read and write operations with one bit at an instance and where each bit of the message is separated by constant time intervals.
- *Input Buffer*: A buffer where an input device puts a byte and the processor reads it later.
- *Output Buffer*: A register buffer from where an output device receives the byte after a processor write operation.
- *Handshaking Signals*: The signals before storing the bits at the port buffer or before accepting the bits from the port buffer.
- *Device*: A unit that connects to the processing unit through the ports. It has fixed pre-assigned port addresses (*device addresses*) according to its interfacing circuit.

- *Control Register*: A register for bits, which controls the actions of a device. It is for a *write* operation only.
- *Status Register*: A register for bits, which reflects the current status at the port buffer. It is for a *read* operation only. The bit may or may not auto-reset on device servicing.
- *I/O Port*: A port for input or output operation at an instant. Handshake input and handshake output ports are also known as I/O ports. For example, a keypad is set up to connect to an I/O port.
- *Half-Duplex*: A serial port having one common I/O line. For example, a telephone line. Message flows one way at an instance.
- *Full-Duplex*: A serial port having two distinct I/O lines. For example, a modem connection to the computer COM port. There are two lines TxD and RxD at a 9-pins or 25-pins connector. A message flows both ways at an instance.
- *Device Decoder*: A circuit to take the address bus signals as the input and generate a chip select signal, CS, for the port address selection.
- *UART*: A standard asynchronous serial input and output port for serial bits. UART (in microcontrollers) usually sends a byte in 10-bits or 11-bits format. The 10-bits format is when a start bit precedes the 8-bit message (character) and a stop bit succeeds the message. An 11-bit format is when a special bit also precedes the stop bit.
- *TxD*: A line used for transmission of UART serial bits. The 0 and 1 signals are at RS232C voltage levels when RS232C COM port is used, or at the TTL levels in microcontrollers.
- *RxD*: A line used for reception of UART serial bits. The 0 and 1 signals are at levels similar to that for TxD line.
- *COM Port*: A port at the computer where a mouse, modem or serial printer connects.
- *RS232C port*: A standard for UART transmission and reception in which TxD and RxD are at different Voltage levels (+ 12V for '0' and – 12 V for '1') and handshaking signals, CTS, RTS, DTR, DCD and RI, are at the TTL levels. [Recall RS232C standard is used at the COM ports].
- *Protocol*: A way of transmitting messages on a network by using a software for adding the additional bits like starting bits, headers, addresses of source and destination, error control bits and ending bits. Each layer or sub-layer uses its protocol before a message transmits on a network.
- *HDLC*: High Level Data Link Control Protocol for synchronous communication between primary (master) and secondary (slave) as per standard defined. It is a bit-oriented protocol.
- *Real Time*: A time which always increments without stopping or resetting.
- *Real Time Clock* (RTC): A clock that continuously generates interrupts at regular intervals endlessly. An RTC interrupt ticks the other timers of the system, for example SWTs.
- *System Clock*: A clock scaled to the processor clock and which always increments without stopping or resetting and generates interrupts at preset time intervals.
- *Hardware Timer*: A timer present in the system as hardware and which gets the inputs from the internal clock with the processor or system clock. A device driver program programs it like any other physical device.
- *Timer Overflow or Time-Out*: A state in which the number of count-inputs exceeded the last acquirable value and, on reaching that state, an interrupt can be generated.
- *Timer Finish*: A state after the timer acquired the preset count-value and stopped. An interrupt generates on completion.

- *Timer Reset*: A state in which the timer shows all bits as 0s or 1s. A reset can also be after overflow in case a timer is programmed for continuous running.
- *Timer Reload*: State in which the timer shows all bits as 0s or 1s. A reload can also occur after finishing in case a timer is programmed to auto-reload and start again.
- *Software Timer*: This is software that executes and increases or decreases a count-variable on an interrupt from a timer output or from a real-time clock interrupt. A software timer can also generate interrupt on overflow of count-value or on finishing value of the count-variable (reaching the predefined value).
- *Counter*: Unit for getting the count-inputs on the occurrence of events that may be at irregular intervals.
- *Free Running Counter*: A counter that starts on power-up and is driven by an internal clock (system clock) and which can neither be stopped nor be reset.
- *Event*: A change of present condition.
- *Event Flag*: A Boolean variable to indicate the event occurrence when it is true.
- *Delay*: An action blocked for a certain pre-defined period.
- *Watchdog Timer*: An important timing device in a system that resets the system after a predefined timeout. This time may be definable within the first few clock cycles after reset.
- *Open Drain Output*: A gate with an internally missing connection between its drain and supply. The advantage is that it pulls up circuit voltage and current levels, which are required when interfacing it. An external pull up circuit is needed when using the output.
- *Quasi Bi-directional Port*: A port with the dual advantage of using pull up circuit as per the voltage and current level required when interfacing it and using no pull up circuit for a short period sufficient to drive a LSTTL circuit.
- *Time Division Multiplexing*: A way in which, in different time slots, the message from different channels can be sent.
- *Demultiplexing*: A way to separate a multiplexed input and direct the messages to multiple channels.
- *On-chip Ports and Devices*: The ports and devices along with the processing unit, for example, in microcontrollers.
- *Synchronous Communication*: Communication in which a constant phase difference is maintained between the clocks that guide the transmitter and receiver. A maximum time interval is pre-fixed and affects the time in which a frame of bytes transmits.
- *Iso-synchronous Communication*: Communication in which a constant phase difference is not maintained between the frames but maintained within a frame. Clocks that guide the transmitter and receiver are not separate. Only the maximum time interval is not pre-fixed in which a frame of bytes transmits i.e., it can be variable. Uses are for transmission on a LAN or between two processors.
- *Asynchronous Communication*: A communication in which a constant phase difference is not maintained and the clocks that guide the transmitter and receiver are separate. Time interval between which a frame of bytes transmits is not pre-fixed and is indeterminate.
- *PISO*: A shift register for a Parallel Input and Serial Output. It is used for serial bits reception in synchronous mode.
- *SIPO*: A shift register for a Serial Parallel Input and Parallel Output. It is used for serial bits transmission in synchronous mode.

- *FSK* modulation: Frequency Shifted Keying. The 0 and 1 logic states are at different frequency levels. For example, 0 at 1050 Hz and 1 at 1250 Hz on a telephone line. It permits use of the telephone line for serial bit transmission and reception.
- *PSK* modulation: Phase Shifted Keying modulation. The 0 and 1 logic have different phases in a high frequency signal. PSK modulation permits use of the telephone line for serial bit transmission and reception.
- *QPSK*: Quadrature Phase Shifted Keying. An example is the pair of bits 00, 01, 10 and 11, which are sent at different quadrant phase differences of a voice frequency signal. It permits the use of the telephone line for serial bit transmission and reception at double the rate. It permits the 56 kbps modem to show a performance equivalent to 112 kbps.
- *Master/slave communication*: A communication between two processors when one processor guides the transmission of the bits to a slave processor after receiving acknowledgement from the address slave.
- I^2C *Bus*: A standard bus that follows a communication protocol and is used between multiple ICs. It permits a system to get data and send data to multiple compatible ICs connected on this bus.
- *PCI Bus*: A standard bus used is the Peripheral Component Interconnect Bus.
- *CAN Bus*: A standard bus used at the Control Area Network in automotive electronics.
- *ISA Bus*: A standard bus based on IBM Standard Architecture Bus.
- *USB Bus*: A standard bus for fast serial transmission and reception.
- *PCI/X Bus*: A standard bus used as PCI Extended Bus.

■ REVIEW QUESTIONS ■

1. What is the advantage of a processor that maps addresses of I/O ports and devices like a memory device?
2. Compare the advantages and disadvantages of data transfers using serial and parallel ports/devices.
3. Explain three modes of serial communication, asynchronous, iso-synchronous and asynchronous, from the serial devices with one example each.
4. How do the following indicate the start and end of a byte or data frame? (a) UART (b) HDLC (c) CAN
5. What are the *internal serial-communication devices* in (a) 8051 and (b) 68HC11? Compare the modes of working of both of these. [Refer to Sections 3.1.3 and G.5.]
6. A port device may have multi-byte data input buffers and data output buffers. What are the advantages of these?
7. Refer to the tables in Section D.4. What are the advantages of DAA and McBSP ports in DSPs?
8. How does the TPU (Timer Processing Unit) in the 68HC16 and 683xx microcontrollers help in an embedded real time controller? [Refer to Section C.2 of Appendix C.]
9. What is meant by the buses for networking of serial devices? What are the buses for networking of parallel devices?
10. Explain the use of each control bit of I^2C bus.
11. What is meant by plug and play devices? What bus protocols (of the buses listed in the exercises below) support plug and play devices?

12. What is hot attachment and detachment? What bus protocols (of buses listed in the exercises) support hot attachment and detachment?

13. What is a timer? How does a counter perform (a) timer functions, (b) prefixed time inititated events generation and (c) time capture functions?

14. Why do you need at least one timer device in an embedded system?

▪ PRACTICE EXERCISES ▪

15. How does the following device's features help in an embedded system? (a) Schmitt trigger input, (b) low Voltage 3.3V IOs, (c) Dynamically controlled impedance matching, (d) PCS subunit, (e) PMA subunit and (f) SerDes. Give one exemplary application of each.

16. PPP protocol for point to point networking has 8 starting flag bits, 8 address bits, 8 protocol specification bits, a variable number of data bits, 16-bit CRC and 8 ending flag bits. The maximum number of bits per PPP frame can be 12064. How many maximum number of bytes can be transferred per PPP frame? What is minimum percentage of overhead in the payload (frame)?

17. Refer to Section G.6. List the applications of a free running counter, regularly interrupting timer and pulse accumulator counter (PACT). How do you get PWM output from a PACT? How do you get DAC output from a PWM device?

18. A 16-bit counter is getting inputs from an internal clock of 12 MHz. There is a prescaling circuit, which prescales by a factor of 16. What are the time intervals at which overflow interrupts will occur from this timer? What will be the period before these interrupts must be serviced?

19. What is meant by a software timer (SWT)? How do the SWTs help in scheduling multiple tasks in real time? Suppose three SWTs are programmed to timeout after the overflow interrupts 1024, 2048 and 4096 times from the overflow interrupts from the timer in Exercise 18. What will be the rate of timeout interrupts from each SWT?

20. What is the advantage and disadvantage of negative acknowledgement?

21. A new generation automobile has about 100 embedded systems. How do the bus arbitration bits, control bits for address and data length, data bits, CRC check bits, acknowledgement bits and ending bits in the CAN bus help the networking devices distributed in an automobile embedded system?

22. How does the USB protocol provide for a device attachment, configuration, reset, reconfiguration, bandwidth sharing with other devices and device detachment (while others are in operation) and reattachment?

23. Refer to Sections 3.3 and F.1. Design a table that compares the maximum operational speeds and bus lengths and give two example of uses of each of the followings serial devices: (a) UART, (b) 1-wire CAN, (c) Industrial I^2C, (d) SM I^2C Bus, (e) SPI of 68 Series Motorola Microcontrollers, (f) Fault tolerant CAN, (g) Standard Serial Port, (h) MicroWire, (i) I^2C, (j) High Speed CAN, (k) IEEE 1284, (l) High Speed I^2C, (m) USB 1.1 Low Speed Channel and High Speed Channel, (n) SCSI parallel, (o) Fast SCSI, (p) Ultra SCSI-3, (q) FireWire/IEEE 1394 and (r) High Speed USB 2.0

24. Refer to Sections 3.4 and F.2 and use a Web search. Design a table that compares the maximum operational speeds and bus lengths and give two example of uses of each of the following parallel devices: (a) ISA, (b) EISA, (c) PCI, (d) PCI-X, (e) COMPACT PCI, (f) GMII, (Gigabit Ethernet MAC Interchange Interface), (g) XGMII (10 Gigabit Ethernet MAC Interchange

Interface), (h) CSIX-1. 6.6 Gbps 32 bit HSTL with 200 MHz performance and (i) RapidIO™ Interconnect Specification v1.1 at 8 Gbps with 500 MBps performance or 250 MHz dual direction registering performance using 8 bit LVDS (Low Voltage Data Bus).

25. Search the Internet and design a table that gives the features of the following latest generation serial buses. (a) IEEE 802.3-2000 [1 Gbps bandwidth Gigabit Ethernet MAC (Media Access Control)] for 125 MHz performance, (b) IEE P802.3oe draft 4.1 [10 Gbps Ethernet MAC] for 156.25 MHz dual direction performance, (c) IEE P802.3oe draft 4.1 [12.5 Gbps Ethernet MAC] for four channel 3.125 Gbps per channel transreceiver performance, (d) XAUI (10 Gigabit Attachment Unit), (e) XSBI (10 Gigabit Serial Bus Interchange), (f) SONET OC-48, OC-192 and OC-768 and (g) ATM OC-12/46/192.

# Chapter 4

Device Drivers and Interrupts Servicing Mechanism

What We Have Learnt

Let us summarize the following facts that have been presented in the previous chapters:

1. Embedded system hardware consists of the processor, memory devices (ROM and RAM), internal timing and I/O port devices and the basic hardware units: power supply, clock and reset circuits.

2. Organization of the processor, memory and devices in the system.

3. A memory-mapped I/O processor accesses the internal devices, devices at the I/O ports, peripheral devices and other off-chip devices using the port addresses analogous to the memory addresses.

4. Each device has addresses for three sets of its registers: data registers (or buffers), control registers and status registers. The processor, when accessing a register, uses its address.

We also learnt about the other important devices and buses for networking of the distributed devices.

5. Synchronous and asynchronous serial devices, for example, HDLC and UART device (serial line device). Internal serial communication devices in microcontrollers. Devices with sophisticated interfacing features for fast I/Os and fast transceiver function devices. Devices for real time voice and video I/Os. Timing and counting devices. Software timers as the virtual timing devices. Devices like keypad, keyboard, LED display unit, LCD display unit, DAC and ADC and pulse dialer.

6. Serial buses for interconnecting distributed devices: I^2C bus between multiple distributed ICs for inter-IC communication; CAN bus (Control Area Network bus) at the control network between distributed devices in automobiles; and USB (Universal Serial Bus) for fast serial transmission and reception between the host embedded system and distributed serial devices like keyboard, printer, scanner and ISDN system.

7. Parallel buses—ISA (IBM Standard Architecture), and PCI (Peripheral Component Interconnect)/PCI-X (PCI Extended) interfacing bus between a host computer or system, and the PC-based devices or systems or cards, for example, NIC (Network Interface Card).

Think about a device in a vending machine. The device is at an input port that collects a coin inserted by a person. *The system awakens and activates on an interrupt* through a hardware or software signal. The system on *interrupt* collects the coin by running a servicing routine. This routine is the *device driver routine for the port.* How does a port device awaken? One way is the 'programmed I/O-transfer', also called 'wait and transfer' or 'busy wait transfer' by which the device waits for the coin continuously, activates on sensing the coin and runs the service routine. Just as a grandmother watches at a window to see whether the school bus has arrived, in the standard method the device should awaken and activate on an interrupt after sensing the coin-inserting event.

Consider a digital camera system. It has an image input device. [It would otherwise be just a toy!] *How does the system awaken and activate, and the device grab an image frame?* Here again, the system awakens and activates on *interrupt* through a hardware or software signal from the device. The system on *interrupt* then runs a servicing routine. This routine reads the device frame-buffer (called frame grabbing) and processes, compresses and stores the image data in a flash memory. The servicing routine is the device driver for the image input device. How does the input frame-grabber device awaken? Again, on an *interrupt* on sensing the image.

The two examples given above clearly show that *interrupts and interrupt-servicing device-drivers (ISRs) play the crucial role in an embedded system.* Think of any system and it will have devices and thus needs device drivers. The embedded software designer must use the codes for device (i) configuring (initializing), (ii) activating (also called opening or attaching), (iii) driving using an ISR, and (iv) resetting (also called deactivating or closing or detaching). Codes are to be designed for each device function as well as for each interrupt of device in the system. [Note: (i) An ISR is often called a device driver when it services the device interrupt. (ii) Remember that device drivers are the memory and processor-sensitive programs.]

Consider the example of a video system. When the system is running, two device-driver ISRs also run. One driver is for the voice device and the other for the image device. Assume that ISRs and other system software designs are such that these two device drivers do not maintain synchronization. The system therefore does not meet the deadlines set for servicing of each system device. The next set of images and the next set of voice signals will be missed. Therefore, a crucial question for the system software designer is how to design the appropriate ISRs for multiple device-interrupts so that all device-interrupt calls are serviced within the stipulated deadlines of each interrupt. *The designer should provide optimum latencies and set appropriate deadlines for each task.*

What are the latencies for each ISR servicing? What are the deadlines? How (and in how much time) is the context of previously running routine *saved* when there is switching to another context of another ISR? How should the saved context be restored on returning to the previous ISR? *I/O devices and timing devices, control and status bits, sequences and speed of actions needed on interrupt from these, and the answers to the questions of latency and deadline, must be clearly understood by the designer.*

What We Will Learn

The objective of this chapter is to learn the following:
1. Necessity of using the interrupt service routines in any system and the working of interrupt mechanisms in the systems with the help of simple examples.
2. Device driver routines to service the device interrupt(s), and the driving of a peripheral unit (device).
3. Device initialization and port accesses.
4. Use of Linux Internals and device drivers.
5. Examples of *device initialization* and **device driver coding** for the parallel ports and serial line UART.
6. Device interrupts and lists of the various possible sources of the software and hardware interrupt.
7. Software instruction–related and runtime error–related interrupts in a system.
8. Interrupt system and individual device interrupt enabling and disabling, interrupt vectors, interrupt-pending registers and status registers, nonmaskable, and maskable, and nonmaskable-only, when so defined within a few clock cycles after reset.
9. Context and context switching on an interrupt.
10. Interrupt latency and deadline for interrupt servicing.
11. New methods for the fast context switching adopted in 80196, 80960 and ARM 7.
12. Classification of processors for an Interrupt service that 'Save' or 'Don't Save' the context other than the program counter.
13. Use of the DMA channel for facilitating the small interrupt-latency period of an interrupt source.
14. Assignment of software and hardware priorities among the multiple sources of interrupts.
15. Methods of servicing in case of simultaneous service demand from multiple sources.

4.1 DEVICE DRIVERS

4.1.1 Device Servicing without Using an ISR

Example 4.1 shows device servicing without using an ISR. It explains the advantages of using the interrupt mechanism and also explains various points for a device servicing mechanism.

Example 4.1   Assume a 64 kbps network. Using a UART that transmits in format of 11-bit per character format, the network transmits at most 5818 characters per second. Before 172 µs, a receiver expects another character assuming that all the received characters are in succession without any time-gap. Let Port A be at an Ethernet interface card in a PC, and Port B be its modem input which puts the characters on the telephone line. Let *In_A_Out_B* be a routine that receives an input character from Port A and re-transmits an output character to Port B. *Without using an interrupt servicing mechanism*, *In_A_Out_B* schedules the following steps *a* to *e* and executes the cycles of functions (*i*) to (*v*), thus ensuring that the modem gets the characters from the card continuously.

Step *a*: Function (*i*): Check for a character at port A. If not available, then wait.
Step *b*: Function (*ii*): Read Port A.
Step *c*: Function (*iii*): Decrypt the Message.
Step *d*: Function (*iv*): Encode the Message.
Step *e*: Function (*v*): Transmit the encoded character to Port B.

Let 150 μs define the minimum delay needed before there has to be a switch to execute the *In_A_Out_B* cycle start. Instead of continuously checking for characters at the port A by executing function (*i*), a delay function can be first called for 150 μs and then function (*i*) may be used. Other codes can be executed during 150 μs. *When the characters are not received in succession, the waiting period during step (a) can be very significant.* A loss of processor time for the waiting periods is the most significant disadvantage of the present approach.

Network Driver Program In_A_Out_B without Interrupts

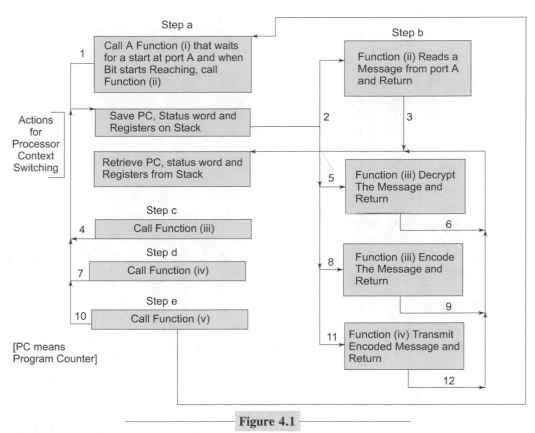

——— **Figure 4.1** ———

Steps (a) to (e) for five function calls in an exemplary network driver program IN_A_OUT_B, also shown is how each called function processes on a call and on a return. Numberings on the arrows show the program running sequences.

When there is a ***call*** to a function (method) in each step *a* to *e*, the function executes by three main steps:

1. Saving the current address of the program counter onto a stack, from where the processor will execute in case there is no *call* to a function.
2. Readjusting the contents of the stack pointer.

3. Getting an address (pointer) from where the function begins, loading the address into the program counter and then executing the called function's instructions. Before executing these instructions, also saved are the current program's status word, registers, and other program contexts if not done automatically by the processor, and if these may be needed by the newly called function.

Figure 4.1 shows the steps *a* to *e* for the five functions called by In_A_Out_B and how each called function processes on a *call* and on a *return*. Numberings on the arrows show the sequences during the program run (flow).

The last instruction (action) of any routine or function is always for a *return*. Steps are as follows during return from the function.

1. Before return, retrieve the previously saved status word, registers and other contexts.
2. Retrieve the saved program counter from the stack and readjust the contents of the stack pointer.
3. Execute the remaining part of the program that called the function.

The important points to remember are:

(i) A function call, after executing any instruction, is a planned (user-programmed) diversion from the current sequence of instructions to another sequence of instructions—and this sequence of instructions executes until the return from that.

(ii) On a function call, the instructions are executed as in a function in the 'C' or in a method in Java.

One approach for servicing (input or output or any other action) is 'programmed I/O transfer', also called 'busy and wait transfer'. When waiting periods are a significant fraction of the total program execution period, loss of processor time in waiting is the most significant disadvantage of this approach.

4.1.2 Device Driver ISR

An example for a device driver is a driver for handling the port inputs, which:

(i) Resets the device-buffer-full flag (in status register) and thus prepares the device for the next read. [Remember: The flag is a register bit for a Boolean variable that sets to signal a need. The need is for executing an interrupt service routine (ISR). The flag must reset when a corresponding ISR starts executing. Also, any flag of a source, if provided for indicating an occurrence, must be distinct for each source.]

(ii) Reads the input buffer(s) by emptying the buffer and storing the byte(s) in memory or using the bytes received as per system requirement.

Remember, before the driving, it is also necessary to configure a device. A device driver routine differs from other ISRs in that respect.

Example 4.2 shows device servicing using an ISR. It explains the advantages of using the system interrupt mechanism and also explains the various points that are involved in any device servicing mechanism.

Example 4.2 Consider how to *use an interrupt mechanism* for the *In_A_Out_B* cycle from (*i*) to (*v*). Use of the interrupt servicing mechanism saves the processor waiting time in the function (*i*). It also saves time when the characters do not reach port *A* continuously in immediate succession. During the saved time the processor becomes free in the system and is available for executing other routines or sets of codes.

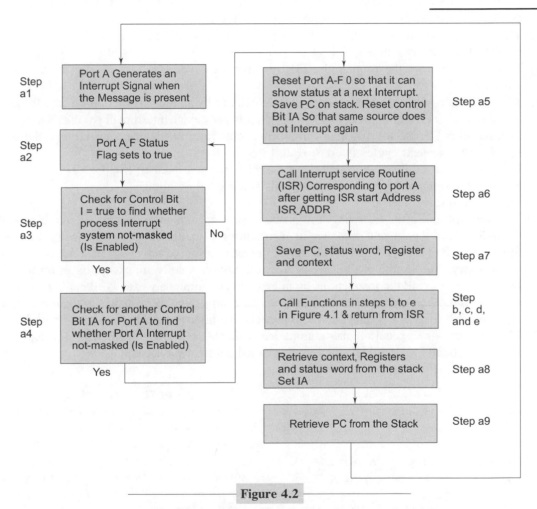

Figure 4.2

Steps on interrupt call in the exemplary network, In_A_Out_B

Figure 4.2 shows the interrupt service calling steps *a1* to *a6* for the above exemplary network in the previous example and shows the driving ISR in steps *a7* to *a9 for the steps (b) to (e) in the previous example.*

When using the interrupt servicing mechanism, the following steps are executed.

(i) Port A generates an interrupt signal when the message arrives. [Step *a1*]

(ii) As soon as a character reaches port A, a status-showing flag, *PORT_A_F*, sets (true) to show the need to activate this interrupting source (an internal or external hardware signal). [Step *a2*]

(iii) When PORT_A_F sets using steps *a3* and *a4 the interrupt* of any other ongoing function or program occurs and the corresponding service routine executes the functions (ii) to (v). When an ISR call initiates, the steps *a5* to *a9* run. If step *a3* shows that a control bit *I* has masked (disabled) the interrupt service mechanism for various device(s) interrupts, then interrupt service is denied. [This control bit is also called primary level mask bit.] If step *a4* shows that a control bit *IA* has masked this interrupt service for port A, then port A interrupt service is denied. [This control bit is also called secondary level mask bit.]

(iv) The last instruction of the ISR is also like a function. It is for a *return*. On return from the routine, the steps continue as follows.

(v) Before return from the routine, the ISR restores the saved status word, registers, and other contexts.

(vi) Also, it enables the further interrupts automatically or by an user instruction, in case these were either (a) disabled automatically by sourcing hardware device interrupt or (b) disabled by a user instruction in step *a5* for disabling further interrupts from the device, respectively. Set control bit *IA* again for next cycle of port A read. [Step *a8*.]

(vii) Retrieve the saved program counter and readjust the stack pointer. [Step *a9*.]

(viii) Execute the remaining part of the previously interrupted program.

Important points to remember are:

(i) An interrupt call after executing any instruction is an unplanned planned (source activated) diversion from the current sequence of instructions to another sequence of instructions—and this sequence of instructions executes until the return. Source activation instances (events) are not a priory. The source can be any device-interrupt or a software exception or error.

(ii) On an interrupt call, the instructions do not execute continuously exactly like a C *function* or a Java *method. These execute as per the interrupt mechanism of the system. For example, 'return'* from an ISR differs in certain important aspects. An interrupt mechanism may be such that an ISR on beginning execution may automatically disable other device interrupt services. These are automatically re-enabled if they were enabled before a service call.

Real-world devices are much slower than the processor. Therefore, the real world approach for servicing (input or output or any other action) is an interrupt-driven IO transfer. Instructions do not continuously execute exactly like a C function method in Java. They execute according to the interrupt mechanism of the system. Driver codes are sensitive to the processor and memory. This is because (i) when device addresses change, the program should also be also be modified and (ii) when a processor changes, the interrupt service mechanism also changes. When waiting periods are a significant fraction of the total program execution period, preventing loss of processor time caused by waiting is the most significant advantage of this approach.

4.1.3 Device Drivers

A system has a number of physical devices (Table 3.1). A device may have multiple functions. ***Each device function requires a driver.*** A timer device performs timing functions as well as counting functions. It also performs the delay function and periodic system calls. [Section 3.2]. A *transceiver* device transmits as well as receives, and may not be just a repeater. It may also do the *jabber control* and *collision control*. [Jabber control means prevention of continuous streams of unnecessary bytes in case of system fault. Collision control means that it must first sense the network bus availability then only transmit.] A voice-data-fax modem device has transmitting as well as receiving functions for voice, fax as well as data. *A common driver or separate drivers for each device function are required*. Why are device drivers important routines in a system? The following points will make it clear.

1. *The driver provides a software layer (interface)* between application and the actual device: When running an application, the devices of the system are to be used. A driver provides a routine that facilitates the use of a device-function in the application. For example, an application for mailing generates a stream of bytes. These are to be sent through a network driver card after packing the stream messages as per the protocol used in the various layers, for example TCP/IP. The network Driver routine will provide the software layer between the application and the network for using the network interface card (device).

2. *The driver facilitates the use of a device*: The driver routine is usually written in such a manner that it can be used like a black box by an application developer. Simple commands from a task or function can then drive the device. *Once a driver is available, the application developer does not need to know anything about the mechanism, addresses, registers, bits and flags used by the device.* For example, consider a case when the system clock is to be set to tick every 100 μs (10,000 times each second). The user OS for the application simply makes a call to function, like OS_Ticks (10,000). It is not necessary for the user of this function to know by which timer device it will be performed. What are the addresses, which will be used by the driver? Which will be the device register where the value 10,000 registers? What are the control bits that will be set or reset? OS_Ticks (10,000) when run, simply interrupts the device. Then the driver which executes takes 10,000 as input and reconfigures the real time clock to let the system clock tick each 100 μs and generate the system clock interrupts continuously every 100 μs to get 10,000 ticks each second.

3. *Usually, there is a common method to use the device driver*: All accesses to all devices are not made directly—but are always made by first interrupting the device and then executing the required driver routine.

4. *Usually, a device (or device function module) is opened (or registered or attached) before using the driver*: Using a user function or an OS function, a device is first initialized and configured by setting and resetting the control bits of device control register and the use of the interrupt servicing mechanism is enabled. This process is also called opening or registering or attaching the device (or device function module). Using a user function or an OS function, a device (or device function module) can also be closed or de-registered or detached by another process. After executing that process, device driver is not accessible till device is reopened (re-registered or re-attached).

Drivers for port, keypad, display, timer and network devices (Sections 3.1 to 3.4) are most commonly used in the systems. Drivers for PCS and PMA are required in media devices for most voice and video systems. It becomes impractical for a software designer to write the codes for each function of a device. *For commonly used devices, a designer most often relies on drivers that are readily available in the thoroughly tested and debugged operating systems*. [Refer to Chapter 10 for μCOS II and VxWorks operating system.] *The Linux operating system is a thoroughly tested and debugged operating system and is used throughout the world*. It has a large number of drivers that are, moreover, in the public domain. [Public domain means *nonproprietary* and usable by anyone.] The designer may therefore choose Linux when the embedded system being designed has the devices that have the drivers available in Linux. [Refer http://www.linuxdoc.org].

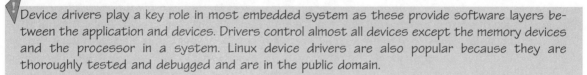

Device drivers play a key role in most embedded system as these provide software layers between the application and devices. Drivers control almost all devices except the memory devices and the processor in a system. Linux device drivers are also popular because they are thoroughly tested and debugged and are in the public domain.

4.1.4   Linux Internals as Device Drivers and Network Functions

Linux has internal functions called *Internals*. Internals exist for device drivers and network management functions. Examples of useful *Linux drivers* for the embedded system are listed in Table 4.1.

Table 4.1
Useful Linux Device Drivers

Driver Type	Explanation
char	Drivers for char devices. A char device is a device for handling a stream of characters (bytes).
block	Drivers for block devices. A block device is a device that handles a block or part of a block of data. For example, 1 kB data handled at a time. [Note: Unix block driver does not facilitate use of a part of the block during read or write.]
net	Drivers for network devices. A net device is a device that handles network interface device (card or adapter) using a line protocols, for example tty or PPP or SLIP.
input	Drivers for the standard input devices. An input device is a device that handles inputs from a device, for example keyboard.
media	Drivers for the voice and video input devices. Examples are video-frame grabber device, teletext device, radio-device (actually a streaming voice, music or speech device).
video	Drivers for the standard video output devices. A video device is a device that handles the frame buffer from the system to other systems as a char device does or UDP network packet sending device does.
sound	Drivers for the standard audio devices. A sound device is a device that handles audio in standard format.
system	Platform-specific drivers. Recently, system-processor specific drivers have also become available in this OS. Examples are drivers for an ARM processor–based system.

The Linux internal functions exist for Sockets, Handling of Socket buffers, firewalls, network Protocols (for examples, NFS, IP, IPv6 and Ethernet) and bridges. These are in the *net* directory. They work separately as drivers and also form a part of the network management function of the operating system. [The reader can refer to a standard textbook for bit-wise meaning of UDP, PPP and SLIP and for socket functions, firewall and network protocols. For example, refer to *Internet and Web Technologies* from Tata McGraw-Hill, 2002, for bit-wise description of PPP, SLIP, TCP, IP, Ethernet and other protocols.]

The Linux operating system has internals and a large number of readily available device drivers for the most common physical and virtual devices and has the functions for the network sockets and protocols.

4.1.5 Writing Physical Device Driving ISRs in a System

Recall Section 1.4.6, which defines and explains the functions of a device driver. Also recall, Section 2.4.4 and Example 2.16, which explain the registers and addresses of a PC serial line device (UART device). Writing the software for the driver is not soft! The following points must be clear before writing a device driver.

1. A device has three sets of device registers—data registers or buffers, control registers and status registers.
2. There may be more than one device register at a device address.
3. A device *initializes* (configures, registers, attaches) by setting the control register bits.
4. A device *closes* (resets, de-registers, detaches) by resetting the control register bits.
5. Control register bits control all actions of the device. A control bit can even control which address corresponds to which data register at an instant. For example, at the very instant that the DLAB control bit is set, the 0x2F8 corresponds to the divisor-latch lower byte. [Example 2.16.]
6. Status register bits reflect flags for the status of the device at an instant and change after performing the actions as per the device driver. A status flag at a status register reflects the present status of the device, for example, an instance between finishing the transmission of bits from a TRH buffer register and obtaining the new bits for next transmission. A transmitter empty flag reflects it. [Example 2.16]
7. Either setting of a status flag (software call or a hardware call by a signal) is used by the system to *initiate* a call for executing an ISR. [An ISR (also called Interrupt Handler Routine) executes if **(i)** it is enabled (not masked at the system) and **(ii)** the interrupt system itself is also enabled.] The following information must therefore be available when writing a device control and configuring and driver codes.
8. ***Addresses for each register***. Physical device hardware and its interfacing circuit fix the addresses for a physical device and they usually cannot be relocated. The device becomes the *owner* of these addresses. For example, IBM PC hardware is designed such that the device addresses are as following: *Timer* addresses between 0x0040-5F, *Keyboard* addresses between 0x00600-6FD, Real time clock (System clock) addresses between 0x0070-7F, Serial COM port 2 addresses between 0x02F8-2F and Serial COM port 1 addresses between 0x03F8-3FF. Notes: *(i)* There may be an input buffer register and an output buffer register at an address. This is because during device write and read instructions at the control bus the different signals are issued. The physical device can thus select the appropriate register when taking action. For example, there is a register SBUF at 8051. It addresses both the output serial buffer and input serial buffer. *(ii)* There may be multiple registers at the same address. Recall Example 2.16, which shows the following: RBR (Receiver Data Buffer Register) and TRH (Transmitter Holding Register) are at the same address (0x2F8) in PC COM2 serial device. This address is also common for lower byte of divisor latch, which is used for presetting the device baud rate. A control bit is made 1 to write this byte when setting the device baud rate and later it is made 0 for using same address as RBR and TRH during the device '*read*' and '*write*' instructions, respectively.
9. Purpose of each bit of the control register.

10. *Purpose of each status flag in the status register.* Which status bit when set and reflects a device interrupt, calls to which ISR.
11. Whether control bits and status flags are at the same address. The processor reads the status from this address during read instructions. The processor writes the control bits at that address during write instructions.
12. Whether both the control bits and flags may be in the same register.
13. Whether the status flag, which sets on a device interrupt, auto resets on executing the ISR or if the ISR needs to reset it.
14. Whether control bits need to be changed, reset or set again before returning to the interrupted process.
15. *List of actions required* by the driver at the data buffers, control registers and status registers.

A device-driving ISR is designed using the device addresses and three sets of device registers— data register(s) or buffer(s), control register(s) and status register(s). A device is configured and controlled by the control bits. The driver initiates and executes on a status flag change. A list of actions required by the driver at the data buffers, control registers and status registers is needed and is prepared before writing the driver codes.

4.1.6 Virtual Devices

Recall that Section 1.4.5 defined virtual devices and gave two examples of virtual devices, file and pipe. Besides the physical devices of a system, drivers are also used in embedded systems for *virtual devices. Physical device drivers and virtual device drivers have analogies*. Like virtual devices, physical device drivers may also have functions for device *open*, *read*, *write* and *close*. Consider the analogies of a file device with a physical device. (i) Just as a *file* needs to be *opened* to enable read and write operations, a device may need to be sent an *interrupt call* for initializing and configuring it (opening, registering or attaching it). [This is done by setting control bits appropriately.] (ii) Just as a file is sent a *read call,* a device must be sent another *interrupt call* when its input buffer is to be read. (iii) Just as a file is sent a *write call*, a device needs to be sent *another interrupt call when its output buffer is to be written.* (iv) Just as a file is sent a *close call* a device needs to be sent *another interrupt call to disable (close or deregister or detach)* it from the system for further read and write operations.

The concept of virtual (software) interrupt device drivers is very important in programming. Examples are as follows:

1. A memory block can have data buffers (in analogy to buffers at an I/O device) and can be accessed from a *char* driver or *block* driver. Device is called *char device* or *block device when driver can access a character or a block of characters, respectively.* [Recall Table 4.1 rows 1 and 2.]
2. A physical device transceiver (with input-output block buffer) or repeater is equivalent to a virtual device called *loop back device*. It stores allocated memory blocks using a block device driver and returns the data back from the memory.
3. A bounded buffer device in memory can be like a printer buffer. A data stream is sent by one routine (driver) and read by another routine (driver). A bounded buffer device is a virtual device, usually called *pipe* device.
4. A program can store in a set of memory blocks called *RAM disk* in the analogous way a file system does at the hard disk. It is a *file device*. RAM disk is a device that consists of multiple internal file devices.

The virtual device is an innovative concept for system software design. Drivers for these are also written like the physical device drivers. Important devices are char device, block device, loop back device, file device, pipe, socket and RAM disk. Device configuring is equivalent to creating a file. Device activation on interrupt is equivalent to opening a file. Device resetting is equivalent to closing a file. Device detaching is equivalent to freeing the memory space allotted for file data.

4.2 PARALLEL PORT DEVICE DRIVERS IN A SYSTEM

Driving a parallel port needs three modules. One module is for initializing by placing appropriate bits at the control register. The second module is for calling on interrupt whenever a status flag sets at the status register. The third module is for interrupt servicing (device driver) programming.

Following are two examples that explain device drivers' (service routines) actions. The chosen microcontroller 68HC11 (or 68HC12) has the memory mapping of the I/O ports. When there is an interrupt, the device driver *interrupt service (handler) routine* is called for an input or output service and the following are the services done by the drivers.

1. An input-port service part of the driver routine, *portA_ISR_Input*, fetches from the input-port A as if it is a *read* from a memory and generates a queue into which the values read adds. This service routine needs two parameters, *queue-tail pointer* and *queue-length* [Figure 2.4(b)]. Example 4.3 explains the driver.

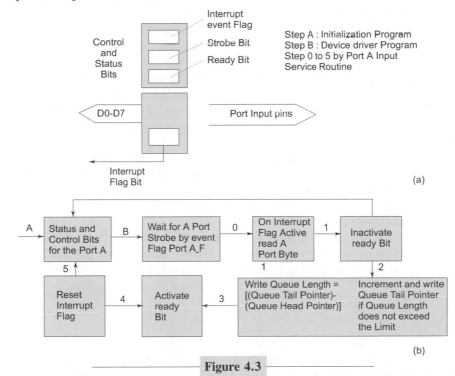

Figure 4.3

(a) Control and Status bits used in the system call, and port pins interface with data bus
(b) Step A for initialization, step B for system call to the driver and steps 0 to 5 for driver Port_ISR_Input. The driver reads a byte from a port and puts it into a queue that builds in memory on successive inputs to the port.

2. An Input-port service part of second driver routine PortC_ISR reads (deletes) from the queue the byte for the port and writes to the output-port C as if it is a *write* to a memory. This service routine also needs the parameters, *queue-head pointer* and *queue-length*. It calls *exception* when the queue is full. Example 4.4 explains the driver.

Example 4.3 Consider Port A of 68HC11 in single chip mode. There are no handshaking signals associated with this port. Figure 4.3(a) shows control and status bits used during the system calls. It also shows port pins interfacing with the data bus. Figure 4.3(b) shows Step *A* of device initialization and *step B for processor call to the driver*. It shows *steps 0 to 5 at the driver portA_ISR_Input routine. The driver reads a byte from a port and puts it into a queue that builds in memory on successive inputs to the port* steps *A*, *B* and steps 0 to 5 of Port A input service routine, *portA_ISR_Input*, in the device driver. These steps are for reading a byte from a port into a queue, which builds in memory on successive inputs to the port. Before the call to the device driver for Port A, two actions are necessary. (*i*) Define Port A address as follows. # define port A 0x1000 /* when memory page address for the 68HC11 registers is 0x1000. [Port memory-address is 0xp000, where page-address is defined by *p* at a processor-reset; *p*, 4 bit maximum significant nibble defined as bits, 0001.] (*ii*) Processor operational system sets the event (status) flag, *portA_F* to '1' and enables a call to the device driver program for Port A input. Device initialization and system call steps before the Driver *portA_ISR_Input* is called, are as following.

  (i) Wait till *portA_F* equals '1'. [Step *A* in Figure 4.3.]

  (ii) If portA_F equals '1' the system calls *portA_ISR_Input*. [Step *B* in Figure 4.3.]

The driver *portA_ISR_Input* service-routine programming will perform the following actions.

  (i) Read Port A.

  (ii) If *portA_Queuelength* is less than the *portA_MaxQueueSize* then store port input byte at a pointer **portA_Queuetail*. (a) **portA_Queuetail* is the queue-tail pointer that points to a memory address where the byte from Port A adds into the queue. (b) *portA_Queuelength* is present queue length at the port. (c) *portA_MaxQueueSize* is the maximum queue length defined for the Port A bytes.

  (iii) If *portA_Queuelength* is not equal to *portA_MaxQueueSize* then increment **portA_Queuetail* to the next address. The pointer now points to where to put the byte on the next call to the routine. If equal, call an error task routine for port A.

  (iv) It is necessary to reset *portA_F* if the operational system does not reset on action for the event (on call to the driver).

Here *portA_Queuelength* = present length of a queue where the bytes from Port A are being saved. *portA_MaxQueueSize* = Maximum permissible queue length. **portA_Queuetail* = Queue-tail pointer to a memory address. portA_F is the flag bit used in Step *B*.

Example 4.4 Now consider another example for reading Port C in single chip mode. 68HC11 Port C uses handshaking signals. Figure 4.4(a) shows handshaking signal to a Port C. Figure 4.4(b) shows control and status bits used in the *call* to the driver. Figure 4.4(c) shows port C as input and its interface with data bus. Figure 4.4(d) shows port C as output. Figure 4.4(e) Step *A* for initialization routine, step *B*

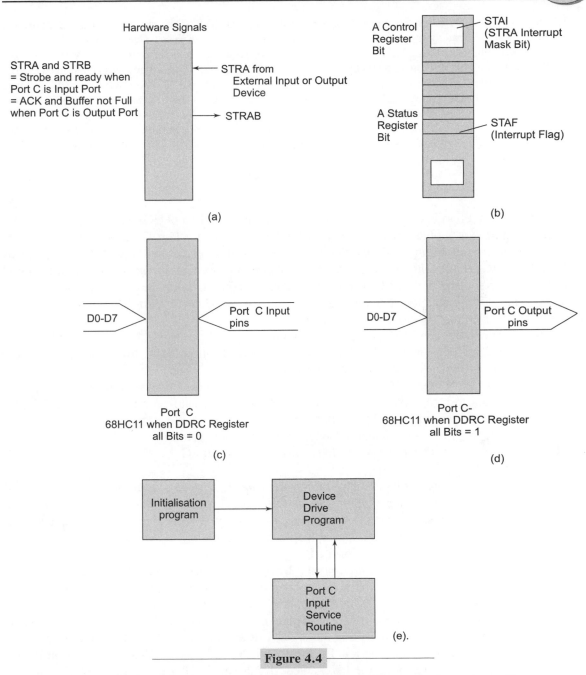

Figure 4.4

(a) Handshaking signal to a port (b) Control and Status bits used in the system calls (c) Port C as input and its interface with data bus (d) Port C as output (e) Steps A initialization, step B for system call to the driver and step C for driver PortC_ISR.

for system call to the driver and steps *C* for the driver PortC_ISR. The driver reads from the port and adds the byte read into a queue. The latter builds in memory on successive inputs to the port. An external device activates STRA pin. The external device requests a transfer of its byte to Port C through STRA. When a STRA pin activates by making it '0', the Port C gives an acknowledgement in case *STAI* (STRA interrupt mask bit) at a control register is not set (*STAI is not at* '1'). A STRB pin sends hardware signal for the ready-status (acknowledgement) from the port C to the external device. When *STAI* is programmed to '0', the external device puts the byte into the buffer as soon as STRB pin sends acknowledgement. As soon as the external device completes putting the byte at the Port C, the *STAF* sets. *STAF* is at a status register. *STAF* is the interrupt flag, which sets when the external device completes putting the byte at the Port C.

Port memory address is 0xp003, when page address configured on 68HC11 is 0xp000. [*p* is 4 bit maximum significant nibble.] Before a call to the device driver for Port C, three actions occur by device initialization program. (*i*) Define Port C address as follows. # define PortC 0x1003 /* p bits as 0x0001 */. (*ii*) Reset all eight bits to 0s at DDRC so that port C becomes an input parallel port. DDRC is data direction register for port C at memory address 0xp007. (*iii*) On an operational system call, *STAI* becomes '0' for enabling interrupting by the device. *STAI* is the sixth bit of PIOC (Port I/O Control register). It is at memory address 0xp002.

A driver program system call will be for the following actions.

1. If *STAI* is set '1' then read *STAF*. [*STAF* is the seventh bit at PIOC. PIOC also provides the status bits. It is for control-cum-status bits.]
2. If *STAF* is set '1' then call *portC_ISR* (Port C service routine), otherwise wait.
3. There is no need to software reset of *STAF* as there is an automatic hardware reset of it by 68HC11 as soon as *portC_ISR* is called.

Driver routine *portC_ISR* programming is done for the following actions.

4. If *quasi_bidir* bit does not equal to false, write 0xFF to Port C.
5. Read Port C.
6. If *portC_Queuelength* is less than the *portC_MaxQueueSize* store port bits at the address defined by **portC_Queuetail*. Abbreviations are as following. (a) **portC_Queuetail* is a pointer that points to a memory address where the byte from Port C adds to the queue. (b) *portC_Queuelength* is present queue length at the port. (c) *portC_MaxQueueSize* is the maximum queue length defined for the Port C bytes.
7. If *portC_Queuelength* is not equal to *portC_MaxQueueSize* than increment **portC_Queuetail* to let it point to the next address. Now this pointer will point where to put the byte on the next call to the routine. When both are equal then call an exception (*error* routine) for port C.

Examples 4.5 to 4.7 are C programs. These explain the device initialization by the preprocessor C commands, system call function for calling the driver ISR and the driver ISR. Required explanations are given as comment lines in the program. Example 4.5 is 'C' language implementation of the device initialization, calling program for device driver and device driver ISR. It uses in-line assembly. [Section 5.1]. Let 68HC11 Port A be configured at the address 0x1000. Flag, STAF, 7th bit in the PIOC (Port I/O Control) register will then be at address 0x1001. Secondary level Port A Interrupt mask is STAI. It is the 6th bit in the PIOC.

Example 4.5 /************* Definitions in Pre-Processor *********************/

```c
# define false 0
# define true 1
# define enable_Maskable_Intr ( )
# define disable_Maskable_Intr ( )
# define enable_PortA_Intr ( )
# define disable_PortA_Intr ( )
# define PortA_ISR_Input (unsigned char* )
# define volatile portA (volatile unsigned char *) 0x1000 /* Let portA be at address 0x1000*/
# define volatile PIOC (volatile unsigned char *) 0x1001 /* Let Port IO Control register PIOC be
at address 0x1001*/
# define STAF equ %10000000 /* Let Port A interrupt flag, STAF be the 7th bit */
# define STAI equ %01000000 /* Let Port A interrupt Mask, STAI be the 6th bit */
/********************************************************************/
void enable_Maskable_Intr ( ) {
asm {
/* Assembly Instruction for I bit setting at the CCR to enable all maskable Interrupts at primary
level */Bit
   STI;
   };
};
/
*********************************************************************/
void disable_Maskable_Intr ( ) {
asm {
/* Assembly Instruction for I bit setting at the CCR to enable all maskable Interrupts at primary
level */Bit
   CLI;
   };
};
/
*********************************************************************/
void enable_PortA_Intr ( ) {
asm {
/* Assembly Instruction for Bit Set (true) of portA_F in PIOC*/
   bset PIOC, STAI;
   };
};
/
*********************************************************************/
void disable_PortA_Intr ( ) {
asm {
/* Assembly Instruction for Bit Reset (false) of portA_F in PIOC*/
   bclr PIOC, STAI;
   };
};
```

```
/*********** Port A Interrupt Service Routine *********************/
void portA_ISR_Input (*portAdata) {
disable_PortA_Intr ( ); /* Disable another interrupt from port A*/
/* Insert Code for reading Port A bits*/
portAdata = &PortA;
}
/* Main Program with wait instruction for interrupt flag setting and then calling Port ISR */
void main (void) {
int n=0; int nMax = 100; /*define an index for a char array */
unsigned char [ ] portAinput; /* Let portAinput is an arrary to hold the data from port A*/
unsigned char *portAdata;
/* Wait till an interrupt occurs and sets STAF*/
while (STAF != 1) { };
/* Execute interrupt service routine for portA*/
if (STAF == 1) {
   portA_ISR_Input   (&portAdata);};
portAinput [n] = portAdata;/* Write port A input array element from the returned data*/
n++; /* Prepare for next byte from port A*/
if (n >nMax) println ("Error! Port A input bytes excceded the Defined Array size") else
enable_PortA_Intr ( ); /* Prepare for another interrupt from port A*/
}
```

Example 4.6 gives the same example coding again in high-level codes without using in-line assembly for coding.

Example 4.6 /*************** Definitions in Pre-Processor ********************/
```
typedef unsigned char int8bit;
# define int8bit boolean
# define false 0
# define true 1
# define volatile boolean IntrEnable
# define enable_Maskable_Intr ( ) {
IntrEnable = true; IntrDisable = false;
};
# define volatile boolean IntrDisable
# define disable_Maskable_Intr ( ) {
IntrEnable = false; IntrDisable = true;
};
# define volatile boolean IntrPortAEnable
#define enable_PortA_Intr ( )
# define volatile boolean IntrPortADisable
#define disable_PortA_Intr ( )
boolean STAF /* Let Port A interrupt flag is variable STAF */
```

```
boolean STAI /* Let Port A interrupt Mask, STAI be the 6th bit */
void enable_PortA_Intr ( ) {
IntrPortAEnable = true; IntrPortADisable = false; STAI = true;
};
void disable_PortA_Intr ( ) {
IntrPortAEnable = false; IntrPortADisable = true; STAI = false;
};
/************ Port A Interrupt Service Routine *********************/
void portA_ISR_Input ( ) {
disable_PortA_Intr ( ); /* Disable another interrupt from port A*/
/* Insert same Code for reading Port A bits as in example 4.3*/
.
.
.
}
/* Main Program with wait instruction for interrupt flag setting and then calling Port ISR */
void main (void) {
/* Insert same Code for reading Port A bits as in example 4.5*/
.
.
.
.
}
```

Example 4.7 gives an exemplary C program *In_A_Out_B* layout for Example 4.6 of a 64 kbps inter-network between the ports A and B defined in Section 4.1. It gives device initialization, device driver calling function and port driver function. The terms 'volatile', infinite loop and use of functions and pointers are explained later in Section 5.3.3.

Example 4.7 typedef unsigned char int8bit;
```
# define int8bit boolean
# define false 0
# define true 1
/* Same preprocessor declaration for Port A as in Example 4.5 */
/* Insert preprocessor declaration for Port B similar to that for the port A */
/* Note: Port B of 68HC11 is not interrupt driven */
# define volatile portB (volatile unsigned char *) 0x1004 /* Let portB be at address 0x1004*/
/*****************************************************************/
/* Main Program with wait instruction for interrupt flag setting and then calling Port ISR */
void main (void) {
/* The Declarations of all variables, pointers, functions here and also initializations here */
unsigned char *portAdata;
unsigned char *portBdata;
void portA_ISR_Input (unsigned char *portAdata);
```

```
void enable_Port A_Intr ();
/* Insert code here similar to the main function in Example 4.5 in loop to continue port A read
before 172 microsecond deadline */
while (true) {
/* Codes that repeatedly execute */
/* The code for a function for availability check of a character at port A*/
/* Wait till an interrupt occurs and sets STAF*/
while (STAF != 1) { };
/* The code for a function for reading a character at port A*/
if (STAF == 1) {
    portA_ISR_Input  (&portAdata);};
enable_PortA_Intr ( ); /* Prepare for another interrupt from port A*/
}
/* The code for a function for deciphering the character read at port A*/
/* .... Insert Here the codes for deciphering portAdata */
.
.
.

/* .... Insert Here the codes for encrypting before sending portAdata to portB */
.
.
.

/* Function for retransmit output to PortB*/
portB = &portAdata;

};
}
```

We learnt writing the driver codes from examples. Parallel port device drivers can be designed in the assembly language of the specific processor or in C language with assembly codes added as in-line assembly when changing or using the device registers.

4.3 SERIAL PORT DEVICE DRIVERS IN A SYSTEM

Recall Example 2.16 for a serial line UART device. All PCs use this device. Example 4.8 gives the device driver codes for the case of IBMPC with 80x86 processor and UART 8250 or new generation UART device that have the FIFO input and output buffers. Portability of the UART drivers in different systems is essential. There is an IEEE standard. It is called POSIX (Portable Operating System Interface) standard. The following driver example uses the POSIX defined functions.

Example 4.8
1. /************* Definitions in Pre-Processor *************************/
define false 0

define true 1

2. #define baud_l data

#define baud_h interrupt_enable

#define SERIALOUT 3

#define SERIALIN 4

#define I_CHAR_IN (1<<0)

#define I_TRANS_EMPTY (1<<1)

#define F_BAUD_LATCH (1<<7)

#define F_NORMAL (0<<7)

#define F_BREAK (1<<6)

#define F_NO_BREAK (0<<6)

#define F_PARITY_NONE (0<<3)

#define F_STOP1 (0<<2)

#define O_LOOP (1<<4) /* Loopback test */

#define O_OUT1 (1<<2) /* Port COM1 unused */

#define O_OUT2 (1<<3) /* Port COM2 Serial Interrupts used /

#define O_RTS (1<<1) / * RTS output signal Used*/

#define O_DTR (1<<0) / * DTR output signal Used*/

#define S_TBE (1<<5) /* Transmitter buffer empty */

#define S_RxRDY (1<<0) /* We have got a character */

#define SPEED 48 /* 2400 baud */

#define F_DATA8 (3)

Device Configuration Definition Codes

3. /* Device Initialization Codes Initialize 8250. Refer Example 2.16. We have to define Output port and input port addresses */

COM2 registers:

2F8 Txbuffer (THR) DLAB=0 (Write) (DLAB is at LCR control register)

2F8 Rxbuffer (RBR) DLAB=0 (Read) (Read Write Addresses are same for THR and TBR)

2F8 Divisor Latch LSB DLAB=1 (Define Baud Rate Divisor Latch when control bit DLAB =1)

2F9 MSB DLAB=1

2F9 Interrupt enable register (IER) (When DLAB = 0)

2FA Interrupt identification register (IIR)

2FB Line control register (LCR)

2FC Modem control register (MCR)

2FD Line status register (LSR)

2FE Modem status register

Device Activation Codes

4. /* Device Initialization Codes—Set Control Bits for Interrupt Mask and for Allocate Interrupt Vector Addresses for the Device Interrupts. We can define Mask serial lines to interrupt IRQ3 or IRQ4. Set interrupt vector and set interrupt mask: COM2 interrupt is at IRQ3 in Interrupt controller of 80x86. Recall Interrupt vector is a memory address where the driver ISR is located. 80x86 can be programmed to run in two modes. Recall that the number of real mode processor interrupts = 8 and the number of protected mode processor interrupts = 32. */

/* Codes for Real Mode of Processor */

setvect (0xB, serial_interrupt); /*0xB is Serial Interrupt Vector at 80x86 in Real Mode /

5. /* Define function outportb (q, p) by Assembly Language instructions, move variable p in 80x86 AL register and send AL to the output at address q */

```
Outportb (q, p) means:
mov al, p
out q, al
outportb (0x21, inportb (0x21) & 0xE7);
outportb (0x20, 0x20); /* End of Interrupt processing =EOI*/
6. enable ( ); /* Enable Interrupts */
7. /* Declare Eight Character data structure for data at the device registers *
struct sio {
char data; /* Data register RBR THR*/
char interrupt_enable; /* Interrup enable register IER*/
char interrupt_id; /* Interrupt identification register IIR*/
char format; /* Line control register LCR */
char out_control; /* Modem control register MCR */
char status; /* Line status register LSR */
char i_status; /* Modem status register MSR /
char temp; / * Dummy character for non existent 0x2FF address */
};
8. /* Dcfine Memory address*/
#define COM ( (struct sio near *) 0x2F8)
9. /* Set Serial line Interrupts Set */
10. outportb ( (int) & COM- > interrupt_enable,I_CHAR_IN | I_TRANS_EMPTY);
11. /* Set line parameters: ( DLAB=1)
outportb ((int)&COM->format,F_BAUD_LATCH|F_NO_BREAK|
F_PARITY_NONE|F_STOP1| F_DATA8);
12. /* Set Baud Rate */
outportb ((int)&COM->baud_l,SPEED&0xff);
outportb ((int) & COM->baud_h,SPEED >>8);
13./******************** Initialize Normal Serial Data Output
***********************/
outportb ((int) &COM->format,
F_NORMAL|F_NO_BREAK|F_PARITY_NONE|F_STOP1|F_DATA8);
outportb ((int) &COM-> out_control,O_OUT1|O_OUT2|O_RTS|O_DTR);
14. /* Reset the Serial Line Device */
(void ) inportb ((int) &COM->data);
 (void ) inportb ((int) &COM -> interrupt_enable);
 (void ) inportb ((int) &COM -> interrupt_id);
(void ) inportb ((int) &COM -> status);
15. /* Enable Interrupt Controller as initialization of device is finished by above steps. */
outportb (0x20, 0x20);
enable( );
16. /* For Mailbox concept refer Section 5.7.3 */
/* Transfer character to mailbox circular */
int GetSerial ( )
{ int x;
disable ( );
x = GetBuff(&MailBox [SERIALIN]); /* Read char from serial port */
enable ( );
return (x);
```

```
}
Device Driver Codes
17. /* Transfer a character to Serial Line Device by writing to serial port at 0x2F8 */
int transferring;
int PutSerial (char c) /* Write char to serial port */
{
disable ( );
if (Going)
{
if ( (PutBuff (&MailBox [SERIALOUT], (int) c)) == -1)
{enable ( );return (-1);}
else { enable ( ); return (1);} }
else {
transferring=1;
outportb ((int)&COM->data,c); /* Start I/O */ }
enable ( );
return (1);
}
/
**************************************************************************/
void interrupt serial_interrupt ( )
{
int  int_status;
disable( );
int_status = inportb ((int) &COM >status);
inportb ((int) &COM->interrupt_enable);
inportb ((int) &COM->interrupt_id);
if ( (int_status & S_RxRDY) != 0)
{PutBuff (&MailBox[SERIALIN],inportb ((int)&COM->data) & 0x7f); }/* Receive char */
if ((int_status & S_TBE ) != 0)
{ if ( MailBox[SERIALOUT].Head == MailBox[SERIALOUT].Tail) transferring= 0;
else  outportb((int)&COM->data,
(char)GetBuff (&MailBox[SERIALOUT]));
}
outportb( 0x20, 0x20);
enable ( );
} /* End of the Driver ISR Codes */
/
**************************************************************************/
```

We learnt about writing the serial device driver coding from an example of a PC 80x86 processor system with an UART 8250 device.

4.4 DEVICE DRIVERS FOR INTERNAL PROGRAMMABLE TIMING DEVICES

Recall the following points presented in Section 3.2. (i) Generally, there is at least one hardware timer as an internal device in systems needing timers. (ii) Using the time-outs (ticks) from it as many software timers as needed can be driven. The 80x86, 80960 and PowerPC-based embedded systems need off-chip hardware timers as there is no on-chip timer in the processor chip. The 8051, 68HC11 and 80196 microcontroller families (Appendix C) have on-chip timers.

The timer device driver programming needs an understanding of the programming of each bit of timer control registers and status registers. The timer device driver programming's important step is the programming of each bit of one or two control registers present and the status register. The driver programmer must also take into account the following. Instead of interrupt enable, a device may a have a mask bit. Mask bit means interrupt disables on set and enables on reset. Its actions are opposite to that of enable bit. The driver programmer must also remember that a certain interrupt cannot disable (cannot mask). [Non-Maskable Interrupt.] These enable or mask bits are the secondary bits. There is an overall interrupt system enable bit, which is like a master key (primary level bit) for all maskable interrupt sources. The driver must set that bit also.

Step I Write in a register that holds the timer reset value, the number of count-inputs, *numTicks* for the RTC.

Step II Write in status register the timer status flag(s) = reset (in case the device does reset automatically this flag(s) on a read of the status flag(s).

Step III Write each bit present in the control register(s). Write interrupt secondary and primary level enable bits = true in control register, write other bits according to their uses. It is essential to write the device enable bit to let the device work. Definitions of each of the bits in mode register if present are also essential.

Each bit of one or two control-registers present, status register, and output compare register(s) is to be programmed by the driver. A Free Running Counter is used as a timing device. The device driver programming steps in 68HC11 will be as in the following example:

Example 4.9
   (*i*) *Step a*: Define the output compare register(s) that holds an instance(s) of the FRC for the OC flag(s) setting and OC interrupt(s) occurrence(s).
  (*ii*) *Step b*: Flag(s) on its read from status register must be reset in case the device does reset automatically. The flags that may be present are the FRC overflow status flag, OC flag(s), ICAP_F flag(s), RTC flag, and SWT flag(s). These are to be reset on a read of the status register.
 (*iii*) *Step c*: Define control register(s) bits. Here, the definition for each bit present is essential. The bits may be as follows. Prescaling bits for count-input clock, Overflow interrupt enable bit, RTC interrupt enable, OC interrupt enable bit(s), OC enable bit(s), OC output level bit(s), ICAP enable bit(s), ICAP input edge bit, ICAP input bit(s), ICAP interrupt(s) enable bit, SWT enable (bits), and SWT interrupts enable bit(s).
  (*iv*) *Step d*: Also enable the primary level interrupt enable bit if already not enabled.

4.5 INTERRUPT SERVICING (HANDLING) MECHANISM

4.5.1 Hardware and Software Related Interrupt Sources

Hardware interrupts are processor specific. ISRs are needed not only for the device interrupts but also for the software related interrupts. One example of software interrupt is the '*exception*' interrupt generated by the processor. Example 4.10 clarifies this.

Example 4.10   Assume a division by zero occurs during execution of a certain instruction of a program. An ISR is needed which must execute whenever the division by zero occurs. This ISR could be to display 'A division by zero error at ....' on the screen and then terminate an ongoing program.

A user program under execution currently by the processor does not know when its ALU (arithmetic logic unit) will issue this internal error flag (a hardware signal). Using an interrupt mechanism, that service routine executes, which is meant for service on a zero-division error-signal. On setting of the message, an interrupt of the ongoing program happens just after completing the current instruction being executed, and then the ISR executes for post zero division task after resetting the flag.

A hardware source can be internal or external. Each of the interrupt sources (when not masked) demands a temporary transfer of control from the presently executed program to an ISR corresponding to the source. The internal sources and devices differ in different versions and families. Table 4.2 gives five classifications of hardware and software interrupts from multiple sources. Not all of the given type of sources in the table may be present or enabled in a given system. Further, there may be some other special type of the sources provided in the system.

4.5.2   Software-Error-Related Hardware Interrupts

A software error may be due to an illegal or not-implemented opcode. The examples are as follows. There is an *illegal opcode trap* in 68HC11. This error causes an interrupt to a vector address. A not-implemented opcode error causes an interrupt to a vector address in 80196. There are the software-error-related interrupts in 80x86. Examples are the division by zero (type 0) and overflow (type 2). [Refer to Section 4.5.3 for meaning of type.] These two generate by the hardware in the ALU part of the processor. Each processor has a specific instruction set. It is designed for that set only. An illegal code (instruction in the software) is an instruction which does not correspond to any instruction in this set. Whenever such a code fetches, an interrupt occurs in certain processors. The error-related interrupts are also called *software-traps (or software exceptions)*. [Note: A flag, 'Trap,' is an interrupt source of a different type in case of the 8086 and 80x86 processors. In this instance, trap has a different meaning altogether. It means exceptional run time condition, which causes the trapping of the presently running routine.] Software exception is a run-time exceptional condition, which, if it occurs (sets) during execution, causes a diversion to another routine called *exception*, which is called on occurrence of that condition. For example, a *queue* [figure 2.4(b)] getting full is an exceptional run-time condition. It should cause the diversion to routine called *exception* that initiates the appropriate action. *Exceptions* are important routines for handling the run-time errors. Action on encountering the exception is called *throwing an exception*.]

Table 4.2

Five Classifications and Twenty-Eight Types of Sources of the Interrupts

Software-error-related Sources (exceptions or SW traps)	Software Codes Related Sources	Internal Hardware Device Sources	External Hardware Devices with Internal Vector Address Generation	External Hardware Devices providing the ISR Address or Vector Address or Type externally
1. Overflow	11. Breakpoint Instructions [INT n, INT0 and type 3 in 80x86, swi in 68HC11.]	21. Parallel Ports and UART Serial Receiver Port- [Noise, Overrun, Frame-Error, IDLE, RDRF in 68HC11.	25. Within first few clock cycles Non-Maskable declarable Pin [XIRQ in 68HC11]	28. INTR [in 8086 and 80x86]
2. Division by zero	12. Debugging Trap Flag [TF in 8086]	22. Synchronous Receiver byte complete	26. Non-Maskable Pin [NMI in 8086 and 80x86]	
3. Underflow	13. An instruction for an exception. [For example, to enter supervisory mode in ARM 7]	23. UART Serial Transmit Port-Transmission Complete, TDRF Empty Synchronous Transmission of byte completed	27. Maskable Pin [INT0 and 1 in 8051, IRQ in 68HC11.]	
4. Illegal opcode	14. RTC-driven Software Timer			
5. Programmer$^\$$ defined *exceptions*[#] (*signals*)	15. Input-driven Software Timer	24. ADC Start of Conversion, End of Conversion		
6. Task Blocking	16. Pulse-Accumulator Overflow			
7. Task Finished	17. Real Time Clock time-outs			
8. Event Related	18. Watchdog Timer Resets			
9. Semaphore Take and Release	19. Timers Overflows on time-out			
10. A Boolean variable used as the status flag to define software-interrupt source.	20. Timer comparisons with Output compare Registers or Timer captures on inputs			

Note: (1) Processor specific example are in bracket (2) [#] *Signal* is a term sometimes used for an *exception*. [For example, in VxWorks RTOS. Refer to Section 10.3.] *Signal* or *exception* is an interrupt on setting of certain conditions or on obtaining certain results or output during a program run. $^\$$ There are two types of *exceptions*. One is internally generated by the processor. An example is *division by zero* in 80x86. The second is user defined.

(1) Software-error-related hardware interrupts are needed to respond to errors such as division by zero or an illegal opcode. (2) Exceptions are essential for handling the run-time errors. An 'exception' is like an ISR. The ISR executes on interrupt and services the interrupt. The exception executes on occurrence (throwing an exception) of the exceptional condition during a software run and it services (handles) the exceptional condition encountered therein.

4.5.3　Software-Instruction-Related Interrupts Sources

There are certain software instructions for interrupting. These differ from a *call* instruction as follows. *Software interrupt* in 68HC11 is caused by an instruction, SWI. There is a single-byte instruction INT0 in 80x86. It generates a type 0 interrupt, which means the interrupt should be generated with corresponding vector address 0x00000H. Instead of the type 0 interrupt which 8086 and 80x86 hardware generates on a division by zero, the instruction INT 0 does exactly that. Consider another single-byte 8086 and 80x86 instruction TYPE3 (corresponding vector address 0x00004*3=0x0000CH). This generates an interrupt of type 3, called break point interrupt. This instruction is like a PAUSE instruction. PAUSE is a temporary stoppage. It enables a user to do some housekeeping and return to instruction after the break point by pressing any key. Consider another 8086 and 80x86 two-byte instruction INT n, where n represents type and is the second byte. This means 'generate type n interrupt' and get ISR address from the vector address 0x00004* n. When n =1, it represents single step in 8086 and 80x86. There is a flag called trap and denoted by TF. This flag is at the FLAG register and EFLAG register of 8086 and 80x86, respectively. This means when TF sets (written '1') automatically after every instruction, the processor action causes an interrupt of type 1 repeatedly. The processor fetches each time the ISR address from the vector address 0x00004 (same as type 1 interrupt address). INT 1 software instruction will also cause type 1 interrupt once. The instruction is identical to the one caused at the end of an instruction on a TF flag setting. There is an instruction called Trap for the debugging software in 80196. Until the next instruction after the Trap is executed, no interrupt source can interrupt and service.

There may be a need to run a number of routines concurrently just as the need to run ISRs from the multiple devices and sources run concurrently. A Boolean variable can then be used instead of status flag, which, when set, signals a request to run a software interrupt service routine (SWISR). There can be a set of Boolean variables instead of the status register or for the interrupt-pending register to have multiple requests to run the SWISRs. Instruction INT n_{type} is used in 80x86 processors to run an SWISR. It is done by defining the appropriate value of n_{type} to be executed whenever a flag sets. Another set of Boolean variables can be used for the interrupt controlling bits [masking (disabling) bits]. An alternative interrupt mechanism can be programmed as follows. Just as the CPU looks at the status register or interrupt-pending registers at the end of each instruction, in order to provide an interrupt mechanism, a software interrupt (SWINT) scheduler (or foreground program) must then periodically check the above sets of variables periodically to let the SWISRs run concurrently.

Note: A system may have complex software interrupt sources also. Complex means unrelated to an instruction like INT n_{type} in 80x86 but related to a set of instructions (SWISR). The exemplary uses of these are in task scheduling and time slicing of the tasks [Sections 9.4.4]. A software timer may interrupt on an event during the program execution. Its use lies in counting the number of events from a source and then generating the interrupt form it. There can be interrupts on a task blocking, task finishing, semaphore taking and releasing [Section 9.4.3].

Software-instruction related or software-defined-condition related software interrupts are used in the embedded system. They are essential to design ISRs like error-handling ISRs and software-timer-driving ISRs. These interrupts are also called exceptions or signals. Besides device drivers, system operating systems must provide interrupt-servicing mechanisms for these also.

4.5.4 Hardware Interrupts Related to Internal Devices

In Table 4.2, the third column gives the usual hardware interrupts that relate to the internal devices.

4.5.5 Interrupt Vector

1. A system has internal devices like the on-chip timer and on-chip A/D converter. A very commonly used method is an internal device (interrupt source or interrupt source group) which auto generates an interrupt vector address, ISR_VECTADDR as per the device. These vector addresses are specific to a specific microcontroller or processor with that internal device. When an interrupt source is identified by the setting of a status flag, the processor finds the ISR address from the bytes at the ISR_VECTADDR. The system software designer must *place the bytes at ISR_VECTADDR for each ISR_ADDR using the locator and device programmer.* ISR_ADDR is the starting address of the device-driving ISR. In fact, in the ROM image, there exists a table for the interrupt vectors for the multiple sources of interrupts in the system. This table *facilitates the servicing of the multiple interrupting sources for each internal device.* Each row for an ISR_VECTADDR gives the bytes to provide the corresponding ISR_ADDR from the columns. [Vector table location in memory depends on the processor. It is located at the higher memory addresses, 0xFFC0 to 0xFFFB in 68HC11. It is located at the lowest memory addresses 0x00000 to 0x003FF in 80x86 processors. It starts from lowest memory addresses 0x0000 in ARM7.] An external device may also send to the processor the ISR_VECTADDR through the data bus.
2. A mechanism when using a vector address is present in 80x86 processors. The mechanism is such that only a short type number of the address, called ISR_type is needed in case of the external search for ISR_VECTADDR. [ISR_type multiplied by 0x00004 gives the address of the bytes. From these bytes the ISR_VECTADDR generates. From there, the ISR_ADDR is fetched.]
3. There is great ease of using the multiple sources of the interrupts even from a single device. This is when the device does not have to send ISR_VECTADDR.
4. *A source group has the same* ISR_VECTADDR. For example, in 8051, TI and RI are the sources in same group having identical ISR_VECTADDR. [TI is an interrupt that is generated when the serial buffer register for transmission completes serial transmission, and RI is when buffer receives a byte from the serial receiver. ISR at the ISR_ADDR must first identify the

interrupt source in case of the identical vector address or ISR address for a group of sources. Identification is from a flag or otherwise.]

The interrupt vector table is an important part of an interrupts-servicing mechanism, which associates a processor. It provides the ISR address to the processor for an interrupt source or a group of sources.

4.5.6 Classification of All interrupts as Nonmaskable and Maskable Interrupts

There are three types of interrupt sources in a system. (i) *Nonmaskable.* Examples are RAM parity error in a PC and error interrupts like division by zero. These must be serviced. (ii) *Maskable.* These are the interrupts for which the service may be temporarily disabled to let higher priority functions be executed uninterruptedly. (iii) *Nonmaskable only when defined so within few clock cycles after reset.* Certain processors like 68HC11 contain this provision. For example, an external interrupt pin, XIRQ interrupt, in 68HC11. XIRQ interrupt is *nonmaskable only when defined so within few clock cycles after 68HC11 reset.*

4.5.7 Enabling (Unmasking) and Disabling (Masking) of All the Maskable Interrupt Sources

There can be control bits in the devices. There may be one bit called the primary level bit for enabling or disabling the complete interrupt system of the maskable interrupts. A set of bits may enable or disable a specific source or source group in the interrupt system. *By appropriate instructions in the user software, a write to the primary and secondary levels enables (or the opposite of it, masks) either all or a part of the total maskable interrupt sources.*

4.5.8 Interrupt Pending Register or Status Register

An identification of a previously occurred interrupt from a source is done by one of the following:
(*i*) A processor pending-flag (boolean variable) in an *interrupt-pending register (IPR)* which sets by the source (setting by hardware) and auto-resets immediately by the internal hardware when at a later instant, the corresponding source service starts on diversion to the corresponding ISR. (*ii*) A local level flag (bit) in a *status register,* which can hold one or more status-flags for the one or several of the interrupt sources or groups of sources. The IPR and status register differ as follows. The status register is read only. (*i*) A status register bit (an identification flag) is *read only*, and is cleared (auto-reset) during the *read*. An IPR bit either clears (auto-resets) on servicing of the corresponding ISR or clears by a *write* instruction for resetting the corresponding bit. (*ii*) An IPR bit can be set by a write instruction as well as by an interrupt occurrence that waits for the service. A status register bit is set by the interrupting source hardware only. (*iii*) An IPR bit can correspond to a pending interrupt from a group of interrupt sources, but identification flags (bits) are separate for each source among the multiple interrupts.

Properties of the interrupt flags are as follows. A separate flag for every identification of an occurrence from each of the interrupt sources must exist in case of multiple interrupt sources. The flag sets on the occurrence of an interrupt. (*i*) It is present either in the internal hardware circuit of the processor or in the IPR or in status register. (*ii*) It is used for a *read* by an instruction and for a *write* by the interrupting source hardware only. (*iii*) It *resets* (becomes inactive) as soon as it is *read*. This is an auto-reset characteristic provided for in certain hardware designs in order to let this flag indicate the next occurrence from the same interrupt source. (*iv*) If set at once, it does not necessarily mean that it will be recognized or serviced later. When a mask bit corresponding to its source exists, even if the flag sets, the processor may ignore it unless the mask (or enable) bit modifies later. This makes it possible to prevent an unwanted interrupt from being serviced by a diversion to a corresponding ISR.

4.6 CONTEXT AND THE PERIODS FOR CONTEXT-SWITCHING, DEADLINE AND INTERRUPT LATENCY

4.6.1 Context, Latency and Deadline

A software system designer must learn the following if there is multitasking programming involved.

1. *A context of a program includes a program counter as well as the program status word, stack pointer and processor registers.* The context may be at a register set or at the separate memory block for the stack. The set or block holds information regarding the program counter (to indicate the address of the next instruction to be executed for this task), memory block allotted to a particular foreground program or ISR, ID of the process to be interrupted and CPU state (registers etc.) [Section 8.1]. What should constitute the context? It depends on the processor or the operating system supervising the program.

2. Context switching means saving the context of the interrupted routine (or function) or task and retrieving or loading the new context of the called routine or task. The *time taken in context switching* has to be included in a period, called interrupt latency period, T_{lat}. Recall Example 4.2; implementation of saving and retrieving steps in Figure 4.1; and steps a6, a7, a8 and a9 in Figure 4.2 on a function or ISR call and return is faster by a special hardware in certain processors. The examples are given below for the fast context switching facility in certain processors.

The 80960 embedded processor program allocates a set of local registers, *p0, *p1, ... for use later whenever a call instruction is executed. Each set then stores the various local variables of a routine or foreground program. When the call instruction is executed, an internal stack frame is created, and the set de-allocates from the previous routine or foreground program. [Internal stack means the stack is not at the external RAM but internally associated with the processor.] All local registers of the set save onto the stack frame before de-allocation. Local registers of another set now allocate to the called program. On the return, the stack frame returns the variables in the reallocated local register set and the frame de-allocates. Saving and retrieving are fast. This is due the fact that firstly, the external RAM is not being used as stack and, secondly, that it is only the pointers for the context in a stack frame whose values only change or interchange when there is program diversion from one routine to another.

Figure 4.5 explains the allocation and re-allocation processes as in the 80196 and 80960 processors. This figure also shows an exemplary context switching from task1 to task2. Only the references interchange from a register window (set) to another register set (or stack frames). The context switching then takes a few clock cycles of time once the programmer preloads the context in the register sets for the different functions, ISRs and *tasks*. [Refer to Section 8.1 for definition of the *task*.]

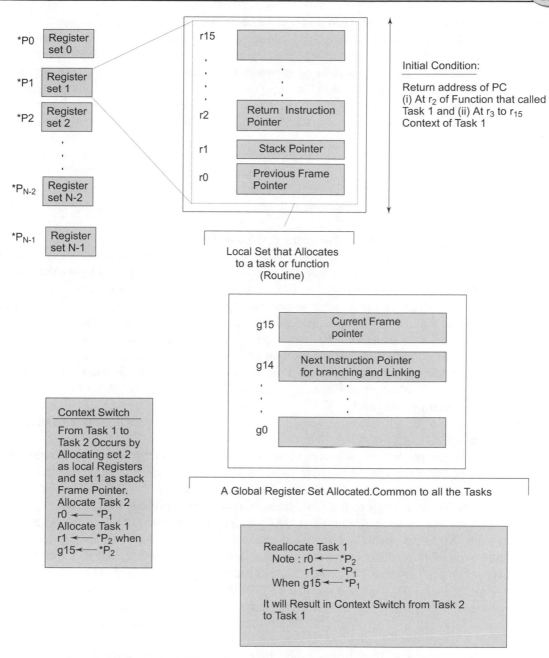

Note : As the Codes Execute r1 changes but g15 remains constant so that saved variables can be pushed and popped when retrieving.

Figure 4.5

Context Switching using the register sets and stack frames allocation and re-allocation in 80196 and 80960-based systems.

Context switching is as follows in ARM 7 processor on an *ISR call*. (i) The interrupt mask (disable) flags are set. [Disable low priority interrupts.] (ii) Program Counter (PC) is saved. (iii) Current Program Status Register, CPSR copies into one of the saved program status register (SPSR) and CPSR stores the new status (interrupt source data or information). (iv) PC gets the new value as per the interrupt source from the vector table. *An ISR return context switching back to previous context is* as follows. (i) Program Counter (PC) is retrieved. (ii) The corresponding SPSR copies back into CPSR. (iii) The interrupt mask (disable) flags are reset. [It re-enables the earlier disabled low priority interrupts.]

3. *For every source there may be a maximum period only up to which the service of its ISR instructions can be kept pending.* This period defines the *deadline* during which the service must be completed. [Refer to the example of a video processing system described earlier. Also refer to Section 1.3.1 for an application of deadline information for reducing power dissipation in the embedded systems.]

An *Interrupt-latency* period, T_{la}, is the sum of the periods as follows.

(i) Time to be taken is for the response and initiating the ISR instructions. This includes the time to save or switch the context (including the program counter and registers) plus the time to restore its context. For example, in the ARM 7 processor, this period equals two clock cycles plus zero to twenty clock cycles for finishing an ongoing instruction plus zero to three cycles for aborting the data. [The variation is because an interrupt can occur just before the end or just after the start of an instruction. The longest instruction duration takes 20 cycles of clock in ARM 7. Aborting the data (CPSR coping into an SPCR) in a specific case may or may not be needed. The three clock cycles are needed for the latter.] The minimum period is thus four and the maximum is 27 clock cycles in this processor. Thus for a T_{la} calculation 27 clock cycle periods are taken into account.

(ii) This includes the periods needed to service all the interrupts of higher priority than that of the present source.

(iii) This may or may not be included depending upon the programmer's approach: the maximum period of disabling the interrupt service in any set of codes of any source or function. [A set may consist of the codes for the critical region instructions.]

Figure 4.6(a) shows the periods of deadline, latency, servicing of ISR for an interrupt.

Consider the case of a serial synchronous or UART communication port at which the buffer is not sufficient to store any further bytes and service is necessary on interrupt before the next interrupt at the port. If the principle that ISRs be as short as possible is not followed, an ISR currently under execution may not return before that *deadline*, T_{max}. Recall Example 4.1. In In A_Out_B given above, the bytes transmit at each 170 µs interval, and the T_{max} for the inter-network interrupt service is 170µs. If the interrupt latency in the given embedded system is T_{lat}, then the maximum time for its ISR instructions available will be T_{max} minus T_{lat}. Let T_{ISR} be the time for servicing the instruction in In_A_Out_B. T_{ISR} should be less than the T_{max} minus T_{lat}.

The interrupt latency and deadline of any source help in allocating the time for the various routines and functions. Figure 4.6(b) shows five exemplary interrupts; RTI, RI, STATUS_I, FIFI_4thEntry and FIFO-Full, in order of servicing priorities. They are explained by Examples 4.11 and 4.12.

Example 4.11   Assume that a serial input from a network reads a byte and on full buffer it generates an interrupt. Let a FIFO of eight bytes (0th to 7th) be present. Any byte from the receiving data buffer transfers to FIFO in case the buffer is not *read* within the deadline in an exemplary processor 80196. Let there be three sources of interrupts, when buffer receives a byte, when FIFO is half full and when FIFO is full. Let the three source names be RI, FIFI_4thEntry and FIFO_FULL, respectively. Deadline T_{max} for a service for these sources will be the reciprocal of the rate. Using an SWT the status of the network must be checked for the port status. Let us name this interrupt source STATUS_I.

Column 2 of Table 4.3 shows when a network reads at the rate of 1,000 bytes per second. Assume $T_{lat} = 340$ μs for each of three sources. Also assume that the T_{ISR} for RI, FIFI_4thEntry and FIFO_FULL source to service are 20 μs, 90 μs and 170 μs, respectively. Let there be an interrupt, RTI (real time clock interrupt), which obviously has a higher priority. The time available in-between for servicing of an RTI will be T_{max}- T_{lat} - T_{ISR}. [Column 3].

Example 4.12   Let latency, T'_{lat} for RTI be 120 μs. If ISR_RTI takes $T''_{ISR} = 200$ μs to give count input to four SWTs when none of the timer time-outs, and takes T'''_{ISR} 300 μs when one timer time-outs. [Refer to Example 3.5.] We have to choose the bigger period between the two. What is then the minimum time left $(T'^v_{ISR})_{max}$ for other higher priority ISR instructions of interrupt STATUS_I. to be serviced from the network for its status information? Assume $T''_{lat} = 80$ μs for STATUS_I source. $(T'^v_{ISR})_{max} = T_{max}$- T_{lat} - T_{ISR} - T'_{lat} - T''_{lat}. [Column 4]

Table 4.3

Deadline Source with Two Other Higher Priority Interrupts	T_{max}(μs) at 1000 B/s	Time available for rendering the service of RTI (μs)	Time available for instructions of ISR of network-status service at 1 ms rates
RI	1	640	160
FIFI_4thEntry	5	4570	4070
FIFO_Full Maximum Latency	9	8490	7990

A good software design principle for multiple interrupt sources is to keep the ISR as short as possible. Why? So as to service the in-between pending interrupts and leave the functions that can be executed afterwards for a later time. When this principle is not adhered to, a specific interrupting source may not be serviced within the *deadline* (maximum permissible pending time). Example 5.6 explains the use of queuing of the functions of the ISRs for a later period execution. This keeps the ISR execution periods short. Deadlines are thus met.

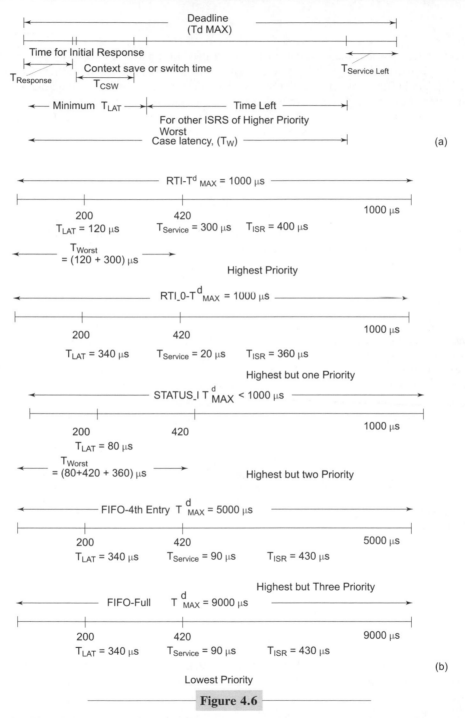

Figure 4.6

(a) The deadline, latency, servicing and ISR execution periods for an interrupt. (b) Five exemplary interrupts; RTI, RI, STATUS_I , FIFI_4thEntry and FIFO–Full, in order of servicing priorities.

Each running program has a context at an instant. Context reflects a CPU state—program counter, stack pointer(s), registers and program state (variables that should not be modified by another routine). Context saving on the call of another ISR or task or routine is essential before switching to another context. Certain processors provide for fast context switching by providing internal stack frames or sets of local registers for the contexts. Fast context switching reduces the interrupt latencies and enables the meeting of each task deadline for service. The operating system program provides for memory blocks to be used as stack frames if internal stack frames are not available at the processor being used in the system.

4.6.2 Classification of Processors Interrupt Service Mechanism from Context Saving Angle

The 8051 interrupt mechanism is such that on occurrence of an interrupt service, the processor pushes the processor registers PCH and PCL on to the memory stack. The 8051 family processors do not save the context of the program (other than the absolutely essential program counter) and a context can save only by using the specific set of instructions. For example, use Push instructions. It speeds up the start of ISR and returns from ISR but at a cost. The onus of context saving is on the programmer in case the context is to be modified on servicing or the function calls during execution of the remaining ISR instructions.

The 68HC11 interrupt mechanism is such that processor registers save onto the stack whenever an interrupt service occurs. These are in the order of PCL, PCH, IYL, IYH, IXL, IXH, ACCA, ACCB, and CCR. The 68HC11 thus does automatically save the processor context (not program variables context) of the program without being so instructed in the user program. As context saving takes processor time, it slows a little at the start of ISR and return from ISR but at the great advantage that the onus of processor registers context saving is not on the programmer and there is no risk in case the context modifies on servicing or function calls.

Context-switching process in 80960 is very powerful. [Figure 4.5.] ARM 7 also provides a mechanism for fast context switching.

4.6.3 Use of DMA Channel for Facilitating the Small Interrupt-Latency Period Sources

A good feature of servicing with small latency periods is that use a DMA channel when multiple I/O interrupt sources exist. The interrupt service routine period from start to end can now be very small as the ISR that initiates the DMA (Direct Memory Access) to the interrupting source, simply programs the DMA registers for the command, data-count, memory block address and I/O bus start address. [Section 2.6]

The use of DMA channels for the I/O services in place of processor interrupt driven ISRs provides an efficient method. This is because a DMA transfer uses the periods when the system buses are free.

4.6.4 Assignment of Priorities to Meet Service Deadlines

There is a need for assigning the level numbers and assigning a priority order. This can be understood from the following scenario. Assume that there are seven interrupt sources: watchdog timer, timer 1 input capture, timer 2 input capture, timer 1 output compare, timer 2 output compare, serial port input and A/D conversion. The A/D conversion should have the lowest priority since the timer and serial port interrupt need servicing faster than A/D. The watchdog timer should have the highest priority among these. Therefore, after a programmed time there should be the program *initialization* again and a *run* again. Also, the internal timers need faster servicing than the external serial port. There can be nested interrupt calls in case the higher priority interrupt source activates in succession. A return from any of the ISRs is to the lower priority ISR. When there are the multiple interrupt sources, an occurrence of each interrupt source (or source-group) is identifiable from a bit or bits in the status register and/or in the IPR. Case 1: When there are multiple sources of interrupts from the multiple devices, the processor hardware assigns to each source (including traps or exceptions) or source-group a pre-assumed priority (or level or type). Let us assume a number, p_{hw} represents the hardware presumed priority for the source (or group). Let the number be among 0, 1, 2, ..., k, ..., m-1. Let $p_{hw} = 0$ mean the lowest; $p_{hw} = 1$ next to lowest;........; $p_{hw} = m$-1 assigned the highest. Why does the hardware assign the presumed priority? Several interrupts occur at the same time during the execution of a set of instructions, and all or some interrupts are enabled for servicing. The servicing using the source corresponding ISRs can only be done in a certain order of priority. [There is only one processor.]

The processor's hardware can assign $p_{hw} = 6, 5, 4, 3, 2, 1, 0$, respectively for seven interrupts. Take the example of the 80x86 family processor. Consider its six interrupt sources; division by zero, single step, NMI (nonmaskable interrupt from RAM parity error, etc.), break point, overflow, and print screen. These interrupts can be assumed to be of $p_{hw} = 0, -1, -2, -3, -4$ and -5, respectively. The hardware processor assigns highest priority for a *division by zero*. This is so because it is an exceptional condition found in user software. The processor assigns the *single stepping* as the next priority as the user enables this source of interrupt because of the need to have a break point at the end of each instruction whenever a debugging of the software is to be done. NMI is the next priority because an external memory *read* error needs urgent attention. Print screen has lowest priority.

ARM 7 provides for two types of interrupt sources (requests), IRQs (interrupt requests) and FIQs (fast interrupt requests).

Which is the interrupt to be serviced first among those pending? Some method of polling resolves this question. The 8086 has a 'vectored priority polling method'. A processor interrupt mechanism may internally provide for the number of vectors, ISR_VECTADDRs. The *vectored priority method* means interrupt mechanism assigns the ISR_VECTADDR as well as p_{hw}. There is a call at the end of each instruction cycle (or at the return from an ISR) for a highest priority source among those enabled and pending. Vectored priorities in 80x86 are as per the n_{type}. The $n_{type} = 0$ highest priority and $n_{type} = 0xFF (=255)$ lowest priority.

The 80196 is another exemplary system for interrupts control. There are distinct flag bits for each source. But there is a grouping of the sources. The 80196 processor divides all maskable interrupts in 8 groups. Each group has a common enabling bit at the primary level. There are 8 secondary level enable bits of the processor in-between servicing of an ISR. There is in-between diversion to a higher priority ISR than the present.

4.6.5   Software Overriding of Hardware Priorities

This can be understood from some examples. The 8051 and 80960 internal interrupt mechanisms are as follows. Which source or source group has higher priority with respect to others? *It is first decided among the ISRs that have been assigned higher priority in the user software. If user-assigned priorities are equal then among that highest priority, which is pre-assigned at the processor internal hardware.* For example, there are the interrupt registers at 8051 in which there are five priority bits for the 5 interrupt sources in 8051 at IP register. [Also there are the five interrupt-enable bits. These are the secondary level enable bits of the processor's in-between servicing of an ISR.] When a priority bit at IP is set, the corresponding interrupt source gets a high priority, and if reset, it gets a lower priority. The 8051 first selects by polling among the high priority according to the bits at the IP register. The concept of the IP register is not necessary because the secondary level masking bits are available. IP registers do not exist in 68HC11 and 80196. Supervisory mode instructions of 80960—and not user instructions—facilitate p_{sw}, which is definable by a set of five bits.

> When there are multiple device drivers and hardware and software interrupts, assignment of the priorities for each source or source group is essential. Either the hardware-defined priority can be used as such or software-assigned priorities are used. These override the hardware priorities. Operating systems of the system being used assigns the software priorities to each routine or task.

▪ SUMMARY ▪

The following is a summary of the important points we learned in this chapter.

- Physical device drivers, virtual device drivers and ISRs for the software instruction, software defined condition and error condition are used by the software designer for a system.
- It is essential to use the interrupt service routines (ISRs) in any system.
- Every system has an interrupt servicing mechanism. This has been explained by using simple examples.
- Device driver routines service the device interrupt(s) and drive a peripheral unit (device). A device is configured and initialized by using the control bits at its control register(s). The device driver executes on interrupts as reflected by flags in the status register(s).
- Linux has a large number of drivers.
- We learnt by using examples about *device initialization* and *device driver coding* for the parallel ports, serial line UART and internal timing device in 68HC11.
- Virtual devices like char device, block device or file device, RAM disk, socket, pipe, and loop back device, are used during system software design. These are treated in a way analogous to physical devices.
- There are the device interrupts as well as other interrupt sources, the drivers for which must be written by the software designer. A list of the various possible sources of software and hardware interrupts is given. A software instruction or a condition during running-related or runtime-error-related interrupts are important mechanisms in the systems.
- The interrupt servicing mechanism in a system provides for individual device interrupt enabling and disabling, interrupt vectors and vector table, interrupt pending registers and status registers, and nonmaskable and maskable interrupts.
- Each running-program has a *context* at an instant. Context means a CPU state—program counter, stack pointer(s), registers and program state (variables that should not be modified by another routine). The context must be saved on a *call* to another ISR or *task* or routine. It must be done before processor switching to another context. Certain processors like those from the

ARM family and 80960 provide for fast context switching. These have the internal stack frames or sets of local registers for the contexts. Fast context switching reduces the interrupt latencies and enables the meeting of each task deadline. The operating system program provides for memory blocks (allocates the blocks) to be used as stack frames if internal stack frames are not available at the processor being used in a system.

- The software design should be such that interrupt latencies are as short as possible. This helps in meeting the deadlines for each interrupt service. The use of a DMA channel facilitates the small interrupt-latency periods of an I/O interrupt source.
- There may be simultaneous service demands from multiple sources. Assignment of software priorities among the multiple sources of interrupts, keeping in mind the available hardware priorities, is essential in a software design.

■ LIST OF KEYWORDS AND THEIR DEFINITIONS ■

- *Device Initialization Codes*: Codes for programming the device control register of a device.
- *Device Driver Codes*: Codes for read and write operations at the device addresses and for reading device status and initiating an interrupt.
- *Device Attaching (Adding)*: Configuring a device and enabling the use of its driver.
- *Device Detaching (Removing)*: Disabling the use of a device driver by the system.
- *Device Opening*: Resetting the device control bits and preparing it for the use of its driver.
- *Device Closing*: Resetting the device control bits and its next time use is then possible only by opening it again.
- *Linux*: An open source operating system. It has a large number of device drivers and network management functions.
- *Linux Device Drivers*: Device drivers taken from the Linux source.
- *Interrupt*: CPU on an interrupt message may initiate a further action by calling an interrupt service routine (ISR) or else it continues with the current process (task).
- *Interrupt Service Routine (ISR)*: A program that is executed on interrupt after saving necessary parameters onto the stack so that the same can be retrieved on return from the routine's last instruction. An ISR is also called a device driver when it services a device interrupt.
- *Hardware Timer*: A timer present in the system as hardware and which gets inputs from the internal clock with the processor, called system clock. A device driver program programs it like any other physical device.
- *Interrupt Mechanism*: A mechanism for interrupt driven servicing of the devices and ports. It saves the processor waiting time, because it lets the processor process the multiple devices and virtual devices. The mechanism also sets the priorities and provides for enabling and disabling the services.
- *Foreground Program*: A foreground program is one that is executed when no interrupt call is being serviced.
- *Interrupt Enable Bit*: If set as true it enables the interrupts from a source(s).
- *Interrupt Mask Bit*: When a bit is reset (= false) the request for the initiation of interrupt service is responded, otherwise it is not responded.
- *Interrupt Flag*: A register bit for a Boolean variable that sets to signal a need for executing an interrupt service routine (ISR). It resets when a corresponding ISR starts executing.
- *Nonmaskable Interrupt*: A source that cannot be disabled and is used for most needed service cases.

- *Maskable Interrupt*: A source that can be disabled or masked.
- *Primary Level Enable Bit*: A bit that enables or disables any service on interrupt by all the maskable sources. It helps in executing critical section codes by preventing service to any other maskable source.
- *Secondary Level Mask Bit*: It disables service from an individual source or source group.
- *Interrupt Vector*: A memory address where there are bytes to provide the corresponding ISR address. The system has the specific vector addresses assigned by the hardware for the interrupting sources for each internal device.
- *Interrupt Vector Table*: A table for the interrupt vectors in the memory. The table facilitates the servicing of the multiple interrupting sources for each internal device. In each row for an interrupt vector address, there are bytes to provide the corresponding interrupt service routine address.
- *Interrupt-Pending Register*: A register to show the interrupt sources or source groups from various devices that are pending for servicing by executing the corresponding ISRs. It is a read and write register. A bit auto-resets in it when the corresponding interrupt service starts. A user instruction can also reset a bit in the register.
- *Status Register*: A read-only register for a device to set a flag on arising of an interrupt. A user instruction can also reset a bit in it. If a device has a number of sources, the status register has that number of flags—a distinct source for each source. When it is read by a processor instruction, the flag resets.
- *Context*: The program counter and stack pointer as well as the program status word and processor registers for a foreground program or ISR or task. It can also include memory block addresses allotted to the program or routine. It can include ID.
- *Context Switching*: Saving the interrupted routine (or function) context and retrieving or loading the new context of the called routine. The time taken in context switching is included in the interrupt latency period.
- *Stack Frame*: A set of registers or a memory block that stores the context for a program or ISR.
- *Interrupt Latency*: A period for waiting for the service after a service demand is raised (source status flag sets).
- *Worst-case Latency*: Maximum interrupt latency is found in the worst possible case.
- *Deadline*: A period during which service to an interrupt must start.
- *Software Interrupt*: An interrupt by an instruction, by a software timer or by an error condition trap or illegal opcode.
- *Exception*: Setting of a condition that may be defined by the programmer, in which the programmer also defines the ISR for handling servicing for that condition. Error conditions are handled by the exceptions.
- *Signal*: An *exception* may also sometimes be called a *signal*. The processor may also signal an exception. For example, on division by zero in 80x86.
- *Hardware Interrupt*: Interrupt to devices or ports at the system.
- *Hardware Assigned Priority*: Priority assigned by the processor itself to service a source when several interrupts need the interrupt service.
- *Software Assigned Priority*: A priority for a source or source group. It is defined at a register called interrupt priority register. When several interrupts occur at the same time, software assigned priorities override the hardware priorities.
- *Polling*: A method by which, at the end of an instruction or at the end of an ISR, the pending interrupts are searched by the processor to service the one with the highest priority.

■ REVIEW QUESTIONS ■

1. What are the disadvantages and advantages of *busy and wait transfer* mode for the I/O devices?
2. What are the advantages and disadvantages of an interrupt driven data transfer?
3. What are the advantages of *DMA*-based or peripheral-transaction-server-(80960)-based data transfer over the *interrupt-driven data transfer*?
4. How does a programmer use the vector address for an interrupt source?
5. Interrupt vector address are pre-fixed in the interrupt mechanism for the known internal peripherals in a microcontroller. How are the vector addresses assigned for exceptions and user-defined interrupts?
6. An interrupt mechanism in each processor differs from a processor family to another. Explain why the device drivers are processor-sensitive programs.
7. How do you intialize and configure a device? Take an example of a serial line driver at a COM port of PC.
8. How is a *file* at the memory act handled as a device?
9. What are the advantages of a RAM disk?
10. Make a list of Linux internal *net* directory functions for sockets, handling of socket buffers, firewalls, network protocols (for examples, NFS, IP, IPv6 and Ethernet) and bridges. Why are these device drivers assigned in a separate directory of the network management function of the Linux operating system?
11. Define *context*, *interrupt latency* and *interrupt service deadline*.
12. Why is the context switching in an embedded processor faster than saving the pointers and variables on the stack using a stack pointer? Use the example of 80960.
13. How does the context switching handled in ARM 7?
14. DMA helps in reducing the processor load by providing direct access for the I/Os. How does it help in faster task execution in a multitasking system by the reduced interrupt-servicing latencies?
15. What is meant by throwing an exception? How is the exception condition during the execution of a function (routine) handled?
16. Refer to Sections G.7, G.8 and G.9. How do the device drivers and ISRs differ in the 80x86 microprocessor and 68HC11 microcontroller?
17. How does one assign service priority to the multiple device drivers of a system? How does one assign priorities to the timer devices and ADC device?
18. What are the uses of hardware assigned priorities in an interrupt mechanism?
19. What are the uses of software assigned priorities in an interrupt mechanism?
20. How is a break point interrupt important for debugging embedded software?
21. What is meant by POSIX function?

■ PRACTICE EXERCISES ■

22. How does one write device driver? List the steps involved in writing a device driver.
23. Search the Web and design a table to show features in device driver modules of Embedded Linux OS and Red Hat eCOS (Embedded Configurable Operating System) systems.
24. Explain using examples each of the following: a char device, a block device and a block device configurable as char device. A UART is a char device. Why is it a char device?
25. Give software related interrupt examples. What are the interrupts in 8086, which generate on software error?

# Chapter 5

Programming Concepts and Embedded Programming in C and C++

What We Have Learnt

Let us recapitulate the following facts that have been presented in the previous Chapters:

1. System hardware consists of processors, memory devices (ROM and RAM) and internal, I/O ports, physical internal and external devices, timing devices and basic hardware units—power supply, clock circuit and reset circuit.

2. A system has the buses and interfacing circuits for connecting the physical devices at the ports.

3. A system has the processor and memory-sensitive software for the device drivers as per interrupt servicing mechanism for handling the interrupts from the devices, drivers for the virtual devices and software-interrupts, exceptions and errors.

Besides the above software, the software is also needed in the system for the simple or sophisticated hardware application(s). Except for certain processor and memory-sensitive instructions where program codes may be written in assembly, the codes are written in a high-level language. ***Programming is the most essential part of any embedded system design***.

What We Will Learn

The objective of this Chapter is to explain **(i)** the programming concepts, elements and data structures in detail, **(ii)** the use of objected oriented programming approach, **(iii)** the use of source code engineering tools and **(iv)** the use of memory optimization methods. The objective is achieved by the explanation of the following concepts and practices.

1. Programming in the *assembly language* vs. *high-level language* and the powerful features of C for embedded systems

2. Program elements: Preprocessor directives and the header files *include* files and source files that provide a program for an application
3. Program elements: *Macros* and *functions* and their uses in a C program
4. Program elements: *Data types*, *data structures*, modifiers, conditional statements and loops
5. Program elements: *Pointers*, function calls, multiple functions, *function pointers*, function queues and service-routine queues
6. Important data structures: Arrays, queues, stacks, lists and trees
7. The vital role played by queues in network communication or client-server communication
8. Program details for using 'queue' an important data structure used in a program
9. Queues for *implementing a protocol* for a network
10. Queuing of Functions on Interrupts for an efficient use of the interrupt mechanism
11. A new form of queue data structure, which facilitates flow control by a *first in provisionally out* queue and how it is used in a network protocol like TCP for *data flow control*
12. Programming details for using 'stack'—another important data structure used in a program
13. Programming details for using '*list*' and priority wise '*ordered list*' as these are also the important data structures used in an embedded system program.
14. Example of a *list* for running real time clock interrupts driven *software timers* (RTCSWTs)
15. Example of a *list* of *ready tasks for multitasking* by operating system functions
16. Object Oriented Programming concepts; Embedded programming in C / C++ and advantages and disadvantages of programming in *C++* and *Java*
17. Use of *compiler, cross compilers and source code engineering tools*
18. Steps needed to *optimise the memory* needs of embedded software

Program modeling concepts and software engineering practices adopted for designing an embedded system program are dealt with in Chapters 6 and 7. Definition of a process and inter-process synchronization, RTOSs and RTOS programming using an RTOS are described in Chapters 8 to 11.

5.1 SOFTWARE PROGRAMMING IN ASSEMBLY LANGUAGE (ALP) AND IN HIGH-LEVEL LANGUAGE 'C'

Assembly language coding of an application has the following advantages:

1. *It gives a precise control* of the processor internal devices and *full use of processor specific features* in its instruction set and its addressing modes.
2. The machine codes are compact. This is because the codes for declaring the conditions, rules, and data type do not exist. The system thus needs a smaller memory. Excess memory needed does not depend on the programmer data type selection and rule-declarations. It is also not compiler or specific library-functions specific.
3. Device driver codes may need only a few assembly instructions. For example, consider a small embedded system, such as a timer device in a microwave oven, a washing machine or a vending machine. Assembly codes for these can be compact and precise, and are conveniently written.

It becomes convenient to develop the *source files* in C or C++ or Java for complex systems because of the following advantages of high-level languages for such systems.

1. ***The development cycle is short for complex systems*** due to the use of functions (procedures), standard library functions, modular programming approach and top-down design. Application programs are structured to ensure that the software is based on sound software engineering principles.

   (a) Let us recall Example 4.8 of a UART serial line device driver. Direct use of this function makes the repetitive coding redundant as this device is used in many systems. We simply change some of the arguments (for the variables) passed when needed and use it at another instance of the device use.

   (b) Should the square root codes be written again whenever the square root of another value (argument) is to be taken? The *use of the standard library function*, square root ( ), saves the programmer time for coding. New sets of library functions exist in an embedded system specific C or C++ compiler. Exemplary functions are the delay ( ), wait ( ) and sleep ( ).

   (c) *Modular programming approach* is an approach in which the building blocks are reusable software components. Consider an analogy to an IC (Integrated Circuit). Just as an IC has several circuits integrated into one, similarly a building block may call several functions and library functions. A module should however, be well tested. It must have a well-defined goal and the well-defined data inputs and outputs. It should have only one calling procedure. There should be one return point from it. It should not affect any data other than that which is targeted. [Data Encapsulation.] It must return (report) error conditions encountered during its execution.

   (d) Bottom-up design is a design approach in which programming is first done for the sub-modules of the specific and distinct sets of actions. An example of the modules for specific sets of actions is a program for a software timer, RTCSWT:: run. Programs for delay, counting, finding time intervals and many applications can be written. Then the final program is designed. The approach to this way of designing a program is to first code the basic functional modules and then use these to build a bigger module.

   (e) Top-down design is another programming approach in which the *main* program is first designed, then its modules, sub-modules, and finally, the functions.

2. *Data type declarations provide programming ease.* For example, there are four types of integers, *int*, *unsigned int*, *short* and *long*. When dealing with only positive values, we declare a variable as *unsigned int*. For example, numTicks (Number of Ticks of a clock before the timeout) has to be unsigned. We need a signed integer, *int* (32 bit) in arithmetical calculations. An integer can also be declared as data type, *short* (16 bit) or *long* (64 bit). To manipulate the text and strings for a character, another data type is *char*. *Each data type is an abstraction for the methods to use, to manipulate and to represent, and for a set of permissible operations.*

3. *Type checking* makes the program less prone to error. For example, type checking does not permit subtraction, multiplication and division on the *char* data types. Further, it lets + be used for concatenation. [For example, micro + controller concatenates into microcontroller, where micro is an array of *char* values and controller is another array of *char* values.]

4. *Control Structures* (for examples, *while*, *do-while*, *break* and *for*) and *Conditional Statements* (for examples, *if*, *if-else*, *else-if* and *switch-case*) make the program-flow-path design tasks simple.

5. *Portability* of non-processor-specific codes exists. Therefore, when the hardware changes, only the modules for the device drivers and device management, initialization and locator modules [Section 2.5.2] and initial boot-up record data need modifications.

Additional advantages of C as a high-level language are as follows:

1. It is a language between low (assembly) and high-level language. Inserting the assembly language codes in between is called in-line assembly. A direct hardware control is thus also feasible by in-line assembly, and the complex part of the program can be in high-level language. Example 4.8 shows the use of in-line assembly codes in C for a Port A Driver Program.

High level language programming makes the program development cycle short, enables use of the modular programming approach and lets us follow sound software engineering principles. It facilitates the program development with 'bottom-up design' and 'top-down design' approaches. Embedded system programmers have long preferred C for the following reasons: (i) The feature of embedding assembly codes using in-line assembly. (ii) Readily available modules in C compilers for the embedded system and library codes that can directly port into the system-programmer codes.

5.2 'C' PROGRAM ELEMENTS: HEADER AND SOURCE FILES AND PREPROCESSOR DIRECTIVES

The 'C' program elements—header and source files and preprocessor directives—are as follows:

5.2.1 Include Directive for the Inclusion of Files

Any C program first includes the header and source files that are readily available. A case study of sending a stream of bytes through a network driver card using a TCP/IP protocol is given in Example 11.2. Its program starts with the codes given in Example 4.5 and Example 5.1. The purpose of each included file is mentioned in the comments within the * symbols as per 'C' practice.

Example 5.1
```
# include "vxWorks.h" /* Include VxWorks functions */
# include "semLib.h" /* Include Semaphore functions Library */
# include "taskLib.h" /* Include multitasking functions Library */
# include "msgQLib.h" /* Include Message Queue functions Library */
# include "fioLib.h" /* Include File-Device Input-Output functions Library */
# include "sysLib.c" /* Include system library for system functions */
# include "netDrvConfig.txt" /* Include a text file that provides the 'Network Driver Configuration'. It
provides the frame format protocol (SLIP or PPP or Ethernet) description, card description/make, ad-
dress at the system, IP address(s) of the node(s) that drive the card for transmitting or receiving from the
network. */
# include "prctlHandlers.c" /* Include file for the codes for handling and actions as per the protocols
used for driving streams to the network. */
```

Included is a preprocessor directive that includes the contents (codes or data) of a file. The files that can be included are given below. Inclusion of all files and specific header files has to be as per the requirements.

(*i*) *Including codes files*: These are the files for the codes already available. For example, # include "***prctlHandlers.c***".

(*ii*) *Including constant data files:* These are the files for the codes and may have the extension '.const'.

(*iii*) *Including stings data files:* These are the files for the strings and may have the extension '.strings' or '.str.' or '.txt. For example, # include "*netDrvConfig*.txt" in Example 5.1.

(*iv*) *Including initial data files:* Recall Section 2.4.3 and Examples 2.13 to 2.15. There are files for the initial or default data for the shadow ROM of the embedded system. The boot-up program is copied later into the RAM and may have the extension '.init'. On the other hand, RAM data files have the extension '.data'.

(*v*) *Including basic variables files:* These are the files for the local or global static variables that are stored in the RAM because they do not posses the initial (default) values. The term static means that there is a common but not more than one instance of that variable address and it has a static memory allocation. There is only one real time clock, and therefore only one instance of that variable address. [Refer to case (iv) Section 5.4.3.] These basic variables are stored in the files with the extension '.bss'.

(*vi*) *Including header files:* It is a preprocessor directive, which includes the contents (codes or data) of a set of source files. These are the files of a specific module. A header file has the extension '.h'. Examples are as follows. The string manipulation functions are needed in a program using strings. These become available once a header file called "*string.h*" is included. The mathematical functions, *square root, sin, cos, tan, atan* and so on are needed in programs using mathematical expressions. These become available by including a header file, called "*math.h*". The preprocessor directives will be '# include <*string.h*>' and '# include <*math.h*>'.

Also included are the header files for the codes in assembly, and for the I/O operations (conio.h), for the OS functions and RTOS functions. # include "VxWorks.h" in Example 5.1 is a directive to the compiler, which includes VxWorks RTOS functions.

Note: Certain compilers provide for *conio*.h in place of *stdio*.h. This is because embedded systems usually do not need the file functions for opening, closing, read and write. Including stdio.h makes the code too big.

What is the difference between inclusion of a header file, and a text file or data file or constants file? Consider the inclusion of *netDrvConfig*.txt.*txt* and *math.h*. (*i*) The header files are well-tested and debugged modules. (*ii*) The header files provide access to standard libraries. (*iii*) The header file can include several text file or C files. (*iv*) A text file is a description of the texts that contain specific information.

5.2.2　Source Files

Source files are program files for the functions of application software. The source files need to be compiled. A source file will also possess the preprocessor directives of the application and have the *first function from where the processing will start*. This function is called *main* function. Its codes start with *void main ( )*. The *main* calls other functions. A source file holds the codes as like the ones given earlier in Examples 4.3 to 4.5.

5.2.3　Configuration Files

Configuration files are the files for the configuration of the system. Recall the codes in Line 3 of Example 4.8. Device configuration codes can be put in a file of basic variables and included when needed. If these codes are in the file "serialLine_cfg.h" then # include *"serialLine_cfg.h" will be preprocessor directive.* Consider another example. *# include "os_cfg.h".* It will include the os_cfg header file.

5.2.4 Preprocessor Directives

A preprocessor directive starts with a sharp (hash) sign. These commands are for the following directives to the compiler for processing.

1. *Preprocessor Global Variables*: "# define volatile boolean IntrEnable" is a preprocessor directive in Example 4.6. It means it is a directive before processing to consider IntrEnable as a global variable of boolean data type and is volatile. [Volatile is a directive to the compiler not to take this variable into account while compacting and optimising the codes.] IntrDisable, IntrPortAEnable, IntrPortADisable, STAF and STAI are the other global variables in Example 4.5.

2. *Preprocessor Constants*: "# *define* false 0" is a preprocessor directive in example 4.3. It means it is a directive before processing to assume 'false' as 0. The directive *'define'* is for allocating pointer value(s) in the program. Consider # define portA (volatile unsigned char *) 0x1000 and # define PIOC (volatile unsigned char *) 0x1001. [Refer to Section 4.2.] 0x1000 and 0x1000 are for the fixed addresses of portA and PIOC. These are the constants defined here for these 68HC11 registers.

Strings can also be defined. Strings are the constants, for example, those used for an initial display on the screen in a mobile system. For example, # define *welcome* "Welcome To ABC Telecom".

Preprocessor constants, variables, and inclusion of configuration files, text files, header files and library functions are used in C programs.

5.3 PROGRAM ELEMENTS: MACROS AND FUNCTIONS

Table 5.1 lists these elements and gives their uses.

Table 5.1

Uses of the Various Sets of Instructions as the Program Elements

Program Element	Uses	Saves context on the stack before its start and retrieves them on return	Feasibility of nesting one within another
Macro	Executes a named small collection of codes.	No	None
function	Executes a named set of codes with values passed by the calling program through its arguments. Also returns a data object when it is not declared as void. It has the context saving and retrieving overheads.	Yes	Yes, can call another function and can also be interrupted.
Main function	Declarations of functions and data types, typedef and either (i) executes a named set of codes, calls a set of functions, and calls on the Interrupts the ISRs or (ii) starts an OS Kernel.	No	None

Program Element	Uses	Saves context on the stack before its start and retrieves them on return	Feasibility of nesting one within another
Reentrant function	Refer to Section 5.4. 6 (ii)	Yes	Yes to another reentrant function only
Interrupt Service Routine or Device Driver	Declarations of functions and data types, typedef, and executes a named set of codes. Must be short so that other sources of interrupts are also serviced within the deadlines. Must be either a reentrant routine or must have a solution to the shared data problem.	Yes	To higher priority sources
Task	Refer to Section 8.2. Must either be a reentrant routine or must have a solution to the shared data problem.	Yes	None
Recursive function	A function that calls itself. It must be a reentrant function also. Most often its use is avoided in embedded systems due to memory constraints. [The stack grows after each recursive call and its usage may choke the memory space availability.]	Yes	Yes

Preprocessor Macros: A macro is a collection of codes that is defined in a program by a name. It differs from a function in the sense that once a macro is defined by a name, the compiler puts the corresponding codes for it at every place where that macro name appears. The 'enable_Maskable_Intr ( )' and 'disable_Maskable_Intr ( )' are the macros in Example 4.5. [The pair of parentheses is optional. If it is present, it improves readability as it distinguishes a macro from a constant]. Whenever the name enable_Maskable_Intr appears, the compiler places the codes designed for it. Macros, called test macros or test vectors, are also designed and used for debugging a system. [Refer to Section 7.6.3.]

How does a macro differ from a function? The codes for a function are compiled once only. On calling that function, the processor has to save the context, and on return restore the context. Further, a function may return nothing (*void* declaration case) or return a Boolean value, or an integer or any primitive or reference type of data. [Primitive means similar to an integer or character. Reference type means similar to an array or structure.] The enable_PortA_Intr ( ) and disable_PortA_Intr ( ) are the function calls in Example 4.5. [The brackets are now not optional]. Macros are used for short codes only. This is because, if a function call is used instead of macro, the overheads (context saving and other actions on function call and return) will take a time, $T_{overheads}$ that is the same order of magnitude as the time, T_{exec} for execution of short codes within a function. We use a function when the $T_{overheads}$ $<< T_{exec}$, and a macro when $T_{overheads}$ $\sim=$ or $> T_{exec}$.

Macros and functions are used in C programs. Functions are used when the requirement is that the codes should be compiled once only. However, on calling a function, the processor has to save the context and, on return, restore the context. Further, a function may return nothing (void declaration case) or return a Boolean value, or an integer or any primitive or reference type of data. Macros are used when short functional codes are to be inserted in a number of places or functions.

5.4 PROGRAM ELEMENTS: DATA TYPES, DATA STRUCTURES, MODIFIERS, STATEMENTS, LOOPS AND POINTERS

5.4.1 Use of Data Types

Whenever a type of data is named, it will have the address(es) allocated at the memory. The number of addresses allocated depends upon the data type. C allows the following primitive data types: *char* (8 bit) for characters, *byte* (8 bit), *unsigned short* (16 bit), *short* (16 bit), *unsigned int* (32 bit), *int* (32 bit), *long double* (64 bit), *float* (32 bit) and *double* (64 bit). [Certain compilers do not take the 'byte' as a data type definition. The 'char' is then used instead of 'byte'. Most C compilers do not take a Boolean variable as data type. As in the second line of Example 4.6, *typedef* is used to create a Boolean type variable in the C program.]

A data type appropriate for the hardware is used. For example, a 16-bit timer can have only the unsigned short data type, and its range can be from 0 to 65535 only. The typedef is also used. It is made clear by the following example. A compiler version may not process the declaration as an unsigned byte. The 'unsigned character' can then be used as a data type. It can then be declared as follows:

    typedef unsigned character portAdata
    #define Pbyte portAdata Pbyte = 0xF1

5.4.2 Use of Data Structures: Queues, Stacks, Lists and Trees

Marks (or grades) of a student in the different subjects studied in a semester are put in a proper table. The table in the mark-sheet shows them in an organised way. When there is a large amount of data, it must be organised properly. A data structure is a way of organising large amounts of data. A data element can then be identified and accessed with the help of a few pointers and/or indices and/or functions. [The reader may refer to a standard textbook for the data structure algorithms in C and C++. For example, *Data Structures and Algorithms in C++* by Adam Drozdek from Brooks/Cole Thomson Learning (2001).]

A data structure is an important element of any program. Section 2.5 defines and describes a few important data structures, *stack*, *one-dimensional array*, *queue*, *circular queue*, *pipe*, a *table* (two dimensional array), *lookup table*, *hash table* and *list*. Figures 2.3 to 2.5 show different data structures and how they are put in the memory blocks in an organised way. Any data structure element can be retrieved.

Sections 5.5, 5.6 and 5.7 describe separately three data structures in detail: the queues, stacks and lists, respectively. Table 5.2 gives the uses and show exemplary uses of queues, stacks, arrays, lists and trees.

Table 5.2

Uses of the Various Data Structures in a Program Element

Data Structure	Definition and when used	Example(s) of its use
Queue	It is a structure with a series of elements with the first element waiting for an operation. An operation can be done only in the first in first out (FIFO) mode. It is used when an element is not to be accessible by any index and pointer directly, but only through the FIFO. An element can be inserted only at the end in the series of elements waiting for an operation. There are two pointers, one for deleting after the operation and other for inserting. Both increment after an operation.	(1) Print buffer. Each character is to be printed in FIFO mode. (2) Frames on a network. [Each frame also has a queue of a stream of bytes.] Each byte has to be sent for receiving as a FIFO. (3) Image frames in a sequence. [These have to be processed as a FIFO.]
Stack	It is a structure with a series of elements with its last element waiting for an operation. An operation can be done only in the last in first out (LIFO) mode. It is used when an element is not to be accessible by any index or pointer directly, but only through the LIFO. An element can be pushed (inserted) only at the top in the series of elements still waiting for an operation. There is only one pointer used for pop (deleting) after the operation as well as for push (inserting). Pointers increment or decrement after an operation. It depends on insertion or deletion.	(1) Pushing of variables on interrupt or call to another function. (2) Retrieving the pushed data onto a stack.
Array (one dimensional vector)	It is a structure with a series of elements with each element accessible by an identifier name and an index. Its element can be used and operated easily. It is used when each element of the structure is to be given a distinct identity by an index for easy operation. Index starts from 0 and is +ve integers.	$ts = 12 * s(1)$; Total salary, ts is 12 times the first month salary. marks_weight [4] = marks_weight [0]; Weight of marks in the subject with index 4 is assigned the same as in the subject with index 0.
Multi-dimensional array	It is a structure with a series of elements each having another subseries of elements. Each element is accessible by identifier name and two or more indices. It is used when every element of the structure is to be given a distinct identity by two or more indices for easy operation. The dimension of an array equals the number of indices that are needed to distinctly identify an array-element. Indices start from 0 and are +ve integers.	Handling a matrix or tensor. Consider a pixel in an image frame. Consider a Quarter-CIF image pixel in 144 x 176 size image frame. [Recall Section 1.2.7.] A *pixel* [108, 88] will represent a pixel at 108th horizontal row and 88th vertical column. [#]See following note also.
List	Each element has a pointer to its next element. Only the first element is identifiable and it is done by list-top pointer (Header). No other element is identifiable and hence is not accessible directly. By going through the first element, and then consecutively through all the succeeding elements, an element can be read, or read and deleted, or can be added to a neighbouring element or replaced by another element.	A series of tasks which are active. Each task has pointer for the next task. Another example is a menu that point to a submenu.

Data Structure	Definition and when used	Example(s) of its use
Tree	There is a root element. It has two or more branches and each has a daughter element. Each daughter element has two or more daughter elements. The last one does not have daughters. Only the root element is identifiable and it is done by the treetop pointer (Header). No other element is identifiable and hence is not accessible directly. By traversing the root element, then proceeding continuously through all the succeeding daughters, a tree element can be read, or read and deleted, or can be added to another daughter or replaced by another element. A tree has data elements arranged as branches. The last daughter, called node, has no further daughters. A binary tree is a tree with a maximum of two daughters (branches) in each element.	An example is a directory. It has number of file folders. Each file folder has a number of other file folders and so on. In the end is a file.

Note: #pixel [0,0] represents the pixel at the left corner on top and pixel [144, 176], the right bottom. *Pixel* [10,108, 88] is a pixel data element in a three-dimensional array form. It represents pixels at same position (108 x 88) in the 10th frame.

5.4.3 Use of Modifiers

The actions of modifiers are as follows:

Case (*i*): Modifier '***auto***' or No modifier, if *outside* a function block, means that there is ROM allocation for the variable by the locator if it is initialised in the program. RAM is allocated by the locator, if it is not initialised in the program.

Case (*ii*): Modifier '***auto***' or No modifier, if *inside* the function block, means there is ROM allocation for the variable by the locator if it is initialised in the program. There is no RAM allocation by the locator.

Case (*iii*): Modifier '*unsigned*' is modifier for a short or int or long data type. It is a directive to permit only the positive values of 16, 32 or 64 bits, respectively.

Case (*iv*): Modifier '*static*' declaration is inside a function block. Static declaration is a directive to the compiler that the variable should be accessible outside the function block also, and there is to be a reserved memory space for it. It then does not save on a stack on context switching to another task. When several tasks are executed in cooperation, the declaration *static* helps. Consider this example declaration, 'private: static void interrupt ISR_RTI ( ). The static declaration here is for the directive to the compiler that the ISR_RTI ( ) function codes limit to the memory block for ISR_RTI ( ) function. The private declaration here means that there are no other instances of that method in any other object. It then does not save on the stack. There is ROM allocation by the locator if it is initialised in the program. There is RAM allocation by the locator if it is not initialised in the program.

Case (*v*): Modifier *static* declaration is outside a function block. It is not usable outside the class in which declared or outside the module in which declared. There is ROM allocation by the locator for the function codes.

Case (*vi*): Modifier *const* declaration is outside a function block. It must be initialised by a program. For example, #define *const* Welcome_Message "There is a mail for you". There is ROM allocation by the locator.

Case (*vii*): Modifier *register* declaration is inside a function block. It must be initialised by a program. For example, '*register* CX'. A CPU register is temporarily allocated when needed. There is no ROM or RAM allocation.

Case (*viii*): Modifier *interrupt*. It directs the compiler to save all processor registers on entry to the function codes and to restore them on return from that function. [This modifier is prefixed by an underscore, '_interrupt' in certain compilers.]

Case (*ix*): Modifier *extern*. It directs the compiler to look for the data type declaration or the function in a module other than the one currently in use.

Case (*x*): Modifier *volatile* outside a function block is a warning to the compiler that an event can change its value or that its change represents an event. An event example is an interrupt event, hardware event or inter-task communication event. For example, consider a declaration: '*volatile* Boolean IntrEnable;' in Example 4.6. It changes to false at the start of service by a service routine, if previously true. The compiler does not perform optimization for a *volatile* variable. Let a variable be assigned, $c = 0$. Later, it is assigned $c = 1$. The compiler will ignore statement $c = 0$ during code optimisation and will take $c = 1$. But if c is an event variable, it should not be optimised. IntrEnable = 0 is at the beginning of the service routine in case an interrupt enable variable is used for disabling any interrupt during the period of execution of the ISR. IntrEnable = 1 is executed before return from the ISR. This re-enables the interrupts at the system. Declaration of IntrEnbale as volatile directs the compiler not to optimise two assignment statements in the same function. There is no ROM or RAM allocation by the locator.

Case (*xi*): Modifier *volatile static* declaration is inside a function block. Examples are (a) '*volatile static* boolean RTIEnable = true;' (b) '*volatile static* boolean RTISWTEnable;' and (c) '*volatile static* boolean RTCSWT_F;' The static declaration is for the directive to the compiler so that the variable should also be accessible outside the function block; there is to be a reserved memory space for it; and volatile means a directive should not optimise an event that it can modify. It then does not save on the stack on context switching to another task. When several tasks are executed in cooperation, the declaration static helps. The compiler does not optimise the code due to declaration volatile. There is no ROM or RAM allocation by the locator.

5.4.4 Use of Conditions, Loops and Infinite Loops

Conditional statements are used often. If a defined condition(s) is fulfilled, the statements within the curly braces after the condition (or a statement without the braces) are executed, otherwise the program proceeds to the next statement or to the next set of statements. Sometimes a set of statements is repeated in a loop. Generally, in case of array, the index changes and the same set is repeated for another element of the array.

Infinite loops are never desired in normal programming. Why? The program will never end, never exit or never proceed further to the codes after the loop. However, an infinite loop is a feature in embedded system programming! This is clarified by the following examples. (i) What about switching off the telephone? The system software in the telephone always has to be in a waiting loop that finds the ring on the line. An exit from the loop will make the system hardware redundant. (ii) Recall Example 4.1 of inter-networking program for In_A_Out_B. Port A may give the input at any instance. The system has to execute codes up to the point where there is output at port B, and then return to receive and wait for another input. The hardware equivalent of an infinite loop is a ticking system clock (real time clock) or a free running counter.

Example 5.2 gives a C program design in which the program starts executing from the main ( ) function. There are calls to the functions and calls on the interrupts in between. It has to return to the start. The system main program is never in a halt state. Therefore, the main ( ) is in an infinite loop within the start and end.

```
Example 5.2
# define false 0
# define true 1
/*****************************************************************/
void main (void) {
/* The Declarations here and initialization here */
.
.
.
/* Infinite while loop follows. Since the condition set for the while loop is always true, the statements
within the curly braces continue to execute */
while (true) {
/* Codes that repeatedly execute */
.
.
.
}
/*****************************************************************/
```

Assume that the function *main* does not have a waiting loop and simply passes the control to an RTOS. Consider a multitasking program. The OS can create a task. The OS can insert a task into the *list*. It can delete from the list. Let an OS kernel *preemptively schedule* the running of the various listed tasks. Each task will then have the codes in an infinite loop. [Refer to Chapters 9 and 10 for understanding the various terms used here.] Example 5.3 demonstrates the infinite loops within the tasks.

How does more than one infinite loop co-exist? The code inside waits for a signal or event or a set of events that the kernel transfers to it to run the waiting task. The code inside the loop generates a message that transfers to the kernel. It is detected by the OS kernel, which passes another task message and generates another signal for that task, and preempts the previously running task. Let an event be the setting of a flag—and the flag setting is to trigger the running of a task whenever the kernel passes it to the waiting task. The instruction, 'if (flag1) {...};' is to execute the task function for a service if flag1 is true.

```
Example 5.3
# define false 0
# define true 1
/*****************************************************************/
void main (void) {
```

```
/* Call RTOS run here */
rtos.run ( );
/* Infinite while loops follows in each task. So never there is return from the RTOS. */
}
/*************************************************************/
void task1 (....) {
/* Declarations */
.
.
while (true) {
/* Codes that repeatedly execute */
.
.
/* Codes that execute on an event*/
if (flag1) {....;}; flag1 =0;
/* Codes that execute for message to the kernel */
message1 ( );
}
}
/*************************************************************/
void task2 (...) {
/* Declarations */
.
.
while (true) {
/* Codes that repeatedly execute */
.
.
/* Codes that execute on an event*/
if (flag2) {.......;}; flag2 =0;
/* Codes that execute for message to the kernel */
message2 ( );
};
}
/*************************************************************/
.
.
/*************************************************************/
void taskN (...) {
/* Declarations */
.
.
while (true) {
/* Codes that repeatedly execute */
.
.
/* Codes that execute on an event*/
if (flagN) {....;}; flagN =0;
```

```
/* Codes that execute for message to the kernel */
message2 ( );
};
}
/*******************************************************************/
```

5.4.5   Use of Pointers and NULL Pointers

Pointers are powerful tools when used correctly and according to certain basic principles. Exemplary uses are as follows. Let a byte each be stored at a memory address.

1. Let port *A* in a system have a buffer register that stores a byte. Now a program using a pointer declares the byte at port A as follows: 'unsigned byte *portA'. [or Pbyte *portA.] The * means 'the contents at'. This declaration means that there is a pointer and an unsigned byte for portA, The compiler will reserve one memory address for that byte. Consider 'unsigned short *timer1'. A pointer *timer1* will point to two bytes, and the compiler will reserve two memory addresses for contents of *timer1*.

2. Consider declarations as follows: void *portAdata—The void means the undefined data type for portAdata. The compiler will allocate for the *portAdata without any type check.

3. A pointer can be assigned a constant fixed address as in Example 4.5. Recall two preprocessor directives: '# define portA (volatile unsigned byte *) 0x1000' and '# define PIOC (volatile unsigned byte *) 0x1001'. Alternatively, the addresses in a function can be assigned as follows. 'volatile unsigned byte * portA = (unsigned byte *) 0x1000' and 'volatile unsigned byte *PIOC = (unsigned byte *) 0x1001'. An instruction, 'portA ++;' will make the portA pointer point to the next address and to which is the PIOC.

4. Consider two items: unsigned byte portAdata; unsigned byte *portA = &portAdata. The first statement directs the compiler to allocate one memory address for portAdata because there is a byte each at an address. The & (ampersand sign) means 'at the address of'. This declaration means the positive number of 8 bits (byte) pointed by portA is replaced by the byte at the address of portAdata. The right side of the expression evaluates the contained byte from the address, and the left side puts that byte at the pointed address. Since the right side variable portAdata is not a declared pointer, the ampersand sign is kept to point to its address so that the right side pointer gets the contents (bits) from that address. [Note: The equal sign in a program statement means 'is replaced by'].

5. Consider two statements, '*unsigned short *timer1;*' and '*timer1++;*'. The second statement adds 0x0002 in the address of timer1. Why? timer1 ++ means point to next address, and unsigned short declaration allocated two addresses for timer1. [timer1 ++; or timer1 +=1 or timer = timer +1; will have identical actions.] Therefore, the next address is 0x0002 more than the address of timer1 that was originally defined. Had the declaration been 'unsigned int' (in case of a 32-bit timer), the second statement would have incremented the address by 0x0004.

    When the index increments by 1 in the case of an array of characters, the pointer to the previous element actually increments by 1, and thus the address will increment by 0x0004 in the case of an array of integers. For array data type, * is never put before the identifier name, but an index is put within a pair of square brackets after the identifier. Consider a declaration, '*unsigned char portAMessageString [80];*'. The port A message is a string, which is an array of 80 characters. Now, portAMessageString is itself a pointer to an address without the star sign before it. [Note: Array is therefore known as a reference data type.] However, *portAMessageString will now refer to all of the 80 characters in the string. portAMessageString [20] will refer to the twentieth element (character) in the string. Assume that there is a list of RTCSWT (Real Time Clock interrupts triggered Software Timers) timers that are active at an instant. The top of the list can be pointed as '*RTCSWT_List.top' using the pointer. RTCSWT_List.top is now the pointer to the top of the contents in a memory for a list of the active RTCSWTs. Consider the statement '*RTCSWT_List.top ++;'* It increments this pointer in a loop. It *will not point* to the next top of another object in the list (another RTCSWT) but to some address that depends on the memory addresses allocated to an item in the RTCSWT_List. Let *ListNow* be a pointer within the memory block of the list top element. A statement '*RTCSWT_List. ListNow = *RTCSWT_List.top;'* will do the following. RTCSWT_List pointer is now replaced by RTCSWT list-top pointer and now points to the next list element (object). [Note: RTCSWT_List.top ++ for pointer to the next list-object can only be used when RTCSWT_List elements are placed in an array. This is because an array is analogous to consecutively located elements of the list at the memory. Recall Table 5.2.]

6. A NULL pointer declares as following: '*#define NULL (void*) 0x0000'*. [We can assign any address, instead of 0x0000, that is not in use in a given hardware.] NULL pointer is very useful. Consider a statement: '*while (* RTCSWT_List. ListNow -> state != NULL) { numRunning ++;'*. When a pointer to ListNow in a list of software timers that are running at present is not NULL, then it will only execute the set of statements in the given pair of opening and closing curly brackets. One of the important uses of the NULL pointer is in a list—it is the last element to point to the end of a list, or to no more contents in a queue or empty stack, queue or list.

5.4.6 Use of Function Calls

Table 5.1 gives the meanings of the various sets of instructions in the C program. There are functions and a special function for starting the program execution, '*void main (void)'*. Given below are the steps to be followed when using a function in the program.

1. *Declaring a function:* Just as each variable has to have a declaration, each function must be declared. Consider an example. Declare a function as follows: '*int* **run** (int *indexRTCSWT*, unsigned int *maxLength*, unsigned int *numTicks*, SWT_Type *swtType*, SWT_Action *swtAction*, boolean *loadEnable*);'. Here *int* specifies the returned data type. The **run** is the function name. There are arguments inside the brackets. Data type of each argument is also declared. A modifier is needed to specify the data type of the returned element (variable or object) from any function. Here, the data type is specified as an integer. [A modifier for specifying the returned element may be also be *static*, *volatile*, *interrupt* and *extern*.]

2. *Defining the statements in the function:* Just as each variable has to be given the contents or value, each function must have its statements. Consider the statements of the function 'run'. These are *within a pair of curly braces* as follows: '*int RTCSWT:: run (int indexRTCSWT, unsigned int maxLength, unsigned int numTicks, SWT_Type swtType, SWT_Action swtAction, boolean loadEnable) {...};'*. The last statement in a function is for the *return* and may also be for returning an element.

3. *Call to a function:* Consider an example: 'if (delay_F = = true & & SWTDelayIEnable = = true) ISR_Delay ( );'. There is a call on fulfilling a condition. The call can occur several times and can be made repeatedly. On each *call*, the values of the arguments given within the pair of bracket pass for use in the function statements.

(i) Passing the Values (elements)

The values are copied into the arguments of the functions. When the function is executed in this way, it does not change a variable's value at the *called* program. A function can only use the copied values in its own variables through the arguments. Consider a statement, 'run (int *indexRTCSWT*, unsigned int *maxLength*, unsigned int *numTicks*, SWT_Type *swtType*, SWT_Action *swtAction*, boolean *loadEnable*) {...}'. Function 'run' arguments *indexRTCSWT*, *maxLength*, *numTick*, *swtType*, and *loadEnable* original values in the calling program during execution of the codes will remain unchanged. The advantage is that the same values are present on return from the function. The arguments that are *passed by the values* are saved temporarily on a stack and retrieved on return from the function.

(ii) Reentrant Function

Reentrant function is usable by the several tasks and routines synchronously (at the same time). This is because all of its argument values are retrievable from the stack. A function is called ***reentrant function*** when the following three conditions are satisfied.

1. *All the arguments pass the values and none of the argument is a pointer (address) whenever a calling function calls that function.* There is no pointer as an argument in the above example of function 'run'.

2. *When an operation is not atomic, that function should not operate on any variable, which is declared outside the function or which an interrupt service routine uses or which is a global variable but passed by reference and not passed by value as an argument into the function.* [The value of such a variable or variables, which is not local, does not save on the stack when there is call to another program.]

   Recall Section 2.1 for understanding *atomic operation*. The following is an example that clarifies it further. Assume that at a server (software), there is a 32-bit variable *count* to count the number of clients (software) needing service. There is no option except to declare the *count* as a global variable that shares with all clients. Each client on a connection to a server sends a call to increment the *count*. The implementation by the assembly code for increment at that memory location is non-atomic when **(i)** the processor is of eight bits, and **(ii)** the server-compiler design is such that it does not account for the possibility of interrupt in-between the four instructions that implement the increment of 32-bit count on 8-bit processor. There will be a wrong value with the server after an instance when interrupt occurs midway during implementing an increment of *count*.

3. *That function does not call any other function that is not itself Reentrant.* Let RTI_Count be a global declaration. Consider an ISR, *ISR_RTI*. Let an '*RTI_Count ++;*' instruction be where the RTI_Count is variable for counts on a real-time clock interrupt. Here *ISR_RTI* is a not a Reentrant routine because the second condition may not be fulfilled in the given processor hardware. There is no precaution that may be taken here by the programmer against shared data problems at the address of the *RTI_Count* because there may be no operation that modifies RTI_Counts in any other routine or function than the *IST_RTI*. But if there is another operation that modifies the RTI_Count the shared-data problem will arise. [Refer to Section 8.2.1 for a solution.]

(iii) Passing the References

When an argument value to a function passes through a pointer, that function can change this value. On returning from that function, the new value will be available in the calling program or another function called by this function. [There is no saving on a stack a value that (*a*) *passes through a pointer in the function-arguments* or (*b*) *operates in the function as a global variable* or (*c*) *operates through a variable declared outside the function block.*]

5.4.7 Multiple Function Calls in Cyclic Order in the Main

One of the most common methods is for the multiple function-calls to be made in a cyclic order in an infinite loop of the *main*. Recall the 64 kbps network problem of Example 4.1. Let us design the C codes given in Example 5.3 for an infinite loop for this problem. Example 5.4 shows how the multiple function *calls* are defined in the main for execution in the cyclic orders. Figure 5.1 shows the model adopted here.

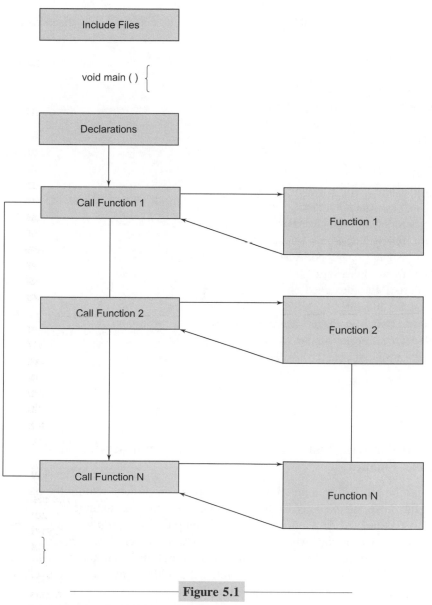

Figure 5.1

Programming Model for multiple function calls in 'main ( )' function.

Example 5.4

```
typedef unsigned char int8bit;
# define int8bit boolean
# define false 0
# define true 1
void main (void) {
/* The Declarations of all variables, pointers, functions here and also initializations here */
unsigned char *portAdata;
boolean charAFlag;
boolean checkPortAChar ( );
void inPortA (unsigned char *);
void decipherPortAData (unsigned char *);
void encryptPortAData (unsigned char *);
void outPortB (unsigned char *);

.
while (true) {
/* Codes that repeatedly execute */
/* Function for availability check of a character at port A*/
  while (charAFlag != true) checkPortAChar ( );
/* Function for reading PortA character*/
  inPortA (unsigned char *portAdata);
/* Function for deciphering */
  decipherPortAData (unsigned char *portAdata);
/* Function for encoding */
  encryptPortAData (unsigned char *portAdata);
/* Function for retransmit output to PortB*/
  outPort B (unsigned char *portAdata);
  };
}
```

5.4.8 Function Pointers, Function Queues and Interrupt Service Routines Queues

Let the * sign not be put before a function name; there are arguments within the pair of brackets; and the statements for those are executed on a call for the function. The statements are inside a pair of the curly braces. Consider a declaration in the Example 5.4, '*boolean checkPortAChar_ ( );*'. '*checkPortAChar*' is a function, which returns a *Boolean* value. Now, *checkPortAChar* is itself a pointer to the starting address of the statements of the function inside the curly brackets without the star sign before it. The program counter will fetch the address of *checkPortAChar*, and the CPU sequentially executes the function-statements from here.

Now, let the * sign be put before the function. '* *checkPortAChar*' will now refer to all the compiled form statements in the memory that are specified within the curly brackets.

Consider a declaration in the example, 'void *inPortA* (unsigned char *);'.

1. *inPortA* means a pointer to the statements of the function. Inside the parentheses, there is an unsigned character pointed by some pointer.
2. **inPortA* will refer to all the compiled form statements of *inPortA*.
3. (* *inPortA*) will refer to calls to the statements of *inPortA*.
4. What will a statement, 'void *create* (void (**inPortA*) (unsigned char *), void *portAStack, unsigned char port Apriority);' mean?
   (a) First modifier 'void' means *create* function does not return any thing.
   (b) '*create*' is another function.
   (c) Consider the argument of this function 'void (**inPortA*) (*unsigned char *portAdata*)'. (**inPortA*) means call the statements of *inportA* the argument of which is 'unsigned char *portAdata'.
   (d) The second argument of *create* function is a pointer for the portA stack at the memory.
   (e) The third argument of *create* function is a byte that defines the portA priority.

An important lesson to be remembered from earlier discussions is that a returning data type specification (for example, void) followed by '*(*functionName) (functionArguments)*' calls the statements of the *functionName* using the *functionArguments*, and on a return, it returns the specified data object. We can thus use the function pointer for invoking a call to the function.

When there are multiple ISRs, a high priority interrupt service routine is executed first and the lowest priority is last. [Refer to Section 4.6.4.] It is possible that function calls and statements in any of the higher-priority interrupts may block the execution of a low-priority ISR within the deadline. How is the deadline problem for low-priority routines to be solved? One solution is by using the function pointers in the routines, and forming a queue for them. The functions are then executed at a later stage. [Refer to exemplary codes in Section 5.5.3].

> Pointers are needed in number of situations, for example, port bit manipulation and read or write. Software designers must learn the uses of pointers in depth. An innovative concept is the use of function queues and the queues of the function pointers built by the ISRs. It reduces significantly the ISR latency periods. Each device ISR is therefore able to execute within its stipulated deadline.

5.5 QUEUES

5.5.1 Queue

Figures 5.2 (a) and (b) show a *queue* and a *circular queue*, respectively. Let us implement a queue consisting of general types of elements waiting for operations. Let us assume that each element of the

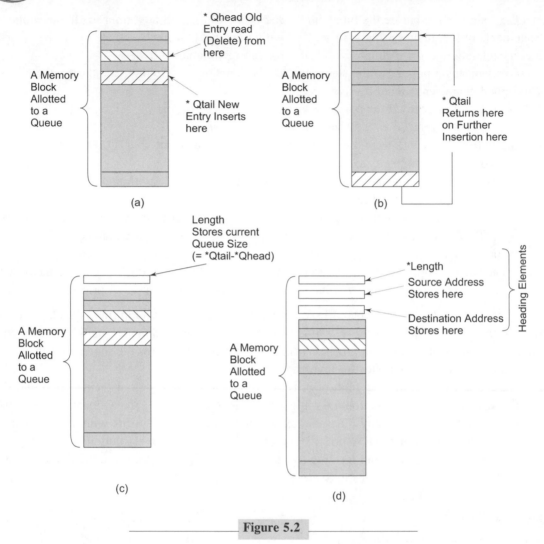

Figure 5.2

(a) A queue. (b) A circular queue. (c) A queue with its current size (length) as its heading element before the data in the queue. (d) A queue with its length, source address and destination address as its heading elements before the data in the queue follows.

queue is stored in the memory as an array. What then is the difference between a queue and an array? The difference lies in the accessibility and read of an element. It is in the FIFO (First In First Out) mode in a queue. An element is accessible and read by its index in the array. A queue can also be assumed as a list of elements from which an element is accessible and read as a FIFO list in which write (insertion) is not feasible in-between the list but at the last step only.

The queue in array implementation is also accessible through a pointer called *head* (front) and is inserted into it by a pointer called *tail* (back).

A queue can be restricted to 256 elements in a system that has a small memory. When needed, the *unsigned byte* can be replaced in the programming example with either *short* or *int* in a system with a larger memory. The queue size can also be enlarged from 256 to the maximum limit. [Use *char* in case the compiler does not define the *byte* as a data type.]

Example 5.5 gives the codes of a C++ class QueueElArray and assumes queue-size maximum 65,536 elements. [The reader may refer to a standard textbook for C ++ for in-depth study. One example is *Standard C++ with Object Oriented Programming* by Paul S. Wang from Brooks/Cole Thomson Learning (2001). Object oriented programming concepts are described in Section 5.8.]

A design concept that can be very useful in an embedded networking system is used for coding. It differs from a conventional application coding for a queue as follows.

1. There are five flags: *Qerrorflag, headerFlag, trailingFlag, CirQuFlag,* and *PolyQuFlag.*
   (a) The *Qerrorflag* equals to true when there is an error. A service function for error returns a string according to the error.
   (b) The *headerFlag equals to true when there are the header bytes* before the queue elements. [Refer to examples shown in Figures 5.2(c) and (d).] The service functions insert and return the header bytes.
   (c) The *trailingFlag* equals to true when there are trailing bytes after the queue elements. The service functions inserts and returns the trailing bytes. [Example of the trailing bytes at a queue is placing a check-sum, which receiver subsequently uses for the error check. Fields after the data in Tables 3.2, 3.3, 3.9 and 3.10 are the other examples of using the trailing bytes in the *queue*.]
   (d) If the *CirQuFlag* equals to true then a service function does the circular queuing of the bytes at the queue. [Data to a print buffer is an example of circular queuing method.]
   (e) When the *PolyQuFlag* equals to true a service function does the queuing of the bytes at the different blocks (or frames) of the queues. Polygon queuing means when a memory-block holding a queue fills and tail pointer reaches the block-end, another empty block starts inserting the elements. There can be multiple blocks of queues. [Data from the data link layer at an Ethernet LAN is an example of polygon queuing method. Data split into the different frames (blocks) with each frame of minimum 64 bytes and maximum 1,518 bytes. Transmission of image frames in video is another example of using the polygon queuing method. Each frame forms one queue block.]

2. There are four Boolean variables: *headerEnable, trailingEnable, CirQuEnable,* and *PolyQuEnable.*
   (a) If *headerEnable* equals to true, then the first few bytes are for the header with the queue.
   (b) If *trailingEnable* equals to true, then the first last few bytes are for the trailing byte with the queue elements.
   (c) If *CirQuEnable* equals to true, then only the circular queuing is permitted.
   (d) If *PolyQuEnable* equals to true, then only the bytes are permitted in a next block when a queue becomes full.

3. There are four unsigned short variables: *headerNumByte, trailingNumByte, CirQuNumByte,* and *PolyQuBlockNum.*

(a) The *headerNumByte* equals to the number of bytes at the queue header.

(b) The *trailingNumByte* equals to the number of bytes at the queue tail.

(c) The *CirQuNumByte* equals to the number of circular queuing permitted.

(d) The *PolyQuBlockNum* equals to the block number in the polygon queues (the blocks of queues) permitted.

Example 5.5

```
# define false 0
# define true 1
typedef unsigned char int8bit;
# define int8bit boolean
/*Declare a constant of Assumed Size of the queue = 65536. */
static const AssumedQSize = 65536;
/* ............ Insert Codes that the type of a queue element, QElType */;
class QueueElArray {
private:
```

/* Define three numbers, qhead, qtail and qsize of the queue. The qsize means the number of elements in a queue. The qhead means the first element. The qtail means the last element. Any new inserted element will be at the qtail. Any deleted element will be from the qhead*/

```
 unsigned short qhead, qtail, qsize; unsigned short qfull;
 boolean headerEnable, trailingEnable, CirQuEnable, PolyQuEnable;
 unsigned short headerNumByte, trailingNumByte, CirQuNumByte, PolyQuBlockNum;
 boolean headerFlag, trailingFlag, CirQuFlag, PolyQuFlag;
 void incCirc (int & item); /* inserting and deleting the element at the increasing indices circularly */
 QueueElArray (const QueueElArray & Qelement); /* Prevent calling a queue element using Qelement*/
/*Define Queue Error Handling variable and function */
 boolean volatile Qerrorflag;
 static void interrupt ISR_Qerror (volatile boolean Qerrorflag, unsigned short [ ] );};
public :
/* A constructor for the QueueElArray */
 QueueElArray (QElType * QelementsArray, unsigned short maxSize = AssumedQSize);    /* The
function delete in C++ is equivalent to function free in C. A destructor for the QueueElArray */
 ~QueueElArray {delete [ ] QelementsArray;};
/* An operator for the QueueElArray*/
 const QueueElArray & operator = (const QueueElArray & Qelement);
 boolean isQNotEmpty ( ) const {return (qsize > 0);};
 void Qempty ( );
 void QElinsert (const QElType & item); /* A function for inserting an element at tail*/
 boolean isQNotFull ( ) const {return (qsize < qfull);};
 QElType QElReturn ( ); /* A function for returning an element from head*/
}; /* End of class Queue */
/*****************  Constructor  for  Queue  *********************************/
QueueElArray (QElType * QelementsArray, unsigned short maxSize) {qfull = maxSize; Qerrorflag =
false;
```

```
Qempty ( );/* Construct Empty Queue */
QelementsArray = new QElType [maxSize];
/* Now handle the errors */
If (QelementsArray = = NULL) {Qerrorflag = true; ISR_Qerror (Qerrorflag, "Error! Queue Space
Not Available");
}
/
*********************************************************************/
void QueueElArray :: Qempty ( ) {qhead =1; Qtail = 0; qsize = 0;}
/*******************************************************************/
void QueueElArray :: QElinsert (const QElType & item) {
if (isQNotFull ( )) {
qsize ++;
incCirc (qtail);
QelementsArray [qtail] = item; Qerrorflag = false;}
else {Qerrorflag = true; ISR_Qerror (Qerrorflag, "Error! Queue Found Full");};
} /* End of insertion into the QueueElArray */
/
*********************************************************************/
QElType QueueElArray :: QElReturn ( ) {
QElType = temp;
if (isQNotEmpty ( )) {temp = QelementsArray [qhead]; qsize —; incCirc (qhead); Qerrorflag =
false;
return (Qelement);}
else {Qerrorflag = true; ISR_Qerror (Qerrorflag, "Error! Queue Found Empty);}; /*
} /* End of a deletion from the QueueElArray */
/
*********************************************************************/
void QueueElArray :: incCirc (unsigned byte & item) {
if (++item = = qfull) {item = 0;};
} /* End of incCirc and Next Queue Element pointer back to start*/
/
*********************************************************************/
/*
Place here codes for ISR_Qerror
*/
```

5.5.2 Use of the Queues for Implementing the Protocol for a Network

Networking applications need the specialised formations of a queue. Figure 5.2(c) shows a queue with its current size (length) as its heading element before the data in the queue. Figure 5.2(d) shows a queue that has the length, source address and destination address as its heading elements before the data in the queue follows.

On a network, the bits are transmitted in a sequence and retrieved at the other end in a sequence. To separate the bits at the different blocks or frames or packets, there are header bytes. [A packet differs from a block in the sense that a packet can follow different routes (paths or pipes) to reach the destination. A block may have several frames. A frame always succeeds in the same path (pipes) to reach the destination port. A block may have the frames that may be for the different ports but at same destination address.] The header with the queue elements follows a protocol. A protocol may also provide for appending the bytes at the queue tail. These may be the CRC (Cyclic Redundancy Check) bytes at the tail. Figure 5.3(a) shows a pipe from a queue. Figure 5.3(b) shows a queue between the sockets. Figure 5.3(c) shows three queues of the packets on a network. Table 5.3 gives the various cases and shows how the codes have to be modified from those in Example 5.5, in each case.

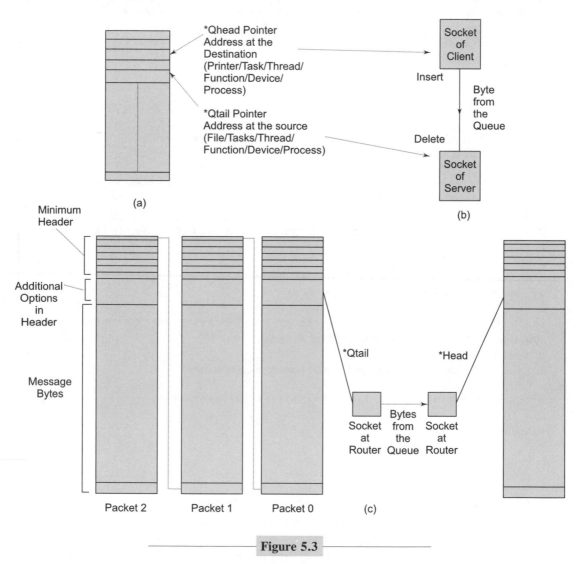

Figure 5.3

(a) A pipe from a queue. (b) A queue between two sockets.
(c) The queues of the packets on a network.

Table 5.3
Use of Queues for Implementing a Networking Protocol

S.No.	Additional Header Bytes	Additional Bytes at Tail	Queue MaxSize constant	Formation of Another Queue When a Queue is Full	Changes in the Codes of Example 5.5	Example of an Application
1.	Yes, Length Two bytes	No	Yes	No	headerEnable = 1; trailingEnable = 0; CirQuEnable = 1; PolyQuEnable = 0.	String transmission on a socket
2.	Yes, length plus source and destination addresses	No	Yes	Yes	headerEnable = 1; trailingEnable = 0; CirQuEnable = 0; PolyQuEnable = 1.	A network layer protocol, UDP
3.	Yes, length plus source and destination addresses	Yes with CRC bytes	Yes	No	headerEnable = 1; trailingEnable = 1; CirQuEnable = 1; PolyQuEnable = 0.	Ethernet Data link layer
4.	Yes,	Yes	Yes	Yes	headerEnable = 1; trailingEnable = 1; CirQuEnable = 0; PolyQuEnable = 1.	Transmission through the routers

5.5.3 Queuing of Functions on Interrupts

Let us redesign the codes given in Example 5.4. Figure 5.4 shows a programming model used for the design. The model is as follows. Multiple function pointers are queued by the ISRs and device driving ISRs. Each ISR is designed with a short set of codes. It does not execute any nonessential codes within the ISR. These are executed at the functions later by 'operationPortA ( )', which is called the '*main*'.

Example 5.6 shows how the called functions are not executed in the interrupt service routines. Only the pointers for these functions are placed in a queue by the ISR. This gives the very important feature that the ISRs now have only a few codes. Now, assume that there are multiple sources. A source, with a smaller deadline for its service, can be serviced without missing its service. On a return from a service routine, the operation function '*operationPortA ( )*' gets the function pointers from the queue and then executes the pointed functions.

Example 5.6
```
/*
```
From Example 4.5 Insert here all preprocessor directives, commands and functions except the main and portA_ISR_Input ( ) functions.
```
*/
void main (void) {
/* The Declarations of all variables, pointers, functions here and also initializations here */
```

```
while (true) { operationPortAFunctionQueues ( ); /*Call Functions from a Queue in cyclic (Round
Robin) Mode*/
 };
}
/*******************************************************************/
```

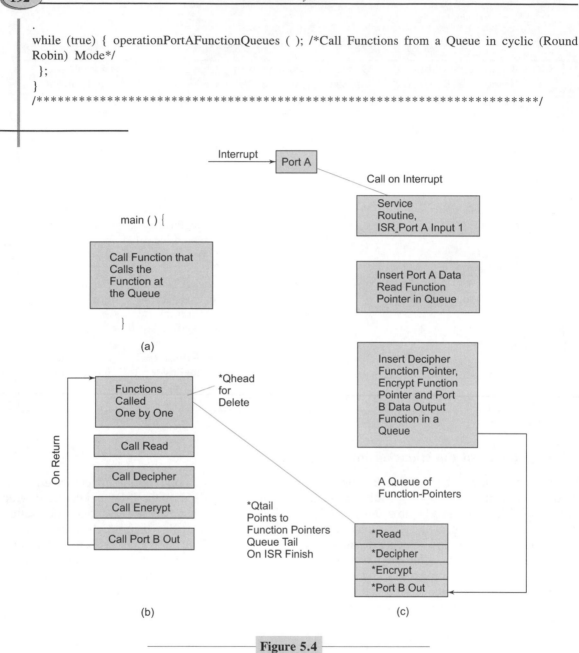

Figure 5.4

*A programming model for the multiple function pointers that are queued
by the interrupt service routine. These functions are executed later by operationPortA ( )
(a) Main ( ) function, (b) Function 'opertionPortAFunctionQueues ( )', (c) Creation of a queue
of the function pointers by the ISRs.*

```
void operationPortAFunctionQueues ( );
unsigned char *portAdata;
boolean checkPortAChar ( );
void inPortA (unsigned char *);
void decipherPortAData (unsigned char *);
void encryptPortAData (unsigned char *);
void outPortB (unsigned char *);
void checkPortAChar ( );
QueueElArray In_A_Out_B = new QueueElArray (QElType * QelementsArray, 65536);
portAIF = false; portAIEnable = true;
/* Codes that repeatedly execute */
while (portAFlag != true) checkPortAChar (In_A_Out_B, STAF);
In_A_Out_B.QElReturn ( );
In_A_Out_B.QElReturn ( );
In_A_Out_B.QElReturn ( );
In_A_Out_B.QElReturn ( );
};
void checkPortAChar (QueueElArray In_A_Out_B, volatile boolean portAIF) {
while (portAIF != true) { }; /*Wait till the occurrence of Port A Interrupt */
/* Call ISR_PortAInputI, an Interrupt Service Routine on a real time clock interrupt. */
void interrupt ISR_PortAInputI (QueueElArray In_A_Out_B);
}
/********************************************************************/
void interrupt ISR_PortAInputI (QueueElArray In_A_Out_B) {
disable_PortA_Intr ( ); /* Disable another interrupt from port A*/
void inPortA (unsigned char *portAdata); /* Function for retransmit output to Port B*/
void decipherPortAData (unsigned char *portAdata); /* Function for deciphering */
void encryptPortAData (unsigned char *portAdata); /* Function for encrypting */
void outPort B (unsigned char *portAdata); /* Function for Sending Output to Port B*/
/* Insert the function pointers into the queue */
In_A_Out_B.QElinsert (const          inPortA & *portAdata);
In_A_Out_B.QElinsert (const decipherPortAData & *portAdata);
In_A_Out_B.QElinsert (const encryptPortAData & *portAdata);
In_A_Out_B.QElinsert (const outPort B & *portAdata);
enable_PortA_Intr ( ); /* Enable another interrupt from port A*/
}
/********************************************************************/
```

5.5.4 Use of the FIPO (First-In Provisionally-Out) Queues for Flow Control on a Network

A commonly used network transport protocol is '*Go back to N*'. It is used in case of a point-to-point network. Receiver acknowledgment occurs at successive but irregular intervals of time. Bytes transmit from the network driver (transmitter) and queue up to a certain limiting number or up to the occurrence of a time-out, whichever is earlier.

Flow control of 'bytes' or 'packets' or 'frames' in many network protocols is done by taking into account the acknowledgments from the receiving entity.

1. If there is no acknowledgment within the limit or time-out, there is complete retransmission of the bytes from the queue.
2. If there is an acknowledgment for any byte that was sent in a sequence, there is retransmission of the bytes that remained unacknowledged at that instance of N-th sequence. [It is therefore called *Go back to N* and also *sliding window protocol*. Window means fragments or frames queued up during an interval for the receiver to accept in its buffer.]

There has to be three pointers, one for the *front* (*QHEAD), a second for the *back* (*QTAIL) and a third pointer is *tempfront* (*QACK). Two pointers are the same as in every queue. The third pointer defines a point up to which an acknowledgment has been received. The acknowledgment is for a byte inserted (placed) at the queue back.

The insertion into the queue is at the *back* (*QTAIL). There is a predefined limiting difference between *front* and *back* (*QTAIL). There is a predefined time-interval up to which insertions can occur at the *back* (*QTAIL). There is a predefined limiting maximum permitted difference between *tempfront* (*QACK) and *front* (*QHEAD).

This design gives a necessary feature. There can be a variable amount of delays in transmitting a byte a well as in receiving its or its successor acknowledgment. The receiver does not acknowledge every byte. There is acknowledgment only at successive predefined time-intervals. The design can be called FIPO (First In Provisionally Out). Figure 5.5 shows a FIPO queue for accounting the acknowledgments on the networks. It also shows at the bottom, the pointer addresses at three instances, at the beginning of transmission, on acknowledgment and after acknowledgment. Note that the window—that is between N-th sequence pointed by QACK and waiting sequence pointed by QTAIL as a function of time—is shown as sliding after receipt of QACK from the receiver for N-th sequence. [Refer to the left-to-right changes in Figure 5.5.]

1. *front* (*QHEAD) *equals back* (*QTAIL) *as well as tempfront* (*QACK) at the beginning of the transmission.
2. When there is an acknowledgment, *front* (*QHEAD) *resets and equals tempfront* (*QACK).
3. The transmission starts from the *tempfront* (*QACK) again.
4. There is a limiting maximum time interval difference between transmission from *tempfront* (*QACK) and after that time if *tempfront* (*QACK) is not equal to *front* (*QHEAD) then *front* (*QHEAD) *resets and equals tempfront* (*QACK) *again.* It means that after the limit, the *tempfront* (*QACK) will be forced to be equal to *front* (*QHEAD). This is because the receiver did not acknowledge within the stipulated time interval.

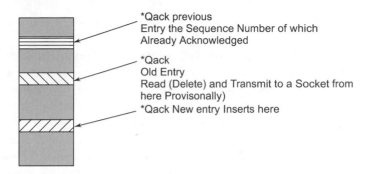

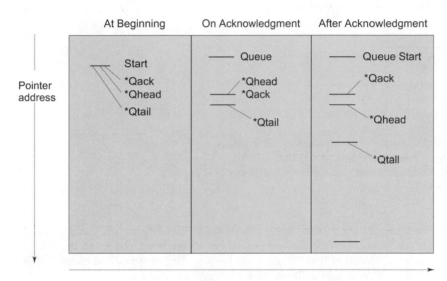

Time as Transmission Through a Socket Progresses

Figure 5.5

FIPO queue for accounting for the acknowledgments on the Networks with Go back to N (sliding window protocol) for transmission flow control. Note the three pointer addresses at three instances: at the beginning of transmission, on acknowledgment and after acknowledgment.

Queues and pipes are data structures used in the extremely useful program elements for embedded networking systems. A network protocol stack is implemented by creating a queue in specialized format with headers and trailing bytes. Creating a queue of function pointers is used to solve the problem of deadline misses by the low priority ISR.

5.6 STACKS

Let us implement a stack of a general type of elements. Each element is inserted last at the stack. What is then the difference between the stack and queue? The difference is that when the insertion of an element is at a pointer, *top*, from the stack, an element returns only from the *top*. When the insertions are at the *tail* in a queue, the return is from the *head*. The accessibility and read mode is LIFO (Last In First Out) in a stack, the while in the queue, it is FIFO.

A stack can be restricted to 256 elements in a system that has a small memory. The *unsigned byte* is replaced with either *short* or *int* when there is a larger memory system. This enlarges the stack size. [Use *char* in case the compiler does not define the *byte* as a data type.] Further, we can then have more elements. Let us restrict the stack to 65,536 elements. Example 5.7 gives the C++ codes for a class with a maximum of 65,536 elements.

Example 5.7

```
# define false 0
# define true 1
typedef unsigned char int8bit;
# define int8bit boolean
/*Declare a constant of Assumed Size of the stack = 65536. */
static const AssumedS_Size = 65536;
/* Insert Codes that the type of a stack element, SElType */
class Stack {
private:
```

/* Define two variables, s_top and ssize for the stack. The size means the number of elements in the stack. The s_top means the first element pointer before the beginning of an insertion and it also means the last element as insertion into the stack progresses. Any new inserted element will be at the s_top. Any deleted element will also be from the s_top*/

```
 unsigned short s_top, ssize; unsigned short sfull;
 void dec (int & item); /* On returning the elements decrease the indices */
 void inc (int & item); /* On inserting the elements increase the indices */
 Stack (const Stack & Selement); /* Prevent calling a stacked item using Selement*/
 /*Define Stack Error Handling variable and function */
 boolean volatile Serrorflag;
 static void interrupt ISR_Serror (volatile boolean Serrorflag, unsigned short [ ] );};
public :
 /* A constructor for the Stack */
 Stack (SElType * SelementsArray, unsigned short maxSize = AssumedS_Size);
 /* The function delete is in C++ equivalent of free in C. A destructor for the Stack */
 ~Stack {delete [ ] SelementsArray;};
 /* An operator for the Stack*/
 const Stack & operator = (const Stack & Selement);
 boolean isSNotEmpty ( ) const {return (ssize > 0);};
 void Sempty ( );
 void SElinsert (const SElType & item); /* A function for inserting an element at s_top*/
 boolean isSNotFull ( ) const {return (ssize < sfull);};
```

```
  SElType SElReturn ( ); /* A function for returning an element from head*/
}; /* End of class Stack */
/************** Constructor for Stack ********************************/
Stack (SElType * SelementsArray, unsigned short maxSize) {
sfull = maxSize; Serrorflag = false;
SEmpty ( );/* Construct Empty Stack */
SelementsArray = new SElType [maxSize];
/* Handle Errors */
If (SelementsArray = = NULL) {Serrorflag = true; ISR_Serror (Serrorflag, "Error! Stack Space Not
Available");
}
/
*****************************************************************/
void Stack :: Sempty ( ) { s_top = 0; ssize = 0; }
/*****************************************************************/
void Stack :: SElinsert (const SElType & item) {
if (isSNotFull ( )) {
ssize ++;
inc  (s_top);
SelementsArray [s_top] = item; Serrorflag = false;}
else {Serrorflag = true; ISR_Serror (Serrorflag, "Error! Stack Found Full" );};
} /* End of insertion into the Stack */
/*****************************************************************/
SEltype Stack :: SElReturn ( ) {
SElType = temp;
if (isSNotEmpty ( )) {temp = SelementsArray [s_top]; ssize —; dec (s_top); Serrorflag = false;
return (Selement); }
else {Serrorflag = true; ISR_Serror (Serrorflag, "Error! Stack Found Empty);};
} /* End of a deletion from the Stack */
/
*****************************************************************/
void Stack :: inc (unsigned byte & item) {(++item;
if (item < sfull) {Serrorflag = false;};
if (item > = sfull) {Serrorflag = true; ISR_Serror (Serrorflag, "Error! Stack Found Out Of Bound"
);};
} /* End of increment of top of the stack*/
/*****************************************************************/
void Stack :: dec (unsigned byte & item) { item —;
if (item < 0) {Serrorflag = true; ISR_Serror (Serrorflag, "Error! Stack Found Out Of Bound" );}
else Serrorflag = false;
} /* End of dec and Next Stack Element pointer back to s_top*/
/
*****************************************************************/
/*
Place here codes for ISR_Serror
*****************************************************************/
```

Stacks are data structures used in the program elements of an embedded system for LIFO accesses.

5.7 LISTS AND ORDERED LISTS

5.7.1 List

Consider a *list* of a general type of elements (objects). A *list* differs from an *array* as follows. *(i)* In an *array*, the memory allocation is as per the index assigned to an element. Each element is at the consecutive memory addresses starting from the 0th element address. Each element value (or object) in an *array* is read or replaced or written by using two values, 0th element address pointer and index only. Each element in the *array* usually has the same memory size if *array* is of primitive data types. *(ii)* In a *list*, each element must include an item as well as a pointer, LIST_NEXT. This is because each element is at the different memory address to which only the predecessor element points. LIST_NEXT points to the next element in the *list*. LIST_NEXT points to NULL in the element at the end of the list. The memory-size of an item of an element can also vary. The address of the list element at the top is a pointer, LIST_TOP. Only by using the LIST_TOP and traversing through all the LIST_NEXT values for the preceding elements can an element be deleted or replaced or inserted between the two elements.

A *list* can be an *ordered list*. All the elements rearrange according to a priority parameter in the *ordered list* on creation, on any new insertion or on deletion. The parameter can be an unsigned *byte* or *short* or *int* or a character ASCII code (alphabetical order). Along with an item and LIST_NEXT, the priority parameter is also stored in each list-element. Each element of the *ordered list* is always as per the order of the priority assigned to its items. Figures 5.6(a) to (c) show the arrangement of the items in an ordered list, rearrangement on an insertion after its first item and on a deletion of its first item, respectively.

A *list* differs from a *queue* as follows: A *queue* can be called a *list*, which is accessible and is readable as FIFO only. An insertion of an element in the *list* can be done anywhere within it by traversing through LIST_NEXT pointers in the preceding elements. An insertion is always at the *tail* in the *queue*. Also, an element can be read and deleted from anywhere in the *list*. It is always from the *head* in the *queue*.

Let us restrict a *list* to 256 elements in a system with a small memory and let us also order the list while inserting an element into it. We replace *unsigned byte* with the either *short* or *int* in a larger memory system to enlarge the *list* size.

Example 5.8 gives a class *OrderedList* codes for a maximum of 256 elements. The advantage of the ordered list is that the search for an item of a higher priority takes less time by the function LElSearch ( ). This is because that list item is near the ListTop. This can be explained as follows.

Let LIST_NOW be the address of the beginning of a list-element. When there is a search for the last item pointer, a '*for*' loop is used. The *ListNow* is varied from beginning (LIST_TOP) to end (LIST_NEXT = NULL). Therefore the search takes little time only in the case of the first item.

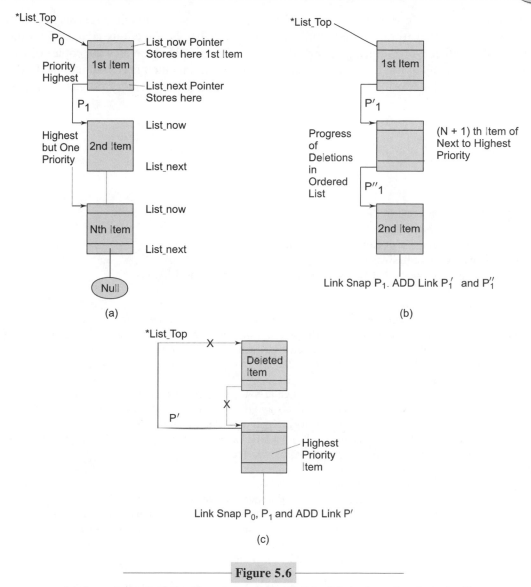

Figure 5.6

(a) Arrangement of the items in an ordered list. (b) An insertion into the list after its first item. (c) A deletion in the list of its first item.

Deletion of the first item using the function LEldeleteFirst ( ) saves execution time compared to LEldeleteLast ( ). Why? *ListNow* varies from beginning to end in case there is a search for last-item pointer, using a '*for*' loop, while in case of the first item, the deletion takes little time. [Figure 5.6(c).] C++ object property of polymorphism is used in creating the list of objects in Example 5.8. A C++ object is stored as a list-element. LElType depends on how the list is constructed by the constructor function. It can be *char*, it can be a function pointer, and it can be an array pointer. The constructor declares the LElType. There is declaration for the search and insert functions as '*virtual*' functions.

This causes the dynamic run time binding. The compiler uses the functions corresponding to the objects that are created from the class OrderedList. Otherwise the functions created from the class will be used. In other words, during compilation or linking, the object creates the copies of functions for the ROM only, and at run time that function which is to be executed is copied into the RAM. [The only disadvantage of declaring *virtual* is that there is an additional memory pointer lookup when a virtual function calls.]

Example 5.8
```
# define false 0
# define true 1
typedef unsigned char int8bit;
# define int8bit boolean
/*Declare a constant of Assumed Size of the ordered list = 256. */
static const AssumedLSize = 256;
/* ............ Insert Codes that the type of an ordered list element, LElType */;
class OrderedList {
protected:
/* Define three numbers, lhead, ltail and lsize of the ordered list. The lsize means that the number of
elements in an ordered list. The lhead means the first element. The ltail means last element. Any new
inserted element will at the ltail. Any deleted element will be from the lhead*/
  unsigned byte lfull; lsize, priorityOld =255;
  struc ListItem {
  LElType Lelement; ListItem *pNext; unsigned byte priority, unsigned byte itemID,
/* Constructor ListItem defines the arguments and default initialises the items in the list as 0, pointer to
next ListItem as NULL, item id =0, priority assignment = 255 maximum*/
  ListItem (LElType Lel = 0, ListItem *pTemp = NULL, unsigned byte prnew = 255, unsigned byte itId =0)
: Lelement (Lel), pNext (pTemp), priority (prnew), itemID (itId) { }
  };
  ListItem *ListTop; ListItem *ListNow;
  void inc (int & item); /* inserting the list item increases an index when the list grows */
  void dec (int & item); /* deleting the list item decrease the index when the list reduces */
  OrderedList (OrderedList & anElement); /* Prevent calling a list item using anElement*/
  Void deleteAll ( );
/*Define Ordered listErrorHandling variable and function */
  boolean volatile Lerrorflag;
  static void interrupt ISR_Lerror (volatile boolean Lerrorflag, unsigned char [ ]);};
/* ——————————————————————————————————————————————
——————————————————————————*/
public :
/* A constructor for the OrderedList. Constructed is definition of the Lelement type as a pointer,
maxSize is AssumedLSize; List top is as new ListItem constructor and List current position now accord-
ing to the previous List top. */
  OrderedList (LElType Lelement, unsigned byte maxSize = AssumedLSize) : ListTop (new ListItem),
ListNow (ListTop) { };
/* Destructor calls the function deleteAll for freeing the memory used by an ordered list. Declare virtual
because there are the virtual declared functions in this class*/
```

```
virtual ~OrderedList ( ){deleteAll ( );};
/* The operators for the Ordered List Note: Before const we may use the inline modifier so that
the compiler inserts the actual codes at all the places where we use these operators. This
reduces time and stack overheads in the function call and return. But this is at the cost more
ROM codes. */
 const OrderedList & operator = (OrderedList & anElement);
 const OrderedList & operator ++ ( ) {
 if (ListNow != NULL) ListNow = ListNow -> pNext;
 return *this;}
 const LElType & operator ( ) ( ) const {
 if (ListNow != NULL) return (ListNow -> LElement);
 else return (ListTop -> LElement); };
 boolean int OrderedList & operator ! ( ) const {return (ListNow != NULL);};
/* Empty the list */
 void Lempty ( );
 /* A function for getting to the list top */
 void top ( ) {ListNow = ListTop;};
 /* A function for assigning List first Element */
 void firstElement ( ){if ListTop -> pNext ! = NULL) {
 ListNow = ListTop -> pNext; };
/*Test List items not than the full list*/
 boolean isLNotEmpty ( ) const {return ((ListTop -> pNext) != NULL || (lsize > 0));};
 virtual boolcan LElSearch (boolean present, const LElType & item); /* Refer text for its use */
 virtual boolean LElprioritySearch (boolean present, const unsigned byte & priority); /* Refer
text for its use */
 virtual boolean LElitemIDSearch (boolean present, const unsigned byte & itemID); /* Refer
text for its use */
 /* This function if for searching the priority of the List items viz a viz the item to be inserted. The
inserted item is after the higher priority item. */
 virtual boolean LElprioritySearchThenInsert (boolean present, const unsigned byte & prior-
ity);
 virtual LElinsert (const LElType & item, unsigned byte & priority); /* Refer text for its use */
 virtual LElinsertLast (const LElType & item, unsigned byte &priority) /* Refer text for its use
*/
 virtual LElinsertPrev (const LElType & item, unsigned byte &priority); /* Refer text for its use
*/
 virtual LElinsertFirst (const LElType & item, unsigned byte &priority); /* Refer text for its use
*/
/*Test List items not than the full list*/
boolean isLNotFull ( ) const {return (lsize < lfull);};
/* Check in a ListItem */
 boolean checkInList (const LElType & item);
/* Delete a ListItem from anywhere. Return false if not successful else true */
 boolean LEldelete (const LElType & item);
/* Delete the last element */
void LEldeleteLast ( );
}; /* End of class OrderedList */
/********* Constructor for Ordered List ************************/
```

```
OrderedList (LElType *Lelement, unsigned byte maxSize = AssumedLSize) : ListTop (new
ListItem), ListNow (ListTop) {
lfull = maxSize; Lerrorflag = false;
/*Construct Empty Ordered List */
Lempty ( );
/* Handle Errors */
If ((ListTop == NULL) || (lfull == 0)) {Lerrorflag = true; ISR_Lerror (Lerrorflag, "Error! Ordered List Space
Not Available");
}
/***********************************************************************/
void OrderedList :: Lempty ( ) {
ListTop -> pNext = NULL; ListTop -> itemID = 0; ListTop -> priority = 255, lsize = 0;
}
/***********************************************************************/
boolean LElType OrderedList :: checkInList (const LElType & item) {
ListItem *pTemp;
if (isLNotEmpty ( )) {
for (pTemp = ListTop -> pNext; pTemp != NULL; pTemp = pTemp -> pNext ) {
if (pTemp -> Lelement = = item) {return true;} else return false;};};
return false;};
}
/***********************************************************************/
boolean LElType OrderedList :: LElSearch (boolean present, const LElType & item) {
ListItem *pTemp;
if (isLNotEmpty ( )) {Lerrorflag = false;
 if (present) {
 for (pTemp = ListTop -> pNext; pTemp != NULL; pTemp = pTemp -> pNext ) {
 if (pTemp -> Lelement = = item) {ListNow = pTemp; return true;}};};
 else {
 for (pTemp = ListTop -> pNext; pTemp -> pNext != NULL; pTemp = pTemp -> pNext) {
 if (pTemp -> Lelement = = item) {ListNow = pTemp; return true;}};}
 } else {Lerrorflag = true; ISR_Lerror (Lerrorflag, "Error! Ordered List Found Empty);};
return false;};
}
/***********************************************************************/
boolean LElType OrderedList :: LElprioritySearch (boolean present, const unsigned byte & priority) {
ListItem *pTemp;
if (isLNotEmpty ( )) {
 if (present) {Lerrorflag = false;
 for (pTemp = ListTop -> pNext; pTemp != NULL; pTemp = pTemp -> pNext) {
 priority = pTemp -> priority; return true;};
 }
 else {
 for (pTemp = ListTop -> pNext; pTemp -> pNext != NULL; pTemp = pTemp -> pNext) {
 priority = pTemp -> priority; return true;};}
else {Lerrorflag = true; ISR_Lerror (Lerrorflag, "Error! Ordered list Found Empty);}; return false;
};
}
```

```
/********************************************************************/
boolean LElType OrderedList :: LElitemIDSearch (boolean present, const unsigned byte &
itemID) {
ListItem *pTemp;
if (isLNotEmpty ( )) {
 if (present) { Lerrorflag = false;
 for (pTemp = ListTop -> pNext; pTemp = NULL; pTemp = pTemp -> pNext) {
 itemID = pTemp -> itemID; return true;};}
 else {
 for (pTemp = ListTop -> pNext; pTemp -> pNext != NULL; pTemp = pTemp -> pNext ) {
 itemID = pTemp -> itemID; return true;};}
else {Lerrorflag = true; ISR_Lerror (Lerrorflag, "Error! Ordered list Found Empty);}; return false;
}
/********************************************************************/
void OrderedList :: LElinsert (const LElType & item, unsigned byte priority) {
unsigned byte prnow;
if (isLNotFull ( )) {
if (! isLNotEmpty ( )) {LElinsertFirst (const LElType & item, unsigned byte priority); return;};
Lerrorflag = false;
LElpprioritySearch (present = true, & prnow); /* Get priority of Last List Item */
if (priority < = prnew) { LElinsertLast (const LElType & item, unsigned byte priority); return;}
else{LElinsertPrev (const LElType & item, unsigned byte priority); return;}
};} clsc {Lerrorflag = true; ISR_Lerror (Lerrorflag, "Error! OrderedList Found Full"); return;};
}
/
/********************************************************************/
void OrderedList :: LElinsertLast (const LElType & item, unsigned byte priority, unsigned
priority) {
/* Find item id in case needed for the new last List Item in the last using the function call
LElitemIDSearch (true, & itemID); */
ListItem *pLast;
ListItem *pTemp;
lsize++;
firstElement ( ); pLast = ListNow;
for (pLast; pLast != NULL; pLast = pLast -> pNext) {
 ListNow = pLast;};
ListNow -> pNext = pTemp;
pTemp = new (item, ListNow -> pNext, priority, itemID); pTemp -> pNext= NULL;
}
/* End of insertion at the last into the Ordered List */
/
/********************************************************************/
void OrderedList :: LElinsertPrev (const LElType & item, unsigned byte priority, unsigned
priority) {
/* When needed find item id for the new last List Item in the last using the function call */
LElitemIDSearch (false, & itemID); */
ListItem *pbefore;
ListItem *pTemp;
```

```
lsize++;
/* Retrieve the pointer of last but one list item */
firstElement ( ); pbefore = ListNow;
for (pbefore; pbefore -> pNext != NULL; pbefore = pLast -> pNext ) {
ListNow = pbefore;};
ListNow -> pNext = pTemp;
pTemp = new (item, ListNow -> pNext, priority, itemID);
pTemp -> pNext= pbefore;
pbefore -> pNext = NULL; pbefore -> itemID = itemID + 1;
}
/* End of insertion at the last but one into the OrderedList */
/*********************************************************************/
void OrderedList :: LElinsertFirst (const LElType & item, unsigned byte priority) {
ListItem *pTemp;
itemID = 1; lsize = 1;
ListTop-> pNext = pTemp;
pTemp = new (item, ListTop -> pNext, priority, itemID);
ListNow = pTemp;
pTemp -> pNext= NULL;
}
/* End of insertion at the first element into the OrderedList when a list is empty*/
/*********************************************************************/
boolean OrderedList :: LEldelete (const LElType & item) {
ListItem *pTemp;
/*Find the ListNow of the previous ListItem to present one where the item is found, then delete */
if (LElSearch (false, const LElType & item) ) {
pTemp = ListNow -> pNext;
ListNow -> pNext = pTemp -> pNext;
delete pTemp; lsize—; return true;};
return false;
}
/*********************************************************************/
void OrderedList :: void deleteAll ( ) {
/* This function frees the memory space of all the pointers in a list */
ListItem *pTemp;
ListItem *pdel;
pTemp = ListTop -> pNext;
while (pTemp ! = NULL) {
pdel = pTemp -> pNext;
delete pTemp;
pTemp = pdel;};
delete ListTop;
}
/*********************************************************************/
void OrderedList :: LEldeleteLast ( ) {
ListItem *pdel;
ListItem *pbefore;
if (isLNotEmpty ( )) {return;};
```

```
firstElement ( ); pbefore = ListNow;
/* Find the last but one List Item position pointer ListNow */
for (pbefore; pbefore -> pNext != NULL; pbefore = pbefore -> pNext ) {ListNow = pbefore; };
pdel = pbefore -> pNext; delete (pdel);
/* Define ListNow next pointer as NULL*/
ListNow -> pNext = NULL;
lsize—; }
}
/* End of deletion at the last into the OrderedList and also freeing the memory */
/**********************************************************************/
void OrderedList :: LEldeleteFirst ( ) {
ListItem *pdel;
ListItem *plater;
if (!isLNotEmpty ( )) {return;};
/* Find the pointer ListNow for the List Item at first position */
firstElement ( );
pdel = ListNow;
plater = pdel -> pNext;
ListTop -> pNext = plater; delete (pdel);
lsize—; }
}
/* End of deletion at the first into the OrderedList and also freeing the memory for that element.
*/
/
***********************************************************************/
/* Place here codes for
boolean OrderedList : : LElprioritySearchAndInsert (boolean present, const unsigned byte &
priority);
*/
/**********************************************************************/
/* Insert here codes for ISR_Lerror
*/
/**********************************************************************/
```

The explanation is as follows for the functions that are not yet explained within the *comments* in Example 5.8.

- A function LElSearch retrieves a fresh pointer ListNow, for a ListItem when the item is present and when the boolean variable *present* passes *true*. The function returns 'true' if the search is successful. When the boolean variable *present* passes 'false', after a ListItem is found, the function retrieves the pointer 'ListNow' not of this but of the previous *ListItem*. The function returns 'true' if the search for the item is successful.

- A function LElprioritySearch retrieves a priority for a present ListItem at ListNow when *present* (a boolean variable argument) is passed value = *true* on calling the LElprioritySearch. The function returns 'true' when the search is successful and 'false' if unsuccessful. When *present* is passed value = *false* on calling the LElprioritySearch, function retrieves a priority of the preceding ListItem to the ListNow. The function returns 'true' if the search for the preceding List item is successful.

- A function LElitemIDSearch retrieves an itemID for a present ListItem at ListNow when the boolean variable *present* is passed *true*. The function returns 'true' if the search is successful. When boolean variable *present* is passed *false*, the itemID for the element preceding to the list item at ListNow is retrieved. The function returns 'true' if the search for the item is successful.
- Three functions, LElinsertLast, LElinsertPrev and LElinsertFirst, insert an element as last list item, last but one list item, or first ListItem, respectively.
- Function LElinsert tests the list for not full and for not empty: if empty, insert as first list item; if not full and not empty, insert as last item; if priority is lower insert as last but one item; if priority is higher than the present, insert at the end. LElinsert is smart! It measures its priority with respect to the last one and then decides whether insertion is to be at last or last but one.
- 'LElprioritySearchAndInsert' function is useful in case there is a need for re-ordering or for ordering a disordered list.

5.7.2 Uses of a List of Active Device Drivers (Software Timers)

A system may have a number of devices and thus, device drivers. Consider timing device drivers, called software timers (RTCSWTs). An insert function is used in the active RTCSWTs to generate a list. The function may be denoted by RTCSWT_List.LElinsert (this). All timers which start running (getting count-inputs from the real time clock interrupts) are inserted into the list. All the timers inserted in the list in the codes then get the count-input on each real time clock (or system clock) interrupt. An up counter in the list on an overflow or a down-counter on a time-out, is deleted from this list on reaching the finished state. The finished state occurs on overflow or time-out of non-periodic (one-shot) timer. Each of the activated and running RTCSWTs is in the ordered list at an instant. The ordering is in order of counts left for reaching the finished state. The priority is high if the counts left for finishing are few. An RTCSWT finishes earlier when it is one shot, down counter and has lesser numTicks to be counted down. A counter having fewer counts left at any juncture is nearer to the top of the list.

Figures 5.7(a) and (b) show a programming model for a list that has five initiated RTCSWTs as the elements. [Initiated RTCSWT means defined RTCSWT object.] An initiated RTCSWT at an instant can either be active (running) or inactive (finished). Each element at an instance stores the followings five program-variables. (*i*) Priority. (*ii*) ItemID. (*iii*) State (= active or inactive). (*iv*) C, counts that are remaining for finish. (*v*) Pointer for the next element, *pNext. Figure 5.7(a) shows the case of three software timers that are in an active list, RTCSWT_list out of a total of five RTCSWTs. Figure 5.7(b) shows a case of none of the software timers at the active ordered list, and therefore each of the pointers at the elements in the list points to NULL.

5.7.3 Uses of a List of Tasks in a Ready List

Instead of the function pointers queue (Example 5.6), the multiple tasks are used. [Refer to Section 8.1 for *task* definition.] Each function is designed as a *task* that is monitored by a multitasking OS software. A task has reentrancy or, alternatively, the methods to solve the shared data problem. [Refer to Section 5.4.6(ii) for reentrant function definition and uses]. Creation and deletion from the ordered list of ready tasks is done as follows by the OS.

1. The OS creates an ordered list of initiated (ready)-state tasks. These tasks are the ones that have been flagged to be prepared for execution by the CPU in an order.
2. When the OS receives a message for initiating a task into the ready state, the task from the idle state is inserted into a list of initiated (ready) tasks.

3. The codes in each listed task are then executed as directed (scheduled) by the OS.
4. A task, which finishes (completes execution of its codes), is deleted from the listed tasks on a direction from the OS.

A programming model for a multitasking operations example can be as follows. Let there be four ISRs. Each ISR has the minimum required codes, and its functions that can wait for execution are the tasks. Let there be six tasks that the OS supervises. Each element at an instance stores the following five program-variables. (*i*) Priority. (*ii*) ID for the task. (*iii*) State (= running, ready or idle). (*iv*) Codes for the task that has to be executed when readied. (*v*) Pointer for the next list-element, *pNext. Figure 5.8 shows a programming model for six tasks and four ISRs. Three tasks are in the initiated *task-list* with a task either in the running or ready state. Three steps follow during execution. Step A: The ISR sends an event message (or flag) for initiating the needed task(s) into the ready state(s). Step B: The OS inserts that task into the initiated ready tasks *list*, and monitors the execution. The OS uses a strategy for execution of the tasks at the *list*. One strategy can be cyclic (round robin). Step C: The OS deletes that task into the initiated task list when the execution of all its codes finish and task state = idle.

> List and ordered list are extensively used data structures in Embedded C/C++ for a number of applications. Examples are list of active software timers, list of active tasks and list of ISRs (including attached device drivers).

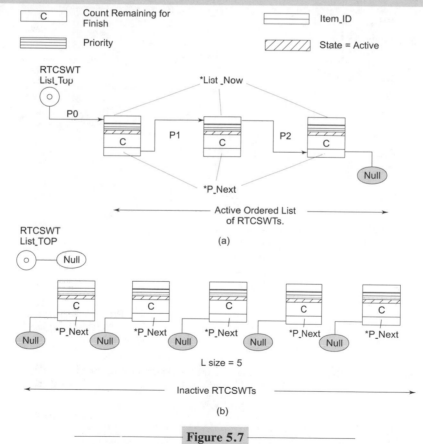

Figure 5.7

*A programming model in which there are three software timers in an active list, RTCSWT_
List, out of a total five of RTCSWTs.*

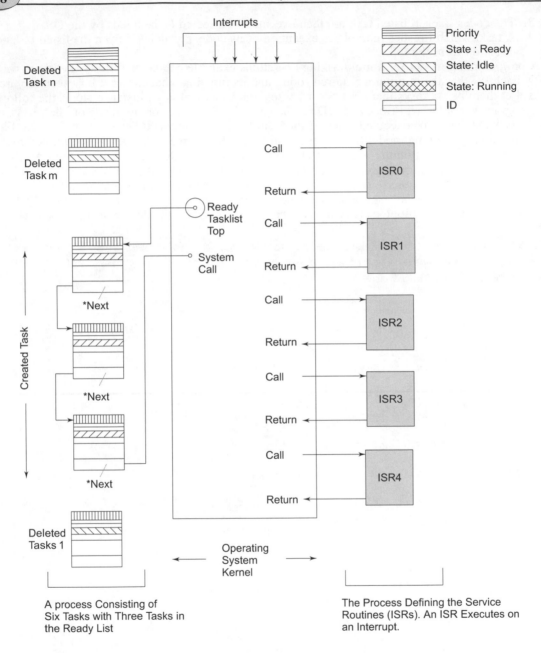

In Single CPU System, at an Instant only one process is Directed by the OS to Run.
Process can be of a task or of an ISR.

Figure 5.8

Six tasks and four ISRs in a programming model for multitasking operations. Three tasks in an initiated task list with state = ready or running.

5.8 EMBEDDED PROGRAMMING IN C++

5.8.1 Objected-Oriented Programming

An *objected-oriented language* is used when there is a need for reusability of the defined object or set of objects that are common within a program or between the many *applications*. When a large program is to be made, an object-oriented language offers many advantages. *Data encapsulation*, *design of reusable software components* and *inheritance* are the advantages derived from the OOPs.

An object-oriented language provides for defining the objects and methods that manipulate the objects without modifying their definitions. It provides for the data and methods for encapsulation. An object can be characterised by the following:

1. An *identity* (a reference to a memory block that holds its state and behavior).
2. A *state* (its data, property, fields and attributes).
3. A *behavior* (method or methods that can manipulate the *state* of the object).

In a procedure-based language, like FORTRAN, COBOL, Pascal and C, large programs are split into simpler functional blocks and statements. In an object-oriented language like Smalltalk, C++ or Java, logical groups (also known as *classes*) are made first. Each group defines the data and the methods of using the data. A set of these groups then gives an application program. Each group has internal user-level fields for the data and the methods of processing that data at these fields. Each group can then create many objects by copying the group and making it functional. Each object is functional. Each object can interact with other objects to process the user's data. The language provides for formation of classes by the definition of a group of objects having similar attributes and common behavior. *A class creates the objects. An object is an instance of a class.*

5.8.2 Embedded Programming in C++

1. What are the programming advantages of C++?

C++ is an *object-oriented program (OOP) language, which supports the procedure-oriented codes of C*. Program coding in C++ codes provides the advantage of objected-oriented programming as well as the advantage of C and in-line assembly. Programming concepts for embedded programming in C++ are as follows:

(*i*) A class binds all the member functions together for creating objects. The objects will have memory allocation as well as default assignments to its variables that are not declared *static*. Let us assume that each software timer that gets the count input from a real time clock is an object. Now consider the codes for a C++ *class RTCSWT*. A number of software timer objects can be created as the instances of *RTCSWT*.

(*ii*) A class can derive (inherit) from another class also. Creating a *child* class from RTCSWT as a *parent* class creates a new application of the RTCSWT.

(*iii*) Methods (C functions) can have the same name in the inherited class. This is called *method overloading*. Methods can have the same name as well as the same number and type of arguments in the inherited class. This is called *method overriding*. These are the two significant features that are extremely useful in a large program.

(*iv*) Operators in C++ can be overloaded like in method overloading. Recall the following statements and expressions in Example 5.8. The operators ++ and ! are overloaded to perform a set of operations. [Usually the ++ operator is used for post-increment and pre-increment and the ! operator is used for a *not* operation.]

const OrderedList & operator ++ ( ) {*if* (ListNow != NULL) ListNow = ListNow -> pNext; return *this;}

boolean int OrderedList & operator ! ( ) const {return (ListNow != NULL) ;};

[Java does not support operator overloading, except for the + operator. It is used for summation as well as string-concatenation.]

There is *struct* that binds all the member functions together in C. But a C++ *class* has object features. It can be extended and child classes can be derived from it. A number of child classes can be derived from a common class. This feature is called polymorphism. A class can be declared as public or private. The data and methods access is restricted when a class is declared private. *Struct* does not have these features.

2. **What are the disadvantages of C++ ?**

Program codes become lengthy, particularly when certain features of the standard C++ are used. Examples of these features are as follows:

(a) Template.
(b) Multiple Inheritance (Deriving a class from many parents).
(c) Exceptional handling.
(d) Virtual base classes.
(e) Classes for I/O Streams. [Two library functions are *cin* (for character(s) in) and *cout* (for character(s) out). The I/O stream class library provides for the input and output streams of characters (bytes). It supports *pipes*, *sockets* and *file management features*. Refer to Section 8.3 for the use of these in inter-task communications.]

3. **Can optimization codes be used in Embedded C++ programs to eliminate the disadvantages?**

Embedded system codes can be optimised when using an OOP language by the following:

(a) Declare private as many classes as possible. It helps in optimising the generated codes.
(b) Use *char*, *int* and *boolean* (scalar data types) in place of the objects (reference data types) as arguments and use local variables as much as feasible.
(c) Recover memory already used once by changing the reference to an object to NULL.

A *special compiler for an embedded system* can facilitate the disabling of specific features provided in C++. Embedded C++ is a version of C++ that provides for a selective disabling of the above features so that there is a less runtime overhead and less runtime library. The solutions for the library functions are available and ported in C directly. The I/O stream library functions in an embedded C++ compiler are also reentrant. So using embedded C++ compilers or the special compilers make the C++ a significantly more powerful coding language than C for embedded systems.

GNU C/C++ compilers (called *gcc*) find extensive use in the C++ environment in embedded software development. Embedded C++ is a new programming tool with a compiler that provides a small runtime library. It satisfies small runtime RAM needs by selectively de-configuring features—like template, multiple inheritance, virtual base class, etc.—when there is a smaller runtime overhead and when fewer runtime library–using solutions are available. Selectively removed (de-configured) features could be template, run time type identification, multiple inheritance, exceptional handling, virtual base classes, I/O streams and foundation classes. [Examples of foundation classes are GUIs (graphic user interfaces). Exemplary GUIs are the buttons, checkboxes or radios.]

An embedded system C++ compiler (other than *gcc*) is Diab compiler from Diab Data. It also provides the target (embedded system processor) specific optimisation of the codes [Section 5.12]. The run-time analysis tools check the expected run-time error and give a profile that is visually interactive.

> Embedded C++ is a C++ version, which makes large program development simpler by providing object-oriented programming (OOP) features of using an object, which binds state and behavior and which is defined by an instance of a class. We use objects in a way that minimises memory needs and run-time overheads in the system. Embedded system programmers use C ++ due to the OOP features of software reusability, extendibility, polymorphism, function overriding and overloading along portability of C codes and in-line assembly codes. C++ also provides for overloading of operators. A compiler, gcc, is popularly used for embedded C++ codes compilation. Diab compiler has two special features: (i) processor specific code optimisation and (ii) run-time-analysis tools for finding expected run-time errors.

5.9 EMBEDDED PROGRAMMING IN JAVA

5.9.1 When Do We Program in Java?

Java has advantages for embedded programming as follows:

1. Java is completely an OOP language.
2. Java has in-built support for creating multiple threads. [For the definition of thread and its similarity in certain respects to task refer to Section 8.1.] It obviates the need for an operating system-based scheduler [Appendix I.5.6] for handling the tasks.
3. Java is the language for most Web applications and allows machines of different types to communicate on the Web.
4. There is a huge class library on the network that makes program development quick.
5. Platform independence in hosting the compiled codes on the network is because Java generates the byte codes. These are executed on an installed JVM (Java Virtual Machine) on a machine. [Virtual machines (VM) in embedded systems are stored at the ROM.] Platform independence gives *portability* with respect to the processor used.
6. Java does not permit pointer manipulation instructions, so it is robust in the sense that memory leaks and memory-related errors do not occur. A memory leak occurs, for example, when attempting to write to the end of a bounded array.
7. Java byte codes that are generated need a larger memory when a method has more than 3 or 4 local variables.
8. Java, being platform independent, is expected to run on a machine with an RISC-like instruction execution with few addressing modes only. [Refer to Table A.1.1 for a few addressing-mode features in RISC.]

5.9.2 What are the disadvantages of Java?

An embedded Java system may need a minimum of 512 kB ROM and 512 kB RAM because of the need to first install JVM and run the application.

Use of J2ME (Java 2 Micro Edition) or JavaCard or EmbeddedJava helps in reducing the code size to 8 kB for the usual applications like a smart card. How? Through one of the following methods.

1. Use core classes only. Classes for basic run-time environment form the VM internal format and only the programmer's new Java classes are not in internal format.
2. Provide for configuring the run-time environment. Examples of configuring are *deleting the exception handling classes*, *user defined class loaders*, *file classes*, *AWT classes*, *synchronized threads*, *thread groups*, *multidimensional arrays*, and *long and floating data types*. Other configuring examples are adding the specific classes for connections when needed, such as datagrams, input output and streams.
3. Create one object at a time when running the multiple threads.
4. Reuse the objects instead of using a larger number of objects.
5. Use scalar types only as long as feasible.

A smart card (Section 11.4) is an electronic circuit with a memory and CPU or a synthesised VLSI circuit. It is packed like an ATM card. For smart cards, there is JavaCard technology. [Refer to *http://www.java.sun.com/products/javacard.*] Internal formats for the run-time environments are available mainly for the few classes in JavaCard technology. Java classes used are the connections, datagrams, input output and streams, security and cryptography.

Described above are the advantages and disadvantages of Java applications in the embedded system. JavaCard, EmbeddedJava and J2ME (Java 2 Micro Edition) are three versions of Java that generate a reduced code size.

Consider an embedded system such as a smart card. It is a simple application that uses a running JavaCard. The Java advantage of platform independency in byte codes is an asset. The smart card connects to a remote server. The card stores the user account past balance and user details for the remote server information in an encrypted format. It deciphers and communicates to the server the user needs after identifying and certifying the user. The intensive codes for the complex application run at the server. A *restricted run-time environment exists in Java classes* for connections, datagrams, character-input output and streams, security and cryptography only.

[For EmbeddedJava, refer to *http://www.sun.java.com/embeddedjava*. It provides an embedded run-time environment and a closed exclusive system. Every method, class and run-time library is optional].

J2ME provides the optimised run-time environment. Instead of the use of packages, J2ME provides for the codes for the core classes only. These codes are stored at the ROM of the embedded system. It provides for two alternative configurations, Connected Device Configuration (CDC) and Connected Limited Device Configurations (CLDC). CDC inherits a few classes from packages for *net*, *security*, *io*, *reflect*, *security.cert*, *text*, *text.resources*, *util*, *jar* and *zip*. CLDC does not provide for the applets—awt, beans, math, net, rmi, security and sql and text packages—in java.lang. There is a separate javax.mircoedition.io package in CLDC configuration. A PDA (personal digital assistant) uses CDC or CLDC.

There is a scaleable OS feature in J2ME. There is a new virtual machine, KVM as an alternative to JVM. When using the KVM, the system needs a 64 kB instead of 512 kB run-time environment. KVM features are as follows:

1. Use of the following data types is optional. (*i*) Multidimensional arrays, (*ii*) long 64-bit integer and (*iii*) floating points.
2. Errors are handled by the program classes, which inherit only a few needed error-handling classes from the java I/O package for the Exceptions.
3. Use of a separate set of APIs (application program interfaces) instead of JINI. JINI is portable. But in the embedded system, the ROM has the application already ported and the user does not change it.
4. There is no verification of the classes. KVM presumes the classes are already validated.
5. There is no object finalization. The garbage collector does not have to do time-consuming changes in the object for finalization.
6. The class loader is not available to the user program. The KVM provides the loader.
7. Thread groups are not available.
8. There is no use of java.lang.reflection. Thus, there are no interfaces that do the object serialization, debugging and profiling.

J2ME need not be restricted to configure the JVM to limit the classes. The configuration can be augmented by Profiler classes. For example, MIDP (Mobile Information Device Profiler). A profile defines the support of Java to a device family. The profiler is a layer between the application and the configuration. For example, MIDP is between CLDC and application. Between the device and configuration, there is an OS, which is specific to the device needs.

A mobile information device has the following:

1. A touch screen or keypad.
2. A minimum of 96 x 54 pixel color or monochrome display.
3. Wireless networking.
4. A minimum of 32 kB RAM, 8 kB EEPROM or flash for data and 128 kB ROM.
5. MIDP used as in PDAs, mobile phones and pagers.

MIDP classes describe the displaying text. It describes the network connectivity, such as for HTTP. [Internet Hypertext Transfer Protocol.] It provides support for small databases stored in EEPROM or flash memory. It schedules the applications and supports the timers. [Recall RTCSWTs.]

An RMI profiler is an exemplary profiler for use in distributed environments.

Java objects bind state and behavior by the instance of a Java class. EmbeddedJava is a Java version, which makes large program development simpler by providing complete object-oriented programming (OOP) features in Java. JVM is configured to minimise memory needs and run-time overheads in the system. Embedded system programmers use Java in a large number of readily available classes for the I/O stream, network and security. Java programs posses the ability to run under restricted permissions. JavaCard is a technology for the smart cards and it is based on Java.

5.10 'C' PROGRAM COMPILER AND CROSS-COMPILER

5.10.1 Compiled, Executable and Locator Files

Two compilers are needed. One compiler is for the host computer which does the development and design and also the testing and debugging. The second compiler is a *cross-compiler*. The cross-compiler runs on a host, but develops the machine codes for a targeted system (processor of the embedded system). There is a popular freeware called GNU C/C++ compiler and free AS11M assembler for 68HC11. A *GNU compiler* is configurable both as host compiler as well as cross-compiler. It supports 80x86, Window 95/NT, 80x86 Red Hat Linux and several other platforms. It supports 80x86, 68HC11, 80960, PowerPC and several other target system processors.

A compiler generates an object file. For compilation for the host alone, the compiler can be turbo C, turbo C++ or Borland C and Borland C++. The target system-specific or multiple-choice cross-compilers that are available commercially may be used. These are available for most embedded system microprocessors and microcontrollers. The IAR System, Sweden, offers cross-compilers for many targets. The targets can be of the PIC family or 8051 family, 68HC11 family or 80196 family. These compilers can switch on its configuring back from the embedded system-specific cross-compiler to the ANSI C standard compiler. PCM is another cross compiler for the PIC (Programmable Interrupt Controller) microcontroller family, PIC 16F84 or 16C76, 16F876.

The host also runs the cross-compiler that offers an integrated development environment. It means that a target system can emulate and simulate the application system on the host.

Note: Figure 1.6 showed the process of converting a C program into machine implementable software. In an embedded system design, as the final step, the bytes must be placed at the ROM after compilation. A 'C' program helps to achieve that goal. The object file is generated on compilation of a program while an executable file is required, which has the source codes having the absolute addresses. The executable file is the file that a device program uses to put (store or burn in) the initial data, constants, vectors, tables, and strings, and the source codes in the ROM. A locator file has the final information of memory allocation to the codes, data, initialization data and so on. The locator then uses the allocation map file, and generates the source code within the allocated addresses. [Section G.2 describes two formats (from Intel and Motorola) of these file output records. A device programmer (Section G.2) uses one of the two formats as the input.] The ROM has the following sections. (*i*) Machine (Executable) codes for the bootstrap (reset) program. (*ii*) Initialization (default) data at shadow ROM for copying into the RAM during execution. (*iii*) Codes for the application and interrupt service routines. (*iv*) System configuration data needed for the execution the codes. [For example, port addresses.] (*v*) Standard data or vectors and tables. (*vi*) Machine codes for the device manager and device drivers.

> Use of appropriate compilers and cross-compilers is essential in any embedded software development.

5.11 SOURCE CODE ENGINEERING TOOLS FOR EMBEDDED C/C++

A source code engineering tool is of great help for source-code development, compiling and cross compiling. The tools are commercially available for embedded C/C++ code engineering, testing and debugging.

The features of a typical tool are comprehension, navigation and browsing, editing, debugging, configuring (disabling and enabling the C++ features) and compiling. A tool for C and C++ is SNiFF+. It is from WindRiver® Systems. A version, SNiFF+ PRO has full SNiFF+ code as well as a debug module. Main features of the tool are as follows:

1. It searches and lists the definitions, symbols, hierarchy of the classes, and class inheritance trees. [The symbols include the class members. A tree is a data structure. A data structure tree has a root. From the roots, the branches emerge and from the branches more branches emerge. On the branches, finally there are the leaves (terminating nodes)].

2. It searches and lists the dependencies of symbols and defined symbols, variables, functions (methods) and other symbols.

3. It monitors, enables and disables the implementation of virtual functions. [Refer to Section 5.7.1 for the use of virtual functions. These are for dynamic run-time binding.]

4. It finds the full effect of any code change on the source code.

5. It searches and lists the dependencies and hierarchy of included header files.

6. It navigates between the implementation and symbol declaration.

7. It navigates between the overridden and overriding methods. [Overriding method is a method in a daughter class with the same name and same number and types of arguments as in the parent class. Overridden method is the method of the parent class, which has been redefined at the daughter class.]

8. It browses through information regarding instantiation (object creation) of a class.

9. It browses through the encapsulation of variables among the members and browses through the public, private and protected visibility of the members.

10. It browses through object component relationships.

11. It automatically removes error-prone and unused tasks.

12. It provides easy and automated search and replacement.

The embedded software designer for sophisticated applications uses a source-code engineering tool for program coding, profiling, testing, and debugging of embedded system software.

5.12 OPTIMISATION OF MEMORY NEEDS

When codes are made compact and fitted in small memory areas without affecting the code performance, it is called memory optimisation. It also reduces the total number of CPU cycles and, thus, the total energy requirements. The following are used for optimising the use of memory in a system.

1. Use declaration as unsigned byte if there is a variable, which always has a value between 0 and 255. When using data structures, limit the maximum size of queues, lists and stacks size to 256. Byte arithmetic takes less time than integer arithmetic.

Follow a rule that uses unsigned bytes for a short and a short for an integer if possible, to optimise use of the RAM and ROM available in the system. Avoid if possible the use of 'long' integers and 'double' precision floating point values.

2. Avoid use of library functions if a simpler coding is possible. Library functions are the general functions. Use of general function needs more memory in several cases.

   Follow a rule that avoids use of library functions in case a generalised function is expected to take more memory when its coding is simple.

3. When the software designer knows fully the instruction set of the target processor, assembly codes must be used. This also allows the efficient use of memory. The device driver programs in assembly especially provide efficiency due to the need to use the bit set-reset instructions for the control and status registers. Only the few assembly codes for using the device I/O port addresses, control and status registers are needed. The best use is made of available features for the given applications. Assembly coding also helps in coding for atomic operations. A modifier *register* can be used in the C program for fast access to a frequently used variable. If portAdata is frequently employed, it is used as follows, 'register unsigned byte *portAdata*'. The modifier register directs the compiler to place portAdata in a general-purpose register of the processor.

   As a rule, use the assembly codes for simple functions like configuring the device control register, port addresses and bit manipulations if the instruction set is clearly understood. Use assembly codes for the atomic operations for increment and addition. Use modifier 'register' for a frequently used variable.

4. Calling a function causes context saving on a memory stack and on return the context is retrieved. This involves time and can increase the worst-case interrupt-latency. There is a modifier *inline*. When the *inline* modifier is used, the compiler inserts the actual codes at all the places where these operators are used. This reduces the time and stack overheads in the function call and return. But, this is at the cost of more ROM being needed for the codes. If used, it increases the size of the program but gives a faster speed. Using the modifier directs the compiler to put the codes for the function (in curly braces) instead of calling that function.

   As a rule, use inline modifiers for all frequently used small sets of codes in the function or the operator overloading functions if the ROM is available in the system. A vacant ROM memory is an unused resource. Why not use it for reducing the worst-case interrupt-latencies by eliminating the time taken in frequent save and retrieval of the program context?

5. As long as shared data problem does not arise, the use of global variables can be optimised. These are not used as the arguments for passing the values. A good function is one that has no arguments to be passed. The passed values are saved on the stacks in case of interrupt service calls and other function calls. Besides obviating the need for repeated declarations, the use of global variables will thus reduce the worst-case interrupt-latency and the time and stack overheads in the function call and return. But this is at the cost of the codes for eliminating a shared data problem. When a variable is declared static, the processor accesses with less instruction than from the stack.

   As a rule, use global variables if shared data problems are tackled and use static variables in case it needs saving frequently on the stack.

6. Combine two functions if possible. For example, LElSearch (boolean *present*, const LElType & *item*) is a combined function in Example 5.8. The search functions for finding pointers to a list

item and pointers of previous list items combine into one. If present is *false* the pointer of the previous list item retrieves the one that has the *item*.

  As a rule, whenever feasible combine two functions of more of less similar codes.

7. Recall the use of a list of running timers and list of initiated tasks in Sections 5.7.2 and 5.7.3. All the timers and a conditional statement (that changes the count input in case of a running count and does not change it in case of idle state timers) could have also been used. A greater number of calls will however be needed and not once, but repeatedly on each real-time clock interrupt tick. The RAM memory needed will be more. Therefore, creating a list of running counters is a more efficient way. Similarly, bringing the tasks first into an initiated task list (Figure 5.8) will reduce the frequent interactions with the OS and context savings and retrievals stack, and time overheads. Optimise the RAM use for the stacks, which is done by reducing the number of tasks that interact with the OS. One function calling another function and that calling the third and so on means nested calls. Reduce the number of nested calls and call, at best, one more function from a function. This optimises the use of the stack.

  As a rule reduce the use of frequent function calls and nested calls and thus reduce the time and RAM memory needed for the stacks, respectively.

8. Use, if feasible, alternatives to the *switch* statements with a table of pointers to the functions. This saves processor time in deciding which set of statements to execute while performing the conditional tests all down a chain.

9. Use the delete function when there is no longer a need for a set of statements after that execute.

  As a rule, to free the RAM used by a set of statements, use the delete function and destructor functions.

10. When using C++, configure the compiler for not permitting the multi-inheritance, templates, exceptional handling, new style casts, *virtual* base classes, and *namespaces*.

  As a rule, for using C++, use the classes without multiple inheritance, without template, with runtime identification and with throwable exceptions.

Embedded software designers must use various standard ways for optimising the memory needs in the system.

▪ SUMMARY ▪

- Programming in the assembly language gives the important benefits of precise control of the processor's internal devices and full use of processor-specific features in its instruction set and addressing modes.
- Programming in a high-level language gives the important benefits of a short development cycle for a complex system and portability to system hardware modifications. It easily makes larger program development feasible.
- C language support to in-line assembly (fragments of codes in assembly) gives the benefits of both.
- C++ provides all the advantages of C and of object-oriented programming and is suitable for embedded systems when: (*i*) Declaring private as many classes as possible. It helps in optimising

the generated codes. (*ii*) Using *char*, *int* and *boolean* (scalar data types) in place of objects (reference data types) as arguments and use local variables as much as feasible. (*iii*) Recovering memory already used by changing reference to an object to NULL. (*iv*) Selectively deconfiguring certain C++ features to get less run-time overhead and less run-time library use. (*v*) Selectively removing the features of template, run-time type identification, multiple inheritances, exceptional handling, virtual base classes, I/O streams and foundation classes.

- Java provides the benefits of extensive class libraries and their availability, modularity, robustness, portability and platform independence.
- The C program uses various instruction elements, preprocessor directives, macro, and constants, including files and header files and functions. Basic *C programming elements* are the data types, data structures, modifiers, conditional statements and loops, function calls, multiple functions, function queues and service-routine queues.
- Infinite looping is a greatly used feature in embedded systems, as it keeps a task or system ready for execution whenever called to run.
- The C program uses passing the variables values by reference to the functions, pointers, NULL pointers and function pointers.
- Queue is an important data structure used in a program. The queue data structure related functions are 'constructing' a queue, 'inserting' an element into it, deleting an element from it and 'destruction' of queue. A queue is a first-in first-out data structure. Queues of bytes in a stream play a vital role in a network communication or client-server communication also. A new form of queue data structure, which facilitates flow control by a *first in provisionally out* queue, is frequently used in embedded networking systems. Deletion is by first in, which has been the provisionally out byte(s). It restarts insertions in the queue from the acknowledged sequence onwards, as and when it is received.   .
- Queuing of pointers to the function on interrupts and later on calling the functions from this queue is a better approach (programming model) as it provides the use of short execution time interrupt-service routines.
- Using a '*stack*' is very common for saving the data in case of interrupts or function calls. Stack related functions are 'constructing' a stack, 'pushing' an element into it, popping an element from it and 'destruction' of a stack.
- The 'list' and priority-wise 'ordered list' related functions are 'constructing' a list, 'inserting' an element into it, finding an element from it, deleting an element from it and 'destruction' of the list. One exemplary application is a list of real-time clock interrupts–driven software timers. Another is the list of ready tasks for scheduling the multiple tasks.
- Once a source code is ready, the compiler and cross-compilers are used for enabling the locator to direct device programmers to store the machine codes in the system ROM.
- Source code engineering tools help in debugging and performance analysis of the codes written in high-level languages.
- Embedded system programs should be optimised in terms of memory requirements. There are many steps that are followed to get the optimised program. Besides reducing the memory size needed, this also reduces the total number of CPU cycles and the total energy requirements.

■ LIST OF NEW KEYWORDS AND THEIR DEFINITIONS ■

- *High-level language*: Programming language in which it is easier to write codes than in the assembly language, and which also gives the important benefits of a short development cycle for a complex system and portability to system hardware modifications.
- *Development cycle*: A cycle of coding, testing, and debugging. A number of cycles may be needed before finalising the source codes for porting in the embedded system ROM.
- *In-line assembly*: A fragment of codes in assembly language in a high-level language that gives the benefits of processor-specific instructions and addressing modes.
- *Object-oriented programming*: A programming method in which instead of operations on data types and structures, and variables and functions as individuals, the operations are done on the objects. The classes create the objects.
- *Class*: A named set of codes that has a number of members—variables, functions, etc., so that objects can be created from it. The operations are done on the objects by passing the messages to the objects in object-oriented programming. Each class is a logical group with an identity, a state and a behavior specification.
- *Scalar data types*: The character, integer, unsigned integer, floating point numbers, long and double are called scalar data type. Unlike an array, data consist of one single element.
- *Private*: A variable belonging to a specific class and not usable outside that class.
- *Reference data types*: Arrays and strings are examples of reference data types.
- *Local variable*: A variable defined within a function and which no other function can modify.
- *NULL*: When a pointer points to NULL, it means there is no reference to the memory. A memory occupied by an element or object or data structure can be freed by pointing it to the NULL.
- *Run-time overhead*: Use of RAM for data and stack is called run-time overhead.
- *Run-time library*: A library function that links dynamically at the run-time. Run-time links increase run-time overheads and out-of-memory errors can arise.
- *Template*: A set of classes using the new classes that are built.
- *Multiple inheritance*: A daughter (derived class) inheriting the member functions from more than one class.
- *Exception handling*: A way of calling the functions on handle development of an exceptional condition. For example, a buffer unable to store any further byte. A programmer thinks of the exceptional conditions and provides for the functions and their calling on occurrence of the *exception*.
- *Virtual base classes*: A special type of classes provided in C++.
- *I/O stream*: A memory buffer created by sending the bytes or characters from a source to a destination so that the destination acts as a sink and accepts them in the sequence that they are sent. An I/O stream object does the writing to a file, printer, queue or network device.
- *Foundation classes*: Classes meant for GUIs (graphic user interfaces—for example, the button, checkbox, menu, etc.).
- *Class libraries*: Classes for a number of applications like encryption and security may be provided after thorough debugging and testing for using these in the requirements. Use of class libraries speeds up program development cycles.

- *Modularity*: A set of codes is said to be modular if it is usable in multiple applications.
- *Robustness*: A program is said to be robust if it can function without errors like stack overflow and out-of-memory errors. Avoiding pointer manipulation instructions, frequently freeing the memory if not needed later and using exceptions, makes a code robust.
- *Portable*: A code that can be ported in another program by suitable configuration changes.
- *Platform independence*: A code that can port on different machines and operating systems.
- *Preprocessor directives*: Program statements and directives for the compiler before the main function to define global variable, global macro (section of code), new data type and global constants.
- *Include file*: File that is included along with the user source code before the compilation by the compiler.
- *Header file*: File containing codes (mostly standard functions) for the user. For example, a file "math.h" containing codes for the mathematical functions.
- *Data type*: Type of data for a variable, for example, an integer.
- *Data structure*: A multi-element structure that can be referenced by a common name (identity).
- *Function queue*: A queue of pointers for the functions awaiting later execution.
- *Infinite loop*: A loop from the program that cannot exit except on interrupt or a change in certain parameters used by it.
- *Passing the value*: From a function, a value is transferred to another function but the same value is reassigned to the original function after return from the called function. Before passing, the argument values save on the stack and retrieve back on return from the function.
- *Passing the reference*: From a function, an address of the argument value is passed from the called function during transfer processing to another calling function. The called function argument becomes modified when operated, and the function may get a different argument value back after return to the calling function. The argument value does not save on the stack on passing by its reference.
- *Queue*: A data structure into which elements can be sequentially inserted and retrieved in a first-in first-out (FIFO) mode. It needs two pointers, one for the queue tail (back) for insertion and the other for the queue head (front) for deletion (read and point to next element).
- *Stack*: A data structure in which elements can be pushed for saving in certain memory blocks and can be retrieved in a last-in first-out (LIFO) mode. A stack needs one pointer for the stack head (top) for deletion (read and point to next element) and one pointer for insertion.
- *List*: A data structure into which elements can be sequentially inserted and retrieved, not necessarily in first-in first-out mode or last-in last-out mode. Each element has a pointer which also points to the address of the next element at the list. The last element points to NULL. A top (head) pointer points to its first element.
- *Ordered list*: A priority-wise ordered list in which it is easy to delete operations (read and then set the pointer to next). It is done sequentially, starting from the top.
- *Source code engineering tool*: A power tool to engineer source codes and also to help in debugging and performance analysis of the codes in high-level languages.
- *Optimisation of memory*: Certain steps changed to reduce the need for memory and having a compact code. It reduces the total size of the memory needed. It also reduces the total number of CPU cycles and, thus, the total energy requirements.

■ REVIEW QUESTIONS ■

1. What are the criteria by which an appropriate programming language is chosen for embedded software of a given system?
2. What is the most important feature in C that makes it a popular high-level language for an embedded system?
3. What is the most important feature in Java that makes it a highly useful high-level language for an embedded system in many network-related applications?
4. What is the advantage of polymorphism, when programming using C++?
5. Why does one break a program into header files, configuration files, modules and functions?
6. Design a table to give the features of top-down design and bottom-up design of a program.
7. Explain the importance of the following declarations: static, volatile and interrupt in embedded C.
8. How and when are the following used in a C program? (a) # define, (b) typedef, (c) null pointer, (d) passing the reference and (e) recursive function.
9. What are the advantages of using freeware and GNU C/C++ compiler?
10. Why do you need a cross-compiler?
11. Why do you use an infinite loop in embedded system software?
12. What are the advantages of reentrant functions in embedded system software?
13. What are the advantages of using multiple function calls in cyclic order in the Main?
14. What are the advantages of building ISR queues?
15. What are the advantages of having short ISRs that build the function queues for processing at a later time?
16. How are the queues used for a network?

■ PRACTICE EXERCISES ■

17. Why do the features in C++ make the code lengthy when using Template, Multiple inheritance (deriving a class from many parents), Exceptional handling, Virtual base classes and I/O streams? Tabulate the reasons.
18. Write a device driver for a COM serial line port in C including in-line assembly codes.
19. What are the most commonly used preprocessor directives? Give four examples of each.
20. How does the use of a macro differ from a function? Explain with exemplary codes.
21. Write program C codes for a loop for summing ten integers with odd indices only. Each integer is 32 bits. Now unroll the loop and write C codes afresh. Compare the code length in both cases.
22. A set of images in a video frame are to be processed. Which data structure will be best suited for storing the inputs before compressing in appropriate format?
23. How does combining two functions reduce the memory requirement? Explain with four examples.
24. Refer to Exercise 16 in Chapter 3. It gives the format of PPP. Write a C program to transmit PPP data frames encapsulating 4096 data bits, where the bits are to be transmitted in a sequence of 32-bit integers stored in memory as in big-endian format.
25. Give two programming examples of each of these data structures in an embedded software: (a) array, (b) queue, (c) stack, (d) list, (e) ordered list and (f) binary tree.

Chapter

6

Program Modeling Concepts in Single and Multiprocessor Systems Software-Development Process

What We Have Learnt

Here is a summary of the following topics related to the software development process that have been presented in the previous chapters.

1. Embedded system software concepts—the most essential parts of a embedded system design.
2. Code generation or source code engineering tool, compiler, cross-compiler and locator for generating the final code as ROM image.
3. Procedure-oriented and object-oriented programming language concepts, coding of the device drivers and the applications using C/C++ programming elements: modifiers, conditional statements and loops, pointers, function calls, multiple functions, function pointers, function queues and service-routines.
4. Programs in assembly languages for certain processor- and memory-sensitive instructions. The coding for these is mostly in C inline assembly.
5. Use of data of various types and with various structures: arrays, queues, stacks, lists and trees and coding examples of uses of data structures.
6. Methods to optimise system memory requirements in programs.

What We Will Learn

A standard-design practice adopted by engineers is to use a model when solutions to problems are to be found. Can modeling be used for programs? The first objective of this chapter is to learn the important concepts of program modeling before software implementation. The following important concepts of program modeling before software implementation are explained.

1. Data flow graphs for program-design and analysis.
2. Controlled data flow graphs for program-design and analysis.

Embedded systems most often need real-time programming. A real-time program is for programming the processes or instruction set with constraints of time for its response, processes with latencies, and processes with deadlines. Complex issues of real-time software design are described in the next chapter after the software development process models and activities in the analysis, design, implementation and testing phases are explained. Real-time operating systems and their working will be described in depth in a subsequent chapter. The second objective of this chapter is to learn the following important concepts of program modeling for real-time programs.

1. Finite State Machine(s) for program models and analysis of real-time programs.
2. Petri Net(s) for program models and analysis of real-time programs.

When using two or more processors, (*i*) program functions (Section 5.3), (*ii*) tasks (Section 8.1.2), (*iii*) single-instruction multiple-data (SIMD) instructions, (*iv*) multiple-instructions multiple-data (MIMD) instructions and (*v*) very long instruction words (VLIW) of DSP instructions can be finished at high speed when a system runs these processes concurrently. [An instruction word means *opcode* plus its operands.] Many sophisticated embedded systems are multiprocessor systems so that processes have short latencies and can thus meet the deadlines. The third objective of this chapter is to explain the following concepts, which are used in modeling a multiprocessor system.

1. SDF graph
2. HSDF representation after unfolding the SDF graph
3. APEG
4. Timed Petri Nets
5. Extended Predicate/Transition Net
6. Multi-Thread Graph (MTG) System
7. Application of models to partitioning, load balancing, scheduling, synchronisation and resynchronisation during the program flow on the multiple processors.

A powerful modeling language is UML. It is important to understand that apart from these modeling processes, software engineering approaches and practices are also necessary. The need for a unified language, which can model many types of processes, activities, designs and development process approaches, should also be understood. It is dealt with in the next chapter.

6.1 MODELING PROCESSES FOR SOFTWARE ANALYSIS BEFORE SOFTWARE IMPLEMENTATION

A software analysis is (i) a description of the system requirement, (ii) a provision of a basic frame before the software is designed and (iii) a definition of a set of requirements, through which the designed software can be validated. Data flow graphs, abbreviated as DFGs, and control data flow graphs, abbreviated as CDFGs, are used for modeling the data paths and program flows of software during a software analysis.

6.1.1 Use of Data Flow Graph for Program Analysis

A data flow means that a program flow and all program execution steps are determined specifically only by the data. The software designer predetermines the data inputs and designs the programming steps to generate the data output. For example, a program for finding an *average* of grades in various subjects will have the data inputs of grades and data output of the *average*. The program executes a function to generate the appropriate output. The data flow graph model is appropriate to model the program for the average.

How does data flow in a program? Data that is input becomes data that is output after a data flow. A diagram called the data flow graph (DFG) represents this graphically. A DFG does not have any conditions within it so that the program has one data entry point and one data output point. There is only one independent path for program flow when the program is executed.

A circle represents each process in DFG. An arrow directed towards the circle represents the data input (or set of inputs) and an arrow originating from the circle represents a data output (or a set of outputs).

When there is only one set of values of each of the inputs and only one set of values of the outputs for the given input, a DFG is also be known as an ADFG, (Acrylic Data Flow Graph). Examples of non-acrylic data input are as follows: (i) An event, (ii) A status flag setting in a device and (iii) Input as per output condition of the previous process.

Example 6.1 gives a DFG during a DSP algorithm.

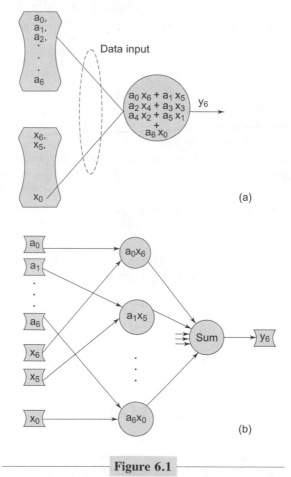

Figure 6.1

(a) DFG for a process for the 6th FIR sequence. (b) DFGs for a set of processes of the same sequence for an FIR filter.

Example 6.1   Figure 6.1 shows a DFG of the following expression for an output sequence y_6 of a 'Finite Impulse Response (FIR) filter'. [Recall Section 2.1. An n-th filtered output sequence, y_n = Σ $(a_i.x_{n-i})$ where the sum is made for i = 0, 1, 2, ..., N-1.] Figure 6.1(a) shows the DFG for a process for the 6th FIR sequence and Figure 6.1(b) shows the DFG for a set of processes of the same sequence. Following are the points notable for the process of calculating $y_6 = a_0.x_6 + a_1.x_5$ + $a_2.x_4$ +$a_3.x_3$ + $a_4.x_2$ + $a_5.x_1$ +$a_6.x_0$.

1. There is one input point to the process represented by the circle for calculating y_6.
2. There is one output point for y_6.
3. There is only one memory address and variable for each coefficient and each filter input. There is only one value of each of the 6 inputs for x and there is only one value of each of the coefficients, a. [This DFG is therefore also an ADFG.].

The order in which inputs are obtained and summation is done is also immaterial.

It must be noted in Example 6.1 that there is no complexity in the process for y_6. Data flow graph models help in a simple code design. A simple code design can be defined as that in which the program mostly breaks into DFGs. A DFG patterns a fundamental program element having an independent path. It gives that unit of a system, which has no control conditions, a single path for the program flow. A unit gives the program context and helps in analyzing a program in terms of complexity. A more complex program would have a smaller amount of DFG processes than a simple program. [Refer to Section 7.2 for the *measures of software complexity*.]

Software implementation becomes greatly simplified when using the data flow graphs because in the DFG model, there is a single data-in point and a single data-out point, with a process or set of processes that are represented by a circle(s). Programming tasks are simplified by representing the code for each process by a circle, using the data input from incoming arrow(s) and generating data output along outgoing arrow(s). When the assignment to an input is fixed in a DFG, it is also called acrylic DFG (ADFG). Programming complexity is minimised by modeling a program in terms of as many DFGs as possible and the use of as many ADFGs as possible.

6.1.2   Use of Control Data Flow Graph for Program Analysis

Let us assume that body temperature and the results of a stethoscopic examination are the inputs on the basis of which a doctor creates a prescription recommending the medicine needed and at what intervals it is to be taken. If the diagnosing process is represented by a DFG (data output is as per data input and there is only one program path) there is no scope for taking into account the different conditions and rates by which the fever varies. The diagnosis process should take into account the conditions and then the prescription must be recommended in each case.

A control flow means that specifically *only the program determines* all program execution steps and the flow of a program. The software designer programs and predetermines these steps. How does one design a process that incorporates controls for taking decisions during the data operations and data flow into a program? A process may have the statements that control the inputs or outputs. It may have loops or condition statements in-between. [Recall Section 5.4.4.] Data that are input generate the data output after a control data-flow as per the controlling conditions. Output(s) depends on the control statements for various decisions in a process. A CDFG is a diagram that graphically represents the conditions and the program flow along a condition dependent path. A CDFG diagram also represents the effect of events among the processes and shows which processes are activated on each specific event. Here, a variable value changing above or below a limit or falling within a range is also like an event that activates a certain process.

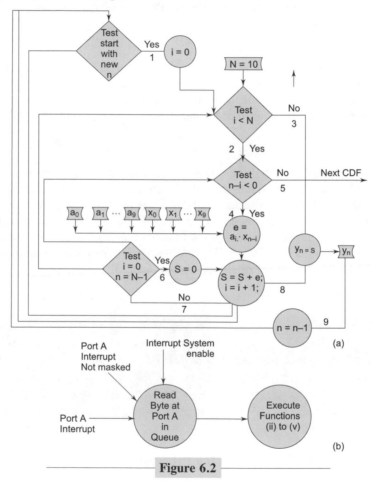

Figure 6.2

(a) Data inputs and Controlling input (decision) nodes shown by test boxes in a CDFG for an FIR filter with ten inputs and ten coefficients. (b) Controlling input conditions marked in the In_A_Out_B programs in Examples 4.1 and 4.2. [Instead of box, a condition is marked at the start of an arc.]

Consider Figure 6.2 and Example 6.2. A circle also represents each process in a CDFG. A directed arrow towards the circle represents the data input (or set of inputs) and a directed arrow from the circle represents a data output (or a set of outputs). A box (square or rectangle set diagonally like a diamond) may represent a condition, for example, in Figure 6.2(a). Alternatively, a condition can be marked (or denoted) at the start of the directed arc or arrow, for example in figure 6.2(b). A directed arrow from the box or a marked starting condition determines the action to be taken when the condition is true.

Example 6.2　Figure 6.2(a) shows the controlling input (decision) nodes by the test condition specifying boxes, and the data inputs to a CDFG for an FIR filter with ten inputs and ten coefficients. [Recall Example 6.1 for meanings of various terms in the n-th filtered output sequence, $y_n = \Sigma(a_i.x_{n-i})$; where the sum is made for $i = 0, 1, 2, ..., 9$.] Following are the points notable for the process of calculating y_n. There is one input point to the process represented by the circle for calculating y_n.

1. There is one output point for y_6. There is only one memory address and variable for each coefficient and each filter input. These are the variables i, n, s and e, which take multiple values during the program flow.
2. The order in which inputs are obtained and summation is done does matter.
   Figure 6.2(b) shows the controlling input in the In_A_Out_B program Examples 4.1 and 4.2. Here, instead of boxes, the condition is marked at the start of the arc.

In Example 6.2, note that there is increased complexity in the process for y_n. The CDFG model helps in understanding all conditions and in determining the number of paths a program may take. It also shows us that software must be tested for each path starting from a decision node, and helps in analyzing the program in terms of complexity [Section 7.2].

> Software implementation becomes simplified when using the specifications of the conditions and decision nodes in the CDFGs that represent the controlled decision at the nodes, and the program paths (DFGs) that are traversed consequently from the nodes after the decisions.

6.2　PROGRAMMING MODELS FOR EVENT-CONTROLLED OR RESPONSE TIME–CONSTRAINED REAL TIME PROGRAMS

6.2.1　Use of Finite States Machine Model

In the case of a washing machine, three states in the process of washing—*Washing*, *Rinsing* and *Drying*, may be defined. The following five *states* may be defined in a telephone process: *Idle*, *Receiving a ring*, *Dialing*, *Connected* and *Exchanging messages*. [A nonfunctioning state is not mentioned since the phone service is assumed to be efficient!] Similarly, the states of a process (function or interrupt service routine (ISR) or task or thread) can be defined.

When is a process modeled using the states? Frequently, there are inputs to a process that change the state of the process to a new state, and generate outputs which may also be the inputs for the next state. Now it can be assumed that in a model the running of the process (program flow) can be considered like the running of a machine generating the states. The program flow can be modeled simply by inter-state transitions (from one state to another).

The steps that model or represent the *states* and *inter-state transitions* are as follows.

1. A transition to a new state occurs from the previous state *on an event (input)*. The event may be the setting a value of a certain parameter or the result of the execution of certain codes. A transition may be also be interrupt flag driven (after a flag sets) or semaphore driven or interrupt-source servicing need driven.

2. A state is identified by a flag condition or a set of codes being executed or a set of values of certain parameters.

3. A state can receive multiple tokens (inputs, messages, flag interrupts or semaphores) from another state. *A token (event) is used here as a general term that means either an input or event-input.* An event-input characteristic is that it is asynchronous (one never knows when an event may happen). An event-input may happen when there is setting or resetting of a flag. It may occur when there is (*i*) a semaphore given or taken or (*ii*) some indication for a resource or signal or data-item generated or (*iii*) completion of execution of a set of codes. [Refer to Sections 8.1.2 and 8.3.1 for meanings of *semaphore* and *signal*, respectively.]

4. A state can generate multiple tokens (outputs, messages, flags interrupts or semaphores).

Let a circle represents a state and let a directed arc (or an arrow) represent the program flow from a state to another. When modeling a process as finite state machine (FSM), the software designer specifies the following for each state.

1. Finite set of inputs (tokens or event-flags or status flags) with their values for reaching the state.

2. Finite actions (for example computations) at the state.

3. Finite set of outputs with their possible values (or tokens or event-flags or status flags) and an output *function* for the state that give the outputs.

4. *State transition function* for each state to take it to the next state.

Real-time system software implementation becomes greatly simplified when using the FSM. Programming tasks are reduced to the following. (i) Coding for each state transition function and output function. (ii) Knowing the time periods taken by the process at each state transition function and between each state, when programming for real-time.

Consider an example of the running state in a timer. The count-input is the clock-input. The changed count value is the output. The output function is the increment in the count value. The state transition function is the time-out on overflow when a predetermined number of count-inputs is reached.

To understand the FSM model representation, Figure 6.3(a) shows *states* of a timer. A timer has four finite states: 'Idle', 'Start', 'Running' and 'Finished'.

1. 'Idle' state: it starts state transition on loading an input, *numTicks* (number of ticks after which the timer finishes), and its transition to 'Start' state occurs.

2. 'Running' state: on each clock input for decrement, the count value decrements.

3. 'Finish' state: program flows to finished state. This is when the count value reaches 0.

Figure 6.3(b) shows how the states of a C *function* can be modeled. The function has four finite states. *'Idle',' Call', 'Executing'* and *'Return'*.

1. *An 'Idle' state to 'Call' state* transition takes place when a function-call occurs in the program. [Refer to Section 5.4.6.] Output from this state is saved onto a stack.

2. Instructions executed in *the 'Executing' state* and the program counter changes with the program flow.

3. The transition to *'Return' state* happens when the instruction reaches the return-instruction.

4. The transition to *'Idle' state* occurs after retrieving the saved status and values from the stack.

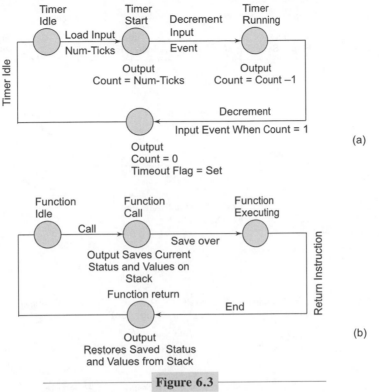

Figure 6.3

(a) States of a timer using FSM (b) States of a C function using FSM.

Figure 6.4 (a) shows how the states of an ***ISR*** can be modeled similar to those in an FSM. An ISR has five finite states, *'Idle', 'Call', 'Executing', 'Blocked',* and *'Return'*.

1. *An 'Idle' state to 'Call' state* transition takes place when an interrupt occurs and a status flag sets. Two of the outputs from this new state consist of (i) the action of saving context on to a stack and (ii) resetting the status flag.
2. Instructions in *the 'Executing' state* are executed and the program counter changes as per the program flow.
3. Transition to *'Blocked' state* occurs on an event when a higher priority interrupt takes place.
4. The program flows to *'Return' state* when the blocked ISR instruction starts running again and reaches the return instruction.
5. The flow to 'Idle' state occurs after retrieving the context from the stack.

Figure 6.4(b) shows how the states of a *task* can be modeled as an FSM. [Refer to Section 8.1 for understanding the concept of a task.] A task has five finite states—*Idle, Ready, Running, Blocked* and *Finished.* For output from one state, which becomes the input to the next state, tokens from the scheduler are *ready flag* and *block flag.* The tokens to the scheduler are *running flag, blocked flag* and *finish flag.*

1. *An 'Idle state' to 'Ready state'* transition occurs when the RTOS schedules this task by sending a token (message) to it. Output from this state consists of saving the scheduler context onto the scheduler stack.

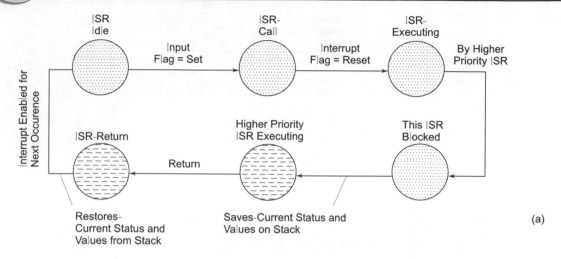

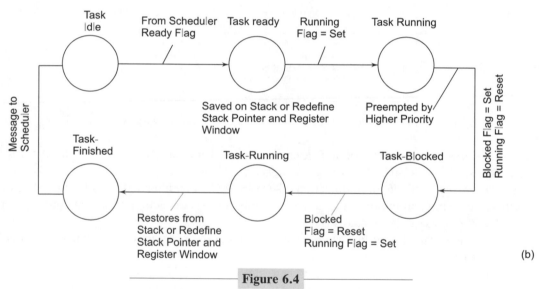

Figure 6.4

(a) States in FSM of an ISR (b) States in the FSM of a task in a multitasking.

2. *The 'Running state'* has instructions being executed and the program counter continuously changes as per program flow.
3. Program flows to *'Blocked' state'* when the scheduler preempts a task. It sends a token (message) to the task. Output from this state consists of saving of the task-context at the task stack.
4. Program flow to *'Running state'* again occurs on a token from the scheduler and after retrieving the values from the stack.
5. The flow to *'Finish state'* happens when the instruction reaches the end-stage. The output is a message to the scheduler.

6. The flow to *Ready state* occurs when the tasks are in an infinite waiting loop.
7. The flow to *Idle state* occurs when a message to the scheduler is sent by the task and the task is deleted from the ready list.

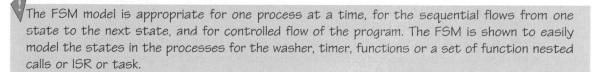

The FSM model is appropriate for one process at a time, for the sequential flows from one state to the next state, and for controlled flow of the program. The FSM is shown to easily model the states in the processes for the washer, timer, functions or a set of function nested calls or ISR or task.

When the FSM model is represented graphically with circles and directed arcs, it becomes complex in the case of a complex process with a large number of states. To design software using the model, a *state table* can then be designed for representation of every state in its rows. The following columns are made for each row.

1. *Present State* name or identification
2. *Action(s)* at the state until some event(s)
3. *The events (tokens)* that cause the execution of state transition function
4. *Output(s)* from the state output function(s)
5. Next State
6. *Expected Time Interval* for finishing the transitions to a new state after the event.

The coding using each row can now be easily done as follows.
'while (*presentState*) {*action* ( ); if (event = ...; token = ...) {output = ...; *stateTransitionFunction* ( ); };}'

Here *presentState* is a Boolean variable, which is true as long as the present state continues and turns false on transition to the next. The *action* ( ) is a function that executes at the state. If certain events occur and tokens are received (for example, clock input in a timer), then a state transition function, *stateTransitionFunction*, is executed which also makes *presentState* = false and transition occurs to the next state by setting *nextState* (a Boolean variable) = true.

When using an FSM model, a state table representation becomes very handy while coding.

6.2.2 Use of Petri Net Model

A. Petri Nets

Petri Nets based modeling has been found to be one of the powerful tools for a real time embedded system. A Petri Net is a *graphical modeling tool* from C.A. Petri. [Refer to Tadao Murata "Petri Nets, Properties, Analysis and Applications," Proceedings of IEEE, <u>77</u>, 541-580, 1989.] Petri Nets based modeling has been used in designing the algorithms for control circuits and computational and communication operators. The model is explained below. The following examples and Figures 6.5 and 6.6 clarify the use of the Petri Net model.

1. *There are the node-places and transitions in the Petri Net model at each Net* instead of the *states* of the FSM model. A computer-network is an interconnection of many types of nodes: computers, servers, printers, hubs, switches and routers. A Petri net interconnection is of two kinds of nodes: *node-places* and *node-transitions*.

2. In a graphically represented Petri Net, a *circle shows a node-place*. A *transition* is shown by a rectangle [Figures 6.5 and 6.6].

3. There is a place called marking place. This shows an initial marking like an initial state in the FSM or 'Idle' place in Figures 6.5 and 6.6.

4. A directed arc (arrow) is between one node-place and node-transition or between a node-transition to the next node-place.

5. A Petri Net base is representation by the directed graphs. These graphs are bi-partite. [Bi-partite means there can be two directions. Note the differences between Figures 6.3 and 6.5 or between Figures 6.4 and 6.6].

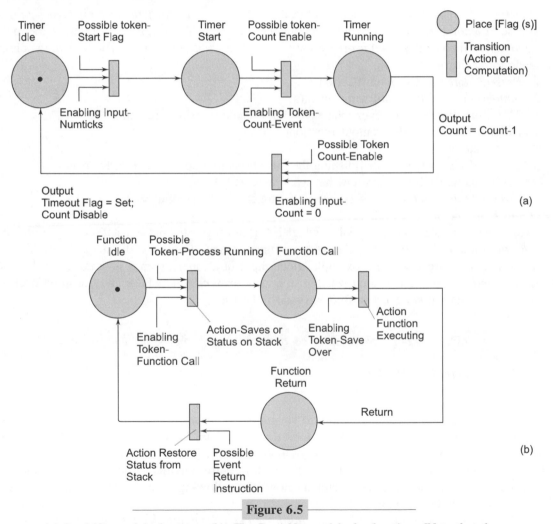

Figure 6.5

(a) Petri Net model of a timer (b): The Petri Net model of a function. [Note that there is only one directed arc from each transition.]

Figure 6.5(a) shows the node places and transitions in the Petri Net model for a timer. The timer is initially at the 'Idle' node-place. There is a possible token, the *start flag* that can start the timer by firing a transition. The transition is fired after an enabling token (number of ticks, *numTicks*) is received. [Firing means leaving a node-place and initiating, which is resulting into the actions or computations or program flow.] After a transition is fired, the timer goes to the 'Start' node-place. (a) There is a possible token, the *count-enable* flag that runs the timer by firing a transition to the 'Running' node-place. (b) There is an enabling token, *count-event* that runs the timer by firing a transition to the *Running* node-place. After a decreasing *count* value, the input is *count*. It enables a transition. A transition is then fired to the 'Idle' place on receiving the possible token, *count enable* and enabling token, and *count* = 0.

Figure 6.5(b) shows the node places and transitions in the Petri net model for a function. Before a function is called, the function is at the 'Idle' node-place. There is a possible token, *process running*. There is an enabling token, *call,* which starts the function by firing a transition from the 'Idle' place. At the transition, the action that takes place is the saving of the processor status onto the stack. Then, the flow is directed to the place of the function call. From this place, a transition is fired after the save-over token (message) from the processor. At the transition, the actions are the computations as per instructions. The program counter changes according to the program flow. The output tokens for the next place(s) are generated. After the transition is fired, the function goes to 'Return' node-place. There is a possible token, *return instruction*. On firing a transition, the actions are the restoring of the saved status. The flow is then directed to the 'Idle' place.

Figure 6.6(a) shows the Petri net model node places and transitions for an ISR. Before the interrupt, the ISR is at the 'Idle' node-place. There is a possible token, *interrupt enable*. There is an enabling token, *Interrupt flag,* which starts the ISR by firing a transition from the 'Idle' place. At the transition, the action done is saving the processor status on to stack. After this, the flow is directed to the *ISR call* place. From this place, a transition is fired after the save-over token (message) from the processor that enables ISR. At the transition, the actions are the computations as per the ISR instructions. The program counter changes according to the program flow. The output tokens for the next place(s) are generated. The ISR is directed to the 'Running' place. A possible token, *p,* can be the input to fire a transition. The *p* is an ISR block message when another ISR is enabled. After the enabling token, 'higher priority ISR enabled', the transition is fired. The actions done are saving the context and execution of the higher priority ISR. The running ISR gets blocked. After the transition is fired on the enabling token, 'return event' from the running higher priority ISR, the ISR is again directed to the 'Running' place. There is an enabling token, q, after a *return instruction*. On firing a transition, the resulting action is context switching. The flow is then directed to the original 'Idle' place of the ISR.

Figure 6.6(b) shows the Petri net model places and transitions for a *task* (Section 8.1.2). Before the interrupt, the task is at the node-place, 'Idle'. There is a possible and enabling token, *task ready flag*. At the transition, the action is taking place to save the processor status (context) on to the stack. After this, the program flow is directed to the task 'Ready' place. From this place, a transition is fired after the possible and enabling token (event), *run*. An output token, *error flag*, is generated in case of an error during the execution of a task. The task is directed to the 'Running' place. The actions at the transition are the computations as per the task instructions. The program counter changes according to the program flow. The output tokens for the next place(s) are generated. The next place may be the queue of IPC messages. [Refer to Section 8.3.3.] The task is directed to the 'Running' place. A possible and enabling token is *p*. The *p* is the task block message when another task is enabled from an RTOS place. The resultant actions are saving the context and blocking for the execution for running the higher priority task. The running task gets blocked. The task is directed to the task place, 'Blocked'. After a transition is fired on an enabling token from a RTOS on a 'return event' from the

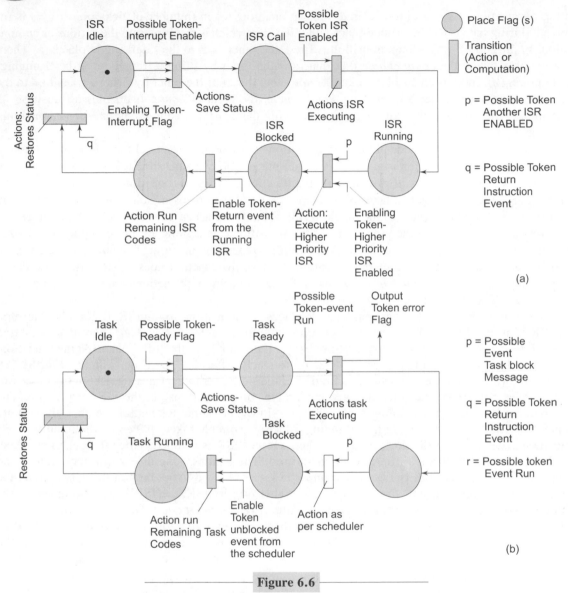

Figure 6.6

(a) Petri Net model for an ISR. (b) Petri Net model for a task.

running higher priority task, the task is again directed by the RTOS to a transition. At the transition, the remaining codes of the task are executed and the task is directed to the *'Running'* place. There is an enabling token, r, after a *return instruction* that fires a transition where the action is context switching. The flow is then directed to the original 'Idle' place of the task.

A directed arc shows a flow from a transition to a place or from a place to a transition. At the transition, the action(s) (for example computations) take place. [Action was at the state in FSM.] A place represents a particular node in the Petri net. From there the transition to which it is directed can be fired only after that transition collects (receives) all the enabling tokens from the place(s). There are possible tokens directed to a transition that enable the transition. A transition thus need not receive tokens from one place only, to be fired. Firing means the action(s) [for example, computations]. From the transition, the output token(s) is generated that can be directed to more than one node-place.

Two models that are of great help in the coding of ISRs, tasks and scheduler are FSM and Petri nets. The FSM is used when there is a finite number of states and Petri nets are used when the number of states is not definite. Petri nets are extremely useful compared to FSM when coding for a scheduler of RTOS.

Real-time system software implementation becomes greatly simplified by the use of Petri Nets in cases where any of several states can lead to any other state, so that it is impossible to know which one will definitely lead to which programming tasks are reduced to the following. (i) Coding for each action (function or set of computations) at every transition and (ii) Knowing the time taken by the actions and processes at each node-transition function (Rectangle) and the time between each fired transition.

A. FSM as a Special Case of Petri Net

When there is only one directed incoming arc from one place and one outgoing arc at each node-transition, a finite state machine equivalent is obtained. *A finite state machine can be considered as a subclass of the Petri net.* Earlier, a washing machine was assumed to be a finite state machine with three sates: *wash*, *rinse* and *dry*. When the machine is modeled as Petri Net, the *wash* state is considered to consist of the *node-transition*, *wash* and node-place, *wash over*. When this transition is fired, the action done is washing. There is one directed input arc, to the *wash-over place*. There will only be one directed arc from *wash-over* to the next node-transition, *rinse*.

A *state* from a process flow in FSM can lead to any one of the several possible subsequent states. This will depend on the output function and state transition function at the state. There can be several directed graphs from a *state*. But there can only be one directed graph towards a state from one definite state. Associated with a node-place in Petri net, there can be more than one node-transition. At a node transition, first the *possible* tokens (semaphores or interrupt enable bits, interrupt flags settings) and then the *enabling* tokens, are sought. A *transition* is fired only after an enabling token.

B. Use of the Petri Table for Real Time Programming

There can be more than one Petri net in a process or in a multitasking scheduler. [Refer to Section 8.1 for definitions of a *process* and a *task*.] The Petri nets model may take into account more than one process at a time. There may be dynamic transitions from one node-place to a node-transition and from one node-transition to the next node-place. There can be a coloured Petri net. A coloured Petri net means more than one type of tokens at the transitions.

A graphical description becomes highly complex when there are many transitions and node-places, coloured Petri nets and dynamic (unpredictable time) transitions. A tabular representation is preferable compared to graphical diagrams in a complex system like a RTOS (Real Time Operating System) for multiple ISRs, and multiple tasks sharing resources. [Resource means CPU(s), variables, memory, files, I/O devices.]

During real-time programming, the use of a row in the Petri table gives a powerful modeling technique that simplifies coding. In a Petri table representation the following columns are defined for each row for each pair of node-places and node-transitions.

1. *Present Node-Place name* for identification of conditions with its associated named node-transition (or input-actions or execution of set of instructions)
2. *Possible event(s) (tokens)* at the node-place
3. *The event(s) (enabling tokens)* that fire the transition
4. Output (tokens)
5. *Output functions and actions* at the transition
6. Next Node-Place
7. *Expected Time Interval* for finishing the transition

The advantage of using Petri tables can be understood from the following exemplary coding scheme. Consider a Petri Table for an exemplary process flow from the present node-place to the next node-places in Example 6.3. The coding in the example is based on the Petri table. Table 6.1 shows this table. Assume the following for the process flow considered here.

Table 6.1

Petri Table for an Exemplary Process Flow from Present Node-place to the Next Node-places

Node-Place	Possible Tokens	Node-transition	Enabling Tokens	Output Tokens	Output Actions	Next Node-Place	Time permitted
Atpresent Node -Place	flagE1, inputE1, semaphoreE1, and eventE1. flagE2, inputE2, semaphoreE2 and eventE2.	Transition1	flagE1, inputE1, semaphoreE1, and eventE1.	error1 from the function output1 ( )	Output-Actions1 ( )	At next Node-Place1	t_1"
Atpresent Node -Place	same as above	Transition2	flagE2, inputE2, semaphoreE2 and eventE2	error2 from the function output2 ( )	Output-Actions1 ( )	At next Node-Place2	t_1"

1. Let there be the possible tokens that can fire a transition 1. These are *flagE1*, *inputE1*, *semaphoreE1*, and *eventE1*.
2. Let there be the possible tokens that can fire a transition 2. These are *flagE2*, *inputE2*, *semaphoreE2* and eventE2.
3. Assume that Boolean variables, *transition1* equals true when the transition 1 is fired and *transition2* equals true when the transition 2 is fired.

4. When transition 1 is fired, assume that the actions are the computations as per the function outputActions1 ( ). When transition 2 is fired, assume that the actions are the computations as per the function outputActions2 ( ).

5. When transition 1 is fired, assume that the output tokens are as per the function output1 ( ). Let tokens be *error1*. When transition 2 is fired, assume that the output tokens are as per the function output2 ( ). Let tokens be *error2*.

6. Assume present node-place of a Petri net is represented by a Boolean variable, *atpresentNodePlace*, and it equals true as long the process is at the present place. The process will be at this place as long as *atpresentNodePlace* does not equal false.

7. Let time spent at a node-place be T. T is t' = *timeatpresentNodePlace*. Let t"$_1$ be the maximum running time permitted for the present node-place to exist. Let t"$_1$ = aTimeInterval. The process remains at present node-place as long as t' < t"$_1$. [Time out does not happen.]

8. Assume that the variable *atpresentNodePlace* will not equal false until either of the transitions, transition 1 or transition 2, is fired.

9. Assume that the transition is to next node-place 1. Boolean variable, *atnextNodePlace1*, represents the next node-place 1 and equals true after transition 1 to the next place. Assume that transition is to next node-place 2. Boolean variable, *atnextNodePlace2*, represents the next node-place 2 and equals true after transition 2 to the next place.

When coding from the table, the first while-loop is used to wait at a place in column 1 until the transition at column 3 is fired or time out happens. The second while-loop is used until a possible token in column 2 is an input from a node place. A transition is fired on collecting the input tokens as per column 4. If conditional statements for the output tokens are as per column 5 and output actions as per column 6, the first is used. The next node-place is as per the column 7. The time out occurs after the time interval in column 8.

Example 6.3 While (*atpresentNodePlace* && (*timeatpresentNodePlace* < *aTimeInterval*)) {
/* A set of possible tokens is at a place as per column 2 of table. Wait for a possible token /
possibleToken = (flagE1 == true || inputE1 == .*input1* || semaphoreE1 = = *taken* || eventE1 = true
|| flagE2 == .true || inputE2 == .*input2* || semaphoreE2 = = *taken* || eventE2 = true);
while (possibleToken) {/* Possible Tokens Expected */
/* Wait for a set of the enabling tokens atPresentNodePlace as per column 4 row 1 or row 2*/
/* Transition 1 fires as per enabling token set in row 1 column 4. */
transition1 = (flagE1 == true) & & (inputE1 == *input1*) &&(semaphoreE1 == *taken*) && (eventE1
= = true);
/* Transition 2 fires as per enabling token set in row 2 column 4. */
transition2 = (flagE2 == true) & & (inputE2 == *input2*) &&(semaphoreE2 == *taken*) && (
eventE2 = = true);
}; /* End of all codes for the present NodePlace possible tokens */
/* If *transition1* = true then transition 1 output tokens as per row 1 column 5, actions as per row
1 column 6 and flow to the next node place as per row 1 column 7. */
if ((atpresentNodePlace) && transition1)) {
output1 ( ); OutputActions1 ( ); atnextNodePlace1 = true; atpresentNodePlace = false;};
/* If *transition2* = true, transition 2 output tokens as per row 2 column 5, actions as per row 2
column 6 and flow to the next node place as per row 2 column 7. */

```
if ((atpresentNodePlace) && transition2)) {
output2 ( ); OutputActions2 ( ); atnextNodePlace2 = true; atpresentNodePlace = false;};
/* End of all codes for two sets of input tokens in two row of the table. These enabled the output of new
tokens and output functions and a transition to another NodePlace */
}; /* End of while loop for present place. Now at the next place*/
```

When using Petri-net models, a tabular representation called the 'Petri table' is very useful for coding.

6.3   MODELING OF MULTIPROCESSOR SYSTEMS

Multiprocessor embedded systems have applications in designing sophisticated embedded systems. [Refer to Section 1.2.7.]

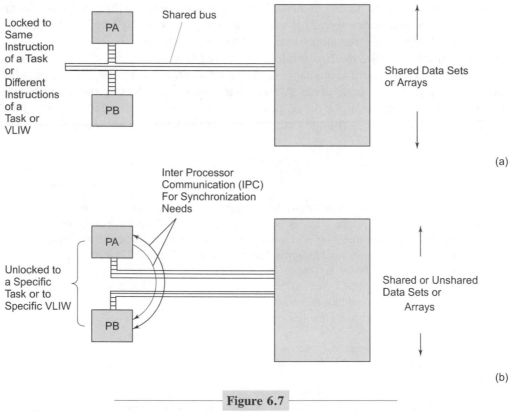

Figure 6.7

(a) Tightly coupled processors sharing the same address space while processing multiple tasks
(b) Loosely coupled processors having separate autonomous address spaces as in a network as well as shared address space for data sets and arrays.

Firstly, how do two processors, *PA* and *PB,* interface with the memory in a system? Case 1: Processors share the same address space through a common bus, called tight coupling between processors. Case 2: Processors have different autonomous address spaces (like in a network) as well as shared data sets and arrays, called loose coupling. Figures 6.7(a) and (b) show both the cases. Case 3: Processors share the memories in alternative bus architecture, for example, three-dimensional mesh, ring, torrid or tree in place of a shared bus between the different tightly coupled processors. Now, how do processors process concurrently?

1. One way of concurrent processing is to schedule each task so that it is executed on different processors and to synchronise the tasks by some inter-processor communication mechanism.
2. The second way is, when an SIMD or MIMD or VLIW instruction has different data (for example, different coefficients in example 6.2), each task is processed on different processors (tightly coupled processing) for different data. This is analogous to the execution of a VLIW in TMS320C6, a recent Texas Instruments DSP series processor. It employs two identical sets of four units and a VLIW (instruction word) can be within four and thirty-two bytes. It has instruction level parallelism when a compiler schedules such that the processors run the different instruction elements into the different units in parallel. Note: The compiler does *static scheduling for VLIWs.* Static scheduling [Section 6.3.6 (c)] is when a compiler compiles the codes to run on different processors or processing units as per the schedule decided; this schedule remains static during the program run even if a processor waits for the others to finish the scheduled processing.
3. An alternate way is that a task-instruction is executed on the same processor but different instructions of a task can be done on different processors (loosely coupled). A compiler schedules the various instructions of the tasks among the processors at an instance.

A multiprocessor system can be tightly or loosely coupled to the memory. It can also be coupled by mesh, ring, torrid or tree. There are several methods of scheduling and synchronising the execution of instructions, SIMDs, MIMDs, and VLIWs in the system.

6.3.1 Issues in Multiprocessor Systems

How is programming done in a multiprocessor system? *The problem is how to partition the program into tasks or sets of instructions between the various processors, and then how to schedule the instructions and data over the available processor times and resources so that there is optimum performance.* Should there be scheduling for running one task on one processor? Then, suppose one processor finishes computations earlier than the other. What is the performance cost [Section 6.3.6(B)]? Performance cost is more if there is idle time left from the available. What is the performance cost if one task needs to send a message to another and the other waits (blocks) until the message is received? Following are the issues in modeling the processing of instructions in a multiprocessor system:

1. Partitioning of processes, instruction sets and instruction(s).
2. Scheduling of the instructions, SIMDs, MIMDs, and VLIWs within each process and scheduling them for each processor.
3. Concurrent processing of *processes* on each processor.
4. Concurrent processing on each superscalar unit and pipeline in the processor. [Refer to Section 2.1 for scheduling of pipelines and super scalar processing.]
5. Static scheduling by compiler, analogous to scheduling in a superscalar processor.

6. Hardware scheduling, for example, whether static scheduling of hardware (processors and memories) is either feasible or not. [It is simpler and its use depends on the types of instructions when it does not affect the system performance.]

7. Static scheduling issue. [For example, when the performance is not affected and when the processing actions are predictable and synchronous.]

8. Synchronising issues during the use of inter-processor or process communications (IPCs) such that there is a definite order (precedence) in which the computations are fired on any processor in multiprocessor system.

9. Dynamic scheduling issues. [For example, the performance is affected when there are interrupts and when the services to the tasks are asynchronous. It is also relevant when there is preemptive scheduling as that is also asynchronous.]

In a multiprocessor system there are a number of issues to be examined during concurrent processing and scheduling of instructions, SIMDs, MIMDs and VLIWs.

6.3.2 Models

Four models that help in programming and scheduling a system are given here. It is necessary to understand the meaning of the directed graph used in these models before the model itself can be understood. Its representation is similar to a DFG. [Section 6.1]. *Vertices (circles) in this graph are called the actors. Actors do the computations.* An actor also represents a complete DFG within itself. *An edge between the vertices (arcs with an arrow for the direction) represents a queue of output values from one vertex and a queue of input values to another vertex. Edges carry the values from one actor to another.*

Let *X* and *Y* be two sets of instructions that once fired (started), do not need any further inputs from any source during the computations. Let *X* generate the output values (tokens/data) *a*, *b* and *c*. Let *Y* get the input values (tokens/data), *a*, *c*, *i* and *j* and let *i* have a delay. The number of inputs to *Y* need not equal the number of outputs from *X*. *Y* gets additional inputs and does not need all the outputs from *X*. These computations and data are now modeled by a directed data flow graph that exists from *X* to *Y*. The number of outputs and inputs are labeled near the arc-origin and arc-end. The actors are equivalent to the transitions and edges are equivalent to the places in the Petri Net model. Figure 6.8(a) shows actors and arcs in a directed graph between *X* and *Y*. It also shows a set of outputs (a, b and c) from *X* and a set of inputs (a, c, i, j) to *Y*. Figures 6.8(b) shows its analogous model using Petri Net (Section 6.2.2).

Multiprocessor system computations and their firing instances can be modeled. Modeling simplifies the programming, scheduling and synchronising of the processes.

6.3.3 Synchronous Data Flow Graph (SDFG) Model

The SDFG model is as follows. [Refer to E. A. Lee and D. G. Messerschmitt, "Static scheduling of synchronous data flow", IEEE Transactions on Computers, Feb. 1987.] Let an arc represent a buffer in physical memory. The arc can contain one or more initial tokens with the delays. A token, until received at the vertex, does not fire the computations at a vertex. Then an initial token may also represent a *delay* that is shown by a dot on the edges of SDFG. [Figure 6.8(a).] For example, *i* and *j* were initial tokens for the vertex *Y* in Figures 6.8(a) and (b) with *i* having a *delay*. The dot on an arc represents the initial token(s) in an SDFG model. If there are more than one initial token the number of initial tokens are mentioned on the dot. [A compiler will subsequently do static scheduling as per the SDFG. Taking into account the tokens and the number of initial tokens the compiler schedules execution at each vertex (firing elements for the computations and creating another set of output tokens).]

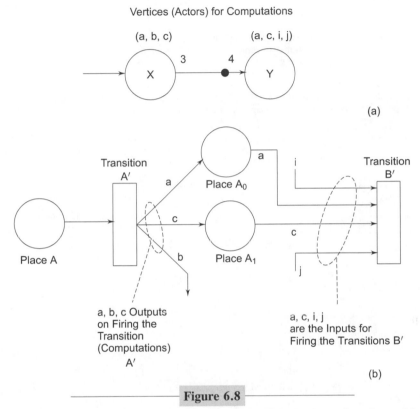

(a) Actors and arcs in a directed graph between X and Y. The outputs a, b and c and inputs a, c, i and j also shown. The i is with a delay (dot). (b) Its analogue Petri Net model in terms of transitions (actors) and places (edges).

!An SDFG model is like a DFG, but also models the delays as well as the number of inputs and outputs.

6.3.4 Homogeneous Synchronous Data Flow Graph (HSDFG) Model

When there is only one token at the input and one at the output, an SDFG is a homogeneous SDFG (HSDFG). For example, if the outputs from vertex X_1 (a set of computations) is *a* and input to Y (another set of computations) is also *a*. An SDFG can therefore unfold into a HSDFG. *An SDF graph can be unfolded into one or more HSDFGs.* Two vertices can be connected by two or more edges in the HSDF graph. An HSDF graph will naturally have more vertices and edges than an SDFG because only one token is permitted at a vertex. Figures 6.9(a), (b) and (c) show an SDFG correspondingly unfolding into a HSDFG graph with three vertices and two edges and the corresponding APEG obtained on removing delays.

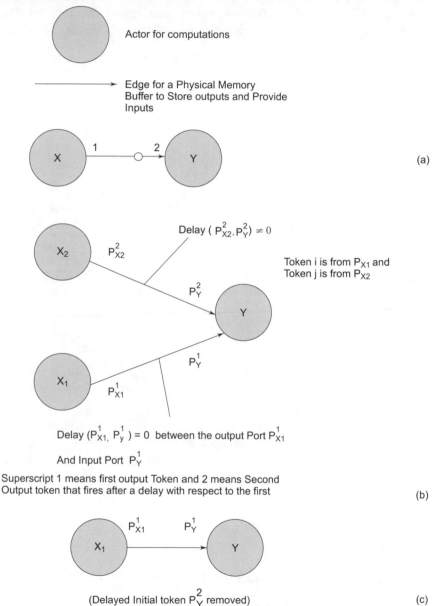

Actor for computations

Edge for a Physical Memory Buffer to Store outputs and Provide Inputs

(a)

Delay (P_{X2}^2, P_Y^2) $\neq 0$

Token i is from P_{X1} and Token j is from P_{X2}

P_{X2}^2

P_Y^2

P_Y^1

P_{X1}^1

Delay (P_{X1}^1, P_y^1) = 0   between the output Port P_{X1}^1

And Input Port P_Y^1

Superscript 1 means first output Token and 2 means Second Output token that fires after a delay with respect to the first

(b)

P_{X1}^1 P_Y^1

(Delayed Initial token P_Y^2 removed)

(c)

Figure 6.9

(a) A modeling of computations by an SDFG. The dot and label over the 2 show delayed two number input tokens at vertex Y. (b) An HSDFG representation after unfolding the SDFG. (c) An APEG representation from an HSDF after removing the delayed edge.

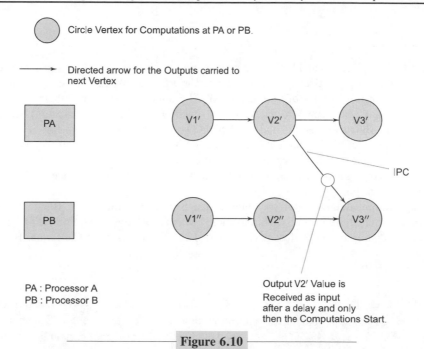

Figure 6.10

A two-processor system with one APEG and one HSDFG with an IPC to PB from PA.

Let there be a sequence of computations that are fired at the vertices. Let *precedence* in a directed graph define the computations order by which the vertices are placed first, then next, and then next to next. A sequence on one processor among the set of processors can be delayed at the arcs. Input from another processor (initial token) can also be delayed.

> HSDFG models are like an SDFG but have the feature that there is only one token that delays along an edge (arc or arrow) because there is only one token at input and one at output.

6.3.5 Acrylic Precedence Expansion Graph (APEG) Model

Acrylic precedence is a precedence of vertices in a directed graph such that there are no delays at the arcs. If initial tokens (delays) are *taken off from an HSDF graph*, an *acrylic precedence expansion graph* (APEG) is obtained. What is the importance of an APEG? An APEG not only has along the arc the starting inputs identical to the output from a previous vertex, but also no delaying tokens. Hence, the execution is smooth along the arc with no inter-processor communication time. An APEG-based algorithm becomes the simplest to schedule such that precedence constraints in the algorithm remain the same as before. Figure 6.9(c) shows a corresponding APEG that is a graph with no *delays*. It drives from a HDFG or SDFG.

A task-level concurrent process as well as an IPC graph can be modeled using APEGs and HSDFGs. A thread running on one processor modeled as APEG can pass a control to another by blocking itself or by sleeping, but the sequence and process flow along the APEG remain intact. An example is given next.

Example 6.4   Let V'_1, V'_2, and V'_3 be the computation vertices assigned to processor *PA*. Let V''_1, V''_2, V''_3 be the computation vertices assigned to processor *PB* concurrently processing with *PA*. An IPC is needed when algorithm (or set of computations) V''_3 cannot proceed until there is a message (token) from V'_2. Let IPC be between V''_3 and V'_2. This synchronises the processes at *PA* and *PB* through the IPCs. Figure 6.10 shows one APEG and one HSDFG with an IPC to *PB* from *PA*.

When there is an indefinitely long data sequence, SDFG-based modeling and the consequent unfolding into the HSDF graph helps. For example, an HSDFG applied to the computations of a fast Fourier transform or for coding a voice data. An HSDF graph can also effectively model an IPC (Inter-Processor Communication) graph.

[For the algorithmic details, refer to Sundararajan Sriram and Shuvra S. Bhattacharya, "Embedded Multiprocessors Scheduling and Synchronization," Marcel Dekker, New York, 2000. Both authors are students of the distinguished Professor Edward A. Lee of University of California, Berkeley.]

APEG models are such that there are no delays during execution at any stage in an APEG or chain of APEGs. Complex problems are therefore first modeled as the SDFGs, then SDFGs are unfolded into HSDFGs and HSDFGs are separated into APEGs. Processing is as per the precedence constraints between the APEGs. APEG-based algorithms become the simplest to schedule but precedence constraints in the algorithm among its APEGs remains constant.

6.3.6   Timed Petri Nets and Extended Predicate/Transition Net Models

Recall Section 6.2 for Petri net model applications. Two models can be used for developing the advanced scheduling algorithms: 'timed Petri Nets' and *probabilistic timed Petri-Nets* (also called Stochastic Petri-nets). These also use *coloured tokens*. A coloured token means a token that has two or more colours. A token is said to be of one colour when it represents a flag (an asynchronous event) and of another colour is when it represents a value(s). A token is said to be of another colour when it represents an inhibiting token. If an inhibiting token is present at a *transition* (Petri-net transition), the transition cannot fire. When this token changes the colour then the transition will fire. [Recall Section 6.2. A *Petri net transition* shows a set of computations that continue without any further input. At a *transition* the actions and computations occur similar to at a *DFG* or *APEG*].

In a 'Timed Petri Nets' model, the times (*delays*) at the node-places and node-transitions are also labeled on the graph like in Example 6.5 and Figure 6.11. When the *delays* cannot be taken as constants but are defined by a probabilistic distribution function, the model is called 'timed Extended Predicate/Transition Net (tEPr/T)'. [Probabilistic distribution function for the times can be understood as follows. There may be 50% certainty that a *delay* is within certain limits, and then may be 70% certainty that the *delay* is within other limits and so on. Predicate means that the delay or distribution function is predictable.]

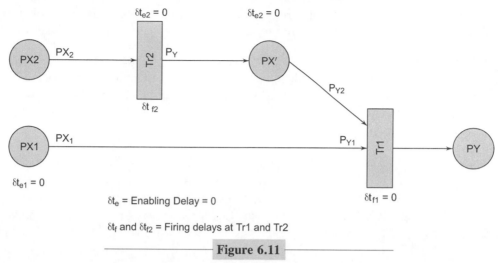

$$\delta t_{e2} = 0 \qquad \delta t_{e2} = 0$$

δt_e = Enabling Delay = 0

δt_f and δt_{f2} = Firing delays at Tr1 and Tr2

————— **Figure 6.11** —————

A timed Extended Predicate/Transition Net (tEPr/T).

Are cooking periods on two identical ovens for two identical recipes, quantities and vessels identical? No! Nearby airflows do not allow the certain predictability that the cooking periods will be identical. Similarly, no two identical electronic gates show exactly equal propagation delays, and so is the behavior of the processes and processors. There will be *probabilistic* distribution of uncertainties in the execution times of two parallel identical processes running on two processors. We, therefore, use a timed Extended Predicate/Transition Net (tEPr/T). [Refer to Example 6.5.]

Timed Extended means time at the node-place times (firing times), and fired computation times are also taken into the model, just like the *delays taken* at the *dots* in an SDFG. Extended Predicate Transition means transition is predictable with a certain degree of uncertainty and is represented by a probability function. A transition connects more than one node-place and can output to more than one node-place. [For details, reader may refer to a *classic* contribution, "Petri Nets, Properties, Analysis and Applications" by Tadao Murata, Proceedings of the IEEE, 27 (4), 541-580, 1989.]

Example 6.5 Figure 6.11 shows an example of a timed Extended Predicate/Transition Net (tEPr/T). There is a node-transition, *Tr1,* that gets the tokens from two node-places, *PX'* and *PX1*. On receiving the coloured tokens from both places, it fires computations. There is another node-transition, *Tr2, which* gets token(s) from one node-place, *PX2*. On receiving all the coloured tokens, it can fire computations like a DFG does. After transitions *Tr1* and *Tr2*, the new node-place is PY. Now each firing may be after a delay, say, δt_{e1} and δt_{e2}, respectively. Let δt_{f1} and δt_{f2} be the computation times, called firing times, at *Tr1* and *Tr2*, respectively. If $(\delta t_{e2} + \delta t_{f2}) < (\delta t_{e1})$, then both *PX1* and *PX2* must make the token(s) available for a duration $= \delta t_{f1}$. After δt_{e1}, only from the node-place *PY*, can further subsequent transitions fire. Using timed Petri net or a Petri table, the modeling can be done as with an SDF. All four delays are fixed in a time Petri-net representation. In tEPr/T model, each *delay* has a *probabilistic* distribution.

Recall the FSM model described in Section 6.2.1. It represents a CDFG approach (Section 6.1.2). In both models, the steps (program paths) are deterministic. The FSM determines the next state by *state transition functions*. The CDFG determines the next path from its decision nodes. Consider a cyclic scheduler. An FSM model, and therefore, the CDFG approach are applicable. An APEG can replace a CDFG approach.

The SDFG and tEPr/T models represent DFG when there are non-acrylic (asynchronous) data inputs and the steps are indeterminists. Unless there is an initial token in the SDFG, the graph does not continue for computations at the next vertex. Unless the timed events occur, the transition does not continue further (does not fire) in timed Petri Nets or tEPr/T. Both can be applied when scheduling a multiprocessor system.

A coloured token means a token that may be an input or an event or an inhibiting event. Its colour can change. For example, form inhibiting event to enabling event. Timed Petri-nets are the Petri-nets with defined times for firing a transition, and for computation times of the fired transitions. Timed Extended Predicate/Transition Net are the timed Petri-nets with times distributed by the probabilistic functions. These two types of nets model a multiprocessing system program flow.

6.3.7   Multi-Thread Graph (MTG) System Model

Let us assume that a task design is that of a thread (Section 8.1.3). Figure 6.12 shows an MTG system model. Multi-threaded system means a system with software consisting of multiple threads with some controlling and scheduling mechanism [Section 9.4.2.] An MTG model of a program has two layers, one of CDFG and other of timed Petri Nets. Modeling by MTG is given below. [For more details, readers may refer to "Modeling, Verification an Exploration of Task-Level Concurrency in Real Time Embedded Systems" by Filip Thoen and Francky Catthor, Kluwer, 2000.]

1. Task level concurrency between different threads models by an FSM. Figure 6.12(a) shows a task modeled as a thread using the FSM. [Recall that FSM is just a subclass of Petri Nets in which a transition gets input only from one node-place and transition to only one specific node-place.]

2. There is concurrent processing of the threads on different processors and processor level concurrency models by timed Petri Nets (as an alternative to the SDFG). Figure 6.12(b) shows timed Petri-nets on two processors, *PA* and *PB*.

Features of MTG are as follows:

(a) MTG provides optimal static and dynamic scheduling at the RTOS level.

(b) Timing analysis formulation and verification approach gives the separations.

(c) Calculation of temporal separations between the MTG entities gives the multiprocessor system *performance matrix*. [For performance definition, refer to Section 6.3.6.]

A Multi-Threaded Graph Model is modeling by using two layers, one of CDFG and the other of timed Petri Nets, for task level and processor level concurrent processing, respectively.

6.3.8   Applications of the Graphs and Petri Nets to Multiprocessor Systems

A. Partitioning and Then Scheduling

Recall Figure 6.10. When there are multiple processors in parallel, partition a program so that the following happens.

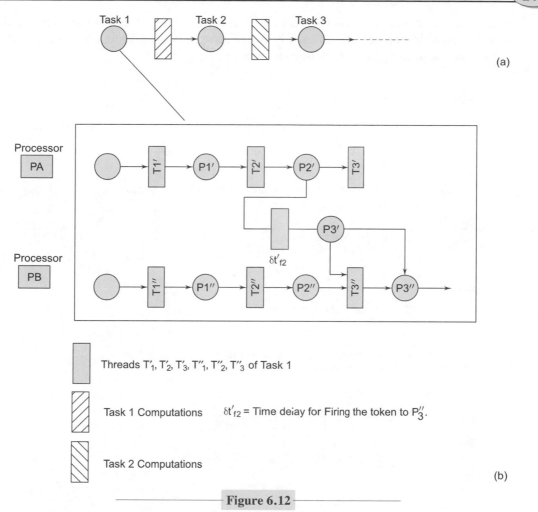

Figure 6.12

An MTG system model. (a) Thread (task) level concurrent processing modeled by FSM
(a subclass of Petri net). (b) For the task 1 Processor level concurrent processing modeled by timed Petri nets.

1. There is a minimum number of IPCs so that the total time of IPC delays (waiting periods) minimises.
2. There is load balancing. Each processor has the least waiting time by sharing the processing load.
3. The performance cost minimises. *Performance cost means the execution time required (a) for computations for the tokens and delays at the edge (communication time), (b) the computation time before firing (computations) by a vertex (transition) and (c) context switch time.*

At each vertex computations occur such that the precedence constraints maintain (remain intact). The graph of a program thus partitions into the functions or tasks or threads. There are three strategies in which a program can be scheduled to run:

1. (a) Each task or function is executed on an assigned processor, (b) Each task or function is executed on the different processors at different periods, (c) Instructions of four different

tasks partitioned on two processors and (d) Instructions of four different tasks partitioned and scheduled on two processors differently in different periods. [Figures 6.13 (a) to (d) show these four partitioning and scheduling strategies.]

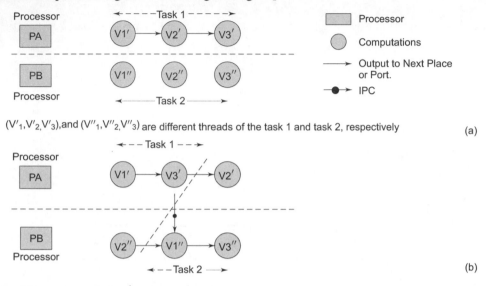

(V'_1, V'_2, V'_3), and (V''_1, V''_2, V''_3) are different threads of the task 1 and task 2, respectively

(a)

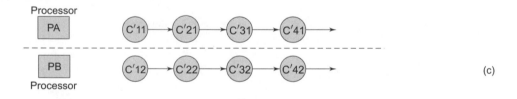

$c'1ij$ is an instruction in $V'1$ to take i-th column and jth row element of matrix A and add bit with the Corresponding Element in B.

(b)

(c)

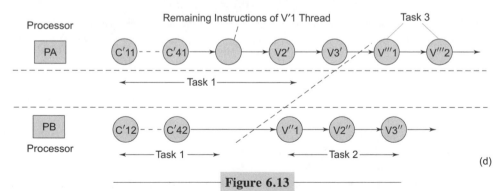

(d)

Figure 6.13

(a) Each task or function is executed on an assigned processor, (b) Each task or function is executed on different processors at different periods, (c) Instructions of four different tasks partitioned on two processors and (d) Instructions of four different tasks partitioned and scheduled on two processors differently in different periods.

2. Each set of data is partitioned in a VLIW instruction and is executed on the different processors, which execute the same program. Consider a matrix addition process. Each row can be added on a different processor when the data of the rows are partitioned among the processors. Such data partitioning is preferred when processing a DSP-VLIW.

3. A combined partitioning is done both at the data level as well as the task (or function) level. Different functions themselves may run concurrently on different processors, but at the micro or atomic level the instructions are run by data partitioning. A Lee multidimensional SDF model or a Printz geometric partitioning model can be used for the graph. [Refer, for the models, to the Sundararajan Sriram and Shuvra S. Bhattacharya book referenced earlier.]

> Partitioning and scheduling of vertices can be done in number of ways: (a) Each task or function is executed on an assigned processor, (b) Each task or function is executed on the different processors at different periods, (c) Instructions of four different tasks partitioned on two processors, (d) Instructions of four different tasks partitioned and scheduled on two processors differently in different periods and (e) data partitioning in case of SIMDs, MIMDs and VLIWs.

B. Method for Minimising Performance Cost by Load Balancing

Let us model all the processes by HSDFG as in Figure 6.14. *Then the following steps are for graph partitioning.*

1. **Step I:** It is shown in Figure 6.15 (a). A graph-partitioning algorithm uses the APEGs as the DFGs. It assigns vertices. By identifying the nested controlling structures, a graph-partitioning tree is obtained. Each nested control structure gives a *micro thread*. Each vertex is then hierarchically organised into a tree of sub-graphs. APEGs are then the leaves of this tree. Consider a multitasking by a scheduler. We can assume each task and ISR and their scheduler to be a macro thread.

2. **Step II:** It is shown in Figure 6.15 (b). The second step is further partitioning of each APEG into the micro threads. Each vertex is now assigned to a distinct micro thread. Now, the pairs of micro threads are successively merged until the following occur. (*i*) It does not result in inter-macro thread communication cycle (inter-branch cyclic communication at the edges between the leaves). (*ii*) It reduces the performance cost.

Average communication time and context switching time in the program add together and give the *performance cost of each partition, and then gives a partitioned program total performance cost.*

Each vertex has a set of computations and the sum of the execution times at each vertex is *static* in any performance cost calculation. An element of a *performance matrix* equals the performance cost of an instruction in a thread. Columns in the matrix correspond to the instructions and rows correspond to the threads.

Reducing the ratio of total performance cost and static cost is the aim of partitioning into multiple processes and macro threads.

> Partitioning is done in such a way so that total performance cost is minimum.

C. Four Scheduling and Synchronisation Strategies during Concurrent Processing

Figures 6.16 (a) to (d) show four scheduling and synchronisation strategies: (a) completely static, (b) self-timed scheduling, (c) quasi-static and (d) completely dynamic

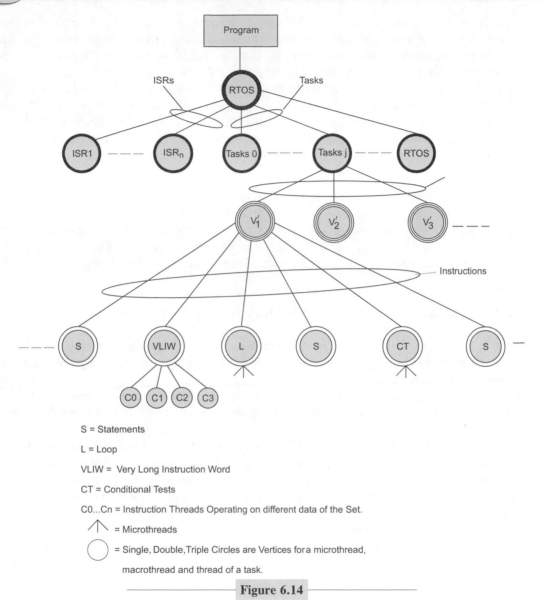

S = Statements

L = Loop

VLIW = Very Long Instruction Word

CT = Conditional Tests

C0...Cn = Instruction Threads Operating on different data of the Set.

∧ = Microthreads

○ = Single, Double, Triple Circles are Vertices for a microthread,
macrothread and thread of a task.

Figure 6.14

*Model for partitioning all the processes running on the processors by HSDFGs and
organisation of each vertex hierarchically as a tree of sub-graphs, each vertex
having a nested controlling structure like a macro thread or task.*

When each vertex computation occurs such that the precedence-constraints are maintained (remain intact), it is called static scheduling. *Static scheduling* is done at the compilation time. The processors are allocated for each vertex computations, inter-vertex communication and ordering of the vertices for processing on an assigned processor in the system.

Suppose the computations of one vertex finish early. A processor will remain idle. Then, instead of a static way there can be a way in which there is dynamic run-time scheduling and synchronisation. It

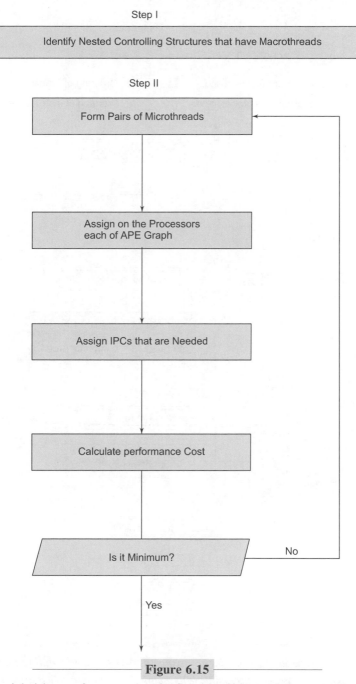

Step I

Identify Nested Controlling Structures that have Macrothreads

Step II

Form Pairs of Microthreads

Assign on the Processors
each of APE Graph

Assign IPCs that are Needed

Calculate performance Cost

Is it Minimum?

No

Yes

Figure 6.15

Steps I and II for minimising performance cost by first identifying and then merging the microthreads.

decides at run time where, after whom, and which vertex to run for performing the computations. Therefore, a *variant* to the completely static method is that the ordering of communication (edge from one vertex to another) between the processors is decided at compilation time but only the predeter-

○ C_1^x, C_2^x, C_3^x, C_4^x,...Instruction 1, 2, 3,.. of Thread x.

◎ V_1', V2, $V3'$, $V4'$, ...Thread 1, 2, 3, 4 of task 1.

▤ Running as per compiler scheduling   ▤ Ordering of Transcations that use the IPC

▥ Assign an APE graph to a processor n.   ▨ Ordering of Instructions as APE graphs.

(a)

(b)

(c)

(d)

----------------------------------- **Figure 6.16** -----------------------------------

(a) Completely static scheduling and synchronising (b) Self-timed scheduling and synchronising
(c) Quasi-static scheduling and synchronising (d) Completely dynamic scheduling and synchronising.

mined transactions are enforced at the run-time target. Suppose there is a compile-time scheduling and it is such that precedence is first V, then V', then V'' computations. At run time, the scheduler schedules that on which processor V should run. After V finishes computations, the scheduler schedules on run time, V' on either the same or another processor.

Another variant to the completely static way is *self-timed scheduling and synchronising*. It is used when the execution times for V, V' and V" are changing at a run. [The changes are due the conditions and control statements encountered during the computations] Now, let V and V" be schedules to run on processor *PA* and V' be scheduled to run on processor *PB*. It is according to a compile-time static scheduling. But at run, *PB* waits until it receives a run-time communication from processor *PA* that V has completed firing. Processor A has a run-time wait until V' completes firing. There is a *self-timing* between A and B. Figure 6.16(b) shows this.

There is flow control at run time. There is run-time *completely dynamic* scheduling and synchronisation. Figure 6.16(d) shows this completely dynamic method of scheduling and synchronisation. For distributed processors forming a network, dynamic scheduling becomes a necessity. A *variant* of dynamic scheduling is that assignments of transactions to the processors are at compile time. Another variant is quasi-static.

Figure 6.16(c) shows this *quasi-static method of scheduling and synchronisation*. It is useful when there are data-dependent firing times at the actors. Only probabilistic assignments and ordering are at the compile time. They can be overridden at run time due to the nature of the data used during the firing by an actor.

D. Resynchronisation

When there are too many IPCs, total performance cost increases. This cost is reduced by appropriate resynchronisation. Example 6.6 explains the resynchronising mechanism.

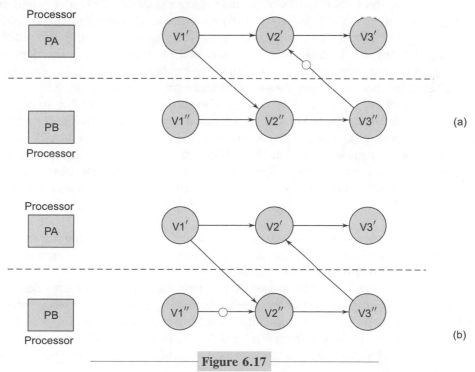

Figure 6.17

(a) Synchronisation of Vertices (b) Resynchronisation on the processors PA and PB.

Example 6.6 Consider a scheduling mechanism shown in Figure 6.17. Let it be that on processor *PA*, the precedence constraints are that V'_1 fires, after that V'_2 fires, and then V'_3 fires. On another sister processor, *PB*, the precedence constraints are that V''_1 fires, after that V''_2 fires, and then that V''_3 fires. Now, assume that from V''_3 there is IPC to V'_2 [IPC is shown by a dot for *delay* in Figure 6.17(a)]. Processor PA cannot fire V'_2 until there is a communication from V''_3. But suppose it is possible to resynchronise such that V'_2 fires by output of V''_2 or V''_1. Then the wait will be reduced. By using appropriate mathematical theorems or analysis, it is also possible to remove the redundant IPCs. Resynchronising is done after estimating how the synchronisation-edge *delay* for an IPC decreases the performance cost when the *delay* position shifts. [Figure 6.17(b) shows the shift of the dot and resynchronisation by replacing the dotted edge between PA and PB with edges with no delays].

The four scheduling and synchronisation strategies are (a) completely static, (b) self-timed scheduling, (c) quasi-static and (d) completely dynamic.

■ SUMMARY ■

- The software engineer must adopt a standard design practice of using a model during the development process for software. Software implementation becomes greatly simplified by the use of data flow graphs (DFGs) because programming tasks are thereby reduced to the following. Coding for each process is represented by a circle using the data input from incoming arrow(s) and generating data output to outgoing arrow(s).

- In a DFG model, there is a single data-in point and a single data-out point with a process or set of processes that are represented by circle(s). When the assignment to an input is fixed in a DFG, it is also called acrylic DFG (ADFG). Programming-complexity is minimised by modeling a program in terms of as many DFGs as possible and the use of as many ADFGs as possible.

- Another important concept of program modeling is the controlled data flow graphs (CDFG) for program design and analysis. The CDFGs represent the controlled decision at the nodes and program paths (DFGs) that are traversed consequently from the nodes after the decisions.

- Program modeling for real time programs is done by the Finite State Machine(s) model or by Petri net(s).

- The FSM model is appropriate for one process at a time, for the sequential flows from one state to the next state, and for a controlled flow of the program. The FSM is found to easily model the states in processes in real time systems.

- Petri Nets–based modeling has been used in designing the algorithms for embedded systems for control circuits and computational and communication operations. Petri Nets–based modeling has been found to be one of the powerful tools for a real-time embedded system. *There are node-places and transitions in Petri Net models at each Net* instead of the *states* in the FSM model. A Petri Net is an interconnection of two kinds of nodes: *node-places* and *node-transitions*.

- When Petri Nets become too large to handle, the use of Petri tables is a powerful real-time programming model.

- The multiprocessor system uses two or more processors for faster execution of the (*i*) program functions, (*ii*) tasks, (*iii*) *single instruction multiple data* instructions, (*iv*) *multiple-instructions multiple-data* instructions or (*v*) *very long instruction words*. The VLIWs in the DSP instructions can be finished at high speed. Modeling of multiprocessor system uses SDFG and HSDFG representations in which there is an unfolding of the SDFG so that there is only one token which delays along its edge and/or an APEG in which there are no delays.
- Timed Petri Nets are used to model a multiprocessor system. It shows at the Nets the transition firing times on the edges and fired computations time.
- Extended Predicate/Transition Net are timed Petri-nets with delays represented by probabilistic distribution functions.
- The Multi-Thread Graph (MTG) System model is used in which there is task level concurrency between the different threads models by an FSM and the processor level concurrency models by timed Petri Nets.
- Models are used for partitioning, load balancing, scheduling, synchronisation and resynchronisation during the program flow on the multiple processors. This gives minimum total performance costs (processing delays).

▣ LIST OF KEYWORDS AND THEIR DEFINITIONS ▣

- *Model*: It is a representation by which a problem, process, design or analysis can be easily understood and the problem becomes simplified after modeling.
- *Data flow graphs (DFG)*: The code for each process is represented by a circle and the data inputs to the process are from incoming arrow(s) and generation of data output is through the outgoing arrow(s).
- *Acrylic DFG (ADFG)*: A DFG model when the assignment to an input is fixed.
- *Controlled data flow graphs (CDFG)*: Modeling by representing the controlled decision at the nodes and program paths (DFGs) that are traversed consequently from the nodes after the decisions.
- *Finite State Machine*: A model in which there are finite states. After a given set of inputs, a state changes according to the state transition function.
- *State transition function*: A process or state of codes that carry a program state from one to another.
- *Petri net*: A model given by C.A. Petri in which there are interconnections of two kinds of nodes: *node-places* and *node-transition. Transitions modeled by a rectangle are the actors (where the computations are fired orinitiated). For firing transitions the tokens are from the node-places. A token is an input or flag or asynchronously occurring event.*
- *Petri Table*: A table to represent the Petri net(s). When Petri nets become too large to handle, the use of Petri tables is a powerful real-time programming model.
- *Multiprocessor system*: A system that uses two or more processors for faster execution of the (*i*) Program functions, (*ii*) tasks or (*iii*) *single instruction multiple data* instructions or (*iv*) *multiple-instructions multiple-data* instructions or (*v*) *very long instruction words*.
- *SDFG*: A DFG representation in which input(s) delays are also shown. Circles (Vertices) are the actors where computations take place like at Petri net transitions or at nodes in a DFG. The edge (arc or arrow) has dots for the delays and labels the number of inputs and outputs.

- *HSDFG*: Representation in which there is unfolding of the SDFG so that there is only one token, with its delay along its edge.
- *APEG*: A SDFG with no delays.
- *Scheduling*: Allocation of different vertices or sub-graphs on different processors.
- *Timed Petri Nets*: Petri-net showing the transition firing times on the edges and fired computations times.
- *Extended Predicate/Transition Net*: Timed Petri-nets with *delays* represented by probabilistic distribution functions.
- *Multi-Thread Graph (MTG) System model*: A model in which there is task level concurrency between the different threads models by an FSM and the processor level concurrency models by timed Petri Nets.
- *Partitioning*: Partitioning the graphs into parts, with each part scheduled on the processors as per scheduling strategy adopted.
- *Load balancing*: Partitioning and scheduling of microthreads and instructions such that each processor shares the processing load in a multiprocessor system.
- *Macrothread*: A thread or its part that has a number of processes and instructions threaded together.
- *Microthread*: A set of instructions or controlled nesting structures in a thread that is the minimum that can be represented by an actor (Vertex or transition or DFG node). Merging of microthreads into macrothreads is used to minimise the performance cost.
- *Performance cost*: Time taken for execution at a vertex or at a sub-graph or microthread.
- *Total Performance Cost*: Total of all performance costs. This will be the minimum if the load on processors is balanced.
- *Resynchronisation*: Repeating synchronisation by suitable mathematical analysis and reducing the number of IPCs, and thus the delays caused at the processors waiting for the IPCs, in a multiprocessor system.

■ REVIEW QUESTIONS ■

1. Why does program complexity increase with a reduced number of DFGs and increasing decision nodes?
2. Explain, using examples: APEG, coloured token, SDFG, HSDFG and resynchronisation.
3. Why do you unfold SDFGs into as many HSDFGs as feasible and then HSDFGs into as many APEGs as possible?
4. How does concurrent processing help in VLIW instruction execution at high speed?
5. How will you schedule a SIMD instruction on two processors?
6. How will you schedule a MIMD instruction on two processors?
7. How will you schedule a VLIW instruction on two processors?
8. How does a Petri table help over a Petri net? Explain with two exemplary problems.
9. What is meant by completely dynamic scheduling and completely static scheduling in a multiprocessor system?
10. What is meant by load balancing? How do you achieve it by combined partitioning?

■ **PRACTICE EXERCISES** ■

11. Recall Table 3.3 for HDLC network protocol format. Draw a data flow graph (DFG) model for a program, which transmits frame start bits, address bits, information frame control bits, data bits, FCS bits and frame ending bits.

12. How will the DFG drawn in Exercise 11 (given above) modify as a control data flow graph (CDFG), which provides decision nodes for bit stuffing (when five 1s transmit for the data, one 0 is stuffed additionally).

13. How will the DFG drawn in Exercise11 (given above) modify as control data flow graph (CDFGs), which provides decision nodes to follow one of the three paths in rows 3a, 3b and 3c in three cases: (a) as per the first bit (b) if the second bit =0 and the first bit is 1(c) if the second bit =1.

14. Draw a CDFG to incorporate decision nodes at the loop start and loop end to limit the summation terms up to n = 10 for Equation D.3 in Appendix D as follows: $y_n = \Sigma a_i.x. (n-i) + \Sigma b_l.y. (n-l)$ used in an IIR filter.

15. Now, minimise the programming-complexity in the resulting CDFGs in Exercise 14 above by modeling a program in terms of as many DFGs as possible and the use of as many ADFGs as possible by converting CDFGs into DFGs, and without using a loop in the program.

16. Draw a FSM model of a vending machine program (Section 11.1). Create the model with these factor: The machine permits only one type of coin, Rs.1; one chocolate vends at a time; and one chocolate costs Rs.8.

17. Draw Petri net model for problem in Exercise 16 above.

18. Recall Example 5.6 for use of the FIPO (First-In Provisionally-Out) Queues for Flow Control on a Network. Model the program as Petri nets with coloured tokens.

19. Using Petri net, model the programs for P and V semaphores as mutex functions in Example 8.1 (Chapter 8).

20. Draw a multiprocessor system for the cases: (a) tightly coupled to the memory, (b) loosely coupled, (c) coupled by mesh, (d) ring coupled, (e) torrid coupled and (f) tree like coupling.

21. Give two exemplary systems in which probabilistic timed Petri Nets can be used.

22. How do you solve the following problem: "How can a program be partitioned into tasks or sets of instructions between the various processors? Then, how can the instructions and data be scheduled over the available processor times and resources so that there is an optimum performance?"

23. A Multi-Threaded Graph Model is modeled by using two layers, one of CDFG and the other of timed Petri Nets for task-level and processor-level concurrent processing, respectively. How will you use it in concurrently processing the image inputs and image compression algorithms?

24. Assume that four processes are scheduled to run on two processors. A program is portioned in such a way that with each 10,000 ns, each process schedules 10 times on each processor. What will be the minimum number of context switching/microsecond?

25. Now, assume in Exercise 24 that there is one inter-process communication per 100ns. Assuming average communication time = 5 ns, what will be the *partitioned program total performance cost.*

Chapter

7

Software Engineering Practices in the Embedded Software Development Process

What We Have Learnt

Let us recap the following points learnt so far for developing embedded system software.

1. Procedure-oriented C and object-oriented programming C++ and Java languages are used in most embedded systems programming. Writing of device drivers and application software codes is done mostly by using C/C++ programming elements: modifiers, conditional statements and loops, pointers, function calls, multiple functions, function pointers, function queues and service-routine queues.

2. Data of various types and with various structures, like the arrays, queues, stacks, lists and trees, are used during coding.

3. Embedded programming is such that methods to optimize the system memory requirements are also used.

4. Embedded system software implementation as final code generation as the ROM image is obtained by using the source code generation tool, source code engineering tool, compiler, cross-compiler, linker and locator.

5. A programmer models the programs during software implementation as DFGs and CDFGs.

6. A programmer models the real-time systems during software implementation by modeling the program flow by FSMs and Petri nets.

7. A programmer models the multiprocessor systems software implementation by using the programming models like SDFG, HSDFG and APEG graphs, Timed Petri Nets, Extended Predicate/Transition Net and Multi-Thread Graph (MTG) Systems.

It is thus clear that embedded software development can be highly complex besides having a large number of data type, structure and algorithm (functions) needs that are to be engineered.

It is also seen that embedded system programs differ from the usual computer programs in the following respects. (i) Several components of vastly different functionalities are found in embedded system software. Functionalities of the physical and virtual device driver routines, application software and operating systems are vastly different and must not only be appropriately engineered, but also system-integrated. (ii) Different software components have response time constraints and have to often meet strict deadlines. (iii) All software components must also use the memory optimally. (iv) Each software-component execution speed must be optimum to enable the minimisation of energy while meeting all deadlines. (v) Software must have controlled complexity and must be thoroughly tested and debugged for errors because the codes will be permanently burned-in at the system ROM. *Therefore, a software algorithm development process must use standard software engineering practice(s).*

Hurrying through the coding directly without following the software engineering approaches and practices should never be done in any system greater than a minimum level of complexity. Research by software engineering experts have shown that on an average, a designer needs to spend about 50% of time for planning, analysis and design, 40 % for testing, validation and debugging and 10–15 % on coding.

There are a number of standard textbooks for in-depth and detailed understanding of software engineering principles, approaches and practices. Software engineering involves:

(i) Developing and using skills and CASE (Computer-Aided Software Engineering) tools so that a system is analysed and specified correctly,
(ii) Software implementation with an appropriate set of source code engineering tools,
(iii) Testing, verification and validation so that quality software is produced.

Standard texts should be studied to develop these skills in depth. [One such textbook is *Software Engineering – A Practitioner's Approach* authored by an international authority on this subject, Roger S. Pressman, McGraw-Hill, 5th Edition 20th Anniversary Edition, 2001. Also see References.]

What We Will Learn

The first objective of this chapter is, therefore, to learn the software engineering concepts and find answers to the following questions:

1. What is the complexity of an algorithm?
2. What are the phases and models of the software development process?
3. What are the processes for implementing, testing, debugging and validation of the developed software?

Real-time programs are needed in embedded systems. This chapter will look for answers to the following questions also:

4. What are the real-time programming and software development process issues?

When software is developed and its design and codes are finalised, it has to be maintained. Another objective of this chapter is to learn, and find answers to, the following:

5. How is a project managed? What are the software management practices and project metrics to estimate efforts at planning stage?
6. How is software maintained?

Unified Modeling Language (UML) is a modeling tool available for use during development or design or process. The last objective of this chapter is to learn:

7. UML basic elements
8. UML diagram
9. Exemplary use of UML for modeling a software implementation

7.1 SOFTWARE ALGORITHM COMPLEXITY

One principle is that a complex system needs greater attention during its development process. Software complexity can be understood by the following:

1. Complexity may be used to predict (a) critical information about reliability and (b) maintainability of software systems from an automatic analysis of source code or for procedural design information.

2. Complexity metrics also provide a numerical feedback during the software development process, to help control the design activity for a project.

3. Complexity provides detailed information about modules and helps to focus on the areas of potential instability during testing and maintenance.

4. Complexity in object-oriented (OO) design is measured in terms of structural characteristics. It is found by examining how classes of an OO design are interrelated to each other.

How can software algorithm complexity be measured? There are at least 18 different categories of software complexity metrics. One of the most popular metrics is from McCabe. [Refer to *Software Engineering—a Practitioners Approach* by Roger S. Pressman, McGraw-Hill, 5th edition, 20th Anniversary Edition, 2001.] It is called Cyclomatic Complexity. When stated in terms of a flow graph, an independent path is the path that moves at least one edge (arc or directed arc) that has not been traversed before the path is defined. An independent path is any path that is traversed due to at least one new set of processing statements or a new condition. Figure 6.2 (a) shows eight paths. Four are independent paths. Cyclomatic complexity, C (g) equals the number of independently traversed paths. The C (g) computation gives us the number of paths to look for, which in this case are 4 (besides the one expected path to facilitate flow). Example 7.1 gives three computation formulas and analysis of C (g) of a CDFG shown in Figure 6.2.

Example 7.1   Three formulas compute C (g) as follows.

1. C (g) = E – N_d +2, where E = the number of edges in the graph and N_d is the number of decision nodes. E = 6. [There are three incoming plus outgoing edges per node. Four nodes have 12 of them. But each edge is shared between two nodes. So the number of edges are taken as = 6.] The number of nodes, N_d = 4. Cyclomatic complexity is again found to be 4 by this formula.

2. C (g) = P + 1, where P is the number of predicate nodes contained in the CDFG. Number of predicate nodes = 3 in the figure. Cyclomatic complexity is again found to be 4 by this criterion. [The node at which there are two predictable paths, one when the test result is true (yes) and the other when the test result is false (no) defines the predicate node.] Figure 6.2(a) shows P = 3 (three predictable paths). The testing node at the test start is simply a decision node and not a predicate node. Calculation of y_n starts on the event of a new *n* value at the input.

3. C (g) = Number of regions in the flow graph. There are four regions. One region is for calculating *e*. The second region is for calculating *s*. The third is to calculate y_n. The fourth region is for testing conditions at the flow start. Cyclomatic complexity is again found to be 4 by this criterion.

Let a program posses a logical complexity. The number of independent paths in a basic set in an algorithm = maximum number of tests that ensures that all the statements have been executed at least once. Figure 6.2(*a*) showed four test conditions and four independent paths. Therefore, cyclomatic complexity in Example 6.2 can be taken as four.

The McCabe cyclomatic complexity provides a quantitative indication of maximum module size. From a number of programming projects, it can be found that a cyclomatic complexity of 10 appears to be a practical upper limit of module size. When cyclomatic complexity exceeds this number, it could become difficult to adequately test a module.

> Software complexity is an important measure. It gives a quantitative indication of maximum algorithm size and its reliability. It measures the number of necessary needed paths that are to be tested. It is also a measure of the difficulties in maintaining software. Cyclomatic complexity is a quantitative measure suggested by McCabe. It gives the number of independent paths that should also be tested whenever software complexity exists.

7.2 SOFTWARE DEVELOPMENT PROCESS LIFECYCLE AND ITS MODELS

Table 7.1 lists the activities in three phases in a software development process.

Table 7.1
Phases in Software Development

Phase	Activities
Definition and Analysis of System Requirement	**(1)** Complete clarity of the specifications of the required system is the aim of the definition phase. **(2)** Specifications are obtained for **(i)** required processing of data, **(ii)** necessary functions and tasks and their expected performance, **(iii)** expected system behavior, **(iv)** constraints of design, **(v)** expected life cycle of the product, **(vi)** human-machine interaction, **(vii)** validation criteria for finally developed systems and **(viii)** time delivery schedule.
Development	Software design, coding and testing are the main activities in the development phase. Conceptual design of the system is evolved during the software design phase for following: **(i)** structuring data, **(ii)** implementing functions, **(iii)** implementing interfaces, **(iv)** algorithms and the languages to be used for them, and **(v)** testing methodologies to be followed. Conceptual design leads to thc development of codes for application software. Testing and validation are essential activities of any software development process.
Support	Embedded software should need little support. Support phase activities are the correction of bugs detected and software enhancement by adding the extra functions necessitated by changing environment and software re-engineering.

There are seven important models from which to choose for the process lifecycle. **(i)** Linear Sequential Lifecycle Model. **(ii)** RAD (Rapid Development) Model. **(iii)** Increment Development model. **(iv)** Concurrent Development. **(v)** Component based model. **(vi)** Fourth generation tool base development (4GL model). **(vii)** Object-Oriented Development Model.

Table 7.2
Stages in the Software Development Process in the Linear Sequential Model

Stage	Activities	Model Deficiency
Stage 1	Modeling and analyzing system requirements Stages 1 and 4 are more prone to blocking	Until a stage is completed by all team members there may be blocking of interdependent activities.
Stage 2	Design of attributes of data structure, software architecture, interfaces and algorithms	

Stage	Activities	Model Deficiency
Stage 3	Code Creation (Software implementation) by translating the design conceptualised	
Stage 4	Testing the internal logic of algorithms; tests for external functions and detecting errors and bugs	

7.2.1 Use of Linear Sequential Model (Waterfall Model or Lifecycle Model) for the Software Development Process

Figure 7.1 shows the model. Stages in development are as per Table 7.2. The linear sequential model is a classic model of the software engineering approach to development processes. Software development progresses through four incremental stages sequentially. Its path is like a waterfall and hence it is

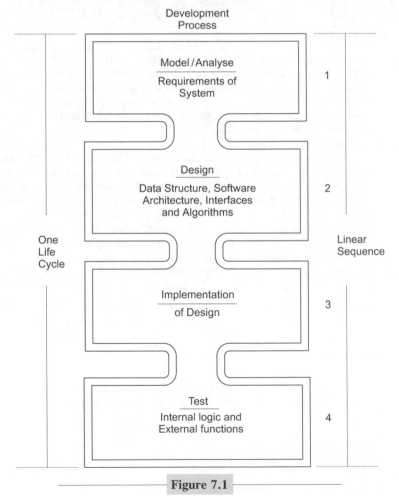

Figure 7.1

Linear Sequential Model (Waterfall Model or Lifecycle Model) for Software Development Process.

also called a waterfall model even though there may be the looping back to the beginning on uncovering of errors, bugs and malfunctions. The model proceeds from a start to an end, and hence it is also called lifecycle model. The development cycle repeats the four stages as in a lifecycle until software is finally validated (qualifies all tests and proven as per required specifications).

7.2.2　Use of the RAD (Rapid Development Phase) Model

This high-speed development model is an adaptation of the linear sequential model for the development of each software component separately by the team. The RAD stages in development are as per Table 7.3.

Table 7.3
Stages in Software Development by RAD Model

Stage	Activities	Model Deficiency
Stage 1	Business Modeling for information flow for business functions	Requirement of strong commitment to rapid development from each member; the necessity for more team members in case of large number of software components; lack of readily available reusable components, which can delay the process; and the risk of bugs in rapid approach
Stage 2	Characteristics and attributes by sets of data structures and objects, data modeling designs of attributes of data structure, software architecture, interfaces and algorithms	
Stage 3	Process modeling creating descriptions of adding, deleting, modifying and retrieving data objects	
Stage 4	Code creation by reuse of available software components	
Stage 5	Testing stage is short due to reuse of previously tested software components. Interfaces the internal logic of algorithms and tests for external functions and detecting errors and bugs.	

7.2.3　Use of the Incremental Model

There are multiple efforts to deliver the software incrementally. First-effort designed software is delivered first and next-effort designed software is then delivered. Successively improved versions are delivered. Incremental efforts for successively delivering improved software are as per Table 7.4.

Table 7.4
Efforts in Software Development by Incremental Model Adapted from Linear Sequential Model

Effort	Activities	Model Deficiency
Stage 1	Same as linear sequential for analyzing, designing, coding and testing	Need of strong commitment to new and incrementally better software with additional functions, and the need for foreseeing future requirements
Stage 2	When first-effort is at designing stage, the second-stage effort cycle starts analyzing for the next incremental software	
Stage 3	When first-effort is at the coding stage and second is at designing f, third stage effort cycle starts analyzing for third incremental software	
Stage 4	When first-effort is at testing stage, second is at coding stage and third is at designing stage, fourth stage effort cycle starts analyzing for fourth incremental software	

7.2.4   Use of the Concurrent Model

The Concurrent model takes all the stages concurrently for each element of the software development process that forms part of the product. For example, consider process analysis stage activities. Each of its elements can be (i) 'under development' (ii) 'awaiting change' (in second and successive cycles) and (iii) 'under revision' (iv) 'done'. Client-server process architecture is well suited for concurrent development during a software process.

7.2.5   Uses of Component Based (Object-Oriented) Software Development Process Model

The component based Object-Oriented Software Development Process Model has stages as per the list in Table 7.5.

Table 7.5
Components Based Object-Oriented Software Development Process

Effort	Activities	Model Deficiency
Stage 1	Components that could be used in software development identified	Need for robust
Stage 2	Selection of available classes (single logically bonded groups) from a software components resource library	interfaces for the components and
Stage 3	Sort available components, components reusable by re-engineering, and unavailable components	slow development in case the reusable
Stage 4	Re-engineer components and create unavailable components	components are not
Stage 5	Construct software from the components and test them	available in required
Stage 6	Repetition of test and construct until final validation of software	numbers

During a development process team members can follow the concurrent engineering approach. Members engineer the components concurrently. For example, one team can engineer the device drivers, another the error handling routines and a third, the application software.

Refer to Section 7.10 for methods of object-oriented design and a modeling language for representing the design.

7.2.6   The Use of Fourth Generation Tools-Based Software Development Process Model

Software tools are available that can generate codes according to higher-level specifications. These tools are called Fourth Generation Tools. Examples are the tools for automatic report generation, automatic high-level graphics generation, creating data base queries and automated HTML code-generation when creating a website. RTOS tools are available for embedded systems.

Use of fourth generation tools can quicken the development processes and provide the tested and debugged solutions available with it. Its drawback is the need for detailed analysis, design and validation.

7.2.7 The Use of Object-Oriented Based and Fourth Generation Tools Based Approach

Currently, software development processes combine object orientation and fourth generation tools. This approach has yet to be perfected for embedded systems.

A successive refinement model is a model in which the process lifecycle is repeated until verification, validation and delivery (or installing the ROM of the system).

Activities during software development are divided into three phases: Definition and Analysis of system requirements, development of designs, code and support. A software development process can be represented by a cycle called lifecycle. From the last stage, the cycle can restart iteratively from the first stage until verified and validated software is completed. Refinement is by successive iterations. There are several software-development process models, which are in use. Activities during seven of these have been discussed.

7.3 SOFTWARE ANALYSIS

Important points in software analysis are as follows:

1. Requirements analysis is the first step in a software development process. [Refer to Table 7.2.] Firstly, general statements are made and then the technical concrete specifications are obtained from them.
2. Any problem can be modeled as consisting of three domains: information, functional and behavioral domains.
3. A problem can be divided into several parts and represented by suitable notations that describe the essence of the problem.
4. Many times, it may not be possible to completely specify the problem at an early stage.
5. Prototype development and prototype specifications can often help.
6. Even with the use of the best analysis and models, problems may change. Therefore communication between the developer team and the customer is essential so that, before the design process starts, both have identical perceptions of the problem, domains, requirements and software and system specifications.

Analyses can be modeled as three-layered structures. (i) Data and its contents and information form the inner layer. (ii) The next layer is 'DFGs', 'CDFGs', 'Entity relationship diagrams' and 'State transition diagrams'. (iii) Data, control and process names, notations and specifications (descriptions) form the highest layer. There may be specifications and descriptions also for the protocols in embedded networking systems.

Implementation specifications—how the messages shall pass between different objects—may also be described by the analysis. Implementation can be viewed at the analysis stage while defining the requirements.

Table 7.6 lists the activities during the use of various models and during obtaining the specifications in the three layers that are used for analyzing the software-requirements. It also gives examples for each stage of activity.

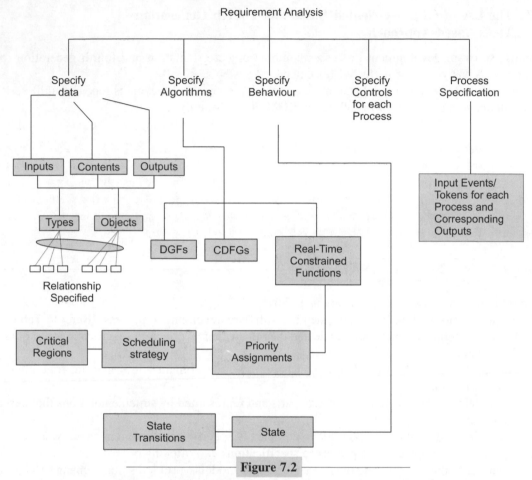

Figure 7.2

Activities for Software Analysis during an Embedded Software-Development Process.

Table 7.6
Analysis of Requirements by Models and Specifications

Modeling	Activities	Example
Data	Data types and objects defined. Attributes known. Relation between the data analysed and required specifications gathered. Entity Relationship modeled.	Network data modeled as a block of queues. Protocol types used at each layer of transmission error checking fields. Transmitter and receiver buffers.
Algorithms (Functions)	**(i)** DFGs for information flow, **(ii)** CDFGs for program flow **(iii)** Real time functions and their priorities, preempting strategies and critical regions	Algorithms for **(i)** application data flow through transport and internet-work layers in TCP/IP networks **(ii)** Control algorithms for establishing network connections, transmission and connection termination **(iii)** Real time functions in automotive cruise control.

Modeling	Activities	Example
Behaviour	States and state transition diagrams	Vending machine behaviour when waiting for coin, when coin is inserted (input event), process activated (internal machine actions) and when chocolate is delivered (output). Also refer to Figures 6.3 to 6.6 for examples.
Data content Specifications	Data content naming, input sources (files), output sources, notations and meanings and specifications	Frames for video processing. Image and sound format names. Input system name. Compression format specifications. Compressed file names and specifications.
Control Specifications	Control specifications analysed for normal and real time functions	Robot motors control specifications and what the controlling functions are specified in real time.
Process Specifications	Process specifications analysed by making lists of inputs on events list, outputs on events, processes activated on each event.	Cruise control process specification, how cruising speed measured, analysed and used for controlling the throttle valve-orifice. [Section 11.3].

Figure 7.2 shows the activities for software analysis during an embedded software-development Process.

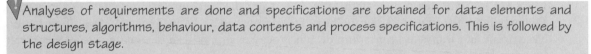

Analyses of requirements are done and specifications are obtained for data elements and structures, algorithms, behaviour, data contents and process specifications. This is followed by the design stage.

7.4 SOFTWARE DESIGN

The reader is referred to Table 7.2 for development phases and a development process lifecycle model. The next stage is design, after software analysis for requirements and specifications is done. Design means a blueprint for software development. It gives an engineering representation of the system to be built. Software design consists of four layers.

1. The first layer is an architectural design. Here, a design of system architecture is developed. The question arises of how the different elements—data structures, databases, algorithms, control functions, state transition functions, process, data and program flow—are to be organised.

2. The second layer consists of data design. Questions at this stage are as follows. What shall be the design of data structures and databases that would be most appropriate for the given problem? Would data organised as a tree-like structure be appropriate? What will be the design of the components in the data? [For example, video information will have two components, image and sound.]

3. The third layer consists of interfaces design. Important questions at this stage are as follows. What shall be the interfaces to integrate the components? What is the design for system integration? What shall be the design of interfaces used for taking inputs from the data objects,

structures and databases and for delivering outputs? What will be the ports structure for receiving inputs and transmitting outputs?

4. The fourth layer is a component-level design. The question at this stage is as follows. What shall be the design of each component? There is an additional requirement in the design of embedded systems, that each component should be optimised for memory usage and power dissipation.

What are the concepts used during the design process? They are as follows.

1. Abstraction. Each problem component is first abstracted. For example, a robotic system's problem abstraction can be in terms of control of arms and motors.

2. Software architectural structural-properties should be understood before the design stage.

3. Extra functional properties from the system being developed should be well understood from the design.

4. Families of related systems developed earlier are taken into consideration during designing.

5. Modular design concepts are used. System designing is the decomposition of software into modules that are to be implemented. Modules should be such that they can be composed (coupled or integrated) later. Effective modular design should ensure effective **(i)** function independence, **(ii)** cohesion and **(iii)** coupling.

6. Modules should be clearly understood and should maintain continuity. For example, in the design for a module for a *list* (Example 5.8) the modules are designed to be inserted into the *list* before as well as after a list element.

7. Also, appropriate protection strategies are necessary for each module. A module is not permitted to change or modify another module functionality. For example, protection from a device driver modifying the configuration of another device.

8. Mapping into various representations is done from the software requirements. For example, DFGs lying in the same path during the program flow can be mapped together as a single entity. Transform and transaction mapping design processes are used in designing. For example, an image is input data to a system; it can have a different number of pixels and colors of each pixel. The system cannot process each pixel and color individually. Transform mapping of image is done by appropriate compression and storage algorithm(s). Transaction mapping is done to define the sequence of the images.

9. User Interface Design is an important part of design. User interfaces are designed as per user requirements, analysis of the environment and system functions. For example, in a vending machine system, the user interface is a LCD matrix display. It can display a welcome message as well as specify the coins needed to be inserted into the machine for each type of chocolate. Interface design has to be validated by the software-customer. For example, the customer must validate message language and messages to be displayed before an interface design can proceed to the implementation stage.

10. Refinement. Each component and module design needs to be refined repeatedly until it becomes the most appropriate for implementation by the software team.

The software design process may require use of Architecture Description Language (ADL). It is used for representing the following: (a) Control Hierarchy, (b) Structural Partitioning, (c) Data Structure and hierarchy and (d) Software procedures.

Figure 7.3 (a) shows the activities for software design during an embedded software-development process.

The software design stage involves abstraction-level to detailed-designing-level design activities. Software design can be assumed to consist of four layers: Architecture design, data design, interfaces design and component level design. There is a need for continuous refinements in the designs by effective communication between designers and implementers.

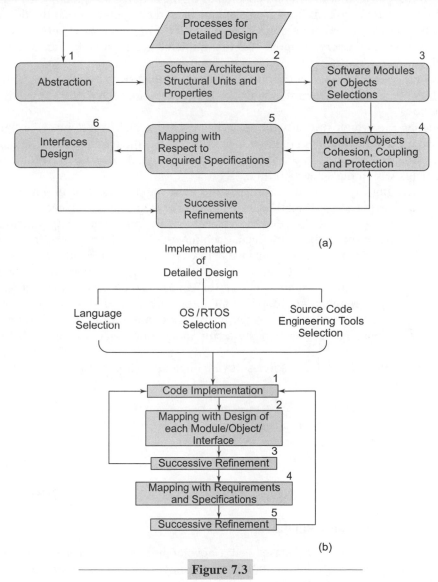

Figure 7.3

(a) Activities for Software Design during an Embedded Software-Development Process
(b) Implementation by Using Detailed Design.

7.5 SOFTWARE IMPLEMENTATION

The reader can refer to Table 7.2 for development phases and a development process lifecycle model. The next stage is software-implementation, after the designing that succeeds the software analysis for requirements and specifications. Implementation is real-world realisation after translating a software design. Figure 7.3(b) shows that the implementation using the detailed design implementation must be according to the software requirements and specifications agreed upon between the developing team and customer. Functions (codes) are written to accomplish information (or events) processing as per the design. Each module of every component is first implemented separately. Later they are coupled and interfaced.

Software is implemented using source code engineering tools. [Refer to Section 5.11 for details.] For software implementation, certain standards and procedures are followed. Here are the principles for implementation.

1. Use language in which the software team has expertise and experience. For example, C++ should be used if a team does not have Java skills but has the required skills and experience in implementing in C++. It is important to remember that good software must fulfil the specified requirements within the scheduled time.
2. Use an operating system or RTOS that is the most appropriate and in which the team has expertise and experience. [Small-scale embedded systems may not need RTOS in simple implementations. This will also keep the code size small without affecting functionality.]
3. Program simplicity should be maintained. Each function should perform a single task and have a simple design. Comments must be added before the control statements. This helps to make a program easily understandable by all team members. It helps in subsequent debugging of the program. Loops should be simple, nested, or concatenated. Unstructured loops should be avoided in any program and one loop intermingling with another should not occur.
4. Program implementation should be as per program structure design specifications.
5. Make the functions general, not tied to specific input conditions or events. This makes reuse easier.
6. Redesigning might be needed to solve software implementation difficulties.

Each component and its interfaces system integration are implemented after the design stage. Program implementation is in a language and may use source-code engineering tools. It should be as per program structure design specifications. Program simplicity should be maintained during the implementation process.

7.6 SOFTWARE TESTING, VALIDATING AND DEBUGGING

7.6.1 Testing, Verifying and Validating

Refer to Table 7.2 for development phases and a development process lifecycle model. The next stage—after analysis, design and implementation—is testing. Testing is an important part of the activity. Embedded software needs thorough testing. This is because the software is to be permanently installed.

The goal of testing is to find errors and to validate that the implemented software is as per the specifications and requirements. Validation means that software must be shown to be according to the specifications and must show behaviour as expected in all the events, conditions and times specified.

What is the most important feature of a good software test? If more errors are discovered at the initial 20% stage of testing, the greater the probability of errors remaining undetected. There will be a higher probability of finding error(s) not yet discovered. When an error is found which was not detected by any previous test, then a test is said to be successful. An important principle of any test is also that the software should meet the customer requirements fully. What is the most effective testing method? The most effective is testing by an independent third party.

Verification and validation have different meanings. **(i)** Verification refers to an activity to ensure that specific functions are correctly implemented. **(ii)** Validation refers to an activity to ensure that the software that has been created is as per the requirements agreed upon at the analysis phase and to ensure its quality.

Tests must be planned long before the testing phase starts. Pareto's principle is that 80% of errors uncovered during testing belong to 20% of the program. Individual components should be tested before the integration into the system.

Table 7.7 lists five test approaches.

Table 7.7
Test Approaches

Approach	Activities	When used	When not needed or feasible
White Box Testing	White box (also called glass box) testing means a close examination of each procedural detail by testing the program flow at the independent paths. Testing is such that **(i)** all decision nodes (Section 7.2) are traversed at least once. It ensures that all conditions are tested. **(ii)** All loops are executed from the starting index to the last. **(iii)** All logical paths are thoroughly tested. Testing is done by test inputs. Test input design sets are crucial and they are designed for the specific sets of loops and conditions. Testing is for control structures, conditions, data flow and loops ('for loops', 'while-do', 'do-until').	White box testing is used in early testing phases and when not reusing the previously developed components. Its need is due to the nature of software defects. Logical error and incorrect assumptions can often be made and these are inversely proportional to the probability of execution of a program path. It could be assumed that a logical path will not be traversed and no situation will arise causing that path to be followed. However, it does arise eventually.	When the number of tests and paths are too large, white box testing is not practical. It should then be applied to a limited number of logical paths that are critical to a system and program logic should be adequately exercised.
Black Box Testing	Black box testing (Behavioural testing) must supplement white box testing. It uncovers a class of errors different from those detected by the white box. Testing	When the focus is on testing the functional requirements and there are tight delivery schedules. Fast but	Early stages of development. Availability of sufficient time for thorough white box testing.

Approach	Activities	When used	When not needed or feasible
	is done by choosing the most appropriate set of input conditions and generating events that test functional validity. Black box testing means testing of behaviour and performance. Black box testing is for: discovering whether the system response is sensitive to particular inputs and system malfunctions at these inputs and events; when the values are beyond the set boundaries in the system, how the system responds; whether there is any specific combination of data objects, inputs or events leading to malfunction; whether the data inputs reaching at faster rates modify the system performance. Black box testing looks only at the relationships between a graphical-collection of nodes. It is for boundary values analyses, not for all in-range inputs. Each set of nodes is examined to describe functions and relationships.	reliable black box test procedure has been designed by the development team.	
Specific Environment Testing	Examples for environment specific testing are as follows. (i) Testing user interfaces and GUIs: the embedded system's LCD matrix display and keypad may be the GUI and user interface, respectively. The testing exercise is for finding the effect of each command from the keys and for proper display of messages. (ii) Testing Client Server architecture: server behaviour for different inputs from each client is tested. Also checked: Whether the response to a client request follows specifications. (iii) Testing Help Utilities and Documents: A system provides Help Utilities for the user. These need testing.	When a system is complete and now has a specific environment to test its behaviour. For example, mobile phone system. It needs to be tested by dialing a number. A robot system needs to be tested by specified reaches of the robot arm at the specified locations.	Early stage and incomplete system.

Approach	Activities	When used	When not needed or feasible
	Documents also need testing to ascertain whether documented functions are working well or even exist.		
Comparison Testing	Several available versions are run in parallel and the functions and behavior are compared.	For example, an automobile's brake system's reliability is absolutely essential. New software is tested for better reliability compared to previous ones.	Comparison systems not available. For example, embedded software is ready but hardware is still under development.
Orthogonal Testing	Testing is done at every combination of values.	Inputs are bounded and every combination can be tested.	When the number of combinations of inputs are large.

How can one ascertain whether the implemented software is testable or not? Testability criteria are as follows: (i) It should be possible to decompose the system so that each system component becomes simple and therefore easier to test. (ii) Observation must be made possible by distinct outputs for each input so that any incorrect output is easily identified and internal errors can then be automatically removed. (iii) The test designing team must understand the designed software well. (iv) Dependencies between internal, external and shared components must be well understood. (v) Changes in design must be communicated to all necessary components. (vi) Technical documents must be well organised, specifically for the detailed tests.

When can a test be more effective? A system can be tested more effectively if it works well. For example, if none of the robotic motors move during a test, more thorough testing is needed; and when a single motor does not move while all others function normally, the test should only be for the module which drives that motor. The test becomes easy and, thus, more effective.

Embedded software needs to be of the highest quality and therefore demands a more systematic approach to testing. What are the strategies adopted for testing?

(i) Testing of each module or unit and component-wise testing.

(ii) Smoke testing exercises all systems from beginning to end such that it is not exhaustive and the software team can access the project regularly at each stage, throughout. As soon as a component is ready, it is tested, and as soon as a set of components are ready they are integrated at periodic intervals and tested periodically.

(iii) Validation testing is validating the software. It is validated after a series of black box tests that show requirement conformity. Validation testing targets the following: (a) Performance is tested. On testing, each required function is shown to possess its characteristics that are in conformity with the requirements. (b) Finding the deviation from the required specification, preparing a list of deficiencies, and creating a deficiency list. (c) Security aspects are tested. This could be done when there are error conditions. For example, when a queue overflows its

size, the software must execute the routine for 'exception'. [Refer to the ISR_Qerror in Example 5.5 and Section 4.5.2 for definition of an *exception*.] There are two types of validation testing: (a) Alpha testing, which is at the developer site, and (b) Beta testing, which is at the embedded system user site.

(iv) Recovery testing is done to determine how a system recovers when software or hardware fails.

(v) System testing means integrated system testing, either by top-down, bottom-up or by regression testing (exercise subset of all test cases).

Testing a system can have several meanings. For example, unit testing, integration testing, design testing and requirements testing.

Embedded system testing is done in two parts: (i) testing during the software development process, on the host machine only, and (ii) testing along with the target system. The latter part is taken up in Chapter 12.

Like software complexity metrics, there exists a software-test metric. It uses (i) Data elements or Objects, (ii) Relationships between the data or objects, (iii) States, (iv) Transitions and (v) A number of functional points. Computations for test metrics can be understood from the following example.

Example 7.2

1. Let (a) k_1 be the number of operators that are distinct and are appearing in n-th software module. (b) k_2 = number of operands that are distinct and are appearing in the module. (c) K_1= Number of times the k_1 operators occur in the module and K_2= Number of times the k_2 operands occur in the module. (d) r = ratio of occurrences of operators per operator appearing in the program and S = size of software.

2. Now, $c = r * k_1 /2$ defines the programming effort coefficient. The logic for so defining is as follows The more times an operator occurs in a module, the greater is the testing effort. If the ratio of occurrences of the operands is larger, the greater will be the efforts needed for testing. Division by 2 is done because the operation is between two operands.

3. Testing effort metric for a module t_e (n) will be = c* S. The logic for this definition is: the greater the module size, the greater the total test efforts.

4. Software testing-efforts metrics, $T_E =\Sigma t_e$ (n), where n = 1, 2, ..., N for N modules in the software.

7.6.2 Debugging

Testing is the activity that detects faults and errors, while debugging means finding the reasons for a fault. Debugging means that once an error (or set of errors) is detected, the problem-creating sources and causes must be identified and the bug present in the software corrected (fixed). The presence of bugs is intolerable in embedded software. Consider a sophisticated embedded system for a rescue spacecraft (Section 1.5). The cost of any bug left uncovered in the system can be imagined!

A fact of life is that a specific software engineer may be able to debug a program easily and another may not. Debugging can be considered an art. Standard guidelines for debugging are given below.

1. Use suitable strategies. Available strategies are as follows: (i) Use early stage testing and debugging. (ii) Debug and locate the problem using a binary portioning method, also called cause-elimination method. It means to first divide the modules in two halves for the expected

sources. Then subdivide into another two halves. For example, in debugging a TV fault, first divide into two sections—video and audio. If an audio problem exists, then divide the audio section circuit into two halves—detector circuits and amplifier circuits—and continue until the problem source is located. (iii) Use backtracking method. Source is found by backtracking along the program flow path. (iv) Use brute force method. A brute force method example is source found by white box testing (Table 7.7).

2. Develop a working hypothesis such that it is possible to predict new values that are to be examined to trace the source during debugging.
3. Develop test modules that can be repeatedly used.
4. Examine previously logged test results.

Table 7.8 gives the various embedded-system debugging techniques during and after the software development process.

Table 7.8
Software Development Process Debugging Techniques

Approach	Activities	When used	When not needed or feasible
Use of breakpoints	The program flow breaks at a certain point and the output at that stage is observed or tested by a well-designed set of inputs. These points are called breakpoints. Breakpoints are added during the software implementation process. The breakpoints are deleted after final verifications and final validation.	During the development stage, when using white box testing and early testing phases and when not reusing the previously developed components	When the addition of breakpoints delays the execution of software unlimitedly
Use of test macros	Results from the macros specially written for the tests help embedded software debugging greatly. [Refer to Section 5.3 for understanding macro in C.]	When test functions are also added in the software development process to ensure efficient debugging	Addition of test macro adds to greater code size beyond the available memory
Use of output files for sampled inputs	A set of sampled input files are designed for the debugging phase and the outputs that the system generates are observed.	Software development process phase and smoke testing activity	Not possible to create inputs such as events
Use of instruction set simulators	Refer to Section 12.3	When testing assembly codes, finding an instruction set throughput, testing device driver codes, examining portability on specific hardware	Software codes that are not CPU-specific codes
Use of laboratory tools	Refer to Section 12.5	Software performance is tested on the target system after embedding it in hardware	Software development process

Caution Bug fixing should not lead to the creation of additional bugs.

Figure 7.4 shows testing, debugging, verification and validation activities for software during an embedded software–development process.

The testing stage follows the implementation stage. Testing approaches are white box testing, black box testing, specific requirement tests, comparison testing and testing by orthogonal data inputs. Tests verify that the software developed after processing through the lifecycle stages is as per required and agreed specifications. Validation of required specifications is done at the last stage before delivering to the customer. The cycles may have to be iterated according to successive refinement models if validation fails. The observed error source is traced during debugging by use of breakpoints, test macros, test vectors, simulators and laboratory tools.

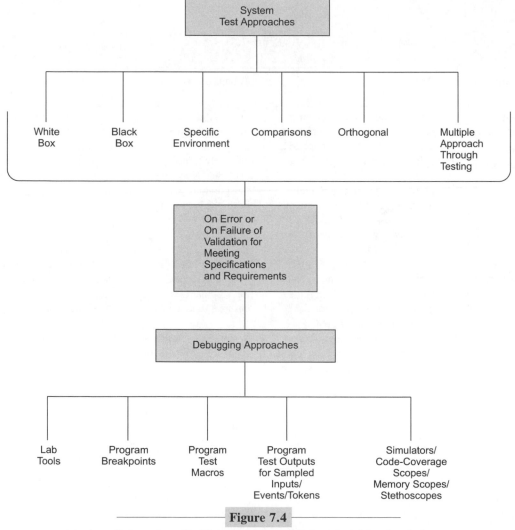

Figure 7.4

Testing, Debugging, Verification and Validation Activities for Software during an Embedded Software–Development Process.

7.7 REAL-TIME PROGRAMMING ISSUES DURING THE SOFTWARE DEVELOPMENT PROCESS

A real-time system is system that has an additional programming requirement that it has to meet response time (deadline) constraints in one or more of its functions. (Section 6.2 for real-time program models). Therefore, real-time programming has the following additional issues.

7.7.1 Issues in Analysis of Requirements and Specifications

(i) Events are asynchronous in nature. What are the maximum latencies and time dependencies of events? What are the program flow–control specifications on occurrences of real-time events?
(ii) What are the specifications of the acceptable maximum response times (time intervals between set of inputs and outputs) of each software function (or component)?
(iii) What are the specifications of the acceptable maximum latencies of the Interrupt Service Routines (ISRs) and what are their servicing deadlines? [Refer to Section 4.6.]
(iv) What are the specifications for the software components (tasks) and interrupts priorities? What are the allocations of priorities to those with shorter deadlines and what are the routines that can await execution at a later stage?
(v) IPCs specifications. [Refer to Chapter 8.]
(vi) What are the risks of failure to meet a deadline and what are the recovery methods in case of failures?
(vii) Reliability and acceptable fault-tolerances. [Refer to the spececraft example in Section 1.5.]

7.7.2 Issues in Design and Implementation

(i) Selection of processor, memories and their sizes, and hardware. [Refer to Sections 1.3.15, 2.2 and 2.4.]
(ii) Selection of single-processor or multiprocessor systems.
(iii) Selection of software design such that there is memory and power dissipation optimisation in the system and hardware-software design is optimised.
(iv) Selection of software language, C or C++ or Java. [Refer to Sections 5.1, 5.8 and 5.9.]
(v) Decision to use a RTOS (real-time operating system) and to design it if needed. [Refer to Chapter 9].
(vi) Decision to design RTOS by the software team or to use the readily tested and debugged RTOS. [Refer to Chapter 10.]
(vii) Design and selection of tests and debugging breakpoints and macros (test-vectors). [Tables 7.7 and 7.8.]
(viii) Selection of uncompressed or compressed software and input data in the system ROM (Section 2.4).
(ix) Decision of whether to use caches and, if used, in what sections and data to manage the power dissipation.
(x) Use of stop-and-wait states in the software to manage the power dissipation.

7.7.3 Issues in System Integration

(i) Integration and development of a unified system by appropriate interfaces.
(ii) Verification of interfaces and the integrated system.

7.7.4 Issues in Testing

As inputs, events and interrupts are time dependent and asynchronous, real-time system software is difficult to test compared to non-real-time software. Beyond white box, black box and smoke testing, other types of testing are required for handling of asynchronous events by the system. Real-time system testing, debugging and verification may offer real challenges. Following are the issues:

(i) Testing whether the real-time data processes properly in one state but in other states fails.

(ii) Selection of testing strategies: (i) Task Testing by testing each task independently so that it not only detects logic errors and functional errors but also errors in meeting timing constraints and in system behavior. (ii) Behavioral testing by simulating the real-time system behavior and studying the behavior(s) on external events.

(iii) Effects of events—testing for each event individually and by pairs and sets of events.

(iv) System testing for hardware and software integration.

(v) Test for proper interrupt-priority assignments.

(vi) Test for each ISR for each of its interrupt source and source groups.

(vii) Testing of throughputs from the tasks and ISRs.

(viii) How does a high number of interrupts arriving at critical times affect the system behavior and performance?

(ix) Testing of timings by performing timing analyses and thorough testing for fulfilling the response time constraints and meeting the deadlines of all ISRs, tasks and functions in all possible sets of events and time intervals between them.

(x) Testing of parallelism and concurrent processing of the tasks, functions and ISRs.

(xi) Testing of IPCs (Chapter 8).

(xii) Testing for memory overflows and stack overflows.

(xiii) Testing of program flows in critical regions for shared data, priority inversion and deadlock problems (Section 8.2).

(xiv) Testing of software portability in hardware.

(xv) Testing of power dissipation optimisation within the response time constraints.

(xvi) Testing for fault tolerances and failure recoveries.

> The real-time systems development process is more complex as there are software response time constraints for the algorithms and ISRs. Testing is to be done for timing analysis, for IPCs and for asynchronous event(s) and their sets at each state of the system. Test for fast-occurring events is also done. Testing has to be thorough; tests for fault tolerances and failure recoveries are also essential.

7.8 SOFTWARE PROJECT MANAGEMENT

7.8.1 Project Management

Pressman defines four Ps in his book—people, product, process and project—for software project management. Table 7.9 lists these four elements and gives their functions. [People in fact means here a team of software engineers or system development engineers.] Software project management principles are as follows: (a) Organisation of the **people**. (b) Development perception correct for analysis, design, implementation and testing when developing the **product**. (c) Monitoring of **process** adopted at each stage. (d) Organise and manage the **project** so as to ensure its success.

Table 7.9
The Four Components of Software Project Management

Component	Roles	What is not advised
PEOPLE Senior Manager	Responsible for effectively facilitating, creating environment, organizing, coordinating and managing all communications and organisational activities. Monitoring and tracking the project and its schedules.	**(i)** Leaving the basic principle that it is the people that matters most in any project **(ii)** Unbalanced approach and unnecessary controls
Project Technical Manager or Team Leader	(i) Selects language, tools and software development process life-cycle model for the process. (ii) Tunes and reorganizes available software specifications, designs and components and existing processes. (iii) Finds new ones that enable initial development perception results into final product within prescribed schedule. (iv) Maintains activity graph for starting, duration, due date and end-points for each activity and control breakdowns. (v) Motivates and encourages implementers. (vi) While executing team-leader guidelines and working within the defined boundaries of the process and its development schedule, lets ideas be generated or new ways be discovered by implementers even when working within bounds of the given product.	Lack of appreciation of the implementer's perceptions, and uncoordinated development.
Implementers	Implements the software (and hardware) development process and uses modeling, source code engineering, testing, simulating and debugging tools.	Not following the agreed and accepted design and lack of coordination among fellow implementers.
Customer of the software (or embedded system)	Specifies the product and its quality requirements and negotiate cost with senior manager(s).	Interference in development process, changing the product specifications after agreeing to them.
End-users	Uses the product within the suggested boundaries.	Not using product as per guidelines. For example, inserting incorrect currency into a vending machine.

Component	Roles	What is not advised
PRODUCT (embedded system)	Correct perceptions of the requirements, needed functions and real-time behavior of the product is necessary. It is done by communication between customer and managers. The scope, context, and objective of the product must be defined correctly and expected performance must be agreed upon.	Lack of correct product specifications.
PROCESS	The development process must be partitioned into a number of problems, components and modules (also data structures and objects) so that activities defined in the activity graph for each activity are finished. How can the process be partitioned such that it adapts to the people and the problem? Selection of the process adopted during development lifecycle must be as per available people and required product. Which is the model to choose? Consider any of the seven given in Section 7.2 or some variation or adaptation from these.	Incorrect partitioning and adaptation of incorrect model.
PROJECT	The software project management goal is to organise the project so as to ensure its success. The success can be ensured by (i) Thorough planning and by estimates (Section 7.7.2) of effort and time taken by each member for each activity for the product. (ii) Clearly defining project test points and quality checkpoints. (iii) Setting effective monitoring and control mechanisms. (iv) Coordinating communication issues and their solutions. Project success is obtained by the following: (i) Working systematically to understand the problem. (ii) Setting realistic activity graphs and clear-cut objects and reasonable expectations from each element. (iii) Maintaining proper development momentum and continuously keeping track of people, products and process activities. (iv) Making smart decisions at the right time. (v) Learning lessons from earlier failures and successes at each element.	Improper planning, incorrect effort estimates and lack of success-oriented focus, keeping people busy in non-project activities.

7.8.2 Project Metrics

Is there a way by which the magnitude of the size of a project and the effort required could be measured in the beginning of a project? Project metrics for software are for **(i)** KLOC (Kilo lines of code), **(ii)** Pages of documentation per KLOC, **(iii)** Errors per KLOC, **(iv)** Defects per KLOC and **(v)** Cost of documentation per page. These give the estimates of (a) Cost per LOC, (b) Error per person-month, and (c) LOC per person-month.

Example 7.3   Consider an embedded system in a car. Assume that following are the external interfacing devices and C functions in the system. Assume a system is being developed to monitor **(i)** Watch, **(ii)** Engine speed, **(iii)** Vehicle speed, **(iv)** Odometer, **(v)** Fuel sensing, **(vi)** Alarm, **(vii)** Illumination, **(viii)** Automatic gearbox, **(ix)** Secured indicators control (recording the car use and malfunctions data for retrieval by mechanics later for servicing and maintenance) and a few other activities. Table 7.10 gives a typical LOC estimates during a project. Table 7.11 gives a format for defects planning and productivity and cost estimates.

Table 7.10
Estimates of Function Size for Coding in C

Function size in Line Of Code (LOC)*			
	Minimum	Average	Maximum
Watch	828	1254	1680
Secured indicators control	120	168	217
Engine Speed	320	399	479
Vehicle Speed	435	557	680
Odometer	958	999	1040
Wake-Up/Sleep-Down	200	252	304
Fuel	360	376	392
Watchdog	160	188	217
Alarms	4640	4781	4922
Coolant temperature	435	537	640
Illumination	1611	1925	2240
Automatic Gear Box	174	207	240
Total	10241	11643	13051

Note: (1) *LOC includes the following:

Commentary	100%
includes ( )	100% !
lines not concluded by semi column	100%
blank lines	0%
special statements like macros	100%

Note: Project metrics failed because of these estimates. Why? It used earlier research estimates and the team did not have any perception of the nature of the efforts needed in assembly code sections of C algorithms in the current project.

Table 7.11

	Function Size in Line Defect Rate Planned in LOC		
	Minimum	Average	Maximum
After First Delivery Bug Contention After Thorough Review			
	Function size in Line Productivity Plan in LOC		
	Minimum	Average	Maximum
LOC Per Hour			
	Estimate Load and Price		
Load Time Per Day Price Per Planned Day			

How does one calculate the estimates for LOC? There is a function-oriented metric, called FP (function points) and called FFP (Full Function Points) for the total of all FPs in the project. Calculation of function points is by counting and then multiplying counts suitably by three weighting factors (simple, average and complex) for the three columns in the Table 7.10. The following function points $F(i)$ are used for finding the i-th counts and then count total. Three weight factors for multiplying a count for an $F(i)$ are given in brackets. These factors are based on the experiences with a large number of projects and have been adapted for the project metrics.

(i) Number of external interfaces (5, 7 and 10)
(ii) Number of user inputs (3, 4 and 6)
(iii) Number of user outputs (4, 5 and 7)
(iv) Number of user enquiries (3, 4 and 6)
(v) Number of files (7, 10 and 15)

FP = count total $*[0.65 + 0.01 * [F(i)]$.

LOC per FP depends on coding language. For C codes, LOC per FP is 128, Assembly = 320 and C++ = 64. It is used in productivity function metrics and for generating a table like Table 7.11.

F must incorporate complexity adjustment values in case of data communication, critical performance, and reusability of code, master file online updating and number of interrelated parameters.

According to Pressman, software project management can be said to consist of management of people, product, process and projects. For planning stage estimation of program efforts, project metrics are available. One such metric is by use of function points and use of FFPs (Full Function Points) to estimate the KLOC (Kilo lines of codes) and productivity per person-month.

7.9 SOFTWARE MAINTENANCE

The lifecycle of software development process ends with delivery, but its life itself does not end there. Maintenance involves changes, removals and additions in the objects or modules after delivery (or installation in ROM in case of embedded systems). The maintenance requirement is inversely proportional to software complexity, as a simpler system needs little maintenance. Table 7.12 gives types of maintenance that may be needed to maintain a system.

Table 7.12

Types of Software Maintenance

Type	Activities	When used	When not needed or feasible
Preventive Maintenance	A system may be checked and maintained periodically.	Wearing of hardware components, for example, a printing device system and hence the need to modify the interface(s).	Difficulties in specifying the necessary preventive actions.
Corrective Maintenance	Correct the deviation noticed under certain specific conditions of field-use of the system.	Product specification from the customer being incomplete and poor customer perception shows the need for corrective maintenance.	System working satisfactorily.
Adaptive	Adapt the software to new conditions. ["Shirt adaptive maintenance" by a tailor creates a slimmer look!]	A robot may have to be adapted to a change in the specifications of its arm reach and range of operations. An automatic chocolate machine may have to adapt to the introduction of new types of coins in the country's currency.	Complex systems may be more difficult or costly to adapt compared to designing a new one.
Enhancement or Perfection Maintenance	Development team delivers a system on schedule and then finds another design capable of working more accurately and effectively.	New design development and availability of new tools with the developer so that better systems can be delivered.	Perfection process can never end and system must have only the acceptable fault tolerances.
System Reengineering	Merely delivering a system never satisfies an innovative team. It does reengineering of previously developed objects and components.	Re-engineering is another way to keep a given system up to date.	Complex systems may be difficult to re-engineer.

When can a software or system be called matured, needing no further maintenance? IEEE has given a standard, which answers this question—the IEEE 962.1 standard, called SMI (Software Maturity Index). When SMI approaches 1.0, the software is matured and needs no further maintenance. Let software have C_c, C_r and C_n—changed, removed and new objects or modules, respectively. Let N_c be total objects or modules in its current release. C_c/N_c, C_r/N_c and C_n/N_c will be the ratios r1, r2 and r3 that reflect the degree of changes, removals and additions per unit objects (or modules). Therefore SMI = (1 − r1 − r2 − r3).

Embedded systems should be engineered to need minimum post-validation and delivery mainte-nance. However, due to improper customer specifications that later require corrections, chang-ing conditions, the need for adaptation, and the need for improved systems or system perfec-tion achievement, software maintenance may be required. Software with greater complexity needs greater effort during maintenance. The IEEE 962.1 standard, called SMI (Software Matu-rity Index) is used to measure maintenance needs. When SMI approaches 1.0, the software is matured and needs no further maintenance.

7.10 UNIFIED MODELING LANGUAGE (UML)

Section 5.8.1 dealt with the following:

1. Object-oriented language is used (i) when there is a need for reusability of the defined object or set of objects that are common within a program or between the many applications, (ii) when there is a need for abstraction and (iii) when, by defining objects by inheritance, new objects can be created. There is data encapsulation within an object.
2. An object is characterised by its identity (a reference to it that holds its state and behaviour), by its state (its data, property, fields and attributes) and by its behaviour (operations, method or methods that can manipulate the state of the object).
3. Objects are created from the instances of a class. Defining the logically related group makes a class. Class defines the state and behaviour. It has internal user-level fields for its state and behaviour. It defines the methods of processing the fields.
4. A class can then create many objects by copying the group and making it functional. Each object is functional. Each object can interact with other objects to process the states as per the defined behaviour.
5. A set of classes then gives an application program.

Object-oriented designing is also done as above.

1. Object-oriented design is done when there is a need for reusability of the defined software components as object or set of objects (reusable components). A new component can be abstracted from the existing. New components and object designs are created by the object inheritances and polymorphs. There is information encapsulation within a designed component or object.
2. A designed-component object is also characterised by its identity (a reference to it that holds its state and behaviour), by its state (its designs for data, property, fields, attributes and algorithms) and by its behaviour (method or methods that can manipulate the state of the design).
3. New object-designs are created from the instances of a designed class.
4. A designed class can then create many component objects (designs) by copying the group and making designs functional. Each design is a functional design. Each object design can interface with other designs to process the states as per the defined behavior.
5. A set of classes then gives the complete software design for a system.

Can we have a unified (common) modeling language for any general system for which object-oriented analysis and design are feasible and which can be abstracted by models? Unification in UML means its common applicability to many designs or processes. We can then model by a similar set of diagrams the following: (i) Software Visualizing, (ii) Data Design(s), (iii) Algorithms design(s), (iv) Software Design(s), (v) Software specifications, (vi) Software Development Process and (vii) An Industrial Process.

UML is a language for modeling. Figures 7.5(a) to (f) show representations of six basic UML elements: class, package, stereotype, object, anonymous object and state. A conceptual design modeling can follow the UML approach. A conceptual design can use the 'User Diagram', 'Object Diagram', 'Sequence Diagram', 'State Diagram'', 'Class Diagram' and 'Activity Diagram'.

Table 7.13 gives its elements. Table 7.14 gives UML diagrams: 'Class diagram', 'State Diagram', 'Object Diagram', 'Sequence Diagram' and 'Collaboration diagram'.

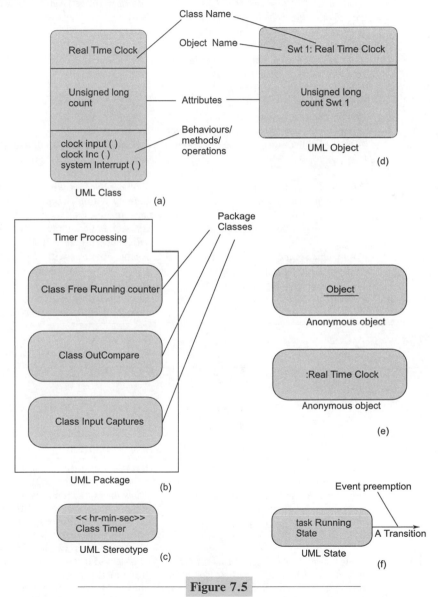

Figure 7.5

Representation of UML Basic Elements (a) class (active class and abstract or inactive class), (b) package, (c) stereotype, (d) object, (e) anonymous object and (f) state.

Table 7.13
UML Basic Elements

Modeling Diagram	What does it model and show?	Exemplary Representation
Class	Class defines the states, attributes and behaviour. A class can also be called active or abstract. An active class means a thread class that has a defined state, behavior and attributes and has the instances of it as objects. A class in general may be abstract when it is in one or more states, when operations or behavior are not completely defined, when in an abstract stage, or when it is not for creating objects but only its inheritances (extensions) can create objects.	Rectangular box with divisions for class names for its identity, attributes, and operations. [Active class identity is by prefixing active before the class identity. Refer to Figure 7.5(a)]
Package	A packed collection of classes and objects	Rectangular box with divisions for class name for its identity, attributes and operations. [Figure 7.5(b)]
Stereo type	An unpacked collection of elements that is repeatedly used	Rectangular box with stereotype identity within two pairs of starting and closing signs followed by class identity. For example, <<SerialLine Driver >> [Figure 7.5(c)]
Object	An instance of a class that is a functional entity formed by copying the states, attributes and behaviour from a class	Rectangular box with object identity followed by semicolon and class identity [Figure 7.5(d)]
Anonymous Object	An object without identity	Rectangular box with no object identity before the semicolon and class identity [Figure 7.5(e)]
State	A state	Rounded rectangle with state name for its identity [Figure 7.5(f)]

Table 7.14
UML Diagrams

Modeling Diagram	What does it model and show?	Representation
Class Diagram	Class diagrams show how the classes and objects of a class relate and hierarchical associations and object interaction between the classes and objects.	Rectangular boxes show the classes and arrows with unfilled triangles at the end show the class hierarchy. Classes can be joined by lines. Start and end numbers on a line show how the number of objects of a class associates with how many objects of other. [Figure 7.6(a)]
State Diagram	States show a model of a structure for its start, end, and in-between associations through the transitions and shows events-labels (or condition) with associated transitions	Dark circular mark shows starting point (refer marking place in Petri net); arrows show transitions. A label over the arrow shows the condition or event, which fires that transition. A dark rectangular mark within a circle shows the end. [Figure 7.6(b)]

Modeling Diagram	What does it model and show?	Representation
Object Diagram	Object diagram defines the static configuration of the system. It also gives the relationship among the consequent objects.	Refer to Figure 7.6(c).
Sequence Diagram[@]	Sequence diagrams visualize the interactions between the objects. Sequence diagrams also specify the sequences of states.	Rounded rectangles for states and rectangular boxes with object identity and class connects by arrows. A vertical axis pointing downward shows the progression of time. [Figures 7.7(a) and (b)]
Collaboration Diagram	Collaboration diagrams visualize the concurrent sequences of states or object-interactions.	Horizontal or vertical axis pointing right or downward shows the progression of time and a parallel set of sequences show concurrency. Conditions or events can be labeled on the arrow. [Figure 7.7(c)]

[@]Figure 7.7 shows the UML sequence diagrams. Figure 7.7(a) shows sequence of interaction between the states, 7.7(b) shows sequence diagram (example, a vending machine sequences of states) and 7.7 (c) shows collaboration diagram (concurrent multiprocessing).

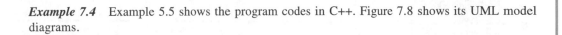

Example 7.4   Example 5.5 shows the program codes in C++. Figure 7.8 shows its UML model diagrams.

Standard texts may be consulted for developing the UML-based representation and modeling skills in depth. [One standard textbook in object-oriented designs is *The Unified Modeling Language User Guide* by Grady Booch, James Rambaugh and Ivar Jacobson, Addison Wesley, 1999.]

Note: UML allows the following:

1. Sequence diagrams may also use the Statechart substrates, or models created by Statechart language. Statechart is a language for implementing the activity diagram, FSM states and state transitions, concurrency, synchronisation, timing and behavioural hierarchy. The message sequence charts are first prepared and from these the Statechart, to show an activity diagram. For example, Statechart can model two concurrent activities of two FSMs.

2. SpecCharts is another language for specifications and charts. Its models, along with its Statecharts-like features, provide implementation of the exception handling (trapping and interrupting) routines easily.

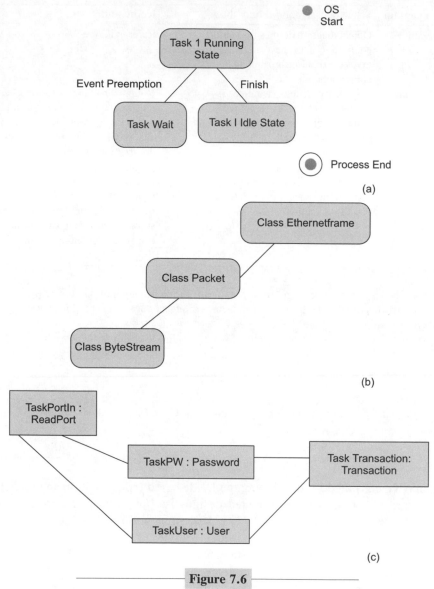

Figure 7.6

UML Diagrams (a) state diagram (b) class diagram (c) object diagram.

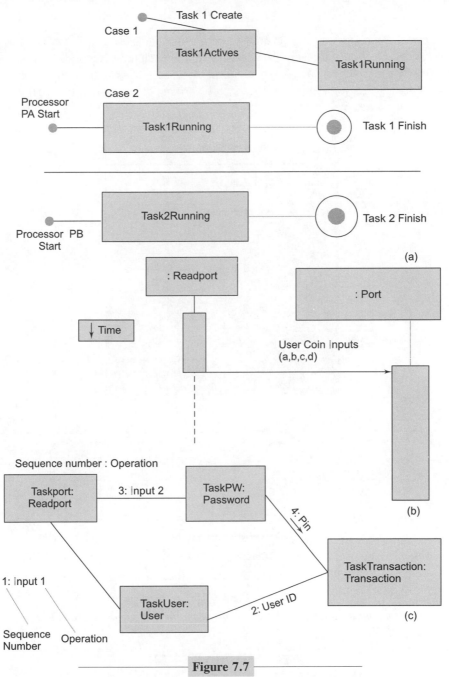

Figure 7.7

*UML Diagrams (a) Sequence of Interaction between the states (b) Sequence diagram
(for example, a vending machine sequences of states) (c) Collaboration
diagram (example, multiprocessing concurrently processing system).*

UML is a powerful modeling language for (i) Software Visualizing, (ii) Data Designs, (iii) Algorithms Design, (iv) Software Designs, (v) Software specifications and (vi) Software Development Process. UML basic elements are class, package, stereotype, object, anonymous objects and state. UML modeling is by class diagrams, state diagrams, object diagrams, sequence diagrams and collaboration diagrams.

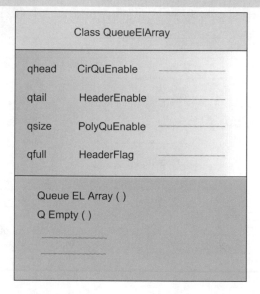

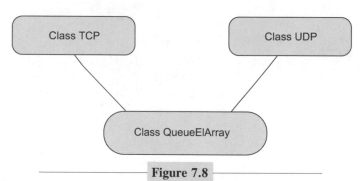

Figure 7.8

UML model diagrams for the C++ program for the queue in Example 5.5.

■ SUMMARY ■

- Software engineering practices must be followed during the development process of any software or system.
- Software complexity gives a quantitative indication of maximum algorithm size and its reliability and a complexity measure is called Cyclomatic complexity. It is the number of needed paths that are to be tested. It is also a measure of the difficulties in maintaining software.

- The software development process can be represented by lifecycle, also called a waterfall or linear incremental model. Analysis, design, implementation and maintenance are four stages of this model. Final product development is by successive refinements.
- There is a choice of seven important models for the process lifecycle. **(i)** Linear sequential Lifecycle model. **(ii)** RAD (rapid development) model. **(iii)** Increment development model, **(iv)** Concurrent development, **(v)** Component based model, **(vi)** Fourth generation object and **(vii)** Oriented Development Model.
- Requirements analysis is the first step in the software development process. General statements are first made and then the concrete technical specifications are obtained from them.
- Any problem can be modeled as consisting of three domains: information, functional and behavioural domains. An important approach to a problem is to first divide it into several parts and then to represent them by suitable notations that describe the problem's essence.
- Often, it may not be possible to completely specify the problem at an early stage. Prototype development and prototype specifications can frequently help. Even with the use of the best analyses and models, the problems may change. Therefore, communication between the developer team and the customer is essential so that before the design process starts both have clear perceptions of the requirements.
- Analyses of the requirements is done and specifications are obtained for data elements and structures, algorithms, behaviour, data contents and process specifications.
- The first designing steps are
  (a) Abstraction. Each problem component is abstracted first. For example, a robotic system problem abstraction can be in terms of the control of the arm and motors.
  (b) Software architectural structural-properties design.
  (c) Extra functional properties design.
  (d) Families of related systems design.
- Modular design concepts are used. System designing is decomposed into modules that are to be implemented. Modules should have the property that allows them to be composed (coupled or integrated) later. Effective modular design should ensure **(i)** function independence, **(ii)** cohesion and **(iii)** coupling.
- Interface design is also essential to integrate the objects and components.
- Software design can be assumed to consist of four layers: Architecture design, data design, interfaces design and component-level design. There is a need for continuous refinements in the designs by effective communication between designers and implementers.
- Implemented programs should be simple. The functions should be made general, not tied to a specific input condition or event. This makes reuse easier. Programs should be implemented as per the program structure design specifications.
- The goal of testing is to find errors and to validate that the implemented software is as per the specifications and requirements. Validation means that the software must be according to the specifications and must show behavior as expected in all the events, conditions and times specified.
- Important testing strategies are white box testing, black box (behavioral) testing, specific environment testing and comparison testing.
- Debugging is completed by using breakpoints, macros, and sample outputs from standard inputs, instruction simulators and laboratory tools.

- The real-time systems development process has complex issues. There are software response-time constraints for the algorithms and ISRs. Testing must be done for timing analysis, for IPCs and for the asynchronous event(s) and their sets at each state of the system. Testing for fast occurring events is also done. Testing has to be thorough; tests for fault tolerances and failure recoveries are essential.

- Software project management issues are the management of people, product, process and project. For planning stage estimation of program efforts, project metrics are available. One project metric is the use of function points and use of FFPs (Full Function Points) to estimate the KLOC (Kilo lines of codes) and productivity per person-month.

- Maintenance means changes, removal and new additions in the objects or modules after delivery (or installation in ROM in the case of embedded systems). The maintenance requirement is inversely proportional to software complexity, as simpler systems need less maintenance.

- Embedded systems should be engineered to need at least post-validation and delivery maintenance. Improper customer specifications need correction maintenance. Changing conditions need adaptation maintenance. Improved system or system perfection achievement needs software perfection maintenance. Preventive maintenance needs perfection maintenance. Software with greater complexity needs greater efforts during maintenance.

- The IEEE 962.1 standard, called SMI (Software Maturity Index), is used to measure the need for maintenance. When SMI approaches 1.0, the software has matured and needs no further maintenance.

- UML is a powerful modeling language for the software development process. Its basic elements are class, package, stereotype, object, anonymous object and state. Modeling in UML is done by developing class diagrams, state diagrams, object diagrams, sequence diagrams and collaboration diagrams.

■ LIST OF KEYWORDS AND THEIR DEFINITIONS ■

- *Software engineering practices*: A practice for using a software engineering approach during the software development process.

- *Software complexity*: A quantitative indication of maximum algorithm size and its reliability and a complexity measure.

- *Cyclomatic Complexity*: A software-complexity measure by finding the number of independent paths a program can follow during execution.

- *Software development process*: A process for developing software, usually by analysis, design, implementation and testing.

- *Lifecycle* (Waterfall model or linear incremental model): A model describing a cycle of analysis, design, implementation and maintenance stages during the life of software at the development stage and then successively refined by the cycles until final verification and validation.

- *Linear sequential lifecycle model*: A linear sequential model for developing software in four sequential stages: analysis, design, implement and test.

- *RAD (Rapid Development) Model*: A model for rapid development of software.

- *Concurrent development model*: Simultaneously developing in parallel a number of processes during the development process.

- *Component-based model*: A way of partitioning (decomposing) software into components and also using reusable components by adapting.
- *Fourth generation tools*: Standard tools and packages for development, for example, Java and use of CASE tools.
- *CASE tools*: Computer Aided Software Engineering tools.
- *Object*: An entity having an identity, state and behaviour and is an instance of a class
- *Class*: A logical group having definitions for data, fields, properties, states, methods, and behaviour.
- *Object-Oriented Development Model*: A process of developing software by first partitioning software into components and also using reusable components by adapting and then each component is composed of objects that are reusable. Reusing an object is by direct adoption, inheritance and overriding states and behavior.
- *Software Analysis*: An analysis for required objects, components, data, data structures, databases and functions, of behavior, control and process specifications.
- *Software Design*: A process of designing software from the abstraction stage to the finer details before implementation is done.
- *Software abstraction*: A visualisation of the software and its development as per requirements and specifications.
- *Problem Abstraction*: Each problem component is abstracted first. For example, a robotic system problem abstraction can be in terms of the control of the arms and motors.
- *Requirements analysis*: The first step in the software development process. First, general statements are made and then from these the technical concrete specifications are obtained. Later, the requirements from all specified software components, data, control and process(es) are ascertained.
- *Modeling domains*: Any problem can be modeled as consisting of three domains: information, functional and behavioural domains.
- *Prototype development*: Developing a prototype, which has functions similar to the final system. For example, first a small laboratory-scale robot prototype is developed and then the industrial-level robot.
- *Modular design*: System designing process is decomposed into modules that are to be implemented. Modules should have the property that allows them to be composed (coupled or integrated) later. Effective modular design should ensure (i) function independence, (ii) cohesion and (iii) coupling.
- *Software design layers*: Architecture design, data design, interfaces design and component level design.
- *Architecture Design*: Describing the structural units and their behaviour and relationships.
- *Data Design*: Design for input and output data elements, structures and databases.
- *Interface design*: Design for integrating the objects and components.
- *Software Implementation*: Coding the software by using a code-engineering tool employing a language like C, C++ and Java.
- *Software Testing*: To find errors until the developed software is verified and validated.
- *Software validation*: Validation means that the software must be according to the specifications and must show behavior as expected in all the events, conditions and times specified, and all components must meet constraints (memory, energy requirement, and response times of functions, routines and drivers).

- *White box testing*: Testing for each program path, each loop and condition, input ranges, events and fast-occurring events and constraints (memory, energy requirement, and response times of functions, routines and drivers).
- *Black box testing*: Testing for behavior and all functional aspects only.
- *Specific environment testing:* Testing for specific environment and components. For example, keyboard, displays, drivers and ISRs.
- *Comparison testing*: Comparison with previously available results.
- *Debugging*: Tracing the source of error by using breakpoints, macros, samples outputs from standard inputs, instruction simulators and laboratory tools.
- *Real time software*: Software with response time constraints for its elements and components.
- *Asynchronous event*: Event occurring at unrelated intervals.
- *Software maintenance*: Maintenance involves changes, removals and additions in the objects or modules after delivery.
- *Software people*: A team of software engineers or system development engineers.
- *Project management*: Managing a project for the development process for a product by the people (team) and monitoring and coordinating each stage of activities.
- *Software planning*: Estimating efforts for a project and developing a suitable strategy for the development process.
- *Program efforts*: Efforts by the people during programming, and it is measured by project metrics.
- *Function points*: A parameter obtained by total counts for a number of function development parameters like external interfaces, user inputs, user outputs, enquiries and files. Appropriate weights are given for each by using the previous results for the software engineering field researchers. Function point is calculated for three types—simple, average and complex (maximum)—to find KLOC-form full function points.
- *Full function points*: FFPs (Full Function Points) are the sum of function points for a simple, average and complex program.
- *KLOC*: Kilo lines of codes.
- *SMI* (Software Maturity Index): A measure of the need for maintenance.
- *UML*: A powerful modeling language, which is extensively used in the software development process, especially for design.

■ REVIEW QUESTIONS ■

1. Why is there a commonality between (i) linear sequential lifecycle model, (ii) software lifecycle model and (iii) waterfall model of a software development process?
2. What are the advantages and disadvantages of using a RAD (Rapid Development) model during the embedded software development process?
3. How does the increment development model differ from the concurrent development model?
4. What are the similarities between the component-based model and the object-oriented development model?
5. How does the use of the fourth generation tool-based development (4GL model) simplify the development process?

6. What is meant by *behavior specification*(s)?
7. How does abstraction of processes help in *OOD* (Objected-Oriented Design)?
8. What is meant by a process specification?
9. How do you implement software from a detailed design of process specifications using programming model, programming language, RTOS and source code engineering tools, while maintaining program simplicity?
10. What are the activities during white box testing, black box testing, orthogonal testing and comparison testing? List these in an exemplary system for each.
11. How does abstract design convert to a detailed design? Explain using the example of the embedded software in a mobile phone.
12. What is the backtracking method for debugging?
13. Why does an embedded system need the highest quality software?
14. Explain a software-testing metric with an example.
15. When is *adaptive maintenance* of an embedded system necessary?
16. When is corrective maintenance necessary?
17. When is perfection maintenance used and when is preventive maintenance?
18. State Pareto's principle.
19. How do uses of macros, test vectors, breakpoints and set of sample output files help in debugging software under development? How does asserting macros in between the program help in debugging and system testing?
20. What are the tests needed in a real-time system? List them.
21. Why is the real-time systems development process more complex?
22. IEEE has a standard, IEEE 962.1, called SMI (Software Maturity Index). Explain the standard.
23. How is an anonymous object denoted in UML?
24. What are the features of UML?

■ PRACTICE EXERCISES ■

25. Compute McCabe's *Cyclomatic Complexity* of the software for Exercise 11 above, and also for Exercise 12 in Chapter 6.
26. Use Table 7.1 and refer to the case study in Section 11.1. Describe three phases in the development process in an automatic vending machine.
27. Specifications are required for data, algorithms, behavior, data contents, controls and processes. List exemplary specifications for the case study of an automobile's cruise control system in Section 11.3.
28. What are the reasons that an embedded system process is programmed with the acceptable maximum latencies, not with minimum latency?
29. When is a brute-force testing method adopted during a software development process?
30. Why should the testing of a real-time system be more stringent? Give an example of a real-time system in which a number of missing deadlines can be within a tolerable limit but variable delays in throughput are not acceptable.
31. Why is the test for fast-occurring events necessary in a real-time video processor or audio processing system?

32. What is meant by fault tolerant systems and fault recovering systems?

33. How can the project specified in Example 7.3 be managed by the 4 Ps described in Section 7.8?

34. Explain, with one example of each, these project metrics: (i) KLOC (Kilo lines of code), (ii) Pages of documentation per KLOC, (iii) Errors per KLOC, (iv) Defects per KLOC, (v) Cost of documentation per page, (vi) Cost per LOC, (vii) Error per person-month and (c) LOC per person-month.

35. Explain with an example the *identity*, *state* and *behaviour* of an object that implements a PPP communication. [Refer to Exercise 16, Chapter 3.6.]

36. Draw the Class diagram, State diagram, Sequence diagram, Collaboration diagram, and Object diagram for the case study in Section 11.1.

# Chapter 8

Inter-Process Communication and Synchronisation of Processes, Tasks and Threads

What We Have Learnt

The following important points have been explained in earlier chapters:

1. Embedded software can be highly complex; there are complex issues in multiprocessor system programming and real-time systems software development.
2. Many times, a program has a number of multiple physical and virtual device drivers, functions, processes (ISRs, tasks and threads), and several program objects that must be concurrently processed on multiple processors or on a single processor.
3. There are issues of response-time constraints, task priorities, latencies and deadlines of vastly different functionalities in embedded-system software.

What We Will Learn

What are the distinctions, meanings and concepts of the following programming entities: functions, processes, tasks and threads? Why and when do embedded software programmers use functions and when the processes (or tasks or threads)?

The first objective of this chapter is to answer these questions by understanding the following:

1. Meanings and concepts of (*i*) *process, task* and *thread* and their states, (*ii*) process control block, task control block and thread control block. Concepts of context, context-switching, multiprocessing, multitasking and multithreading in a system.
2. The distinction between functions, ISRs and tasks in order to understand the finer details of the processing of each during a program run.

How does a data output generated by a process transfer to another? In other words, how does inter-process communication (IPC) take place between two processes? How do we shield data (shared variables) in a critical section of a process before being operated and changed by another higher priority process that starts execution before the first process finishes?

Recall Figures 6.9(a), 6.10 and 6.17. A vertex (set of computations) is waiting for an IPC token to fire computations. Also, recall a transition waiting for input token in a Petri Net (Section 6.2.2). *How does a blocked (waiting) or next process (or task) start running after receiving an IPC (after an event or mail or token) from another process? In other words, how do the processes or tasks synchronize in multiprocessing and multitasking systems?*

The second objective of this chapter is to answer these questions by:

1. Understanding the problem of data that have to be shared between multiple tasks and routines. When one process modifies a variable without another process having been completed, a solution needs to be found.

2. Learning innovative concepts of *semaphores* (as flags, *mutex* or *counting semaphores*) in solving a *shared data problem* and *running critical section codes*.

3. Using *P and V semaphore* functions to solve the classical *producer consumer problem* when using a bounded buffer.

4. Solving the *priority inversion* problem and a *deadlock situation* when using a semaphore.

5. Finding solutions to the shared data problem either (i) by using an *interrupt disabling–enabling mechanism* or (ii) by the *inter-process communication method of using the semaphores.*

6. Programming detailed functions in the RTOS for *inter-process communication* to schedule and synchronize between the multiple tasks (processes) and a scheduler (RTOS).

7. Using a new innovative concept such as spin locks.

8. Having signals (exceptions) for error-handling functions and shortest duration IPCs.

9. Applying semaphores, queues, mailboxes, pipes, sockets, and remote-procedure-calls for scheduling and synchronisation.

Scheduling of tasks by the RTOSs will be described in Chapter 9. The objective of this chapter is to explain the concepts of the IPCs, functions at the RTOS for the IPCs and synchronisation of the processes, tasks and threads. Examples and case studies for a full understanding of the synchronisation and thus concurrent processing of the tasks through the IPCs will be described later in Chapters 10 and 11, respectively.

8.1 MULTIPLE PROCESSES IN AN APPLICATION

8.1.1 Process

Let us first understand the meaning and basic concept of *process*. A *process* defines *a sequentially executing (running) program and its state.* The *state* during running of a process is represented by its status (running, blocked, or finished) and its control block—called process control block (PCB) or *process structure*—its data, objects and resources. A process runs when it is scheduled to run by the OS (kernel), which gives the control of the CPU on a process request (system call). A process runs by executing the instructions; the continuous changes of its state are monitored by the most important parameter, the program counter at the PCB.

Process is that unit of computation that can be modeled by a DFG, CDFG (See Section 6.1), FSM, or Petri Net (Section 6.2) and which is controlled by some *process* at the OS for a *scheduling mechanism* that lets it execute on the CPU and by some *process* at OS for a *resource-management mechanism* that let its use the system memory and other system resources. According to Gary Nutt [*Operating Systems—A Modern Perspective* by Gary Nutt, Addison Wesley, 2nd Edition, 2000] a program can be defined as a "static entity made up of program statements that define process behavior when executed on some set of data". *An application program can be defined as a program consisting of the processes and process behaviors in various states.*

Process is a computational unit that processes on a CPU under the control of a scheduling kernel of an operating system. It has a process structure, called a process control block, at the memory.

8.1.2 Task

An application program can also be defined as a program consisting of the tasks and task behaviours in various states that are controlled by some *scheduling mechanism process* of the system software (called operating system software) that lets it execute on the CPU, and by some *resource-management mechanism process* of the system software that allows tasks to use the system memory.

Embedded software for an *application* may run a number of tasks and each task need a control by a CPU. Assume that there is only one CPU in a system.

1. Each task is *independent,* that is, takes control of the CPU when scheduled by a scheduler at an OS. *The scheduler controls and runs the tasks. No task can call another task.* [It is unlike a C (or C++) function, which can call another function.] *A task is an independent process.* The OS can only block a running task and let another task gain access of CPU to run the servicing codes.

2. Each task has an ID just as each function has a name. The ID, *task ID*, is a byte if it is between 0 and 255. ID is also an index of the task.

3. Each task has its independent (distinct from other tasks) value of the following at an instant: (i) a program counter (memory address from where it runs if granted access to the CPU) and (ii) a virtual stack pointer (memory address from where it gets the saved parameters after the scheduler grants access to the CPU). These two values are the part of its context of a task. Only then does the CPU control switch to the any other process or task. The context must retrieve on transfer of program control to the CPU back for running the task, on the OS unblocking its state and letting it enter the running state again.

4. Each task is recognised by a TCB. TCB is a memory block to hold the current instant program counter information (to indicate the address of the next instruction to be executed for this task), memory map, the signal (message) dispatch table, signal mask, task ID, CPU state (registers, program counter and CPU stack pointer), and a kernel stack (for executing system calls etc.). [Note: The TCB is similar to the Process Control Block (PCB).]

5. Each task may have a priority parameter. The priority, if between 0 and 255, is represented by a byte. [Usually, the higher the value, the higher the priority of that task.]

6. Each task has a context. [Refer to Section 4.6 for definition.] This is a record that reflects the CPU state just before the OS blocks one task and initiates another task into running state. The context thus continuously updates during the running of a task, and the context is saved before switching occurs to another task. Each task also has an initial context, context_init. The context_init includes the initial parameters of a task. The parameters of context_init are as follows: (i) Pointer to a startup function; a function run starts a task from this address. (ii) Pointer to the context data structure. The structure includes the processor registers and status flags. (iii) The task context may also include a pointer to a new task object (function). (iv) It may also include a pointer to the stack of a previous task object (function). The context-switching action must happen each time the scheduler blocks one task and runs another task.

7. Each task may be coded such that it is in endless event-waiting loop to start with. An event loop is one that keeps on waiting for an event to occur. On the start-event, the loop starts from the

first instruction of the loop. Execution of service codes (or setting a flag that is an event for another task) then occurs. At the end, the task returns to the start-event waiting loop.

8. Each task at any instant is in one of the following states:
   (a) *Idle state*: This is the event (or events) waiting loop state before the task initiates (readies for execution). The task in idle state is waiting for this event (or events) to occur. It returns to its idle state after finishing all the codes that are to be executed.
   (b) *Ready State*: The event-waiting loop has been exited, task control has been taken over by the OS and a requisite event has occurred to bring the task into a state where it is ready for the execution of servicing codes from the start (or from where it blocked).
   (c) *Running state*: Executing the servicing codes.
   (d) *Blocked (waiting) state*: Execution of the servicing codes is temporarily suspended after saving needed parameters into its context. For example, a task is pending while it waits for an input from the keyboard or a file. The scheduler then puts it in the blocked state.

9. The task returns to the idle state or ready state after finishing (completion of the running state), that is, when all the servicing codes have been executed.

10. Each task *either* must be a reentrant routine *or* must have a way to solve the shared data problem. [Recall that Section 5.4.4 (ii) explained reentrant function.]

A task is a set of computations or actions that processes on a CPU under the control of a scheduling kernel. It also has a process structure, called a task control block, that saves at the memory. It has a unique ID. It has states in the system as follows: idle ready, running, blocked and finished. It is in ready state again after finish when it has infinite waiting loop—an important feature in embedded system design. Multitasking operations are by context switching between the various tasks.

8.1.3 Threads

A multiprocessing OS runs more than one process. *A process may consist of one or multiple threads that define a minimum unit for a scheduler to schedule the CPU and other system resources.* A thread is a process or sub-process within a process that has its own program counter; its own stack pointer and stack; its own priority-parameter for its scheduling by a thread-scheduler; and its variables that load into the processor registers on context switching. It has its own signal mask at the kernel. The signal mask when unmasked lets the thread activate and run. When masked, the thread is put into a queue of pending threads. *Different threads of a process may share a common structure of the process.* Multiple threads can share the data of the process.

A process structure consists of data for memory mapping, file description and directory. A thread need not possess this data. A process can therefore be considered as a heavyweight process and a kernel-level controlled entity. A process may have process structure with virtual memory map, file descriptors, user-ID, etc. A thread can be considered a lightweight process and a process level controlled entity. [Note: What the structure is, however, depends on the OS.]

How does a task differ from a thread? A thread is a concept used in Java or Unix. A thread can either be a sub-process within a process or a process within an application program. To schedule the multiple processes, there is the concept of forming thread groups and thread libraries. A task is a process and the OS does the multitasking; a task is a kernel-controlled entity while thread is a process-controlled entity. A task is analogous to a thread in most respects. A thread does not call another thread to run. A task also does not directly call another task to run. Both need an appropriate scheduler.

Multithreading needs a thread-scheduler. Multitasking needs a task-scheduler. There may or may not be task groups and task libraries in a given OS.

A thread is a concept in Java and Unix that is a lightweight sub-process or process in an application program. It is controlled by the OS kernel and has a process structure, called thread stack, at the memory. It has a unique ID. It has states in the system as follows: starting, running, blocked and finished.

8.1.4 Clear-cut Distinctions between Functions, ISRs and Tasks by Their Characteristics

Why and when does an embedded software programmer use functions, tasks and threads? When there are multiple devices, functions, ISRs and program objects, the embedded software can be modeled as consisting of multiple tasks—and each task is scheduled by the kernel schedule and uses IPCs for synchronisation. Threads are used in embedded Linux- or Unix-based applications. Functions are subunits of the processes or tasks or ISRs. These do not have analogue of PCB or TCB and have only a stack and no associated scheduler, like a task scheduler or thread scheduler, at the kernel. Table 8.1 summarizes the characteristics of functions, ISRs and tasks.

Table 8.1
Characteristics of the Functions, ISRs and Tasks

Function	ISR	Task
1. Function is an entity used in any routine for performing a specific set of actions as per the arguments passed to it. It may be called from a process or task. Functions are subunits of the processes or tasks or ISRs.	All interrupt-source calls are independent and a call can be from hardware or software. The CPU, during executing an ISR, can let another higher priority ISR execute. It is either the ISR process instructions or RTOS that controls the ISR scheduling on the CPU. It depends on how the kernel manages the ISRs. [Section 9.5]	A task is an independent process, and calls are from the system (RTOS). RTOS can let another higher priority task execute. Only the RTOS (kernel) controls the task scheduling at the CPU.
2. Each function has a program counter and stack that must save before calling another.	Each ISR has a stack for the program counter's instantaneous value and other values that must save before letting another higher priority ISR execute. The stack need not be at a distinct memory block when the different ISRs execute.	Each task has a distinct task stack for the context [program counter instantaneous value and other values (including an ID) in task control block] that must save when blocking from its running state. Each task has a process structure (TCB) for it at distinct memory block.
3. A function calls another function and there may be nesting of one another. There is synchronisation between the functions directly without control of scheduler or OS. [Figure 8.1(b).]	According to the given OS kernel features, there are three alternatives for responding and synchronizing the hardware source calls. [Refer to the next chapter and Figures 9.4(a), (b) and (c).]	There is no synchronisation between the tasks without the kernel facilitating it. Only the RTOS kernel calls a task to run at a time. When a task runs and when it blocks is fully under the control of the RTOS. Depending upon the scheduling by kernel of the RTOS, two common methods are (i) Cooperative Scheduling. (ii) Preemptive Scheduling. [Refer to Section 9.6.]

Figure 8.1(a) shows a characteristic feature of the nested function calls in a program. Figure 8.1(b) shows the program counter assignments at different times on the nested calls.

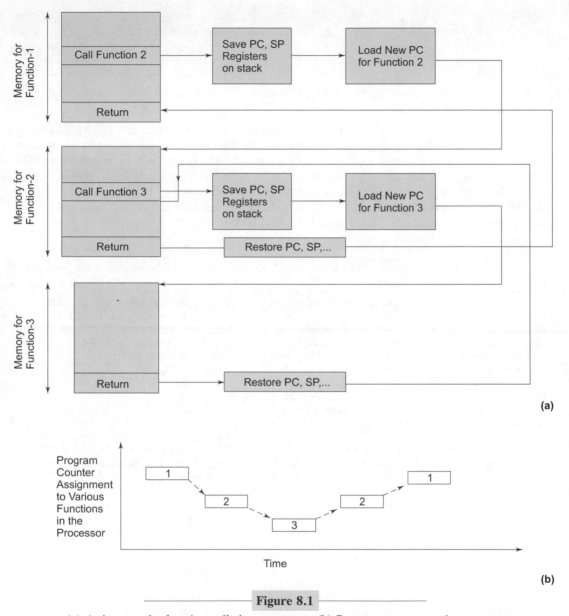

(a)

(b)

Figure 8.1

(a) Actions on the function calls in a program. (b) Program counter assignments to the various functions in a processor on the nested calls.

Function is an entity used in any routine for performing a specific set of actions as per the arguments passed to it. It may be called from a process or task. It may be in an ISR, which executes on interrupts. A task executes on scheduling or on getting an IPC message through the system.

8.2 PROBLEM OF SHARED DATA BY MULTIPLE TASKS AND ROUTINES

8.2.1 Shared Data Problem and Its Solutions

The shared data problem can be explained as follows. Assume that several functions (or ISRs or tasks) share a variable. Let us assume that at an instant the value of that variable operates and during the operations on it, only a part of the operation is completed and a part remains incomplete. At that moment, let us assume that there is an interrupt. Now, if there is another function that also shares the same variable, the value of the variable *may* differ from the value expected if the earlier operation had been completed. The incomplete operation part can occur as follows. Suppose a variable is of 32 bits and the processor of 8 bits. A certain C compiler has to assemble the operations as four 8-bit ALU operations in order to use the 8-bit ALU of this processor. Now, assume that the 32-bit operation is non-atomic when this compiler is used. An interrupt can occur at the end of each 8-bit ALU operation, not necessarily at the end of the 32-bit operation. Therefore, the called ISR or another function can change that variable when it shares with another function. On return, new values of that variable will load into the four registers from the stack. The incomplete operations will now execute with new values in the registers.

Consider another example, the *printf* function in C. It may process non-atomically in certain C compilers for the processor used.

Consider another example. Suppose there is an interrupt service routine that is executing. Let there be a variable that is used in the *condition* test. Let a *while* loop using the test be partially implemented before another interrupt occurs. The same variable returns a new value to the previously interrupted ISR after the return from this interrupt service. Now, the condition test will show a different condition.

Section 2.1.8 describes the use of atomic operation for solving the shared data problem. We need atomic operations because (*i*) an interrupt can occur at the end of an instruction cycle, not at the end of a high-level instruction. (*ii*) A DMA operation can occur at the end of a machine cycle itself and a compiler or program may not take these atomic level details into account. Section 5.4.4(ii) describes the use of reentrant functions as a means to save from the shared data problem.

The following are the steps that, if used together, almost eliminate a likely bug in the program due to the shared data problem.

1. *Use modifier volatile* with a declaration for a variable that returns from the interrupt. This declaration warns the compiler that certain variables can modify because the ISR does not consider the fact that the variable is concurrently shared with a calling function.
2. *Use reentrant functions* with atomic instructions in the part of a function that needs its complete execution before it can be interrupted. This part is called the critical section.

3. *Put a shared variable* in a circular queue. A function that requires the value of this variable always deletes (takes) it from the queue *front*, and another function, which inserts (writes) the value of this variable, always does so at the queue *back*. Now a problem can occur in case there are a large number of functions that send the value into and get the value from the queue, and the queue size is insufficient.

4. *Disable the interrupts before a critical section starts executing* and *enable* the interrupts on its completion. It is a powerful but drastic option. An interrupt, even if of higher priority than the present critical function, gets disabled. The difficulty with this option is that it increases the interrupt latency period and a deadline may be missed for an interrupt service. As an alternative to disabling interrupts, the next subsection describes using semaphores for the shared data problem. A software designer should not use the drastic option of disabling interrupts in all the critical sections. [Note: In the OS for automobile applications, the disabling of interrupts is used before entering any critical section to avert any unintended action due to improper use of semaphores. Refer to Section 11.3.]

The above steps—use of semaphores, use of disabling the switching of task from one to another and other steps—must eliminate the shared data problem completely from a multitasking, multiple ISRs and multiple shared variables cases. Each step has its own inherent benefits in solving the problem. A software designer must utilize the various steps that are optimally suited to solve the problem.

> A shared data problem can arise in a system when another higher priority task finishes an operation and modifies the data or a variable. Disabling the interrupt mechanism, using semaphores and using reentrant functions are some solutions.

8.2.2 Use of Semaphores for a Task or for the Critical Sections of a Task

(i) Use of a Single Semaphore

Suppose that there are two trains that use an identical track. When the first train A is to start on a track, the signal (notification) for A is set (true, taken) and the signal for other train, B is reset (false, released). Similarly, consider using a semaphore, which is a binary Boolean variable when used as an event flag; or it is a signaling variable or notifying variable. [Such a semaphore is called *binary* semaphore when its value is 0 and it is assumed that it has been taken; when its value is 1, it is assumed that no task has taken it and that it has been released.]

The task *A*, when executing a critical section, notifies the OS to take the semaphore (take notice). The OS returns the semaphore as taken (accepted). Now, the task *A* executes the codes of the critical section. The OS, having been notified earlier, does not return the semaphore to another task *B*.

Figure 8.2(a) shows the use of a semaphore between *A* and *B*. It shows the five sequential actions at five different instants, T0, T1, T2, T3 and T4. Figure 8.2(b) shows the timing diagram of the tasks in the running states as a function of time. It marks the five sequential actions at T0, T1, T2, T3 and T4.

An RTOS uses the semaphores for multitasking as follows. Assume *I*, *J*, *K*, *L* and *M* are the tasks. [Refer to Section 9.6.5. It describes the exemplary use of semaphores for the task or for the critical section of a task.]

1. One use of semaphores is to let one task run among many in the initiated into a list with *state* = '*ready*' or '*running*'. The use of semaphore, therefore, permits sharing of a common resource,

the CPU. For example, when a task *K* is to start running, it takes a semaphore. The OS blocks the tasks *I*, *J*, *L* and *M*. These wait for the release of the semaphore by *K*.

2. Another use of the semaphores is as a message (or event flag) from one task to another: inter-task communication through the OS. For example, a task *M* receives the bytes from a port, and another task *I* needs these received values for re-transmission after decoding the message. Task *M* needs inter-task communication. When *I* finishes the re-transmission, it releases the semaphore taken earlier from the OS, and then the OS notifies task *M* so that it can receive another set of bytes from the port.

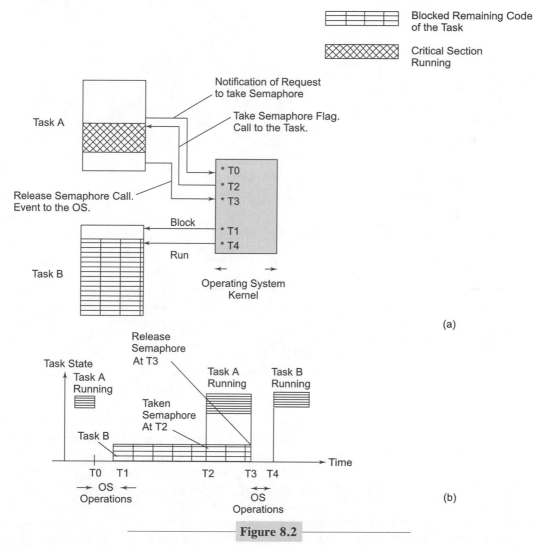

Figure 8.2

(a) Use of a semaphore between tasks, A and B. It shows the five sequential actions at five different instants, T0, T1, T2, T3 and T4. (b) Timing diagram of the tasks in the running states as a function of time. It marks the five sequential actions at five different times, T0, T1, T2, T3 and T4, and shows the use of a semaphore between tasks A and B by the operating system.

A semaphore provides a mechanism to let a task wait until another finishes. It is a way of synchronizing concurrent processing operations. When a semaphore is 'taken' by a task, then that task has access to the necessary resources; when 'given', the resources unlock. A semaphore can be used as an event flag or as a resource key. A resource key is one that permits use of resources, like CPU, memory or other functions or critical section codes.

(ii) Use of Multiple Semaphores

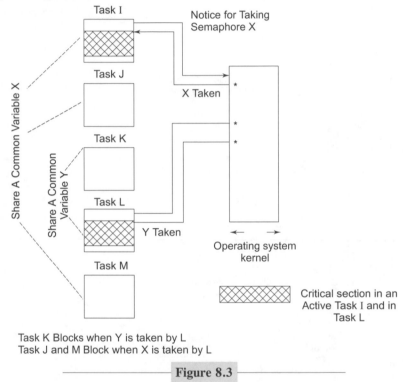

Figure 8.3

Use of two semaphores, x and y, between the tasks I to M.

Let there be five trains *i, j, k, l,* and *m* that use two tracks, *p* and *q,* to different destinations. Assume first, there is only one signal. Let *i, j* and *m* trains share track *p*, and *k* and *l* trains share *q*. When the first train *i* is to start on a track, the signal for *i* is down (released). The other trains *j, k, l* and *m* take notice of it and are not released. Why should the notice be taken by all the trains *j, k, l* and *m*? No. There should be two separate signals for the two tracks. The trains *j* and *m* should take notice when train *i* moves and the train *k* should take notice when *l* moves. Likewise, consider using two semaphores, *x* and *y*. A task *I* when executing a critical section notifies the OS to take the semaphore, *x*. OS returns information that the semaphore has been taken to tasks *J* and *M*. Now, the task *I* executes the codes of the critical section. The OS, having been notified about a taken semaphore *x* from *I*, now does not take and OS does not release any semaphore to task *J* and *M*. But the OS returns another semaphore *y* to either one—task *K* or *L*—at an instance. Figure 8.3 shows the use of two semaphores, *x* and *y* between the tasks, *I* to *M*.

> Multiple semaphores are used and different set of semaphores can share among different set of tasks.

(iii) Use of Mutex

Use of mutex facilitates mutually exclusive access by two or more processes to the resource (CPU). The same variable, sem_m, is shared between the various processes. Let process 1 and process 2 share *sem_m* and its initial value = 1.

1. Process 1 proceeds after sem_m decreases and equals 0 and gets the exclusive access to the CPU.
2. Process 1 ends after sem_m increases and equals 1; process 2 can now gets exclusive access to the CPU.
3. Process 2 proceeds after sem_m decreases and equals 0 and gets exclusive access to CPU.
4. Process 2 ends after sem_m increases and equals 1; process 1 can now gets the exclusive access to the CPU.

The sem_m is like a resource-key. The process which first decreases it to 0 at the start gets the access to and prevents others, who share this key, from running.

> Mutex is a semaphore that gives at an instance two tasks mutually exclusive access to resources.

(iv) Use of P and V Semaphore Functions

A semaphore can also be an integer variable that, apart from initialisation, is accessed only through two standard atomic operations: P and V. [P (for *wait* operation) is derived from a Dutch word 'Proberen', which means 'to test'. V (for *signal* passing operation) is derived from the word 'Verhogen' which means 'to increment'.]

  (i) P semaphore function: requires a resource and, if not available, waits for it.
 (ii) V semaphore function: signal (information) passes to the operating system that the resource is now free to the other users.

In RTOS an efficient synchronisation mechanism—*P and V semaphores*—is used *when using* standard *Posix 1003.1.b*, an *IEEE standard.* [Refer to Section 9.8.] These two functions operate on a mutex-semaphore variable, sem_m (Examples 8.1 and 8.2) and are defined as follows:

Consider P semaphore. It is a function, P (&*sem_1*) which, when called in a process, does the following operations using a semaphore, *sem_1*.

1. /* Decrease the semaphore variable*/
   sem_1 = sem_1 -1;
2. /* If sem_1 is less than 0, send a message to OS by calling a function waitCallToOS. Control of the process transfers to OS, because less than 0 means that some other process has already executed P function on sem_1. Whenever there is return for the OS, it will be to step 1. */
   if (*sem_1* < 0){waitCallToOS (*sem_1*);}.
   Consider V semaphore. It is a function, V(&*sem_2*) which, when called in a process, does the following operations using a semaphore, *sem_2*.
3. /* Increase the semaphore variable*/
   sem_2 = sem_2 + 1;
4. /* If sem_2 is less or equal to 0, send a message to OS by calling a function signalCallToOS. Control of the process transfers to OS, because < or = 0 means that some other process is

already executed P function on sem_2. Whenever there is return for the OS, it will be to step 3.
*/
if (*sem_2* < = 0){signalCallToOS (*sem_2*);}

Example 8.1 Using P and V semaphore functions with a mutex property.

Let sem_1 and sem_2 be the same variable, *sem_m*. The latter functions as a mutex, as follows, when P and V semaphore functions are used in two processes, task 1 and task 2.

Process 1 (Task 1)

while (true) {

/* Codes before a critical region*/

.

.

.

/* Enter Process 1 Critical region codes*/

P (&sem_m);

/* The following codes will execute only when sem_m is not less than 0. */

.

.

.

/* Exit Process 1 critical region codes */

V (&sem_m);

/* Continue Process 1 if sem_m is not equal to 0 or not less than 0. It means that no process is executing at present. */

.

.

.

};

Process 2 (Task 2)

while (true) {

/* Codes before a critical region*/

.

.

.

/* Enter Process 2 Critical region codes*/

P (&sem_m);

/* The following codes will execute only when sem_m is not less than 0. */

.

.

.

/* Exit Process 2 critical region codes */

V (&sem_m);

/* Continue Process 2 if sem_m is not equal to 0 or not less than 0. It means that no process is executing at present. */

.

.

.

};

The same variable, sem_m, is shared between process 1 and process 2. Its use is in making both processes gain mutually exclusive access to the resource (CPU). Either process 1 runs after executing P or process 2 runs after executing P. Also, either process 1 runs after executing V or process 2 runs after executing V.

Figure 8.4(a) shows the use of P and V semaphores at a task, at an ISR and at a scheduler. When a task takes a semaphore P, if sem_m = 'true' (=1) earlier then it becomes 'false' (=0) and task run continues as sem_m is not less than 0. When a task executes V, if sem_m was 'false' earlier then it sets to 'true' and task continues running, else the task blocks and waits for the execution of another task. Figure 8.4(b) shows the program counter assignments of a process or function when using P and V semaphores.

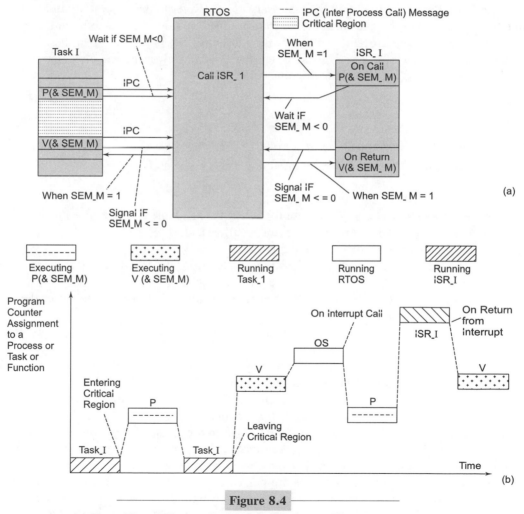

Figure 8.4

(a) Use of P and V semaphores at a task, at an ISR and at a scheduler

(b) The program counter assignments to a process or function when using P and V semaphores.

P and V semaphore functions are in POSIX 1003.1b, an IEEE accepted standard for the IPCs. It can be uses as a mutex; it can also be used as a counting semaphore operation.

(v) Counting Semaphores

Counting semaphores are unsigned integers. A counting semaphore value controls the blocking or running of the codes of a task as well as of an accompanying task(s) with which it shares the value. The counting semaphore counts the number of times it is taken. It increments when taken by a task and decrements when released by a task (same or another). The value of it at an instance reflects the difference in number of times it is taken and number of times released. The use of a semaphore is such that one of the tasks thus waits to execute the codes or waits for a resource until the necessary number of tokens is collected. An exemplary use of a counting semaphore is in a vending machine. It delivers chocolate only on the collection of all the necessary coins. [Refer to this case study in Section 11.1.] Assume there is only one slot and when the machine accepts only one type of coin, until the appropriate number of coins inserted occur, another waiting task will not proceed. When there are multiple types of coins, a number of events is needed and the task for delivering the chocolate will remain blocked, until a requisite set of number of coins collect.

A counting semaphore is a semaphore that can be 'taken' and 'given' a number of times.

(vi) Use of P and V Semaphore Functions with a Counting Semaphore property

Let a task generate the outputs for use by another task. A task can have a separate counting semaphore.

Consider three examples:

  (i) a task transmits bytes to an I/O stream for filling the available places at the stream;

  (ii) a process *'writes'* an I/O stream to a printer buffer; and,

  (iii) a task of producing chocolates is being performed.

In example (i) another task reads the I/O-stream bytes from the filled places and creates empty places.

In example (ii), the print buffer for an I/O stream prints after a buffer-read and, after the printing, more empty places are created at the buffer.

In example (iii), a consumer is consuming the chocolates produced and more empty places (to stock the produced chocolates) are created.

A task blockage operational problem is commonly called producer–consumer problem. A task cannot transmit to the I/O stream if there are no empty places at the stream buffer. The task cannot write from the memory to print buffer if there are no empty places at the print buffer. The *producer cannot produce if there are no empty places at the consumer end.*

A classic program for synchronisation is called program for the *producer–consumer problem program*. It is also called *bounded buffer problem program*. Here, one or more producers (task or thread processes) create data outputs that are then processed by one or more consumers (tasks or processes). The data outputs from the producers are passed for processing by the consumers using some type of IPC (Section 8.8) that uses a shared memory and counting semaphores (or message queues or mailboxes).

Let there be two processes (task 3 and task 4). The P and V functions operate on two shared counting semaphores, sem_c1 and sem_c2, as in Example 8.2.

Example 8.2   Using P and V semaphore functions as the counting semaphores.
Assume two processes using P and V semaphore functions and two tasks, task 3 and task 4. Let *sem_c1* and *sem_c2* be two counting semaphore variables and represent the number of filled places created by the process 3 and number of empty places created by process 4, respectively. P and V functions operate on these as follows.

Process 3 (Task 3)
while (true) {
/* Codes before a producer region*/
.
.

.
/* Enter Process 3 Producing region codes*/
P (&*sem_c2*);
/* The following codes will execute only when sem_c2 is not less than 0. */
.
.

/* Exit Process 3 producing region codes */
V (&*sem_c1*);
/* Continue Process 3 if sem_c1 is not equal to 0 or not less than 0. */
.
.

.
};
Process 4 (Task 4)
while (true) {
/* Codes before a consumer region*/
.
.

.
/* Enter Process 4 Consuming region codes*/
P (&*sem_c1*);
/* The following codes will execute only when sem_m is not less than 0. */
.
.

.
/* Exit Process 4 consuming region codes */
V (&*sem_c2*);
/* Continue Process 4 if sem_c2 is not equal to 0 or not less than 0. It means that no process is executing at present. */
.
.

.
};

Two semaphores, sem_c1 and sem_c2 are shared between processes 3 and 4. When process 3 executes, it first reduces the number of empty places at process 4. When process 3 completes production, it increases the number of filled places at process 3. When process 4 consumes, it first reduces the number of filled places at process 3. When process 4 completes consumption, it increases the number of empty places at process 4. Either process 3 produces output after executing P, or process 4 consumes (uses) inputs after executing P. Also either process 3 proceeds after executing V or process 4 proceeds after executing V.

> P and V semaphore functions can be used for counting semaphore operations and can be used for programs for solving bounded buffer-like problems. These are used for programs having classic situations, including a producer–consumer problem.

(vii) Elimination of Shared Data Problem

The use of semaphores does not eliminate the shared data problem completely. Software designers may not take the drastic option of disabling interrupts in all the critical sections by using semaphores. When using semaphores, the RTOS disables the interrupts. Alternatively, task-switching flags can be used [Section 8.2.3] to avoid the following problems that can arise when using semaphores.

1. Sharing of two semaphores creates a deadlock problem. [Refer to Section 8.2.3.]
2. Suppose the semaphore taken is never released? There should therefore be some time out mechanism after which the error message is generated or an appropriate action taken. There is some degree of similarity with the watchdog timer action on a time out. A watchdog timer on time out resets the processor. Here, after the time out, the OS reports an error and runs an error handling function. Without a time out, an ISR worst-case latency may exceed the deadline.
3. If semaphore is not taken, another task may use a shared variable.
4. What happens when a train takes a signal for a wrong track? When using the multiple semaphores, if an unintended task takes the semaphore, it creates a problem.
5. There may be a priority inversion problem. [Refer to Section 8.2.3.]

8.2.3 Priority Inversion Problem and Deadlock Situations

Let the priorities of tasks be in an order such that task I is of the highest priority, task J is of a lower, and task K of the lowest priority. Assume that only tasks I and K share the data and J does not share data with K. Also let tasks I and K alone share a semaphore and not J. Why do only a few tasks share a semaphore? Can't all share a semaphore? The reason is that the worst-case latency becomes too high and may exceed the deadline if all tasks are blocked when one task takes a semaphore. The worst-case latency will be small only if the time taken by the tasks that share the resources is relevant. Now consider the following situation.

At an instant t_0, suppose task K takes a semaphore, it does not block task J and blocks task I. This happens because only tasks I and K share the data and J does not. Consider the problem that now arises on selective sharing between K and I. At the next instant t_1, let task K become ready first on an interrupt. Now, assume that at the next instant t_2, task I becomes ready on an interrupt. At this instant, K is in the critical section. Therefore, task I cannot start at this instant due to K being in the critical region. Now, if at the next instant t_3, some action (event) causes the unblocked higher than K priority task J to run. After instant t_3, running task J does not allow the highest priority task I to run because K is not running, and therefore K can't release the semaphore that it shares with I. Further, the design of task J may be such that even when the semaphore is released by task K, it still may not let I run. [J runs the codes as if it is in critical section all the time.] The J action is now as if J has higher priority than I. This is because K, after entering the critical section and taking the semaphore when letting J run, did not share the priority information about I—that task I is of higher priority than J. The priority information of another higher-priority task I should have also been inherited by J temporarily, if K waits for I but J does not and J runs when K has still not finished the critical section codes. This did not happen because the given RTOS design was such that it did not provide for temporary priority inheritance in such situations.

The above situation is also called a ***priority inversion problem***. An RTOS must provide for a solution for the priority inversion problem. Some RTOSs provide for priority inheritance in these situations and thus a priority inheritance problem does not occur when using them. Refer to Section 8.2.2(iii) for use of a mutex for resources sharing. A mutex should be a mutually exclusive Boolean flag, by using which the critical section is protected from interruption in such a way that the problem of priority inversion does not arise. Mutex is automatically provided in certain RTOS so that it the priority inversion problem does not arise. Mutex use may also be analogous to a semaphore defined in Section 8.2.2(i) in another RTOS and which does not solve the priority inversion problem.

Consider another problem. Assume the following situation.

1. Let the priorities of tasks be such that task H is of highest priority. Then task I has a lower priority and task J has the lowest.
2. There are two semaphores, *SemFlag1* and *SemFlag2*. This is because the tasks I and H have a shared resource through *SemFlag1* only. Tasks I and J have two shared resources through two semaphores, *SemFlag1* and *SemFlag2*.
3. Let J interrupt at an instant t_0 and first take both the semaphores *SemFlag1* and *SemFlag2* and run.

Assume that at a next instant t_1, being now of a higher priority, the task H interrupts, the task I and J after it takes the semaphore *SemFlag1*, and thus blocks both I and J. In between the time interval t_0 and t_1, the *SemFlag1* was released but *SemFlag2* was not released during the run of task J. But the latter did not matter as the tasks H and J don't share *SemFlag2*. At an instant t_2, if H now releases the *SemFlag1*, lets the task I take it. Even then it cannot run because it is also waiting for task J to release the *SemFlag2*. The task J is waiting at a next instant t_3, for either H or I to release the *SemFlag1* *because it needs this to again enter a critical section in it*. After instant t_3, neither task I can run nor task J. There is a circular dependency established between I and J.

The above situation is also called a *deadlock situation*. On the interrupt by H, the task J, before exiting from the running state, should have been put in queue-front so that later on, it should first take *SemFlag1*, and the task I put in queue next for the same flag, then the deadlock would not have occurred. [Refer to Section 8.3.3 for queuing of messages.]

The use of mutex solves the deadlock problem in certain RTOSs. Its use may be just analogous to a semaphore defined in Sections 6.2.2 (i) and (ii) in other RTOSs, and then the mutex use does not solve the deadlock situation.

Priority becomes inverted and deadlock (circular dependency) develops in certain situations when using semaphores. Certain RTOSs provide the solution to this problem of semaphore use by ensuring that these situations do not arise during the concurrent processing of multitasking operations.

8.3 INTER-PROCESS COMMUNICATION

IPC means that a process (scheduler, task or ISR) generates some information by setting or resetting a flag or value, or generates an output so that it lets another process take note or use it. [Recall Figures 6.9(a), 6.10, 6.17 for a waiting vertex (set of computations) and Section 6.2.2 for a waiting transition.] IPCs in a multiprocessor system are used to generate information about certain sets of computations finishing on one processor and to let the other processors take note of it.

Is it possible to send through the scheduler an output data (a message of a known size with or without a header) for processing by another task? One way is the use of global variables. Use of these now creates two problems. One is the shared data problem (Section 8.2). The other problem is that the global variables do not prevent (encapsulate) a message from being accessed by other tasks. There-fore, the following IPCs may be used:

- *(i)* Signal
- *(ii)* Semaphore (*as flag, mutex*) or counting semaphore for the inter-task communication between tasks sharing a common buffer
- *(iii)* Queue, pipe and mailbox
- *(iv)* Socket
- *(v)* Remote procedure call (RPC) for distributed processes.

A simple example of IPC is seen in a task that runs a *print* function. The scheduler should let the other tasks share this task. The *print* task can be shared among the multiple tasks by using the IPCs. When the printer becomes available, an IPC from the *print* task is generated and the scheduler takes note of it. Other tasks take note of it through the scheduler. Another task generates an IPC to get access to the *print* task.

Consider another example. It is a task for a multi-line display of outputs and another task for displaying current time on the last line. When the multi-line display task finishes the display of the last but one line, an IPC from the *display* task is generated and the scheduler takes note of it. Another task — continuously updating time — can take and generate an IPC output for the current time.

The need for IPC and thus inter-task communication also arises in a client-server network.

The RTOS facilitates the use of IPCs: signals, events, counting, mutex semaphores, mailboxes, queues, pipes, sockets. The RTOS also facilitates the RPCs. The above IPCs are explained in sub-sections 8.3.1 to 8.3.6.

Inter-process communication (IPC) means that a process (scheduler or task or ISR) generates some information by setting or resetting a flag or value, or generates an output so that it lets another process take note or use it under the control of an OS.

8.3.1   Use of the Signals

One method of messaging is to use a *'signal'. A 'signal' provides the shortest message*. The *signal* is a one-bit output from a process for an IPC. *An advantage of using it is that unlike semaphores it takes the shortest possible CPU time*. The *signal*s are the flags that are used for the IPC functions of synchronizing. A signal is the software equivalent of the flag at a register that sets on a hardware interrupt. Refer to the software timers described in the example in Chapter 3. Unless masked by a *signal* mask, the *signal* allows the execution of the *signal* handling function, just as a hardware interrupt allows the execution of an ISR.

A *signal* is unlike the semaphore flag. The semaphore flag has restricted use as an event flag to let another *task process* block, or which locks a resource to a particular *task process* for a given section of the codes. A signal is just a flag that is shared and used by another *interrupt servicing process for signal handling function*. A *signal* raised by one process forces another process to interrupt and catch that *signal* in case the *signal* is not masked. [Use is not disabled.] Signals are to be handled only by very high priority processes as it may disrupt the usual schedule and priority inheritance. It may also cause reentrant problems.

An important use of the *signal*s is to handle *exceptions*. [An exception is a process that is executed on a specific reported run-time condition.] A *signal* reports an error (called 'Exception') during the running of a task and then lets the scheduler initiate an error-handling process or function. An error-handling task may handle the different error login of other task. The handling is through the use of the ISR-handling functions.

Refer to *Operating Systems: A Modern Perspective* by Gary Nutt (Addison Wesley, 2000). Unix and Linux OSs use *signal*s profusely and have thirty-one different types of *signals* for the various events. For the VxWorks *signal* refer to Section 10.3.4.(1).

> The simplest IPC for messaging and synchronizing processes is the use of 'signals'. A 'signal' provides the shortest message. Signals are used for initiating exceptions and error-handling processes.

8.3.2   Uses of Semaphore Flag or Mutex as Resource Key (for Resource Locking and Unlocking to a Process)

A semaphore [Section 8.2(i) and (ii)] describes the use of semaphore flags (tokens). *The semaphore as event flag facilitates inter-task communication for* signaling (through a scheduler) a waiting task M to the running place upon an event at the running task N or at an ISR. *A semaphore token generated at one place is usable at another place*. Before entering the critical section at the running place, a flag, *Sem_NTakenFlag* becomes true.

The semaphores described in Sections 8.2.2.(iii) and (iv) are used as a Mutex (<u>mu</u>tually <u>ex</u>clusive) access to a set of codes (or thread or process). The use of Mutex is such that the priority inversion problem [Section 8.2.3] is not solved in some RTOSs while it is solved in other RTOSs.

A process using a Mutex *locks* on to the critical section in a task. Let there be a critical section in a task. Let *m1_mutex* be a Mutex created by a function to take mutually exclusive value. Two 'C' functions can be written as taskCriticalSec_mutex_lock (&*m1_mutex*) before entering the section and taskCriticalSec_mutex_unlock (&m1_mutex) before exiting the section. Alternatively, P and V semaphores are used with the Mutex. [Refer to Section 6.2.2.(iv).]

The use of a binary semaphore as a flag or as an integer variable Mutex gives a *resource key and facilitates the communication between the task and the scheduler.* The key is to get an access to a resource (a section of the codes or critical section codes or a function). It solves the shared data problem or shared resource problem. It works as a resource locking mechanism if the access to certain resources is blocked. The blocking period of a task during the period when other tasks have taken the semaphore can be limited by defining a timeout value.

Using the key, in a similar manner, a time-consuming ISR can also be blocked after a preset time interval *time-out*. The ISR takes a semaphore and releases it after the time-out to let the other ISR run. This improves the ISR response times of the other ISRs.

Suppose the time taken by a critical section is too long compared to the time that an ISR would have taken. Presetting the timeout for taking a semaphore then helps. In another locking situation, by using the Mutex or P and V semaphores, a resource can lock from task switching on to another task or ISR.

A resource of high priority should not lock the other processes by blocking an already running task in the following situation. Suppose a task is running and a little time is left for its completion. The running time left for it is less compared to the time that would be taken in blocking it and context switching. There is an innovative concept of spin locking in certain schedulers.

A *spin lock* is a powerful tool in the situation described above. [Refer to *Multithreaded Programming with Java* by Bil Lewis and Daniel J. Berg, Sun Microsystems Inc., 2000.] The scheduler locking processor for a task waits to cause the blocking of the running task first for a time-interval t, then for (t - δt), then (t - 2δt) and so on. When this time interval spin downs to 0, the task that requested the lock of the processor now unlocks the running task and blocks it from further running. The request is now granted. *A spin lock does not let a running task be blocked instantly, but first successively tries decreasing the trial periods before finally blocking a task.*

Can a single semaphore be taken multiple times? Yes, in a counting semaphore, it is feasible [Refer to Sections 8.2.(v) and (vi).] The use of a *counting semaphore* is like an event-counting message from one task to another task. It means a semaphore once taken need not be released before being taken again. It is an unsigned integer or byte. *The counting semaphore facilitates multiple inter-task communications in case of producer–consumer problems* [Section 8.2.(vi)].

> ⚠ An RTOS provides the IPC functions for creating and using semaphores as event flags, Mutex, resource keys (for resource locking and unlocking onto a process) and as counting semaphores.

8.3.3 Use of Message Queues

Some RTOSs do not distinguish, or make little distinction, between the use of queues, pipes and mailboxes during the message communication among processes, while other RTOSs regard the use of queues as different.

In the strict sense, a message *queue* is an IPC with the following features.

1. Either one process (task) or a set of (processes) tasks can use the queue.
2. A *read* (deletion) from the queue is in a FIFO mode. Generally, the queue is a circular queue, and the *read* is like a ring of bytes in a buffer. A task that reads the queue (deletes from the queue) may also be of such low priority that execution is deferred. Such a task *reads* the queue and takes actions as per the sequence of *write* operations (insertions) into the queue.

A printer task is an exemplary use of the message queue. The following are the common RTOS features.

1. Each queue for a message may need initialisation before using the functions in the scheduler for the message queue.
2. There may be a provision for multiple queues for the multiple types or destinations of messages. Each queue may have an ID.
3. Each queue either has a user definable size or a fixed pre-defined size assigned by the scheduler.
4. When an RTOS call is to insert into the queue, the bytes are as per the pointed number of bytes. For example, for an integer or float variable as a pointer, there will be four bytes inserted per call. If the pointer is for an array of 8 integers, then 32 bytes will be inserted into the queue.
5. When a queue becomes full, there may be a need for error handling and user codes for blocking the tasks. They may not be self-blocking.

Figure 8.5(a) shows three memory blocks at an RTOS. One block is for initialisation functions. The other two are for queue deleting and various functions. Figure 8.5(b) shows a queue-messages block with the messages and headers. Headers are for the ID of the queue. Message-type (for example, integer, float, array of 8 integers), queue maximum length, present length, and two pointers, *Qfront and *Qback are for queue head and tail memory locations.

The RTOS functions for servicing of a queue can be as follows:

1. The RTOS_QCreate, a queue function that creates a queue and initializes the queue message block and the contents with front and back as queue-top pointers, *Qfront and *Qback, respectively.
2. The RTOS_QWrite (Post) sends a message to the memory as per the queue back pointer, *Qback.
3. The RTOS_QWait (Pend) waits for a queue message at the queue and reads when received.
4. The RTOS_QAccept reads the present queue front pointer after checking its presence yes or no [No wait.]
5. The RTOS_QFlush that reads queue from front to back, and deletes the queue block, as it is not needed later after the *flush*.
6. The RTOS_QQuery just queries the queue message-block when *read* but the message is not needed (used) later.
7. The RTOS_QPostFront sends an urgent message as per the queue front pointer, *Qfront. Refer to Section 5.5.4. Use of this function is made in the following situations. On a network, if a message is not accepted it returns. The FIPO operation is followed (First In and Provisionally Out) [Section 5.5.4].

An RTOS provides the IPC functions for creating and using queues as messages in FIFO or LIFO (for priority message) modes.

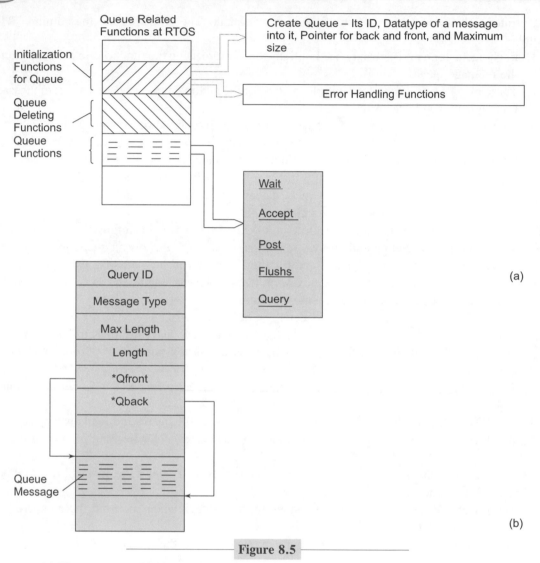

Queue Related Functions at RTOS

Initialization Functions for Queue

Create Queue – Its ID, Datatype of a message into it, Pointer for back and front, and Maximum size

Error Handling Functions

Queue Deleting Functions

Queue Functions

Wait

Accept

Post

Flushs

Query

(a)

Query ID

Message Type

Max Length

Length

*Qfront

*Qback

Queue Message

(b)

Figure 8.5

(a) Three memory-blocks at an RTOS—a block for initialisation functions, two others for queue inserting, deleting and other functions. (b) Queue messages block.

8.3.4　Mailboxes

A message mailbox is an IPC message-block that can be used only by a single destined task. The mailbox message may also include a header to identify the message-type specification. The source (mail sender) is the task that sends the message pointer to a created (initialised) mailbox. [The box initially has the NULL pointer before the sending action.] The destination is the place where the RTOS_BoxWait (Pend) function waits for the mailbox message and reads it when received.

A mobile phone LCD multi-line display task is an example that uses a message mailbox as an IPC. In the mailbox, when the time message from a clock process arrives, the time is displayed at the right corner of the last line. When the message from another task is to display a phone number, it is displayed at the middle line. Another example of using a mailbox is for an error-handling task, which handles the different error logins from other tasks.

Figure 8.6(a) shows three mailbox types at the different RTOSs. Figure 8.6(b) shows the initialisation and other functions for a mailbox at an RTOS. The following may be the provisions at an RTOS for inter-task communication when using the mailbox:

1. A task on an RTOS call puts into the mailbox only a pointer to the mailbox message block.
2. The number of message-bytes sent to the mailbox created by the scheduler is as per a pointer. When an RTOS call is to insert into the box, the bytes inserted at each call are as per the pointed number of bytes.
3. There can be one of the three types of the mailbox provisions at an RTOS [Figure 8.6(a)].
4. A provision for the multiple unlimited messages in one box that queue-up there.
5. A provision for one message-*write* per box and the next message to be accepted only if the previous one is *read*. A mailbox with a provision for only one message per box becomes full with one message.

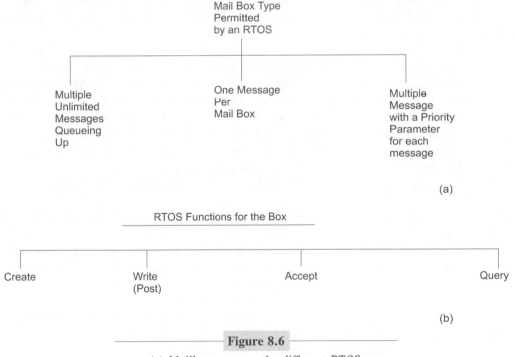

(a)

(b)

Figure 8.6

(a) Mailbox types at the different RTOSs
(b) Initialised and other functions for a mailbox at an RTOS.

A queue may be assumed to be a special case of such a mailbox when there is a provision for a mailbox with multiple messages. Either there is a queue from which a read (deletion) can be on a FIFO basis or there are multiple messages with each message having a priority parameter. The read (deletion) can then only be on priority basis. Even if the messages are accepted in a different priority order, reading is as per the given priority parameter. Such a mailbox is like an ordered list in which writing (insertion) can be in the middle but reading (deletion) is from the top of the list only.

The RTOS functions for the servicing of a mailbox can be as follows.

1. The RTOS_BoxCreate creates a box and initializes the mailbox contents with a NULL pointer.
2. The RTOS_BoxWrite (or Post) sends a message to the box.
3. The RTOS_BoxWait (Pend) waits for a mailbox-message, which is read when received.
4. The RTOS_BoxAccept reads the current message pointer after checking the presence yes or no [No wait.] Deletes the mailbox when read and not needed later.
5. The RTOS_BoxQuery just queries the mailbox when *read* and not needed or used later.

An RTOS provides the IPC functions for creating and using the mailboxes as message pointers or as a message to a destined task.

8.3.5   Pipes

The RTOS pipe functions for servicing are like message-queue functions. The only difference could be in the unlimited or limited pipe size, named or unnamed tasks for read from a byte stream, and named and unnamed tasks for the write into a byte stream.

A message pipe in the strict sense is an IPC queue between two given inter-connected tasks or two sets of tasks. Writing and reading from a pipe is like using a C command *fwrite with a file name* to write into a named file, and C command *fread with a file name* to read into a named file. Pipes are also like Java *PipedInputOutputStreams*. These are the classes that Java defines for input/output streams.

As in the case of the provision for a queue that has multiple messages, a *pipe* also has multiple messages. These are between the two specific tasks (or sets of tasks). A pipe may be assumed to be a special case of a queue from which a *read* (delete) can only be on a FIFO basis and only the specially named sets of tasks can insert and delete through the scheduler.

1. One task in a set of tasks can write through a scheduler *to* a pipe at the back pointer address, *pBACK.
2. One task in a set of tasks can read through a scheduler from a pipe at the front pointer address, *pFRONT.
3. In a pipe there may be no fixed number of bytes per message with an initial pointer for the back and front and there may be no limiting final back pointers. A pipe can therefore be unlimited and have a variable number of bytes between the initial and final pointers.

An example of the need for messaging and thus for inter-task communication using a pipe is a client server network. A client task for an input from a port interconnects with a pipe that reads a byte stream. The port writes into the pipe the byte stream for the server task.

The following are the common RTOS features.

1. Each pipe for a message may need initialisation before using the functions in the scheduler for the message pipe.
2. There may be a provision for multiple connections of the pipes for the multiple destinations of the messages. A number of clients can thus connect with the server.
3. When an RTOS call is to insert into the pipe, a byte is inserted as per the pointed address.
4. A pipe at an RTOS [For example, VxWorks] may be used at a task virtual device like a file device.

Figure 8.7(a) shows three types of functions at an RTOS. One type of functions is for initialisation, creating a pipe, and for defining pipe ID, length, maximum length (not defined in some RTOS) and initial values of two pointers. These are *pFRONT and *pBACK for pipe message-destination (head) and pipe message-source (tail) memory locations, respectively. The second type of functions is for pipe connecting and for defining source ID and destination ID. The third type of functions are for the error-handling functions. Figure 8.7(b) shows pipe-messages in a message-block with its top pointed by *pFRONT and end by *pBACK.

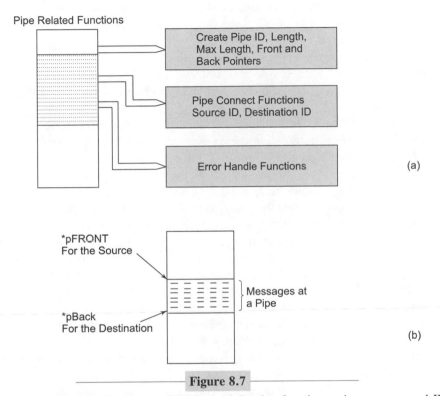

Figure 8.7

*(a) Three types of functions at an RTOS, (initialisation functions, pipe connect and IDs functions, and error handling functions (b) Pipe-messages in a message-block with its top pointed by *pFRONT and end by *pBACK*

An RTOS provides the IPC functions for creating and using the pipes as message queues between source tasks and destination task(s). IPC pipes are used as pipes or file devices. This means a pipe can be opened and closed like a file. It is exactly like a device can be attached or detached from a system.

8.3.6 Virtual (Logical) Sockets

Two physical connector sockets interconnect two systems, for example, a computer COM port with a modem. Similarly, two tasks (or sections of a task) between two sets of the task interconnect through the sockets. [These are virtual (logical), not physical, sockets.]

The use of a virtual (logical) socket for inter-task communication is analogous to the use of virtual sockets for an Internet connection between the browser and the website. Port 80 is conventionally the number at any Web server serving the http websites. The port for the SMTP mail server is 25. The port for the POP3 mail server is 110. Each server or client has an IP address. *The IP address and port number specify a socket on the Internet.*

Assume that using the scheduler, a socket interconnects a byte stream between the source set of tasks *I* and targeted set of tasks *J*. Let there be the four sections or tasks, *a*, *b*, *c* and *d* in task set *I*. There are two sections, *x* and *y* in a set of tasks, *J*. Let the socket be used to send a byte stream from (task set *I*, section *c*) to (task *J*, section *x*). Now, the socket at source (client socket) is specified by the socket at (*I*, *c*) and socket at server by (*J*, *x*).

A socket is a pipe between two specified sections at the specified sets (addresses). Each socket may have the task address (similar to a network or IP address) and a section (similar to a process or port) number. The sections (or ports or tasks) and sets of tasks (addresses) may be on the same computer or on a network.

There has to be a specific protocol in which the messages at the socket interconnect. (I, c) and (J, x). Specification of the section (port) establishes a protocol-based link. The specification of the task (address) numbers establishes a physical link. Figure 8.8 shows the initialised sockets between the client set of tasks and a server set of tasks at an RTOS. Exemplary applications of the sockets are the following. A socket must form before being used for interconnection.

1. One application of sockets is that the tasks in the distributed environment of embedded systems can be interconnected. For example, a network interconnection.
2. A TCP/IP socket is another common application for the Internet. A typical application of a socket is a task receiving a byte stream of TCP/IP protocol at a mobile Internet connection.
3. Yet another application is as following. Consider a mobile phone that has an incoming byte stream, which is a piped input for a certain attending task that receives an SMS message or e-mail through the cellular service.
4. Application as Berkeley Sockets for networking. [It is a standard for a client-server network. An RTOS may provide for POSIX Berkeley Sockets.]
5. An exemplary application is when a task writes into a file at a computer or at a network uses NFS protocol [Network File System Protocol.]
6. Another application of the socket is the interconnection of a task or a section in a set of tasks in an embedded system with another section at a set of tasks in a separate heavyweight process. The scheduler has the socket-connecting functions with the codes specifying the source and destination sections and sets.

A Socket is an IPC for sending a byte stream from (task set-I, section-c) to a (task set-J, section-x). A Socket is a client server or peer-to-peer type of IPC. Now, the socket at the source (client socket) is specified by the socket at (I, c) and the socket at the server by (J, x). A Socket has a number of applications. An Internet socket is for virtual connection between two ports: one port at an IP address to another port at another IP address. An RTOS provides the IPC functions for creating, establishing, using and disconnecting the sockets.

8.3.7 Remote Procedure Calls (RPCs)

There are remote procedure calls (RPCs) in the distributed environment of embedded systems. The RPC provides the inter-task communication when one task is at system 1 and another task is at system 2. Both systems work in the peer-to-peer communication mode, and not in the client-server mode. Each system in peer-to-peer can make RPCs. [The client makes the call that is local or remote and the server response is either remote or local in the client-server calls.]

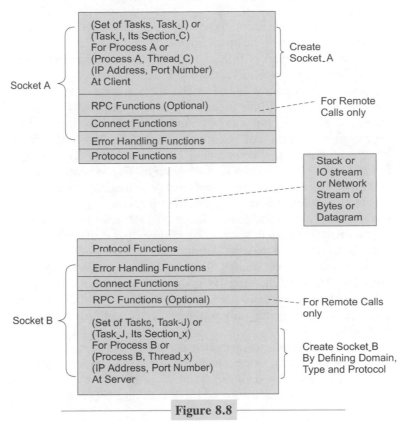

Figure 8.8

Initialised virtual sockets between the client set of tasks and a server set of tasks at an RTOS.

▪ SUMMARY ▪

- A process is a computational unit that processes on a CPU under the control of a scheduling kernel of an operating system.
- A task is a computational unit or set of codes, actions or functions that processes on a CPU under the control of a scheduling kernel of an operating system.
- Each task is an independent process that takes control of the CPU when scheduled by a scheduler at an OS. No task can call another task. Each task is recognised by its TCB (memory block) that holds information of the program counter, memory map, the signal (message) dispatch table, signal mask, task ID, CPU state (registers etc.), and a kernel stack (for executing system calls, etc.). A task is in one of the four states—idle, ready, blocked and running—that are

controlled by the scheduler. [Five in case a task is without infinite-waiting loop—idle, ready, blocked, running and finished.]

- A process may consist of multiple threads that define a minimum unit for a scheduler to schedule the CPU and other system resources. Unix and Java use the threads. Threads are lightweight processes.
- A single CPU system runs one process (or one thread of a process) at a time. A scheduler is essential to schedule a multitasking or multithreading system.
- Often the same data is used in two different tasks (or processes) and if one task interrupts before the operation on that data is completed, then the shared data problem arises. Disabling of interrupts until the completion of the operation by the first task, and then re-enabling interrupts, is one solution. Use of *semaphores (as flags, mutex or counting semaphores)* is another efficient way to solve the shared data problem and run critical section codes.
- A buffer is a memory block for a queue or list of messages or stream of bytes between an output source and input sink. [Examples between the tasks and files, computer and printer, physical devices and network.] It has to be bounded between two limits. It cannot be unlimited or infinite. For example, a print buffer cannot accept unlimited output from a computer. The bounded buffer problem is in synchronizing source and sink. *A producer cannot keep on producing beyond a limit if consumers do not consume. The consumers cannot keep consuming unless the producer keeps producing.* The P and V semaphore functions solve the classical producer–consumer problem when using bounded buffers.
- *The priority inversion* problem and deadlock situation can arise in certain situations when using a semaphore. An operating system should be set up so that it can take care of it by having the appropriate provisions to avoid these situations.
- The RTOS handles inter-process communications between the multiple tasks.
- The RTOS provides for the following IPCs: signals, semaphores, queues, mailboxes, pipes, sockets, and remote procedure calls (RPCs).
- A mailbox may either provide for only one message or multiple messages or for a message pointer in an RTOS.
- A pipe is a queue or stream of messages that connects the two tasks.
- The sockets are used in networks or client servers like communication between the tasks. The RPCs are used for the case of distributed tasks.
- Recently, there has been standardisation of the RTOS and IPC functions.

■ LIST OF KEYWORDS AND THEIR DEFINITIONS ■

- *Process*: A code that has its independent program counter values and an independent stack. A single CPU system runs one process (or one thread of a process) at a time. A **process** is a concept (abstraction). It defines *a sequentially executing (running) program and its state. A state*, during the running of a process, is represented by its status (running, blocked, or finished); its control block, called process control block (PCB) or *process structure*; and its data, objects and resources.
- *Task*: A task is for the service of specific actions and may also correspond to the codes, which execute for an interrupt. A task is an independent process that takes control of the CPU when scheduled by a scheduler at an OS. Every task has a TCB.

- *Task Control Block (TCB)*: A memory block that holds information of program counter, memory map, the signal (message) dispatch table, signal mask, task ID, CPU state (registers etc.), and a kernel stack (for executing system calls, etc.).
- *Task state*: A state of a task that changes on scheduler directions. A task at an instance can be in one of the four states—*idle*, *ready*, *blocked* and *running*—that are controlled by the scheduler.
- *Thread*: A minimum unit for a scheduler to schedule the CPU and other system resources. A process may consist of multiple threads. A thread has an independent process control block like a task control block and a thread also executes codes under the control of a scheduler.
- *Synchronisation*: To let each section of codes, tasks and ISRs run and gain get access to the CPU one after one sequentially, following a scheduling strategy, so that there is a predictable operation at any instance.
- *Inter-process communication*: An output from one task (or process) passed to another task through the scheduler and use of signals, exceptions, semaphores, queues, mailboxes, pipes, sockets, and remote procedure calls (RPCs).
- *Shared Data Problem*: If a variable is used in two different processes (tasks) and if one task interrupts before the operation on that data is completed, then the shared data problem arises.
- *Semaphore*: A special variable (or function) used to take note of certain actions to prevent another task or process from proceeding.
- *Mutex*: A special variable used to take note of certain actions to prevent any task or process from proceeding further and at the same time let another task exclusively proceed further. Mutex helps in mutual exclusion of one task with respect to another by a scheduler in multitasking operations.
- *Counting semaphore*: A semaphore in which the value can be incremented and decremented and which is not a Boolean variable.
- *Buffer*: A memory block for a queue or list of messages or stream of bytes between an output source and input sink, for example between the tasks and files, computer and printer, physical devices and network.
- *P and V semaphores*: The semaphore functions defined in an IEEE standard to be used as mutex or counting semaphore, or to solve the classical producer–consumer problem when using a bounded buffer.
- *Priority inversion*: A problem in which a low priority task inadvertently does not release the process for a higher priority task. An operating system can take care of this it by using appropriate provisions.
- *Deadlock situation:* One task waiting for one semaphore to be released to run a task and another to run different task waiting for another semaphore in order to run. Neither task is able to proceed further. An operating system can take care of this it by appropriate provisions.
- *Message Queue*: A task sending the multiple FIFO or priority messages into a queue for use by another task(s) using queue messages as an input.
- *Mailbox:* A message or message pointer from a task that is addressed to another task.
- *Pipe*: A task sending the messages used by another task using them as input. A pipe can be a device like a file which is also a virtual device.
- *Socket*: It provides the logical link using a protocol between the tasks in a client-server or peer-to-peer environment.
- *Remote Procedure Call:* A method used for connecting two remotely placed methods by first using a protocol for connecting the processes. It is used in the cases of distributed tasks.

■ REVIEW QUESTIONS ■

1. How does a data output generated by a process transfer to another using an IPC (Inter-process Communication)?
2. What are the parameters at a TCB of a task? Why should each task have a distinct TCB?
3. What are the states of a task? Which is the entity controlling (scheduling) the transitions from one state to another in a task?
4. Define a critical section of a task.
5. How is data (shared variables) shielded in a critical section of a process before being operated and changed by anther higher priority process that starts execution before the process finishes?
6. How does use of a counting semaphore differ from a mutex? How is a counting semaphore used?
7. Give an example of a deadlock situation during multiprocessing (multitasking) execution.
8. What are the advantages and disadvantages of disabling interrupts during the running of a critical section of a process?
9. Explain the term multitasking OS and multitasking scheduler.
10. Each process or task has an endless (infinite) loop in a preemptive scheduler. How does the control of resources transfer from one task to another?
11. What is an *exception* and how is an error-handling task executed on throwing the exception?
12. How do functions differ from ISRs, tasks, threads and processes?
13. List the features of P and V semaphores and how these are used as a resource key, as a counting semaphore and as a mutex.
14. What are the situations, which lead to priority inversion problems? How does an OS solve this problem by a priority inheritance mechanism?
15. What is meant by a pipe? How does a pipe differ from a queue?
16. What is meant by a spinning lock? Explain the situation in which the use of the spin lock mechanism would be highly useful to lock the transfer of control to a higher priority task?
17. What is a mailbox? How does a mailbox transfer a message during an IPC?
18. When are sockets used for IPCs? List four examples. When are RPCs used? List two examples.
19. What are the similarities among the process, task and thread? Also list the differences between the process, task and thread.

■ PRACTICE EXERCISES ■

20. Design a table to clearly distinguish the cases when there is concurrent processing of processes with tasks and with threads by using a scheduler.
21. What is the advantage of using a *signal* as an IPC? List the 5 exemplary situations which warrant use of signals.
22. List five exemplary applications of solutions to the bounded buffer problem using P and V mutex semaphores.
23. Every tenth second a burst of 64 kB arrives at 512 kbps in an interval of 100 seconds. Is an input buffer required? If yes, then how much? If yes, then write a program to use the buffer using P and V semaphores.
24. Use a Web search to understand an IEEE-accepted standard, POSIX 1003.1b, in detail.
25. Can different IPCs be used? Given the choice, how would you select an IPC from signal, semaphore, queue or mailbox?

# Chapter 9

Real-Time Operating Systems

What We Have Learnt

The following topics have been covered in the previous chapters:

1. Hardware types—(processor, memories, buses, interfacing circuits, physical ports and virtual devices, timers and real-time clock driven software timers, buses for device networking and their interfacing [Chapters 1 to 3]
2. Details about drivers, writing the device driving interrupt handling routines and details about the interrupt handling [Chapter 4]
3. Procedure-oriented and object-oriented programming concepts, coding of device drivers and application software by using C/C++, using data types and structures [Chapter 5]
4. Program modeling concepts during software implementation in single and multiprocessor systems [Chapter 6]
5. Software Engineering practices in embedded software development processes [Chapter 7]
6. Inter-Process Communication and its use in synchronizing concurrent processes, tasks and threads [Chapter 8]

The following important points that make embedded-system software design challenging were also explained in the previous chapters.

1. There are a number of interrupt sources, physical and virtual. They are processed by interrupt service routines (ISRs) whenever the interrupts occur. Using the ISRs, the system interrupt mechanism lets the processor process a foreground program and an interrupt process. Each ISR has an interrupt latency period after the interrupt source occurrence and before it starts servicing (running) on the processor. An ISR may have a deadline within which it must provide the service and finish the process. Therefore, system software may be such that it assigns higher priorities to few ISRs and sets a priority order for their processing. The system has to

process event-controlled functions and functions have response-time constraints. [Recall Section 4.6.]

2. Modeling of an event-controlled or response-time-constrained function and modeling the scheduling of the real-time functions in the program is by Finite State Machine or Petri Nets. There is inter-process communication (an input for a process on an event, interrupt or finishing certain function or set of instructions) after which a process starts running in a multiprocessor system [Recall Section 6.2.]

3. Modeling of multiprocessor systems and the delays (tokens) between one set of computations and another [Recall Section 6.3.]

4. Inter-process communication between two processes [Recall Section 6.3.]

5. Special issues involved in a real-time system software-development process (analysis, design, implementation and testing) [Recall Section 7.7]

6. Concepts of semaphores (as flags, mutex or counting semaphores) for the IPCs, solving shared data problems and running critical section codes during real-time processing [Recall Section 8.2.]

7. Use of events, signals and exceptions for the error-handling functions as IPCs in order to have the shortest latency period.

8. Use of semaphores, queues, mailboxes, pipes, sockets, and remote-procedure-calls IPCs for scheduling and synchronisation [Recall Section 8.3.]

It is thus clear that embedded software development can be highly complex; there are complex issues in real-time systems software development and there are several components of vastly different functionalities. (i) There is a need for an operating system (OS), which can promptly provide many services like management of the resources, memory, processes, devices, files, I/O subsystems and network subsystems. (ii) Also, there is a need for the OS to be a real-time operating system (RTOS), which can provide the solution of the issues of scheduling and synchronisation, response-time constraints, task priorities, latencies and deadlines and inter-process communications.

An RTOS becomes essential when there is processing and servicing of multiple devices and therefore of multiple tasks with the real-time constraints in a sophisticated application.

Two essential services of RTOS are *inter-process communication* and *scheduling* in addition to other OS services. It has already been seen that there are detailed functions in the RTOS for the inter-process communication mechanism for scheduling and synchronisation between the multiple tasks (processes) and the scheduler in RTOS [Section 8.3].

What We Will Learn

The objective of this chapter is to explain thoroughly the OS and RTOS services and task-states scheduling when there are multiple tasks and ISRs.

In order to fulfil this objective it is necessary to understand the following.

1. (i) Services of kernel, operating system (OS) and real-time operating system (RTOS) and understanding of OS goals, structures, process, memory and devices management, files organisation and management, I/O subsystems and network operating systems (ii) Scheduling Services by three basic strategies for scheduling the multiple tasks—cyclic, preemptive and time slicing

2. Needs of *real time/embedded operating systems* and *real-time task models* and *performance metrics*

3. Handling of interrupts by *Interrupt Service Routines in an RTOS environment*
4. *RTOS* (i) *scheduling* of the tasks and interrupt latencies in cooperative and cooperative with priority ordered list of tasks, time slicing, and preemptive scheduling and (ii) *resource management*
5. Critical Section Handling in priority scheduling cases
6. Fixed real-time scheduling
7. Scheduling of periodic, sporadic and aperiodic tasks by a scheduler
8. Precedence assignment in the scheduling algorithms
9. Advanced scheduling algorithms using the probabilistic Timed Petri nets (Stochastic) and Multi Thread Graph (MTG)
10. Standardisation of the RTOS and IPC functions and list of basic actions in a preemptive scheduler, and providing the relative timings of the various actions to allow optimisation of the processing timings in an application developed using an RTOS
11. Important points to be taken care of during coding for synchronisation between the processes (ISRs, functions, tasks, scheduler functions)
12. OS security issues
13. Embedded LINUX internals
14. Mobile OS

9.1 OPERATING SYSTEM SERVICES

An application programmer for a system can immediately begin using the OS functions that are provided in the given OS without having to write the codes for the services (functions) that follow.

9.1.1 Goals

The OS goals are *'perfection and correctness'* to achieve the following:
1. *Facilitating easy sharing of resources as per schedule and allocations*. [Resources mean processor(s), memory, I/O, devices, virtual devices, system timer, software timers, keyboard, displays, printer and other such resources, which processes (tasks or threads) request from the OS. No processing task or thread uses any resource until it has been allocated by the OS at a given instance.]
2. *Facilitating easy implementation* of the application software with the given system-hardware through the software of the system.
3. Optimally scheduling the processes on one (or more CPUs if available) by providing an appropriate context switching mechanism.
4. *Maximizing the system performance to let different processes (or tasks or threads) share the resources most efficiently with protection and without any security breach*. Examples of a security breach are tasks obtaining illegal access to other task data directly without system-calls, overflow of the stack areas into the memory, and overlaying of PCBs (Section 8.1.1) at the memory.
5. *Providing management functions* for the processes (tasks or threads), memory, devices and I/Os and for other functions for which it is designed.
6. Providing **management and organisation** functions for the I/Os, devices and files and file-like devices.

7. Providing easy interfacing and management **functions for** the network protocols and the **network**.
8. *Providing portability* of the application on different hardware configurations.
9. *Providing interoperability* of the application on different networks.
10. *Providing a common set of interfaces* that integrates various devices and applications through standard and open systems.

The OS goals are perfection, correctness, portability, interoperability, and providing a common set of interfaces for the system, and orderly access and control when managing the processes.

9.1.2 Structures

A system can be assumed to have a structure as per Table 9.1.

Table 9.1
Layered Model of the System

Top-down Structure Layers	Actions
Application Software	—Executes as per the applications run on the given system hardware using the interfaces and the system software
Application Programming Interface (API)	—Provides the interface (for inputs and outputs) between the application software and system software so that it is able to run on the processor using the given system software
System software other than the one provided at the OS	—The OS may not have the functions, for example, for a specific network and for certain device drivers, such as a multi-media device. This layer gives the system software services other than those provided by the OS service functions
OS Interface	—Interface *(for inputs and outputs)* between the above and OS
OS	—Kernel supervisory mode services [Table 9.2], file management and other functions user mode processing services
Hardware–OS Interface	—Interfaces to let the functions be executed on the given hardware (processor, memory, ports and devices) of the system.
Hardware	—Processor(s), memories, buses, interfacing circuits, ports, physical devices, timers, and buses for devices networking.

When using an OS, the processor in the system runs in two modes.
1. *User Mode*: The user process is permitted to run and use only a subset of functions and instructions in the OS.
2. *Supervisory Mode*: The OS runs the privileged functions and instructions in a protected mode and the OS (more specifically, the kernel) only accesses a hardware resource. [The term kernel means nucleus.]

The OS structure consists of kernel and other service functions outside the kernel.

System software includes the kernel.

9.1.3 Kernel

The OS is the middle layer between the application software and system hardware. An OS includes some or all of the following structural units.

1. Kernel.
2. File Management in case it is not a part of the kernel in the given OS [Section 9.1.6] and any other needed functions not provided for at the kernel.

The kernel is the basic structural unit of any OS. It can be defined as a secured unit of an OS that operates in the supervisory mode while the remaining part and the application software operates in the user mode. Table 9.2 gives the functions (services) in the kernel, they are as per the OS design.

Table 9.2
Kernel Services in an OS

Function	Actions
Process Management: Creation to deletion	—Enables process creation, activation, running, blocking, resumption, deactivation and deletion and maintains process structure at a PCB (Process Control Block) [Section 9.1.4]
Process Management: Processing resource requests	—*Processing resource requests* by processes made either by making calls that are known as system calls or by sending message(s) [Section 9.1.4]
Memory Management: Allocation and de-allocation	—*Memory allocation, de-allocation and management*. It also restricts the memory access region for a task. There may be dynamic memory-allocations [Section 9.1.5]
Processes Management: Scheduling	—Processes *scheduling*. For example, in the cyclic scheduling or priority scheduling mode. [Section 9.4]
Process Management Inter-process Communication (IPC) (communication between Tasks, ISRs, OS functions) Services	—*Processes synchronizing* by sending data as messages from one task to another. The OS effectively manages shared memory access by using the IPC signals, exception (error) handling signals, semaphores, queues, mailboxes, pipes and sockets. [Section 8.3]
I/O Management	—Character or block I/Os management. For example, to ensure actions such that a parallel port or serial port obtains access to only one task at a time. [Section 9.2]
Interrupts Control (by handling ISR) Mechanism	—Refer to Section 9.5
Device Management.	—A physical device management is such that it obtains access to one task or process only at an instant. Management of *virtual devices* like pipes and sockets is also provided. [Section 9.1.6] Device manager components are the device drivers and device ISRs (Device Interrupt Handlers) and the resource managers for the devices.
Device Drivers	—Facilitating the use of a number of physical devices like keyboard, display-systems, disk, parallel port, network interface cards, network devices and *virtual devices*.

Note: In models of processes controlled by an OS, *process* also means a task in multitasking models, and a thread in multithreading models. [Refer to Section 8.1.]

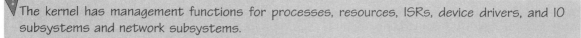

The kernel has management functions for processes, resources, ISRs, device drivers, and IO subsystems and network subsystems.

9.1.4 Process Management

A. Process Creation

The initial process is a process that executes memory instructions at the reset of the processor and then calls the OS. The processor will start executing all the subsequent processes created. Creation means defining address space (memory blocks) for the created process and defining the resources for the process. Processes can be created hierarchically.

The process manager allocates the PCB (or TCB in case a task represents a process) when it creates the process and later manages it. The other OS units can send it queries when necessary. Recall the PCB and TCB described in Section 8.1. (Its structure is explained here again.) It is a *process descriptor* used by the process manager. A PCB or TCB (Section 8.1.2) describes the following:

 (i) Context [Processor status word, program counter, stack pointer and other CPU registers at the instant of the last instruction run executed when the process was left and processor switched to other process)

 (ii) Process stack pointer

 (iii) Current state [Is it created, activated or spawned? Is it running? Is it blocked?]

 (iv) Addresses that are allocated and that are presently in use

 (v) Pointer for the parent process in case there exists a hierarchy of the processes

 (vi) Pointer to a list of daughter processes (processes lower in the hierarchy)

 (vii) Pointer to a list of resources, which are usable (consumed) only once. For example, input data, memory buffer or pipe, mailbox message, semaphore. [There may be producers and consumers of these resources.]

 (viii) Pointer to a list of resource-types usable more than once [A resource type example is a memory block. Another example is an I/O port.] Each resource type will have a count of these types. For example, the number of memory blocks or the number of I/O ports

 (ix) Pointer to a queue of messages. It is considered as a special case of resources that are usable once. It is because messages from the OS also queue up to be controlled by a process

 (x) Pointer to access-permissions descriptor for sharing a set of resources globally and with another process

 (xi) ID which is made by the process manager

B. Management of the created processes

Process management enables process *creation*, *activation*, *running*, *blocking*, *resumption*, *deactivation* and *deletion*. A process manager facilitates the following: Each process of a multiple-process (or multitasking or multithreading) system is executed such that a process-state can switch from one to another. A process does the following sequential execution of the states: 'created', 'ready or activate', 'spawn' (means create and activate), 'running', 'blocked', 'resumed' and 'finished' and 'ready' after 'finish' (when there is an infinite loop in a process) and finally 'deactivated' or 'deleted'. [Blocking and resuming can take place several times in long processes.]

Recall the definitions of process, task and thread in Section 8.1. A process (or task or thread) is considered a unit in which sequential running is feasible only under the control of an OS, with each process having an independent control block (descriptor of the processor at an instant). [Recall the PCB and TCB described in Section 8.1.] Which entity is responsible for controlling a process execution? It is the '*process manager*' unit of the OS.

The process manager executes a process-request for a resource or OS service and then grants that request to let processes share resources. A running process requests by two means, which are explained in Table 9.3.

Table 9.3
Request for a Resource or OS Service by a Running Process

Request Method	Explanation
Message(s)	A process running on user mode generates and puts (sends) a message so that the OS lets the requested resource (for example, input from a device or from a queue) use or run an OS service function (for examples, redefine system-clock rate or define delay period after which the process needs to be run again).
System Call	A call to a function at the OS. First an instruction is issued to trap the processor and switch it to a supervisory mode. The OS then executes a function like a library function (Section 5.1). On finishing the instructions of a called function, the processor switches back from the supervisory mode to user mode and lets the calling process run.

The process manager (i) makes it feasible for a process to sequentially execute or block when needing a resource and to resume when it becomes available, (ii) implements the logical link to the resource manager for resources management (including scheduling of a process on the CPU), (iii) allows specific resources sharing between specified processes only, (iv) allocates the resources as per the resource allocation mechanism of the system and (v) manages the processes and resources of the given system.

A Process Manager creates the processes, allocates a PCB to each, manages access to resources and facilitates switching from one process state to another. The PCB defines the process structure for a process in a state.

9.1.5 Memory Management

A. Memory Allocation

When a process is created, the memory manager allocates the memory addresses (blocks) to it by mapping the process-address space. [Refer to Section 8.1.]

B. Memory Management after Initial Allocation

The memory manager of the OS has to be secure, robust and well protected. There must not be any errors arising out of memory leaks and stack overflows. Memory leaks are attempts to write in the memory block not allocated to a process or data structure. Stack overflow means that the stack exceeds the allocated memory block(s) when there is no provision for additional stack space.

The memory manager manages (i) the use of a memory address space by a process, (ii) specific mechanisms to share the memory space, (iii) specific mechanisms to restrict sharing of a given memory space and (iv) optimisation of the access periods of a memory by using a hierarchy of memories (caches, primary and external secondary magnetic and optical memories). Remember that the access periods are in the following increasing order: caches, primary and external secondary magnetic and then optical. Table 9.4 gives the memory-managing strategy for a system.

Table 9.4
Memory-Managing Strategy for a System

Managing Strategy	Explanation
Fixed-blocks Allocation	Memory address space is divided into blocks—the processes having small address spaces getting a fewer number of blocks, and processes with big address spaces getting a greater number of blocks.
Dynamic-blocks Allocation	Memory address space is divided into blocks—the processes having small address spaces getting a fewer number of blocks, and processes with big address spaces getting a greater number of blocks to start with. The memory manager later allocates variable size blocks (in units of, say, 64 or 256 bytes) dynamically allocated from a free (unused) list of memory blocks description table at the different computation phases of a process.
Dynamic Page Allocation	The memory has fixed sized blocks called pages and the manager allocates the pages dynamically with a page descriptor table.
Dynamic Data Memory Allocation	The manager allocates memory dynamically to different data structures like nodes of a list, queues, and stacks.
Dynamic Address Relocation	The manager dynamically allocates the addresses initially bound to the relative addresses after adding the relative address with a relocation register. The memory manager now dynamically changes only the contents of a relocation register. It takes into account a limit-defining register so that the relocated addresses are within the limit of available addresses. This is also called run-time dynamic address binding.
Multiprocessor Memory Allocation	Refer to Section 6.3 and Figure 6.7. The manager adopts an allocation strategy—the memory is shared with tight coupling between two or more processors, is shared with loose coupling or there are multi-segmented allocations.

The memory manager allocates memory to the processes and manages it with appropriate protection. There may be static and dynamic allocations of memory. This optimizes the memory needs and memory utilization.

9.1.6 Device Management

Recall Section 4.2. There are number of device driver ISRs in a system, each device or device function having a separate driver as per its hardware. A device manager is the software that manages these. Recall Section 1.4.5. An OS device manager provides and executes the modules for managing the devices and their drivers ISRs.

1. It manages the physical as well as virtual devices like the pipes and sockets through a common strategy.
2. Device management has three standard approaches to three types of device drivers: (i) Programmed I/Os by polling the service need from each device. (ii) Interrupt(s) from the device driver ISRs and (iii) DMA operation used by the devices to access the memory. Most common is the use of device driver ISRs.
3. A device manager has the functions given in Table 9.5.

Table 9.5
Functions of a Device Manager

Function	Action(s)
Device Detection and Addition	Provides the codes for detecting the presence of various devices and then initializing or testing them, and configuring them for the use by the OS functions.
Device Deletion	Provides the codes for denying the device resources.
Device Allocation and Registration	Allocates and registers the port (it may be a register or memory) addresses for the various devices at distinctly different addresses and also includes codes for detecting any collision between them.
Detaching and Deregistration	Detaches and deregisters the port (it may be a register or memory) addresses for the various devices at distinctly different addresses and also includes codes for detecting any collision between existing addresses in case of addresses reallocation to the remaining attached (registered) devices.
Restricting Single Device	Restricts a device access to one process (task) only, at an instant.
Device Sharing	Permits sharing of access of a device to the set of processes, but to one process (task) at an instant.
Device Buffer Management	Device hardware may merely have a single byte buffer, or double buffer or 8-byte buffer. A Device Buffer Manager uses a memory manager to buffer the I/O data streams from the device that sends the data and manages such that the computations are at a rate faster than the rate of receiving the data from the device. [A computation example is deciphering the input data. If the deciphering rate is slower than the receiving rate, the buffer(s) will soon choke.] Also used are the multiple buffers and producer-consumer type bounded buffers. [Section 8.2.(vi) and Example 8.2]
Device Queue, Circular-queue or blocks of queues Management	Device IO data streams from the device can be organised as the queues, circular queues and blocks of queues. [Section 5.5.]
Device Driver	Unix Device Driver components are (i) Device ISR, (ii) Device initialisation codes (codes for configuring device control registers) and (iii) System initialisation codes, which run just after the system resets (at bootstrapping). When a device driver is a part of the OS, the device manager effectively operates and provides an optimum performance, after adopting an appropriate strategy. The manager coordinates between application-process, driver and device-controller. A process sends a request to the driver by an interrupt; and the driver provides the actions by executing an ISR. The device manager polls the requests at the devices and the actions occur as per their priorities. The device manager manages I/O Interrupts (requests) queues. The device manager creates an appropriate kernel interface and API and that activates the control register's specific actions of the device. [Activates device controller through the API and kernel interface.]

Function	Action(s)
Device Access Management	(i) Sequential access (ii) Random access (iii) Semi-Random Access and (iv) Serial Communication, which may be by UART or USB [Chapter 3.] The device manager provides the necessary interface.

Table 9.6 gives the set of OS command functions for a device.

Table 9.6

Set of Command Functions for Device Management

Commands	Action(s)
create and open	*Create* is for creating and *open* is for creating (if not created earlier) and configuring and initializing the device.
write	Write into the device buffer or sending output from the device.
read	Write from the device buffer or reading input from the device.
ioctl	Specified device configured for specific functions and given specific parameters. For example, status = ioctl (fd, FIOBAUDRATE, 19200). The *fd* is the device descriptor (an integer returned when the device is opened) and FIOBAUDRATE is the function that takes value = 19200 from the argument. This configures the device for operation at 19200-baud rate.
close and delete	*Close* is for deregistering the device from the system and *delete* is for close (if not closed earlier) and detaching the device.

Note: (1) There are two types of devices: char devices and block devices. [Refer to Table 4.1 for definitions.]
(2) For hardware devices, a device ISR can also be called a system ISR or a system interrupt handler.
(3) ioctl is a powerful function with three arguments. Examples of its use are as follows. (*i*) Accessing specific partition information. (*ii*) Defining commands and control functions of device registers. (*iii*) I/O channel control (*iv*) choosing a command number for a device control (for example, 1 for read, 2 for write). (*v*) Network devices control by defining baud rate or other parameters. What will it be used for? It is as per function defined, as a second argument. Controlled device will be according to the first argument. Values needed by the defined function are at the third argument.

A device driver ISR uses several OS functions. Examples are as follows: *intlock* ( ) to disable device-interrupts systems, intUnlock ( ) to enable device-interrupts. *intConnect* ( ) to connect a C function to an interrupt vector. [The interrupt vector address for a device ISR points to its specified C function.] Function intContext ( ) finds whether an interrupt is called when an ISR was in execution.

UNIX OS makes it feasible for devices and files to have an analogous implementation as far as possible. A device has *open* ( ), *close* ( ), *read* ( ) and *write* ( ) functions analogous to a file *open, close, read* and *write* functions [Section 9.1.7.] APIs and kernel interfaces in BSD (Berkley Sockets for Devices) UNIX are *open, close, read* and *write. There are the following in-kernel commands. (i) select,* which is to first check whether a read or write will succeed. (*ii*) *ioctl* to transfer driver specific information the device driver. [For example, baud rate in Table 9.6.] (*iii*) *stop* to cancel the output activity from the device. (*iv*) *strategy* to permit a block *read or write* or character *read or write.*

The Device Manager initializes, controls, and drives the physical and virtual devices of the system. The main classes of devices are char devices and block devices. Device driver functions may be similar to file functions, open, read, lseek, write and close.

9.1.7 File System Organisation and Implementation

A file is a named entity on a magnetic disc, optical disc, or system memory. A file contains the data, characters and texts. It may also have mix of these. ***Each OS may have differing abstractions of a file***. (*i*) A *file* may be a named entity that is a structured record as on a disk having random access in the system. (*ii*) A *file* may be a structured record on a RAM analogous to a disk and may also be either separately called '*RAM disk*' or simply, a 'file' (virtual device). (*iii*) A *file* may be an unstructured record of bits or bytes. (*iv*) A *file* device may be a pipe-like device for inter-process communication.

It is necessary to organize the files in a systematic way and to have a set of command functions. Table 9.7 gives these functions for POSIX file systems.

Table 9.7
Set of Command Functions in the POSIX File System

Command in POSIX	Action(s)
open	Functions for creating the file
write	Writing the file
read	Reading the file
lseek (List seek) ***or set the file pointer***	Setting the pointer for the appropriate place in the file for the next read or write
close	Closing the file

Note: (1) File devices are block devices in UNIX. Linux permits the use of a block device as a char device also. This is because from *block-device* to *char-device, Linux has an additional* interface. In other words, the kernel interface is identical for the char and block devices in Linux and not in UNIX.

(2) The file on the RAM that is hierarchically organised is known as RAM disk. RAM memory storage is analogous to that on the disk and accessing is also analogous to a disk. [Recall an access from a disk; first directory, then subdirectory, then folder and then subfolder. There is a hierarchy.]

(3) Unix has a structured file system with an unstructured hardware interface. Linux supports different standard file-systems for the system.

Should a file having integers differ from a file having bytes? Should a file having bytes differ from file having characters? Due to the differing approaches to device and file management interfaces, the development of a set of standard interfaces becomes necessary. Only then can systems be portable. A standard set of interfaces is called POSIX, from IEEE. POSIX stands for Portable Operating System Interface standard for coding programs when using the multiple threads. The X refers to the interfaces being similar to ones in UNIX. It is according to the definitions at the AT&T UNIX System V Interface. POSIX defines the functions: open, close, read, write, lseek and fcntl. Function lseek is to move the pointer position in the byte stream. Function fcntl is for file control. The POSIX standard for file operations are as the operations on a linear sequence of bytes.

Window NT assumes a file as the named entity for a record of bytes placed sequentially and the OS has the command functions, CreateFile, ReadFile, WriteFile and SetFilePointer, and CloseHandle for creating a file, reading a file, writing a file and setting the file pointer from the present to new location.

A file in UNIX has *open* ( ), *close* ( ), *read* ( ), *write* ( ) functions analogous to a device, *open*, *close*, *read* and *write* functions. The BSD UNIX interface differs slightly from UNIX.

There are two types of file systems.

- *Block File System:* Its application generates records to be saved into the memory. These are first structured into a suitable format and then translated into block-streams. A file pointer (record) points to a block from the start to the end of the file.

- *Byte Stream File System:* Its application generates record streams. These streams are to be saved into the memory. These are first structured into a suitable format and then translated into byte-streams. A file pointer (byte index) points to a byte from the start index = 0 to N-1 in a file of N bytes.

Just as each process has a processor descriptor (PCB), a file system has a data-structure, called file-descriptor as per Table 9.8. [The structure differs from one file manager to another.] File-descriptor, *fd,* for a file is an integer, which returns on opening a file. *fd* points to the data-structure of the file. *fd* is usable until the closing of the file.

Table 9.8

Data Structure of File-Descriptor in a Typical File System

File-Descriptor	Meaning(s)
Identity	Name by which a file is identified in the *application*
Creator or Owner	Process or program by which it was created
State	A state can be 'closed', 'archived' (saved), 'open executing file' or 'open file for additions'.
Locks and Protection fields	O_RDWR file opens with read and write permissions, O_RDONLY file opens with read only permissions, O_WRONLY file opens with write only permissions.
File Info	Current length, when created, when last modified, when last accessed
Sharing Permission	Can be shared for execution, reading, or writing
Count	Number of Directories referring to it
Storing Media Details	Blocks transferable per access

A file manager creates, opens, reads, seeks a record, writes, and closes a file. A file has a file descriptor.

9.2 I/O SUBSYSTEMS

I/O ports are the subsystems of OS device management systems. Drivers that communicate with the many devices use the subsystems. I/O instructions depend on the hardware platform. I/O systems differ in the different OSs. Subsystems of a typical I/O system are as given in Table 9.9.

Table 9.9

I/O Subsystem in a Typical I/O System in an OS

Subsystems Hierarchy	Action(s) and layers between the subsystems
Application	An application having an I/O system. There may be a sub-layer between the application and I/O basic functions, such as buffered I/O or I/O library functions for read and write functions.
I/O Basic functions	These are device-independent OS functions. There may be a sub-layer between the basic I/O functions and I/O device-driver functions—like file-system functions for read and write.
I/O device driver functions	These are device dependent OS functions. There may be a sub-layer between the I/O's basic functions and I/O device-driver functions [I/O basic functions are buffered I/O or file (block) read and write functions.] A driver may interface with a set of library functions. For example, for serial communication.
Device Hardware or Port or I/O Interface card	Serial Device or Network.

Note: There may be separate functions for synchronous and asynchronous categories. For example, in POSIX, there are the following asynchronous functions: *aio_read* ( ) and *aio_write* for the asynchronous read and write in an I/O system. *aio_list* ( ) is to initiate a list of certain maximum asynchronous I/O port requests. *aio_error* ( ), *aio_cancel, aio_suspend* are functions for asynchronous I/O error status retrieval and for canceling and suspending I/O operations, respectively. Suspension is until the next port device interruption or until a timed out. *aio_return* returns the status of completed operations.

I/O subsystems are an important part of OS services. Examples are the UART access and the parallel port access.

9.3 NETWORK OPERATING SYSTEMS

Refer to the standard texts on Computer Network and Internet and Web Technologies for the meanings of the terms used below. Network means exchange of data frames in the LANs or packets between different host devices (interfaces). Exchange is either in client-server mode or peer-to-peer mode. TCP/IP, Telnet and FTP are important protocols using which byte streams are placed on the net after formatting into frames or packets. Communication between the *char* or block *device*s is different from the network hosts. A case study described in Section 11.2 clarifies the complexity and basic differences between the two. It explains the conversion of byte streams from an application into a protocol stack for the network. Linux or Unix has network subsystems like the I/O subsystems for network interfaces (devices).

Network Operating System (NOS) at Webopedia (Web Encyclopedia) is defined as follows [http://www.webopedia.com/TERM/N/operating_system.htm]:

"An operating system that includes special functions for connecting computers and devices into a local area network (LAN)." Another definition of 'NOS' is "an OS that has functions for protocol-stacks (for example TCP/IP or Ethernet 802.3) plus network device drivers".

'NOS' provides remote logins, file transfers and interconnection through network devices. The kernel of NOS handles the address resolution issues, packet collection, identification and dispatches to the appropriate network device. These functions are process independent and therefore the use of NOS becomes essential for systems (called hosts) on the network.

BSD 4.3 is a 'NOS'. Its kernel, in addition to Unix, supports network computing. It has socket interfaces. [Refer to Figure 8.8.] A process can be referred to by its socket address on a network. [Recall the file device. Each communication system port has separate socket addresses. The socket address is referenced to the specific host just as *fd* is referenced to a file.] *Servers and routers are special purpose NOSs.*

Unix, Mac OS, Linux and OS/2 have *networking functions*. Linux Networking gives firewall services and web services also. OS/2 supports TCP/IP, Peer-to-Peer networks, firewalls and conventional LAN-protocols. Windows OS provides dialup networking, remote access and connection sharing. Is 'NOS' different from the common OS? The basic OS has *networking functions* built-in but a ***Network operating system*** usually has ***software that enhances the basic OS by adding network operation features.***

Examples of NOS are as follows:

1. 3+Share
2. Arcnet systems
3. Artisoft LANtastic
4. AT&T StarGroup (non-LAN Manager)
5. Banyan VINES
6. DEC *Pathworks* and PCSA
7. IBM PC LAN
8. IBM DOS LAN Requester Version 1.30 or Earlier
9. Microsoft LAN Manager and 100-Percent Compatible Networks
10. Microsoft MS-Net and 100-Percent Compatible Networks
11. Net/One PC
12. Novell NetWare for basic server for files and print and web-based networking capabilities
13. PC-NFS
14. TCS 10Net or DCA 10Net

> A network operating system is an OS that includes special functions for connecting computers and devices into a local area network (LAN). It is an OS that has functions for protocol-stacks (for example TCP/IP or Ethernet 802.3) plus network device drivers.

9.4 REAL-TIME AND EMBEDDED SYSTEM OPERATING SYSTEMS

9.4.1 Real-Time Operating System

Real-time operating system (RTOS) software has the operating system structure units given in Table 9.10.

Table 9.10
RTOS Services

Function	Activities
Basic OS Functions	Process Management, Resources Management, Memory Management, Device Management, I/O Devices subsystems and Network Devices and subsystems management
RTOS Main functions	Real-time Task Scheduling and Interrupt-latency control (Section 4.6) and use of timers and system clocks
Time Management	Time allocation and de-allocation to attain efficiency in given timing constraints
Predictability	A predictable timing behaviour of the system and a predictable task-synchronisation
Priorities Management	Priorities Allocation and Priorities Inheritance
IPC Synchronisation	Synchronisation of tasks with IPCs
Time Slicing	Time-slicing of the processes execution
Hard and soft real-time operability	Hard real-time and soft real-time operations [Hard real-time means strict adherence to each task schedule. Soft real-time means that only the precedence and sequence for the task-operations are defined.]

Section 5.7.3 describes the use of a readied task list by the RTOS scheduler. Subsection 9.4.3 describes three scheduling methods for scheduling the multiple tasks from that list.

An RTOS is an OS for response time–controlled and event-controlled processes. An RTOS is an OS for embedded systems, as these have real-time programming issues to solve.

9.4.2 When is an RTOS Necessary and When is it Not Necessary in the Embedded Systems?

Section 8.1.2 introduces the tasks, and Section 9.4.1 the duties, of the RTOS functions. Does an embedded system always need an RTOS? The answer is no, not always. Software for a large number of small-scale embedded systems use no RTOS and these functions are incorporated into the application software.

Recall the list of the functions of an RTOS in Section 9.4.1. Following are the methods used in a small-scale embedded system (An RTOS is not needed).

1. Instead of the memory allocation and de-allocation functions of RTOS, the 'C' function, melloc ( ) and free ( ) can be used for memory allocation and freeing, respectively. ['*delete* ( )' frees the memory in C++ functions.]

2. Instead of RTOS functions for restricting the memory access to various addresses, the use of C++ classes can provide data encapsulation features and restrict the memory access addresses.

3. Instead of using the RTOS kernel for scheduling tasks, function queues can be used. Coding can similarly be done and code design is the simplest. It applies to those cases in which there is sequential use of all the available resources. Recall Examples 5.4 and 5.6 described earlier. [Refer to Section 5.4.6 (iv)]. A 'Function Queue Scheduling' algorithm runs at the main function

in Example 5.4. There is 'Function Queuing on interrupts' in Example 5.6. The queue is later called from the main function for the queue functions (Example 5.5).

4. Software can directly handle the inter-process (task) communication (Section 8.3) also, without recourse to the RTOS. Software codes synchronize and send data from one function to another by effectively managing shared memory access.

5. With a user designed device manager and device driver functions I/O management becomes possible even in the absence of an RTOS [Examples in Chapter 4].

6. Instead of using the RTOS for file management, it can be done by standard 'C' functions, *fopen* ( ), *fread* ( ), *fwrite* ( ) and *fclose* ( ).

7. Device Drivers for keyboard, display, parallel port, network, pipe and sockets can be written easily in a small system.

However, using an RTOS becomes essential in cases of multiple tasks for the following functions that provide the perfection, correctness, protection and security features of any kernel in OS.

1. Memory allocation and de-allocation and restricting the memory accesses for the stack and other critical memory blocks. Allocation of the memory dynamically is also feasible.

2. Effectively scheduling and running and blocking of the tasks in cases of many tasks [Section 9.6].

3. A common and effective way of handling of the hardware source calls from the interrupts [Subsection 9.5.1.]

4. Recall Figure 5.8. It shows the six tasks and four ISRs with three tasks in an initiated (ready) tasks list in a programming model for multitasking. An RTOS provides the effective handling in such a situation.

5. Inter-task (Inter-Process) Communication and its synchronisation [Section 8.3].

6. I/O Management with devices, files, mailboxes, pipes and sockets becomes simple using an RTOS.

7. Effective management of the multiple states of the CPU and internal and external physical or virtual devices. Assume that the following actions are concurrently needed in an application. (i) Physical devices timer, UART and keyboard have issued the interrupts and the service routines are to be executed. (ii) A file is taken as a virtual device. The file also must be opened with its pointer to its first record. (iii) A physical timer is to configure its control register. (iv) Another timer gets a count-input from the system clock. (v) A virtual device, a file, gets the inputs for writing on to it. (vi) A timer status changes on time-out and this generates a need for its service. (vii) A file status is to change on transfer of all needed records to it. (viii) A timer is to execute a service routine on time-out. (ix) A file needs execution of a function, *close* ( ). By effectively using a common method to handle these needs, the RTOS solves the problem.

8. Recall Section 8.2.2. It described the uses of semaphores by tasks or for the critical sections in a task or function. [Critical section indicates a section of codes or a resource or codes that must run without blocking. One critical situation is when there is a shared data or resource with the other routines or tasks.] RTOS provides for effective handling of such a situation.

An RTOS may not be necessary in a small-scale embedded system. An RTOS is necessary when scheduling of multiple processes, ISRs and devices is important. An RTOS is must to monitor the processes that are response time–controlled and event-controlled processes.

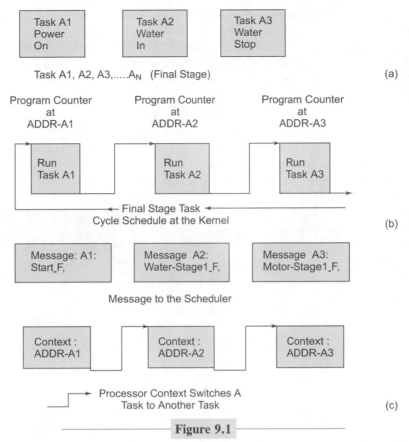

Figure 9.1

(a) First three tasks in a set of tasks A_1 to A_N into which the embedded software is broken for the example in the text (b) Cyclic scheduling (c) Messages from the scheduler and task programs contexts at various instances.

9.4.3 Schedule Management for Multiple Tasks by an RTOS

An RTOS executes the codes for the multiple tasks as per the predefined scheduling strategy. First consider a round robin scheduling by a cyclic scheduler () function by a simple example. Consider an embedded system—an automatic washing machine. There are many tasks and each task is independent. First three tasks are task A1, task A2 and task A3 in a set of tasks A_1 to A_N. Figure 9.1(a) shows the first three tasks of the multiple processes in the embedded-software of the system. The scheduler first starts the task A1 waiting loop and waits for the message A1 from task A1.

1. Task A1: The task is to reset the system and switch on the power if the door of the machine is closed and the power switch is pressed once and released to start the system. The Task1 waiting loop terminates after detection of two events: (i) door closed and (ii) power-switch pressed by the user. At the end, task1 sets a flag start_F, which is a message; *message A1*, to the scheduler for task A2 to start executing its code.

2. Task A2: The scheduler waits for the *message A1* for start_F setting and if that event occurs task A2 starts. A bit is set to signal water into the wash tank and repeatedly checks for the

water level. When the water level is adequate the flag Water-Stage1_F is set, which is a message, *message A2*, to scheduler for task A3 to start executing its code.

3. Task A3: The scheduler waits for the message A2 for the Stage1_F setting and, if that event occurs, task A3 starts. A bit is set to stop filling water and another bit sets to start the wash tank motor. Then a flag, motor-stage1_F is set, which is a message, *message A3*, to the scheduler for the next task to start executing its code.

Figure 9.2(b) shows the scheduler for this example. It is a cyclic scheduler, called a round robin mode scheduler. Figure 9.2(c) shows the task's program contexts at the various instances. Task A1 context has a pointer for task A1, ADDR_A1. Task A2 context has a pointer for task A2, ADDR_A2. Task A3 context has a pointer for task A3, ADDR_A3.

Sections 9.6.1 and 9.6.2 will describe the cyclic scheduling in detail.

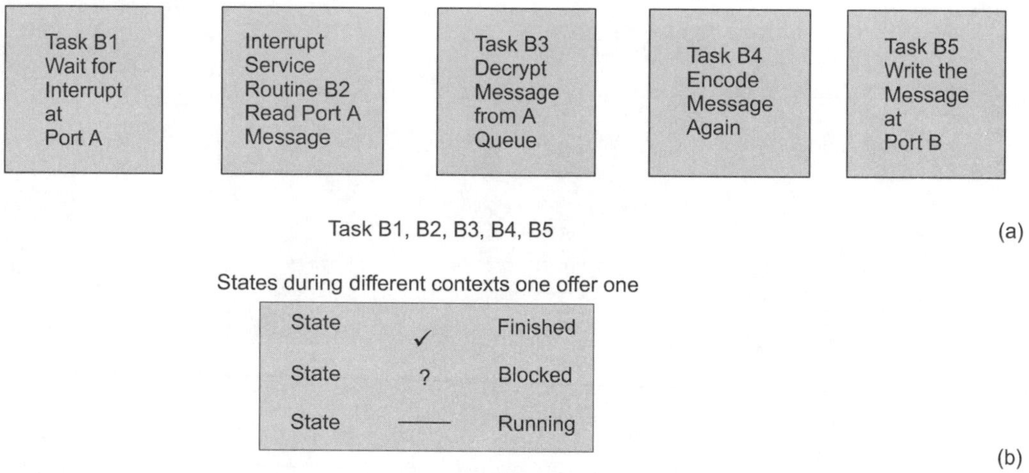

Task B1 Wait for Interrupt at Port A	Interrupt Service Routine B2 Read Port A Message	Task B3 Decrypt Message from A Queue	Task B4 Encode Message Again	Task B5 Write the Message at Port B

Task B1, B2, B3, B4, B5   (a)

States during different contexts one offer one

State	✓	Finished
State	?	Blocked
State	——	Running

(b)

Preemptive scheduling by kernel with preemption on interrupt

Procedure Context	Task B1	Task ISR B2	Task B3	Task B4	Task B5	Saved Context
B3			——			
B1	——		?			B3
B2	✓	——	?			B3
B3	✓	✓	——			
B4	✓	✓	✓	——		
B5	✓	✓	✓	✓	——	
B1	——				?	B5
B2	✓	——			?	B5
B3	✓	✓	——		✓	

(c)

Figure 9.2

(a) First five tasks B1 to B5 (b) The symbols used for the states in a preemptive scheduling
(c) The task programs contexts at the various instances.

Now consider a preemptive scheduler by another simple example. Suppose there is a stream of coded messages arriving to port A of an embedded system. It then decrypts and re-transmits to port B after encoding each decrypted message. [Recall Example 4.1.] Figure 9.2(a) shows the tasks for the multiple processes of this application. Five processes are executed at five tasks, B1, B2, B3, B4 and B5. The order of priorities is as follows.

1. Task B1: Check for a message at port A
2. Task B2: Read Port A.
3. Task B3: Decrypt the Message.
4. Task B4: Encode the Message.
5. Task B5: Transmit the encoded message to Port.

Figure 9.2(b) gives the symbols used to show the preemptive scheduling by the kernel preemptive scheduler in figure 9.2(c). It is a priority-based preemption by the scheduler. A higher priority task takes control from a lower priority task. A higher priority task switches into the running state after blocking the low priority task. The context saves on the preemption. Figure 9.2(c) shows the followings.

1. At the first instance (first row) the context is B3 and task B3 is running.
2. At the second instance (second row) the context switches to B1 as context B3 saves on interrupt at Port A and task B1 is of highest priority. Now task B1 is in a running state and task B3 is in a blocked state. Context B3 is at the task B3 stack.
3. At the third instance (third row) the context switches to B2 on interrupt, which occurs only after task B1 finishes. Task B1 is in a finished state, B2 in running state and task B3 is still in the blocked state. Context B3 is still at the task B3 stack.
4. At the fourth instance (fourth row) context B3 is retrieved and the context switches to B3. Tasks B1 and B2, both of higher priorities than B3, are finished. Tasks B1 and B2 are in finished states. Therefore, Task B3 blocked state changes to running state and B3 is now in running state.
5. At the fifth instance (fifth row) the context switches to B4. Tasks B1, B2 and B3, all of higher priorities than B4, are finished. Tasks, B1, B2 and B3 are in the finished states. B4 is now in a running state.
6. At the sixth instance (sixth row) the context switches to B5. Tasks B1, B2, B3 and B4, all of higher priorities than B5, are finished. Tasks, B1, B2, B3 and B4 are in the finished states. B5 is now in a running state.
7. At the seventh instance (seventh row) the context switches to B1 as context B5 is saved on interrupt at Port A, and task B1 is of highest priority. Now task B1 is in a running state and task B5 is in a blocked state. Context B5 is at the task B5 stack.
8. At the eighth instance (eighth row) the context switches to B2 on interrupt, which occurs only after task B1 finishes. Task B1 is in a finished state, B2 in running state and task B5 is still in the blocked state. Context B5 is still at the task B5 stack.
9. At the last instance (last row) the context is B3 and task B3 is running. The tasks B1, B2 and B5 are in the finished state. There is no context at any task context.

Section 9.6.4 describes preemptive scheduling in detail.

9.4.4 Scheduling of Multiple Tasks in Real Time by RTOS

An RTOS lets the system schedule the various tasks in real time. A real-time system responds to the event within a bound time limit and within an explicit time. A scheduler for the time-constrained tasks can be understood by a simple example.

Task C1 Check Message at Port A Successive 10 mS	Task C2 Read Port A and Place it at Queue	Task C3 Decrypt Queue Messages	Task C4 Encode Queue Messages	Task C5 Transmit by Writing at Port B

Task C1 to Task C5

(a)

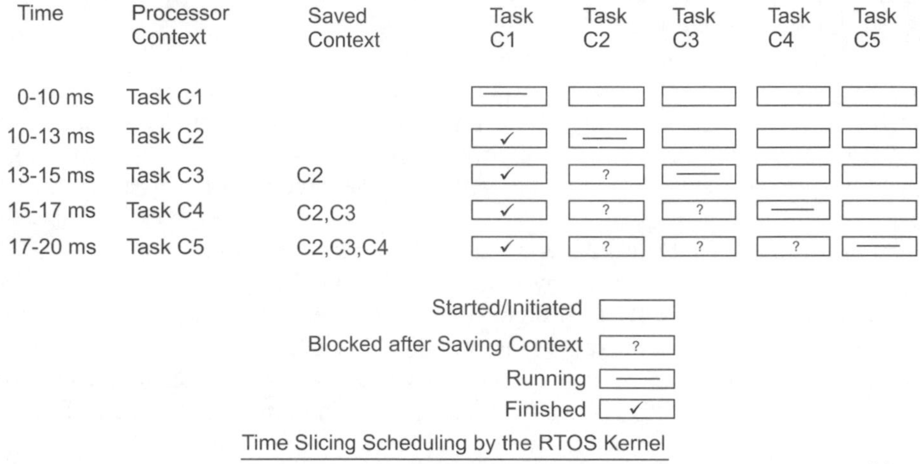

Time	Processor Context	Saved Context	Task C1	Task C2	Task C3	Task C4	Task C5
0-10 ms	Task C1		▭	▭	▭	▭	▭
10-13 ms	Task C2		✓	▭	▭	▭	▭
13-15 ms	Task C3	C2	✓	?	▭	▭	▭
15-17 ms	Task C4	C2,C3	✓	?	?	▭	▭
17-20 ms	Task C5	C2,C3,C4	✓	?	?	?	▭

Started/Initiated ▭
Blocked after Saving Context [?]
Running ▭
Finished [✓]

Time Slicing Scheduling by the RTOS Kernel

(b)

Figure 9.3

(a) The tasks C1 to C5 (b) Tasks programs contexts at the five instances in the Time Scheduling Scheduler for C1 to C5.

Suppose after every ten milliseconds, there is a stream of coded messages arriving to port A of an embedded system, where it is then read and decrypted and retransmitted to the port after encoding each decrypted message. The multiple processes consist of five tasks, C1, C2, C3, C4 and C5, as follows:

1. Task C1: Check for a message at port A at every 10 ms.
2. Task C2: Read Port A and put the message in a message queue.
3. Task C3: Decrypt the Message from the message queue.
4. Task C4: Encode the Message from the queue.
5. Task C5: Transmit the encoded message from the queue to Port B.

   Figure 9.3(a) shows five tasks, C1 to C5, that are to be scheduled. Figure 9.3(b) shows the five contexts in five time schedules, between 0 to 10 ms, 10 to 13 ms, 13 to 15 ms, 15 to 17 ms and 17 to 20 ms, respectively. Let RTOS initiate C1 to C5. Let there be RTC tick interrupts at each ms. Task C1 is scheduled by RTOS to bring it to a running state from its blocked state as soon as a timer triggers an event. If it is known that after every ten millisecond a byte reaches port A, let a timer, RTCSWT, trigger an event every 10 ms. Task C1 finishes after 10 ms, and C2 starts running.

   Figure 9.3(b), at the different time slices, shows the real-time schedules, saved contexts and processor contexts.

1. At the first instance (first row) the context is C1 and task C1 is running.
2. At the second instance (second row) after 10th ms, the RTOS switches the context to C2. Task C1 is finished, C2 is running. As task C1 is finished, nothing is saved on the task C1 stack.
3. At the third instance (third row), the RTOS switches the context to C3 on a timer interrupt, which occurred after 13th ms from start of task C1. Task C1 is finished, C2 is blocked and C3 is running. Context C2 is saved on task C2 stack because C2 is in a blocked state.
4. At the fourth instance (fourth row), the RTOS switches the context to C4 on a timer interrupt, which occurred after 15th ms from start of task C1. Task C1 is finished, C2 and C3 are blocked and C4 is running. Contexts C2 and C3 are at the tasks C2 and C3 stacks, respectively.
5. At the fifth instance (fifth row), the RTOS switches the context switches to C5 on a timer interrupt which occurred after 17th ms from start of task C1. Task C1 is finished, C2, C3 and C4 are blocked and C5 is running. Contexts C2, C3 and C4 are at the tasks C2, C3 and C4 stacks, respectively.
6. On a timer interrupt at the end of 20th ms, the RTOS switches the context to C1. As tasks C5 is finished before 20th ms only the contexts C2, C3, and C4 remain at the stack. Task C1 is in running as per its schedule.

Section 9.6.3 describes time-slice scheduling (each task assigned specific slices of time to run) in a cyclic scheduler in detail.

An RTOS scheduling management may be in cyclic (round robin), preemptive or time-slicing modes when processing multiple tasks.

9.5 INTERRUPT ROUTINES IN A RTOS ENVIRONMENT: HANDLING OF AN INTERRUPT SOURCE CALL BY THE RTOSs

There are three alternative systems for the RTOSs to respond to the hardware source calls from the interrupts. [Recall row 3 of Table 8.1.] Figures 9.4(a), (b) and (c) show the three systems. The following sections explain the *three alternative systems in three RTOSs for responding to a hardware source call on interrupts.*

A. Direct call to ISR by an Interrupting Source

A hardware source calls an ISR directly and the ISR just sends a message to the RTOS. This is shown in Figure 9.4(a). On an interrupt, the process running at the CPU is interrupted and the ISR corresponding to that source starts executing. The RTOS is simply sent a message from the ISR into a mailbox or message queue. It is to inform the RTOS about which ISR has taken control of the CPU. The ISR continues execution of the codes needed for the interrupt service. There are two processes, ISR and RTOS, in two memory-blocks. An i-th interrupt source causes the i-th ISR, ISR_i, to execute. The routine sends a message to the RTOS. The message is stored at the memory allotted for RTOS messages. When ISR finishes, the RTOS returns to the interrupted process or reschedules the processes. RTOS action depends on the messages in the mailbox.

B. Direct call to RTOS by an Interrupting Source and Temporary Suspension of a Scheduled Task

On interrupt, the RTOS intercepts the hardware source call and initiates the corresponding ISR after saving the processor status (or context). The ISR during execution then sends one or more outputs

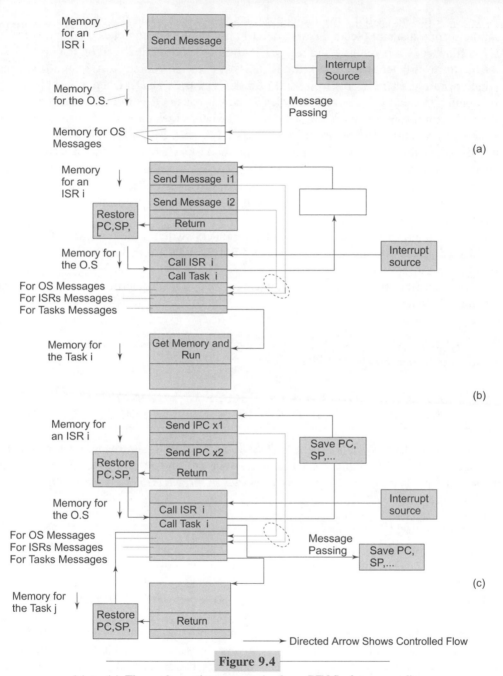

Figure 9.4

*(a) to (c) Three alternative systems in three RTOSs for responding
to a hardware source call on interrupts.*

and messages into the mailboxes or queues. This is shown in Figure 9.4(b). There is a routine i-th ISR
and RTOS and i-th task in three memory blocks. An i-th interrupt source causes the RTOS to call the
i-th ISR, ISR_i, to execute after saving the context on to a stack. The ISR sends a message(s) to the

RTOS for initiating the i-th task and returns after restoring the context. The messages are stored at the memory allotted for RTOS messages. The RTOS now initiates the i-th task to the ready state and to later on run the codes needed for the interrupt service. The ISR must be short and it simply places the messages. It is the task that runs the remaining codes whenever it is scheduled. The RTOS schedules only the tasks (processes) and switches the contexts between the tasks only. The ISR executes only during a temporary suspension of a task.

C. Direct call to RTOS by an Interrupting Source and Scheduling of Tasks as well as ISRs by the RTOS

The RTOS intercepts and executes the task needed on return from the ISR *without any message for initiating the task from the ISR.* This is shown in Figure 9.4(c). There are three i-th ISR, RTOS and j-th task in three memory blocks. An i-th interrupt source causes the RTOS to call the i-th ISR and block the j-th task. ISR_i executes after RTOS switches the context. The routine does not send any message (s) to the RTOS for initiating the i-th task. The routine merely sends the IPCs for the needed parameters (for example, x1 and x2 contents of the device status register and input register) for a task. The parameters are stored at the memory allotted for the RTOS inputs. The RTOS now calls (initiates) on return and restoration of the context and switches the context to later run the codes needed for the j-th or any other task. The ISR need not be short and simply generates and saves as the IPCs the input parameters (device status register and input register). It is the task that runs the codes whenever called. RTOS schedules not only the tasks (processes) but also the ISRs and switches the contexts between the tasks as well as the ISRs.

> An RTOS uses one of three strategies on interrupt source calls: (i) An ISR servicing directly or merely informing the RTOS. (ii) Kernel intercepting the call and calling the corresponding ISR and task. RTOS kernel schedules only the tasks (processes) and ISR executes only during a temporary suspension of the task by the RTOS. (iii) The kernel calling the ISR for the parameters (IPCs) by a context switch and later executing the task corresponding to the source when the RTOS kernel switches back the context when the task is scheduled. RTOS schedules not only the tasks (processes) but also the ISRs (as processes).

9.6 RTOS TASK SCHEDULING MODELS, INTERRUPT LATENCY AND RESPONSE TIMES OF THE TASKS AS PERFORMANCE METRICS

The three common model strategies that a scheduler may adopt are:
1. *Control Flow Strategy:* There is complete control of sequencing of the inputs and outputs. Control steps are deterministic. The worst-case interrupt latencies are well defined. The *scheduler* programs such that its states (or places and transitions) are predetermined by the programming codes. The *Cooperative scheduler* adopts this strategy. Scheduler program can be modeled either by FSM or by simulating FSM by an equivalent subclass of Petri net in which a transition is from only one node place to a specific next place. [Refer to Section 6.2 for FSM and Petri Net models.]
2. *Data Flow Strategy:* Interrupt occurrences are not deterministic. Similarly, the task control steps may not be deterministic. An example is a network. Packets arrival is not deterministic.

When the client demands a service, it is not a known priory. For this case, the worst-case latency is not well defined. The sequencing of the inputs and outputs is not feasible. A *preemptive scheduler* adopts the data flow strategy. A scheduler program can be modeled by Petri Nets, as there are transitions from one or more node places to one or several node places. The generated data from the output function at the transition determine the next sequence in the program. [Refer to Section 6.1.1 for data flow model.]

3. **Control-Data Flow Strategy**: A task design and scheduler functions may provide for the time-out delay features with deterministic delays. Modeling is feasible by the *deterministic timed Petri net* instead of FSM. The worst-case latencies are now deterministic because the maximum delays are deterministic. [Refer to Section 6.1.2 for control data flow model.]

Following are the scheduling models used by RTOS schedulers.

(a) Cyclic Cooperative Scheduling of ready tasks in a circular queue. It closely relates to function queue scheduling.

(b) Cooperative Scheduling with Precedence Constraints

(c) Cyclic Cooperative Scheduling with Time Slicing

(d) Preemptive Scheduling

(e) Fixed Times Scheduling

(f) Scheduling of periodic, sporadic and aperiodic tasks

(g) Dynamic Real Time Scheduling using 'Earliest Deadline First' (EDF) precedence.

(h) Advanced scheduling algorithms using the probabilistic Timed Petri nets (Stochastic) (Section 6.2.2) or Multi Thread Graphs. These are suitable for multiprocessors and for complex distributed systems.

9.6.1 The Cooperative Round Robin Scheduling Using a Circular Queue of Ready Tasks

Figure 8.1 shows the programming model for the multiple function calls. Similarly there can be a scheduler programming-model. Figure 9.5(a) shows a scheduler in which the RTOS inserts into a list the ready tasks for sequential execution in a cooperative round robin model. Program counter changes whenever the CPU starts executing another process. Figure 9.5(b) shows how the program counter changes on switch to another context. The scheduler switches the context such that there is sequential execution of different tasks, which the scheduler calls from the list one by one in a circular queue.

Cooperative means that each ready task cooperates to let a running one finish. None of the tasks does a block anywhere during the ready to finish states. Round robin means that each ready task runs in turn only from the circular queue. The service is in the order in which a task is initiated on interrupt. Each task has the same priority for execution in the round robin mode. We can say that the task-priority parameter sets as per its position in the queue.

Worst-case latency is same for each task. It is t_{cycle}, where t_{cycle} is a period for one round robin in the circular queue of ready tasks. The longer the queue, the greater is the t_{cycle}. If a task is running, all other ready tasks must wait. For an i-th task, let the event detection time when an event is brought into a list be dt_i, switching time from one task to another be st_i and task execution time be et_i. Then if there are n tasks in the ready list, the worst-case latency with this scheduling will be:

$$T_{worst} = \{(dt_i + st_i + et_i)_1 + (dt_i + st_i + et_i)_2 + ... + (dt_i + st_i + et_i)_{n-1} + (dt_i + st_i + et_i)_n\} + t_{ISR}.$$

Here the t_{ISR} is the sum of all execution times for all the ISRs. Remember, the T_{worst} should always be less than the deadline, T_d for any of the task in the list. [Refer to Section 4.6.]

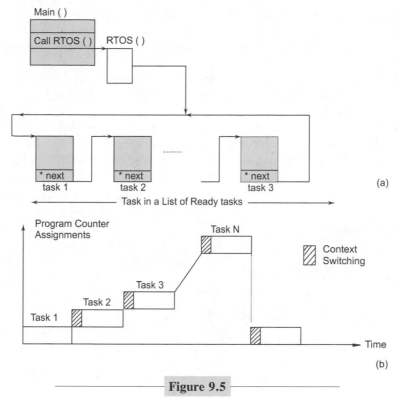

Figure 9.5

(a) A scheduler in which RTOS inserts into a list the ready tasks for a sequential execution in a cooperative round robin mode (b) Program counter assignments switch at different times, when the the scheduler calls the tasks one by one in the circular list.

9.6.2 The Cooperative Scheduling of Ready Tasks Using an Ordered List as per Precedence Constraints

Figure 9.6(a) shows a cooperative *Priority-based* scheduling of the ISRs executed in the first layer (top-right side) and *Priority-based* ready tasks at an ordered list executed in the second layer (bottom-left), respectively. Figure 9.6(b) shows the program counter switch at different times, when the scheduler calls the ISRs and the corresponding tasks at an ordered list one by one. The ordering of list of the tasks is done by the RTOS using a priority parameter, *taskPriority*. [Refer to Sections 5.7.3 for the uses of the ordered list of all the *m* initiated (ready) tasks.]

The RTOS scheduler first executes only the first task at the ordered list and the t_{cycle} equals the period taken by the first task on the list. It is deleted from the list after the first task is executed and the next task becomes the first. Now, if the task in the ordered list is executed in a cyclic order, it is called a cyclic *Priority-based cooperative scheduler.* The insertions and deletions for forming the ordered list are made only at the beginning of each cycle.

At the first layer, an ISR has a set of short codes that have to be executed immediately. [Section 9.5 subsection B.] The ISRs run in first layer (top-right in figure) according to their assigned priorities. It sends a flag or token and its priority parameter for the task to be initiated (serviced). This is the task that is inserted into the ready task list. *There is cooperative scheduling and each ready task cooperates to let the running one finish. None of the tasks does a block anywhere from the start to finish. Here, however, in case of cyclic scheduling, the round robin is among the ready tasks that run in turn only from a priority-wise ordered list. The ordering is according to the precedence of the interrupt source and the task.*

Let P_{em} be the priority of that task which has the maximum execution time. Then worst case latencies for the highest priority and lowest priority tasks will now vary from:

$$\{(dt_i + st_i + et_i)_{P_{em}} + t_{ISR}\}$$

to $\{(dt_i + st_i + et_i)_{p1} + (dt_i + st_i + et_i)_{P2} + ... + (dt_i + st_i + et_i)_{Pm-1} + (dt_i + st_i + et_i)_{Pm} + t_{ISR}\}$.

Here, $P_1, P_2, ..., P_{m-1}$ and P_m are the priorities of the task in the ordered list. Also $P_1 > P_2 > ... > P_m$. With this scheduler, it is easier, but not guaranteed, to meet the requirement that T_{worst} should be $< T_d$ for each task and interrupt source. The programmer assigns the lowest T_d task a highest priority.

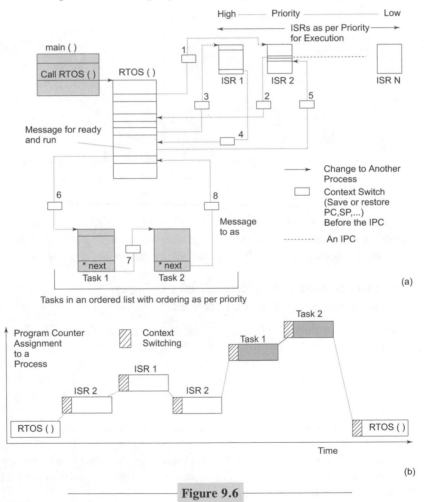

Figure 9.6

(a) Cooperative Priority-based scheduling of the ISRs executed in the first layer (top-right side) and Priority-based ready tasks at an ordered list executed in the second layer (bottom-left) (b) Program counter assignments at different times on the scheduler calls to the ISRs and the corresponding tasks.

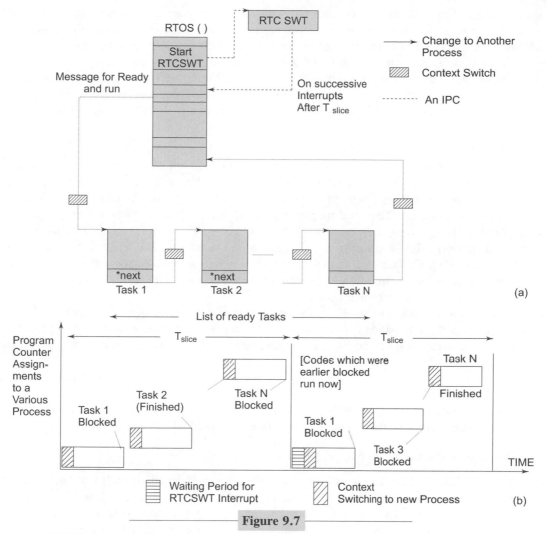

Figure 9.7

(a) The programming model for the cooperative time sliced scheduling of the tasks.
(b) The program counter assignments on the scheduler call to tasks at two consecutive time slices. Each cycle takes time $= t_{slice}$ and $et_2 < et_N < et_1$.

9.6.3 Cyclic Scheduling with Time Slicing (Rate Monotonic Cooperative Scheduling)

When a p-th task has high *execution time, et_p*, the worst-case latency of the lowest priority task can exceed its deadline. To overcome this problem, it is better that the RTOS defines a time slice for each task to finish and if it does not finish, then it blocks it and finishes its remaining codes during the succeeding cycles (next after the lowest priority task is attended and new cycle starts from the first highest assigned priority task) only. Cyclic Scheduling with time slicing is simple and there is no insertion or deletion into the queue or list. Figure 9.7(a) shows a programming model for cooperative

Time sliced scheduling. Figure 9.7(b) shows program counter on context switches when the scheduler call to tasks at two consecutive time slices. Assumed here is that time taken for task 2 < task N < task 1. Each task is allotted a maximum time interval = t_{slice}/N, where t_{slice} = time after which an RTCSWT timer (with the RTOS) interrupts and initiates a new cycle.

The RTOS completes the execution of all ready tasks in one cycle within a time slice, t_{slice} in this mode. The cycle completion rate $(t_{slice})^{-1}$ is constant (for rate monotonic scheduler). Let T_{worst} = sum of the maximum times for the all the tasks if there are *N tasks* in all. Then, when t_{slice} > or = T_{worst}, the T_{worst} equals:

$$\{(dt_i + st_i + et_i)_1 + (dt_i + st_i + et_i)_2 + ... + (dt_i + st_i + et_i)_{N-1} + (dt_i + st_i + et_i)_N\} + t_{ISR}.$$

If t_{slice} equals the sum of the maximum times for each task, then each task is executed once and finishes in one cycle itself. When a task finishes the execution before the maximum time it can take, there is a waiting period between two cycles. The worst-case latency for any task is t_{slice}. A task may periodically need execution. The period for the required repeat execution of a task is an integral multiple of t_{slice}. For each task to run only once, the t_{slice} should also be less that the greatest common factor of all the task periods. The estimation of response time for each task is easy in time slice cyclic cooperative scheduling. Consider a *k*-th task. The task responds within its task period plus the sum of the maximum times taken during a time slice from the task 1 to task (*k*-1). The response time of the *m*-th task at the end of the list is the maximum.

An alternative model strategy can be the *decomposition* of a task that takes an abnormally long time to be executed. The decomposition is *into two or four or more tasks*. Then one set of tasks (or the odd numbered tasks) can run in one time slice, t'_{slice} and another set of tasks (or the even numbered tasks) in another time slice, t''_{slice}.

Another alternative strategy can be the *decomposition of the long-time-taking task* into a number of sequential states or a number of node-places and transitions as in FSM or Petri net. Then one of its states or transitions runs in the first cycle, the next state in the second cycle and so on. This task then reduces the response times of the remaining tasks that are executed after a state.

9.6.4 Preemptive Scheduling Model Strategy by a Scheduler

Cooperative schedulers [described in sub-sections 9.6.1 to 9.6.3] schedule such that each ready task cooperates to let the running task finish. Let there be *N* tasks from task 1 to task *N* and let the assigned order of priority for interrupt servicing be from *1* (highest) to *N* (lowest). Assume now that an interrupt occurs in the time slicing cooperative scheduling just after the cycle starts. It means task 1 misses by a flick the chance of running from start to finish as task *1* will not get serviced until the cycle up to task *N* finishes or until the defined period t_{slice} expires.

The cooperative time slicing scheduler is simpler in design and extremely valuable in many applications where there is a need to use the resources of the embedded systems sequentially, or where none of the tasks has a shorter deadline than the t_{slice} or t_{cycle}. However, *a disadvantage of the cooperative scheduler is that a long execution time of a low-priority task makes a high-priority task wait at least until it finishes*. There is a further disadvantage if the cooperative scheduler is cyclic but without a predefined t_{slice}. It is because then the wait is until all other remaining listed or queued tasks finish. [Refer to Sections 9.6.1 and 2.] Can the higher priority task preempt a lower priority by blocking it? Recall Section 4.6.4. The hardware polls to determine whether an ISR with a higher priority than the present one needs service at the end of an instruction during execution. If yes, then the higher priority ISR is executed. Similarly, the RTOS preemptive scheduler can block a running task at the end of an instruction by sending a message to the task and letting the one with the higher priority take control of the CPU.

Let the priority of task_1 > task_2 > task_3 > task_4.... > task_N. Figure 9.8(a) shows the preemptive scheduling of *N* tasks. Figure 9.8(a) also shows the context switching whenever the process switches from a task to the RTOS and from the RTOS to a task. Figure 9.8(b) shows program counter assignments on the scheduler call to preempt task_2 when the priority of task_1 > task_2 > task_3.

Each task has an infinite loop from start (after the Idle state) up to finish. [Refer to task_1, task_2 and task_N 3 small shaded boxes at the bottom in this figure.] Task_1's last instruction points to the next pointed address, *next. In case of the infinite loop, *next points to the same task_1 start. [Refer to Section 5.4.5.] It is unlike a cooperative scheduler, where it points to the return address to the RTOS and RTOS now initiates and runs the next task in the ready list. In a preemptive scheduler, there is an RTOS message during the running of task_2 to preempt the task. [Figure 9.8(a)]. Note the sequence markings (1), (2) and (3) in the figure. Their meanings are listed in the figure note.

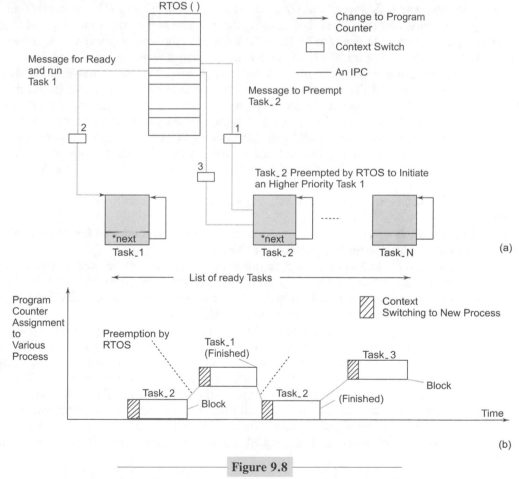

Figure 9.8

(a) Preemptive scheduling of the tasks [A running task is preempted and blocked to let a higher priority task be executed.] Note the sequence markings 1, 2 and 3. For the meanings of these refer to text. (b) Program counter assignments on a scheduler call to preempt task_2. Priority of task_1 > task_2 > task_3.

(1) The higher priority task_1 is initiated as follows. Task 2 blocks and sends a message to the RTOS.

(2) The RTOS now sends a message to task_1 to go to a ready state and run.

(3) A message is sent to RTOS after task_1 finishes and the task_1 context becomes the same as at its start. An RTOS then readies the task_2. Task_2 now runs. A message will be sent to RTOS after task_2 finishes and task_2 context becomes same as at its start. An RTOS message readies the task_3. Task_3 will run now.

Each task design is like an independent program, in an infinite loop between the task ready place and the running task place. [Refer to Examples 5.2 and 5.3 for the infinite loop concept.] The task does not return to the scheduler, as a function does or as a task in a cooperative scheduler does. Within the loop, the actions and transitions are according to the events or flags or tokens. The task design makes a task portable to another preemptive scheduler analogous to a module.

An advantage of using timeout intervals while designing task codes is that worst-case latency estimation is possible. Any task's worst-case latency is the sum of the t_{ISR} and the intervals of all other tasks with higher priority. Another advantage of using the timeouts is the error reporting and handling by the RTOS. Timeouts provide a way to let the RTOS run even the lowest priority task in necessary cases. Refer to the model in Figure 9.9. It shows a Petri net for preemptive scheduling.

Whenever the preemption event takes place, a task switching (a task place transition to its ready place) becomes necessary, and the scheduler searches for the highest priority task at that instance. That task is switched to the running place only by the scheduler. Switching occurs when a taskSwitchFlag is sent to the highest priority task and not to the task that was running previously. The context switching may also occur on an ISR call. [Section 9.5 (C)]. How can the context switching intervals reduce? The context switching intervals are reduced by the static declaration of the variables, as the static variables are RAM resident variables and do not save on the stack on a function call. [Refer to the static declarations in Section 5.4.3.] When this is the case, on a call, the program counter and few must-save registers are saved. Task switching now does not lead to additional stack saving overheads.

The conditions in which an event (token), the *preemptionEvent,* is generated for a task to undergo transition from running place to ready place are as follows:

(*i*) The preemption event takes place when an interrupt occurs and just before the return from the interrupt, there is a service call to the RTOS by the ISR. On this call to the RTOS, a token, the *preemptionEvent*, is set. The task then undergoes transition to the place, *readyTaskPlace*, and runs only when asked by scheduler (by sending *taskSwitchFlag*).

(*ii*) Each RTOS uses a system clock ticking an RTCSWT. The preemption event takes place when an RTCSWT (Real Time Clock–driven Software Timer) interrupt occurs at the RTOS. On this event RTOS takes control of the processor and checks whether it should let the currently executing task continue or to preempt it to make way for the higher priority task. This event makes another higher priority task ready to run, on the switch of the flag to the latter.

(*iii*) The preemption event takes place when any call to the RTOS occurs to enter the critical section or for sending the task message (outputs) to the RTOS, and if another higher priority task then needs to be serviced (take control of the CPU). [Now the preemption is before entering the critical section.]

9.6.5 Critical Section Service by a Preemptive Scheduler

For servicing of a critical section, a preemptive scheduler, on arising out of the service need of the higher priority task, should not preempt a lower priority task in certain cases. The blocking should not occur in the critical section of any lower priority task. This is to prevent the occurrence of a shared data problem. [Refer to Section 8.2.]

Before the critical section, the running task takes a semaphore from the scheduler and releases the semaphore at the exit from the critical section. Figure 9.9 shows a Petri net which models and helps in designing the codes for a task that has a critical section in its running. The figure shows places by the circles and transitions by the rectangles. The following are the places and transitions.

(*i*) The RTOS initiates *idle* to *ready* transition by executing a function, task create ( ). For the present case it is done by executing a function, task_J_create ( ). A transition from the idle state of the task is fired as follows. RTOS sends two tokens, RTOS_CREATE Event and *taskJSwitchFlag*. The output token from the transition is *taskSwitchFlag = true*. [Refer to top-left transition in the figure.] Each task is in idle state (at idleTaskPlace) to start with, and a token to the RTOS is *taskSwitchFlag* = reset.

(*ii*) Consider the *task_J_Idle* place, which currently has highest priority among the ready tasks. After the RTOS creates task_J, the place *task_J_Idle* undergoes a transition to the ready state (to readyTaskPlace), *task_J_Ready* place.

(*iii*) When task J finishes and when it is no longer needed under RTOS control, the RTOS sends a RTOS_DELETE event (a token) the task, it returns to the *task_J_Idle* place and its corresponding *taskJSwitchFlag* resets. [Refer to the bottom-left transition in the figure.]

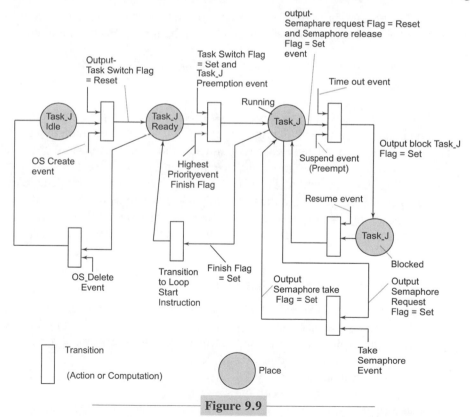

Figure 9.9

The Petri net for the task with a preemptive scheduler and one critical section where it takes a semaphore and release on criticalSectionOver.

(*iv*) At *task_J_Ready* place, the scheduler takes the priority parameter into account. If the current task happens to be of highest priority, the scheduler sets two tokens, *taskJSwitchFlag* = true (sends a token) and *higher Priority Event* = false for the transition to the running task J place, *task_J_Running*. The scheduler also resets and sends the tokens and task switch flags, for all other tasks that are of lesser priority. This is because the system has only one CPU to process at an instant.

(*v*) From the task_J_Running place, the transition to the task_J_Ready place will be fired when the task finish flag sets. [Refer to the bottom-middle transition in the figure.]

(*vi*) At *task_J_Running* place, the codes of the switched task J are executed. [Refer to the top-right most transition in the figure.]

(*vii*) At the runningTaskPlace, the transition for preempting will be fired when RTOS sends a token, *suspendEvent*. Another enabling token if present, is *time_out_event* will also fire the transition. An enabling token for both situations is the semaphore release flag, which must be set. A semaphore release flag is set on finishing the codes of task J critical-sections. On firing, the next place is task_J_Blocked. Blocking is in two situations. One situation is of preemption. It happens when the *suspendEvent* occurs on a call at the runningTaskPlace asking the RTOS to suspend the running. Another situation is a time-out of an SWT that associates with the running task place.

(*viii*) On a *resumeEvent* (a token from RTOS) the transition to *task_J_Running* place occurs. [Refer to the right-side middle transition, which is between the three transitions that are shown in the figure.]

(*ix*) At the *task_J_Running* place, there is another transition that fires so that the task J is back at the to *task_J_Running* place when the RTOS sends a token, take_Semaphore_Event to ask the task J to take the semaphore. [The RTOS sets the semaphore request flag, take_Semaphore_Event. It resets Semaphore release flag. It directs task J to run uninterrupted. Do not block.]

(*x*) There can be none or one or several critical sections. During the execution of a critical section, the RTOS resets the semaphore release flag and sets the take-a-semaphore event token.

Can work be done without the semaphores and/or mutex for the critical sections? Yes, one strategy is disabling and enabling the preemption. The disabling of a preemption means disable only the task Switching flags change, or their passing to the task ready places when using shared data, and enabling the task Switching flags changing and passing again after this. But it should then be ensured that all the ISRs are the reentrant functions. Another strategy could be the use of resources locking semaphore, *mutex*. [Section 8.2.2(iii)]

9.6.6   Fixed (Static) Real-Time Scheduling of Tasks

The scheduling method described in Section 9.6.3 is called 'fixed real-time scheduling'. Let there be *m* tasks and *m* RTCSWTs, the scheduler can thus assign each task a fixed schedule. Each task undergoes a ready place to running place transition on the timeouts of the corresponding timer. The RTOS is supposed to define hard real-time schedules for each task.

Scheduling methods described in Sections 9.6.1, 2, 4 and 5 are also fixed time scheduling methods. This scheduling is called soft real-time scheduling.

A scheduler is said to be using a fixed time scheduling method when the schedule is static and deterministic. The working environment is unaltered when processes are scheduled on the single CPU of the system. Schedules are deterministic as the worst-case latencies for all the interrupts and tasks are pre-determinable. The RTOS scheduler can thus schedule each task at fixed times so that none misses its deadline. [This is when the worst-case latency of each task is less then its deadline for its service.] The 'no deadline miss' advantage is feasible only in deterministic situations. Coding for the tasks are such that execution times do not vary under the different inputs or different conditions.

Schedules once defined remain static in a fixed time scheduling scheduler. Fixed schedules can be defined by one of the three methods.

A. *Simulated annealing* method. Here the different schedules can be fixed and the performance simulated. Now, schedules for the tasks are gradually incremented by changing the RTCSWT timer settings (using a corresponding OS function) until the simulation result shows that none is missing its deadline or that the number of deadline misses are too few to impact the system performance.

B. *Heuristic* method. Here, reasoning or past experience helps to define and fix the schedules.

C. *Dynamic programming* model. A specific running program first determines the schedules for each task and then the RTCSWTs loads the timer settings from the outputs from that program.

If the scheduler cannot fix the schedules, it is a non-deterministic situation. An example is a situation in which a message for a task is expected in a network from another system, and the minimum and maximum time periods for receiving it are unknown. Another example is when the inputs for a task are expected from another system and the minimum and maximum time periods when the inputs will be received are not known.

A dynamic scheduling model is as follows. The software design may be such that the priorities can be rescheduled and fixed times redefined when a message or error message is received during the run.

9.6.7 Precedence Assignment in the Scheduling Algorithms

The best strategy is one which is based on EDF (Earliest Deadline First) precedence. Precedence is made the highest for a task that corresponds to an interrupt source, which occurs at the earliest times, has the earliest deadline and will finish earliest.

How is the precedence assigned in the case of variable CPU loads for the different tasks and EDFs? One method is as follows.

Let t_1 be the instance when task I needs preemption the first time and t_2 be the next instance. A task with minimum $(t_2 - t_1)$ is inserted at the top of the task priority list. It is assigned the highest precedence. The list is dynamically ordered according to $(t_2 - t_1)$.

Firstly, there is deterministic or static assignment of the precedence in advanced scheduling algorithms. It means first there is rate monotonic scheduling (RMS). Later, the scheduler dynamically assigns and fixes the timeout delays afresh, and aligns the precedence as per EDF. The need for the dynamic assignment arises due to the sporadic tasks and the distributed or multiprocessor indeterminate environment.

9.6.8 Advanced Scheduling Algorithms using the Probabilistic Timed Petri nets (Stochastic) and Multi Thread Graph (MTG)

Consider a concurrent tasks system or a *multiprocessor* or *distributed system*. [Also recall Section 6.3.4.] Now, the timeout delays have to be probabilistic and the system is stochastic. When the tasks that are run are of an indeterminate nature, and the input tokens and actions as well as the output tokens and actions are only probabilistic (not definite), the above scheduling model strategies (Section 9.6.1 to 7) become unusable. The scheduling needs to be dynamic as well as stochastic. The scheduler has to use the probabilistic scheduling strategy in a non-deterministic environment. One method in the advanced scheduling algorithm is to use the stochastic timed Petri net model (Section 6.3.4). [For the model details, refer to Tadao Murata, "Petri Nets Properties, Analysis and Applications", Proceedings of IEEE, 77, 541–580, 1989].

An MTG is a two-layered model. It integrates the stochastic timed Petri net and control data flow (CDF) graphs. [For the CDF graph, refer to Section 6.1.2.] It optimizes the performance of the system by integration. It has been found suitable for modeling in real-time embedded systems that have concurrent tasks. The MTG is also used for the verification and performance evaluation of the concurrent tasks scheduler. It is also suitable for integrated real-time digital signal processor systems.

> (1) Scheduling and processing models of the RTOS for tasks are as follows: (i) Cyclic Cooperative Scheduling of ready tasks in a circular queue, (ii) Cooperative Scheduling with Precedence Constraints (iii) Cyclic Cooperative Scheduling with Time Slicing (iv) Preemptive Scheduling. (2) Schedule periods are fixed as follows: (a) Fixed Times Scheduling (b) Scheduling of Periodic, sporadic and aperiodic Tasks (c) Dynamic Real-time Scheduling using 'Earliest Deadline First' (EDF) precedence. (3) Advanced scheduling algorithms using the probabilistic Timed Petri nets (Stochastic) or Multi Thread Graphs are suitable for multiprocessors and for complex distributed systems.

9.7 PERFORMANCE METRIC IN SCHEDULING MODELS FOR PERIODIC, SPORADIC AND APERIODIC TASKS

Different models have been proposed for measuring performances. Three performance metrics are as follows:

 (i) Ratio of the sum of interrupt latencies with respect to the sum of the execution times.
 (ii) CPU load.
 (iii) Worst-Case Execution time with respect to mean execution time.

'Interrupt latencies' in various task models can be used for evaluating performance metrics. The latencies for various task models are described in Section 9.6. The CPU load is another way to look at the performance. It is explained in Section 9.7.1. Worst-case performance calculation for a sporadic task is explained in Section 9.7.2. Refer to *Real Time Systems* by Jane W. S. Liu, Pearson Education, 2000, for details of many models available for evaluating the performances.

9.7.1   CPU Load as Performance Metric

Each task gives a load to the CPU that equals the task execution time divided by the task period. [Task period means period allocated for a task.] Recall In_A_Out_B intra-network Example 4.1. Receiver port A expects another character before 172 µs. Task period is 172 µs. If the task execution time is also 172 µs, the CPU load for this task is 1 (100%). The task execution time when a character is received must be less than 172 µs. The maximum load of the CPU is 1 (less than 100%).

The CPU load or **system load** estimation in the case of multitasking is as follows. Suppose there are *m* tasks. For the multiple tasks, the *sum* of the CPU loads for all the tasks and ISRs should be less than

1. The timeouts and fixed time limit definitions for the tasks reduce the CPU load for the higher priority tasks so that even the lower priority tasks can be run before the deadlines. What does it mean when the sum of the CPU loads equals 0.1 (10%)? It means the CPU is underutilized and spends 90% of its time in a waiting mode. Since the execution times and the task periods vary, the CPU loads can also vary.

When a task needs to run only once, then it is aperiodic (one shot) in an application. Scheduling of the tasks that need to run periodically with the fixed periods can be periodic and can be done with a CPU load very close to 1. An example of a periodic task is as follows. There may be inputs at a port with predetermined periods, and the inputs are in succession without any time-gap.

When a task cannot be scheduled at fixed periods, its schedule is called sporadic. For example, if a task is expected to receive inputs at variable time gaps, then the task schedule is sporadic. An example is the packets from the routers in a network. The variable time gaps must be within defined limits.

A preemptive scheduler must take into account three types of tasks (aperiodic, periodic and sporadic) separately.

  (i) An aperiodic task needs to be preempted only once.

 (ii) A periodic task needs to be preempted after the fixed periods and it must be executed before its next preemption is needed.

(iii) A sporadic task needs to be checked for preemption after a minimum time period of its occurrence. Usually, the strategy employed by the software designer is to keep the CPU load between (0.7 ± 0.25) for sporadic tasks.

9.7.2 Sporadic Task Model

Let us consider the following parameters.

T_{total} = Total length of periods for which sporadic tasks occur

   e = Total Task Execution Time

T_{av} = Mean periods between the sporadic occurrences

T_{min} = Minimum Period between the sporadic occurrences

Worst-Case Execution-time performance metric, p is calculated as follows for the worst case of a task in a model.

$$p = p_{worst} = (e * T_{total}/T_{av}) / (e * T_{total}/T_{min}).$$

It is because the average rate of occurrence of sporadic task = (T_{total}/T_{av}) and maximum rate of sporadic task burst = T_{total}/T_{min}.

There are various models to define a performance metric. Three performance metrics for schedule management by the RTOS are (i) interrupt latencies with respect to the execution times and (ii) CPU load and (iii) Worst-case execution time.

9.8 IEEE STANDARD POSIX 1003.1B FUNCTIONS FOR STANDARDISATION OF RTOS AND INTER-TASK COMMUNICATION FUNCTIONS

A coding method for the processes may not be fixed. For example, (i) an error reporting mechanism may or may not be used and (ii) before entering the critical section, the messaging to the scheduler and use of the semaphores and event flags could be done by different approaches. Either a Boolean variable or the P and V semaphores and a signed byte can be used for a mutex.

The need for standardisation is important. Without this, the interfacing ability and portability is difficult to obtain. IEEE standards give a standard called POSIX. A standard RTOS interface is POSIX 1003.1b since 1996. Earlier, the standard was POSIX 1003.4.

The POSIX 1003.1b defines the standards for the following:

  (i) Portability of the application and system implementation

  (ii) Definitions for the RTOS service interfaces, subroutines, environment variables, headers and structures

 (iii) Ensuring portability, deadlock recovery, error reporting and error recovery

 (iv) Timer setting and reading with a high resolution using POSIX timer and clock interface

  (v) Sharing the memory using POSIX memory management

 (vi) Priority scheduling and priority-based preemptive scheduling

 (vii) Creating and accessing the real-time files and their deterministic performance

(viii) Locking a memory block to a process

 (ix) Synchronizing I/Os

  (x) Asynchronous I/Os using POSIX for asynchronous I/O

 (xi) Defining the performance of a real-time system with a performance matrix

 (xii) Defining the context switch times, interrupt latency [time interval between the interrupt occurrence and first instruction execution] and process dispatch latency [time interval between interrupt occurrence and return to the process.]

(xiii) Language specific services by the RTOS

(xiv) Synchronous and asynchronous inter-process communication using POSIX signals, POSIX semaphores, POSIX message queues.

The RTOS that supports this standard supports the messaging, flow and resources control using the IPCs by the following:

  1. Providing efficient synchronisation using P and V semaphores. The P semaphore accesses a resource and V releases it.

  2. Providing supports for inter-process communication, for example, using a pipe.

  3. Notifying all the synchronous events to the RTOS and for queuing the events with their deterministic delivery.

  4. Supporting spin locks. [Section 8.3.2]

  5. Supporting priority inheritance for solving the priority inversion problems. [Section 8.2.3]

POSIX 1003.1b/1c also contains the definitions for threads and threading mechanisms. Recall Section 8.1.3. Like a task, a thread is also a unit containing a set of instructions and the RTOS only schedules and allocates resources to that unit. An RTOS can schedule the multiple threads (or tasks) only if it knows the internal threads. A lightweight process is one for which the context switch time is less and there is full memory management information at the RTOS. A heavyweight process needs to have separate memory management and resources.

VxWorks (described in Section 10.3) has additional functions for using the POSIX 1003.1b standard interfaces. Its use is made through the POSIX libraries provided in VxWorks for most of the fourteen types of interfaces mentioned above. However, the POSIX memory mapping functions, shared memory functions for open, unlink, protect and un-asynchronous synchronisation, and synchronous communication of files and data may not be supported in a given VxWorks version.

The POSIX 1003.b defines IPC and other standard interfaces. These may be followed in real-time programming and in multitasking and multiprocessing systems.

9.9 LIST OF BASIC ACTIONS IN A PREEMPTIVE SCHEDULER AND EXPECTED TIMES TAKEN AT A PROCESSOR

Table 9.11 gives a list of basic actions in a preemptive RTOS. The expected time in general depends on the specific target processor of the embedded system and the memory access times. However, in order to provide the relative magnitude of the time taken for basic actions at a preemptive scheduler, a new parameter is defined. It defines the time taken for an action by an RTOS scheduler in terms of an assumed scaling parameter; S. S emphasizes the relative magnitudes of execution times for various actions in a typical RTOS.

Let time taken for the simplest instruction be t_{min}. The minimum time is when the semaphores P and V are assigned certain initial values, true or false. Let S be defined in units of T_s. The T_s and t_{min} depend on a specific target processor of the embedded system and the memory access times. For example, a typical value for an 80960 based target processor, t_{min} is 0.6 μs and let $T_s = (4 \text{ μs} \pm 0.6 \text{ μs})$. Let $t_{exec} = (t_{min} + S.T_s)$. This equation defines S as the time over and above the t_{min} in units of a basic time unit, T_s.

If the same RTOS runs on a different processor, the S will therefore remain the same. S is taken as the nearest positive integer for the relative magnitude of execution times for the scheduler actions.

In 80960 with t_{min} 0.6 μs, the S means the following:

1. $S = 1$ will mean t_{exec} between 2.8 μs and 5.2 μs.
2. $S = 2$ will mean 7.4 μs to 9.8 μs.
3. $S = 4$ will mean 14.2 μs to 19.0 μs.
4. $S = 5$ will mean 20.6 μs $\pm$ 3.0 μs.
5. $S = 10$ will mean 40.6 μs $\pm$ 6.0 μs.
6. $S = 15$ will mean 60.6 μs $\pm$ 9.0 μs.

Table 9.11

A list of basic actions in a preemptive RTOS and execution time in terms of a scaling parameter S

Action	S	Action	S	Action	S	Action	S
Context switch	2	Task suspend	1	Sem release/take	1	Message Q Delete	10
Task initiate	12	Task resume	1	Take semaphore flag, mutex, counting semaphore when sem available	1	Q receive Message available	2
Task create and run or delete	28	Task lock when no lock or unlock when lock exists	t_{min}	Take semaphore flag, mutex, counting semaphore when sem available	1	Q receive Message not available	1
Task delete	10	Mem allocate	2	Semaphore flag, mutex, counting semaphore create or delete	6	Message Q Send task pending	5

Contd

Action	S	Action	S	Action	S	Action	S
Task switch flag for running	1	Mem Free	4	Release semaphore flag, mutex, counting semaphore when task in Q	3	Message Q Send task not pending	2
Task create or delete	18	Network byte send	t_{min}	Mutex flush	1	Message Q Send queue full	1
		Semaphore flag or counting semaphore Flush	4	Release semaphore flag, mutex, counting semaphore when no task in Q	1	Message Q create	105

Abbreviations are: Semaphore (Sem), Queue (Q) and Memory (Mem).

A list of basic functions in a preemptive scheduler and IPC functions becomes important when programming using RTOS. Relative speeds for various functions help a programmer in optimizing the response times and enhancing system performance.

9.10　FIFTEEN-POINT STRATEGY FOR SYNCHRONISATION BETWEEN THE PROCESSES, ISRs, OS FUNCTIONS AND TASKS AND FOR RESOURCE MANAGEMENT

An embedded system with a single CPU can run only one process at an instance. The process at any instance may either be an ISR, scheduler or task. [Section 9.5] The RTOS uses one of the three strategies explained in subsections A, B and C of Section 9.5. [Refer to Figure 9.4 for synchronizing the ISRs, RTOS and tasks.] For designing the codes for synchronisation between the processes, the following fifteen points must be taken into account.

1. Use appropriate precedence assignment strategy.
2. The ISR can only write (send) the messages for an RTOS and parameters for the tasks. No ISR instruction should block any task. Only an RTOS must initiate actions according to the variables, semaphores, queues, mailboxes and pipes (Section 8.3) for interactions with the tasks and an RTOS must only control the states of the tasks. The variables and task-switching flags must always be under the RTOS control. No ISR instruction should read and change the task switch flags or messages for the tasks and RTOS.
3. The ISR coding should be like a reentrant function with no shared data problems. The ISR should be short and execute the codes that should not wait for actions by the RTOS and tasks.
4. A task should not call another task as each task has to be under the RTOS control. Such an attempt should generate an error.
5. A task can get the messages and send the messages using the RTOS calls only.
6. While executing the critical section codes, if possible, instead of disabling the interrupts only the task-switching flag changes should be prevented. [It is by using the semaphore.] Thus, only the preemption by RTOS should be prevented. Disabling preemption may be better than disabling interrupts. However, both increase worst-case interrupt latencies. [Recall Section 4.6.]

7. Resource locking using the mutex semaphores may be better than disabling preemption or interrupts. [Refer to Section 8.2.1(v) for mutex semaphores.] Use of mutex semaphore functions, P and V, [Section 8.2.1.(vi).] is a way of resource locking.

8. A task should take the semaphore only during a short period in which the critical section alone is executed. Disabling of running of other tasks for a longer period increases the worst-case interrupt latency periods for all the interrupts.

9. Relative response times improve on multitasking but a limited number of multiple tasks should be preferred. It reduces the context switching intervals and stack needs. Otherwise, the response time advantages are lost and the sum of the stacks for each task needs a larger memory.

10. Remember that the RTOS create ( ) function to create a task takes longer CPU time than writing into a queue and then reading from the queue, and using a semaphore takes the least. [Table 9.3.] Therefore, use semaphores if that suffices. Getting a semaphore takes the least CPU time. Use signals for the most urgent inter-process communications (for example, error reporting by throwing the exceptions). [Section 8.3.1]

11. Create tasks at start-up only and avoid creating and deleting tasks later. The only advantage of deleting is the availability of additional memory space. Suppose a task is deleted by an RTOS delete ( ) function. Now a situation can be that a task is waiting for a semaphore (to let other tasks finish the critical section) or is waiting for a queue (or mailbox) message for a pointer at the RTOS, and that pointer is for a message to the task that has been deleted. A prolonged blocking or a deadly embrace or a deadlock will then occur. An RTOS may not provide protection for these situations.

12. Semaphores, queues, and messages should not be globally shared variables, and each should be shared between a set of tasks only and encapsulated from the rest.

13. Use idle CPU time for internal functions. Often, the CPU may not be running any task. All tasks may be waiting for preemption [for transition from ready place to running place.] The CPU at that instant may associate the RTOS for the following. Read the internal queue. Manage the memory. Search for a free block of memory. Delete or dispatch a task. Perform the internal and inter-process communication functions.

14. If memory allocation and de-allocation are done by the task the number of RTOS functions is reduced. This reduces the interrupt-latency periods as execution of these functions takes significant time by the RTOS whenever the RTOS preempts a task.

15. Use a configurable or hierarchial RTOS, which can put only the needed functions from the scheduler with the rest outside. This is because the preemption scheduling increases the interrupt-latency periods due to time spent in context switching and saving and retrieving pointers for the RTOS functions like memory allocation, and inter-process communications (Section 8.3). The interrupt latency is minimised if the functions for the memory management, file system functions, and inter-process communication (for example, pipe, signal, socket, and RPC) are outside the scheduler at a hierarchical RTOS. Hierarchical RTOS means the RTOS derives the limited functions for the scheduler and the rest of the functions link and bind dynamically when needed.

A fifteen-point strategy is suggested for real-time programming for an embedded system.

9.11 EMBEDDED LINUX INTERNALS: LINUX KERNEL FOR THE DEVICE DRIVERS AND EMBEDDED SYSTEM

Knowledge of Unix and Linux OSs is presumed in the following description. [The OS name Linux is after Linus Torvalds, father of the Linux operating system.] Refer to the standard reference books in the 'References' for more information regarding these OSs. A user application program (tasks) makes the system calls or message passing (Table 9.3) for the following functions at a kernel.

(i) Process Management functions.

(ii) Memory Management functions. [For example, allocation, de-allocation, pointers, creating and deleting the tasks.]

(iii) File System Functions.

(iv) Networking System Functions.

(v) Device Control Functions for any peripheral present into a system (computer).

The following are three important classes of devices:

1. A device type is a char device. An example is a parallel port. Another example is an LCD matrix or a serial port or keypad or mouse. The character access is byte-by-byte and analogous to the access from and to a printer device. System call functions are also like an I/O device, open, close, write and read.

2. Another device type is a block device. It is like a file system (disk). Linux permits a block device to read and write byte-by-byte like a char device or read and write block-wise like a block device. A part of the block can be accessed. [It is always read and written as a block only in Unix.]

3. The protocol-based transmission examples are Ethernet, FTP, TCP, IP, UDP, and Telnet. These devices are called network devices. [Though the FTP and Telnet devices are accessed, like a char device, these devices are also categorised distinctly as network devices.] The network devices permit the exchange of packets (or frames) of data through the protocol based transmissions. Linux has OS functions for the sockets and protocol based transmissions from the network devices. [Sockets logically connect the network transmitter and network receiver devices.] The Linux kernel maps a device to a node of a file system. The network device drivers for the network devices also support the Address Resolution Protocol (ARP) and reverse address resolution support (RARP).

Device control functions are very important in an embedded system. The device driver functions control the keypad in a mobile phone or a TV remote. These functions control the LCD matrix in a mobile phone or video game. Device driver functions are thus extremely important in embedded systems, especially in network-related embedded systems. Recall Section 4.1.4. It describes the Linux drivers in Table 4.1. The important aspects of Linux are explained below.

Earlier, the OS classified devices as char devices and block devices. Network subsystems were added to this to control the network devices and subsystems. [Section 9.3.] Even though final implementation of each device is in terms of stream of characters (bytes) or blocks, internal software implementation is different in these sophisticated devices. Two important questions are:

1. Why is there a need to define other classes for the devices and thus device drivers?

2. Is it possible to define classes of devices other than block devices such as files, in an OS?

The answer to the first question is as follows. Recall Section 3.3.3. The feature of the USB device is that it can be attached, configured and used, reset, reconfigured and used, share the bandwidth with other devices and detached (while others are in operation) and reattached. A file device does not have this feature.

Linux gives the answer to the second question. The Linux OS facilitates definitions of classes of devices and thus device driver modules. A module may be for the USB class of devices or SCSI (Small Computer System Interface). More and more classes of drivers and their modules are now available for Linux. A biggest advantage is that Linux codes are open source.

A Linux kernel can insert a module by registering and remove it by de-registering. [Registering means that it is scheduled later when its turn comes. Deregistering means the module will be ignored. It is similar to task spawning and deleting.] MicroC/OS-II (Section 10.2) registers and deregisters the tasks only. VxWorks (Section 10.3) kernel registers and deregisters the tasks as well as interrupt-service routines (ISRs). Linux kernel provides registering and deregistering of the device-drivers modules as well. Thus, the Linux kernel also permits the scheduling of device drivers and modules. Therefore, the different tasks or programs can send the bytes concurrently or sequentially to a device through its driver registered at the kernel. Further, the Linux kernel enforces the use of sequential accesses, and to specific memory addresses only, by the registering and de-registering mechanism. These features of Linux make the use of Linux OS internals in embedded systems important.

Unix assumes the different device types to be like different tasks. Linux assumes each device type as a module. Linux thus provides reduced memory usage and greater speed, which are the desired features in an embedded system. Each module in Linux implements one driver or one of its functions, and therefore has the same classifications as the device type. Two 8-bit numbers represent the bus number, device number and function number.

There must be a separate device driver for each device. Each device driver is a module that can be inserted into the kernel by registering and removed by de-registering. Can the different programs send the byte to the device concurrently or sequentially? How can a kernel enforce the use of sequential accesses and specific memory addresses only? Does the device accept the ASCII codes only or binary format numbers? A device has the addresses for its ports, addresses, memory space needs, interrupt vector, control register and status register. The kernel must provide a way of registering these. Since each user may have different needs, an OS kernel cannot force any policy for kernel use upon the user. Later, the kernel can perform memory scaling and mapping using its device manager. The kernel must not force anything on the device. It accepts its configurations and needs. Whenever a device permits concurrency, the kernel must give support by executing the device driver module as the tasks. Each device has a security module that the kernel must support. It is risky to implement security functions using the kernel root alone.

The Linux OS supports registering of the driver configurations and functions, and supports de-registering using the functions given in Table 9.12. [Refer to subsection C of Section 9.5 for the RTOSs, which support scheduling of the tasks as well as ISRs. The Linux scheduler not only schedules the tasks and device driving ISRs but all the distinct device modules also.]

Table 9.12
Registering and De-registering and related functions of Linux for Device Driver Modules

Function	Action(s)
insmod	Inserts module into the Linux kernel
rmmod	A module is deleted from the kernel
cleanup	A kernel level *void* function, which performs the action on an rmmod call from the execution of the module
register_capability	A kernel level function for registering
unregister_capability	A kernel level function for deregistering
register_symtab	A symbol table function support, which exists as an alternative to declaring functions and variables static.

The NULL argument will create an empty symbol table. This solves the name space collision problem. The problem source is the use of the same name as in the kernel or by any other module. Linux supports a module initialisation, handling the errors, prevention of unauthorised port accesses, usage-counts, root level security and clean up.

The header file <u>Linux/time.h</u>, if included, supports the timing function in the kernel for scheduling. Delay functions are at <u>Linux/delay.h</u>. Task queues support is provided by including <u>Linux/tqueue.h</u>. The task queues supported are the timers, disk, immediate and scheduler.

Linux device drivers are extremely valuable in a computer system employing embedded systems for handling various additional devices and network interfaces. An example is embedded modem or PCI card (Section 3.4). [PCI is a bus system for a PC. An older alternative is the ISA bus system.] The device driver design can avoid Linux when using any other OS or RTOS. But a software designer must be aware of the functions in the Linux device drivers that, by using a similar code design policy, allows easier coding, which does not need to go through repeated design cycles.

Can the OS and its device driver be integrated with the processor? Yes. An example is the integrated Software Devices Corporation–provided Linux device drivers porting Linux kernel 2.2 into the ARM7 processor.

Linux is increasingly used in embedded systems for reasons besides being a freeware—its powerful device driver features, expandability of the kernel codes at run time and provision of registering and deregistering device driver modules thus facilitating their scheduling like the processes. These are essential in a complex embedded system.

9.12 OS SECURITY ISSUES

When a doctor has to dispense to multiple patients, protection of the patients from any confusion in the medication becomes imperative. When an OS has to supervise multiple processes and their access to the resources, protection of memory and resources from any unauthorised writes into the PCB or resource, or mix up of accesses of one by another, becomes imperative. *The OS security issue is a critical issue.*

Each process determines whether it has a control of a system resource exclusively, whether it is isolated from the other processes, or whether it shares a resource common to a set of processes. For example, a file or memory blocks of a file will have exclusive control over a process and a free memory space will have the access to all the processes. The OS then configures when a resource is isolated from one process and a resource is shared with a defined set of processes.

The OS should also have the flexibility to change this configuration when needed, to fulfill the requirements of all the processes. For example, a process has control of 32 memory blocks at an instance and the OS configures the system accordingly. Later when more processes are created, this can be reconfigured.

The OS should provide protection mechanisms and implement a system administrator(s) defined security policy. For example, a system administrator can define the use of resources to the registered and authorised users (and hence their processes).

What about issues of an *application* changing the OS configuration? The OS needs a protection mechanism for itself. An application-software programmer can find a hole in the protection mechanism and gain an unauthorised access. Thus the implementation of protection mechanisms and enforcement of security policy for resources is a challenging issue presented to any OS software designer. The network environment complicates this issue.

Table 9.13 gives the various activities for implementing important security functions.

Table 9.13
Important Security Functions

Function	Activities
Controlled resource sharing	Controlling read and write of the resources and parameters by user processes. For example, some resources write only for a process and some read only for a set of processes
Confinement Mechanism	Mechanism that restricts sharing of parameters to a set of processes only
Security Policy (Strategy)	Rules for authorizing access to the OS, system and information. A policy example is a communication system having a policy of peer-to-peer communication (connection establishment preceding the data packets flow).
Authentication Mechanism	External authentication mechanism for the user and a mechanism to prevent an application run unless the user is registered and the system administrator (software) has authorised it. Internal authentication for the process is important, and the process should not appear (impersonate) like other processes. User authentication can become difficult if the user disseminates passwords or other authentication methods.
Authorisation Mechanism	User or process(s) allowed using the system resources as per the security policy.
Encryption	A tool to change information to make it unusable by any other user or process without the appropriate key for deciphering it.

The OS security issues are important considerations. Protection of memory and resources from any unauthorised write into the PCB or resource, or mix up of accesses of one by another, becomes imperative from an OS security and protection mechanism.

9.13   MOBILE OS

A new age is dawning. It is the age of Palm computers and Pocket PCs. A mobile OS means the OS of mobile computing or embedded systems (laptop computers or handheld or wireless devices). Examples of mobile OS are Windows CE, Palm OS, PocketPC and Mac OS X Jaguar for connecting the VPN (Virtual Private Networks). WindowsXP, OS and Linux also offer the mobile support. The BlackBerry two-way pager, iPaq PDA, and Motorola StarTac phone use the mobile OSs.

The Windows CE (Windows for C Embedded software) needs are small, 400 kB. Its advantage is the support to Windows API. It makes downloading of Word and Excel files feasible, and Internet browsing feasible. It is a 32-bit scalable RTOS with 256 definable priorities to the processes. In April 2003, Microsoft announced that embedded system designers for devices like handheld PC can use and modify by adding additional functionaltity. No additional royalty for selling modified Windows CE will be payable under shared source Premium Licencing Program.

The Palm OS has a number of industry-specific applications. Palm OS5 is expected to give the features of PocketPC.

PocketPC offers services such as listening to music or detailed navigation using GPS (Geographic Positioning System) when the application is TravRoute using its mobile OS. MicrosoftMoney at PocketPC uses the mobile OS to offer dragging and dropping of Word and Excel files to and from it. PocketPC OS also offers a PC-based personal video recording application interface.

A Linux-based PDA has SharpZarus. There is 'SprintPCS Vision' based *T-Mobile PocketPC phone* (phone with Pocket PC combined). 'Stringer' is a new mobile OS from Microsoft for phones.

> A mobile OS is an OS for mobile systems, handheld devices, PDA, telephone, palm computers and pocket computers. PocketPC, Windows CE, Palm OS are popular mobile OSs. Embedded systems for mobile systems employ the mobile OSs directly or with suitable changes and additions.

■   SUMMARY   ■

- A kernel is a basic unit of any OS that includes the functions for memory allocation and de-allocation, preventing unauthorised memory access, tasks scheduling, inter-process communication, I/O Management, interrupts handling mechanism, and device drivers and management. The OS also controls I/O and network subsystems. An OS kernel may also include file management functions and functions for network subsystems. [These functions may also be separate from the OS kernels in certain OSs.]
- An RTOS has functions for real-time task-scheduling and interrupt-latency control. The RTOS uses the timers and system clocks. Its functions include time allocation and de-allocation to attain the best utilization in giving timing constraints of the tasks. It has a synchronisation mechanism for the tasks using the IPCs and predictable timing and synchronisation behavior in the system.
- Basic strategies for scheduling the multiple tasks are: cyclic as per interrupt sequences, cyclic with cycles arranged in order of priorities to start with, and preemptive and time slicing. Cyclic scheduling means that the tasks from a list of ready tasks are scheduled in sequence. Thus a task that is executed first now becomes a last priority task. Preempting scheduling means that a higher priority task is forced to block by the scheduler to let a higher priority task run. Time slicing means that each task is allotted a time slice after which it blocks and wait for its turn on the next cycle.

- An RTOS has functions for handling interrupts. There are three alternative ways used in the RTOSs for response to hardware source calls from the interrupts.
- The RTOS has functions for critical section handling in priority scheduling cases. RTOS may also provide for fixed real-time scheduling.
- Relative timing of the various actions in a preemptive scheduler helps us to optimize the process timings in an application developed using an RTOS.
- Recently, there has been standardisation (for example, POSIX 1003.b) of the RTOS and IPC functions.
- Linux has a number of device driver functions. Linux assumes each device function of a class of device as a module. Linux thus provides a reduced memory usage with a greater speed and these are the desired features in an embedded system. Each module in Linux implements one driver or one of its functions, and therefore has the same classifications as the device type. Two 8-bit numbers represent the bus number, device number and function number.
- The Linux OS has internal functions for registering and deregistering a device module. Linux OS schedules not only the task for concurrent processing but also facilitates concurrent processing of the modules, devices, ISRs and tasks.
- The popular RTOS MicroC/OS-II kernel registers and deregisters the tasks only. Another popular RTOS, VxWorks, kernel registers and deregisters the tasks as well as interrupt-service routines (ISRs). The Linux kernel provides for scheduling of the device driver modules for a class of drivers as well.
- OS security issues are important considerations in a system and the protection of memory and resources or PCB from any unauthorised write is essential.
- A mobile OS is used in mobile devices like wireless devices, phones, PDAs, Palm and Pocket PCs.

■ LIST OF KEYWORDS AND THEIR DEFINITIONS ■

- *Kernel*: A basic unit of any OS that includes the functions for processes, memory, task scheduling, inter-process communication, management of devices, I/Os, and interrupts and may include the file systems and network subsystems in certain OSs.
- *Operating System*: A system having kernel functions, file management functions and other functions also.
- *Real-Time Operating System*: Operating system with real-time task scheduling, interrupt-latency control, synchronisation of tasks with IPCs, predictable timing and synchronisation behavior of the system.
- *Round Robin or Cyclic scheduling*: A scheduling algorithm in which the tasks are scheduled in sequence from a list of ready tasks.
- *Preempting scheduling*: A scheduling algorithm in which a higher priority task is forced (preempted) to block by the scheduler to let a higher priority task run.
- *Time slicing scheduling:* A scheduling algorithm in which each task is allotted a time slice after which it is blocked and waits for its turn on the next cycle.
- *Cooperative Scheduling*: A waiting task lets another task run until it finishes.
- *Critical section run*: In spite of higher priority pending, a critical section is allowed to run by a scheduler using the semaphore(s).

- *Fixed real-time scheduling*: A scheduling strategy in which the time for each task is fixed.
- *Linux OS*: A powerful operating system with UNIX like features plus special device driver scheduling features and memory management features.
- *Registering a Device*: Making it feasible for RTOS to schedule like a task.
- *Deregistering*: Making it unfeasible for the RTOS to schedule like a task.
- *Protection Mechanism*: A mechanism at an OS to protect against unauthorised accesses to the resources.
- *Mobile OS*: OS for mobile handheld devices, wireless devices, Palm PCs, Pocket PCs, PDAs and phones.

■ REVIEW QUESTIONS ■

1. What should be the goal of an OS?
2. List the layers between *application* and *hardware*.
3. Why does an OS's functions provide two modes, *user mode* and *supervisory mode*?
4. List the functions of a kernel. What can be the functions outside the kernel?
5. Explain the terms *process descriptor* and *process control block* (PCB). What are the analogies in a PCB and TCB?
6. When is a message used and when does a system-call for seeking access to system resources?
7. Process or task creation and management are the most important functions of kernel. Why?
8. A strategy is that the tasks are created at start-up only and creating and deleting tasks later is avoided. Why should it be adopted?
9. Memory allocation and management are the most important functions of kernel. Why?
10. List the advantages and disadvantages of fixed and dynamic block allocations by the OS.
11. The kernel controls the access of system resources, CPU, memory, I/O sub-systems and devices. Why is it needed?
12. What is the importance of device management in an OS for an embedded system?
13. Give examples of I/O subsystems.
14. Define a network operating system. How does a network OS differ from a conventional OS?
15. What are the uses of OS interoperability and portability?
16. How do you choose scheduling strategy for the periodic, aperiodic and sporadic tasks?
17. What is unique feature of Linux device drivers? How does Linux schedule the tasks, ISRs and device drive modules distinctly from other OSs?
18. What is a block file-system?
19. Linux provides for registration and deregistration of device driver modules like the tasks. Why is this feature highly useful in an embedded system?
20. What are the OS units at an RTOS kernel?
21. When do you use cooperative scheduling and when preemptive?
22. Comapre scheduling strategies, real-time scheduling, round robin mode and round robin scheduling.
23. What are the advantages of time slice scheduling by an RTOS?
24. List three ways in which an RTOS handles the ISRs in a multitasking environment.
25. How does a preemption event occur?

26. Real-time system performance metrics are through put, interrupt latencies, avarage response times and deadline misses. Explain the importance of each of these metrics.
27. Why should you estimate worst-case latency?
28. Explain the applications of simulation annealing method.
29. What should be the OS security policy?
30. What is the protection mechanism for the OS?
31. OS security issues are important considerations. Protection of memory and resources from any unauthorised write into PCB or resource or mix up of accesses of one by another becomes imperative from an OS security and protection mechanism. Explain each of these considerations.
32. What is meant by hierarchical RTOS?
33. Why are the mobile OS becoming popular in embedded systems for telephones and PocketPCs?
34. How is the precedence assignment done for the tasks? How is the precedence assignment algorithm used in dynamic programming?
35. List the best strategies for synchronisation between the tasks and ISRs.
36. What is dyanmic program scheduling?

■ PRACTICE EXERCISES ■

37. Give two examples of when the scheduler cannot fix the schedules and there is a non-deterministic situation.
38. How do you estimate CPU load in a multitasking system handling sporadic tasks?
39. When do you use CPU load for performance metrics of a real-time system?
40. Design Petri table for Petri net in Figure 9.9 and then write C codes using the table.

Chapter
10

||||||||||||||||||||||||||||||||||||||||||||||||||||||||||

Real-Time Operating
System Programming Tools:
MicroC/OS-II and VxWorks

What We Have Learnt

We have learnt the following important points in Chapter 8 relating to real-time programming services of RTOS.

1. Process is the computational unit that an OS schedules and, on request either by system call or by message passing, the OS lets the process use the resources: CPU, memory, I/O subsystems, devices, and network subsystems.

2. Process also means 'task' in a multitasking model of the processes and means 'thread' in a multithreading model of the processes, both controlled by an OS.

3. Inter-process synchronisation during concurrent processing of the tasks takes place through signals, semaphores, queues, mailboxes, pipes, sockets and RPCs.

We have also learnt the following important points in Chapter 9 relating to real-time programming services of RTOS.

4. System structure includes application software, Application Programming Interface (API), system software other than the one provided at the OS, OS interface, OS, hardware–OS interface and hardware.

5. The basic functions (services) of the OS are process management from creation to deletion, processing resource requests, memory management (allocation and de-allocation), process scheduling, processing and managing Inter-Process Communication (IPC) (communication between tasks, ISRs, OS functions), I/O subsystems management, managing the Interrupts Control Mechanism (by handling ISR), management of device and device drivers, and, in Linux,

handling even the device driver modules for each class of devices and the open-source functions of the powerful, kernel device driver (Table 4.1).

6. Handling of interrupts and scheduling of tasks is done by the RTOS. RTOS has the basic functions of the OS plus functions for real-time task-scheduling and interrupt-latency control (Section 4.6); it uses the timers and system clocks, time allocation and de-allocation to attain best utilisation under the given timing constraints, a predictable timing behavior of the system and a predictable task-synchronisation, priorities allocation and priorities inheritance, synchronisation of tasks with IPCs, time-slicing of the processes execution, and hard real-time and soft real-time operations.

7. Basic strategies for scheduling the multiple tasks are cyclic (cooperative), preemptive scheduling and time slicing.

8. POSIX Standard.

9. Relative timings for different RTOS functions.

10. Important points that should be taken care of during coding for synchronisation between the processes (ISRs, functions, tasks and scheduler functions).

The goals of any embedded software, and hence of RTOS, are perfection and correctness. The reader must have now realised that there is a great deal of complexity in real-time programming of an embedded system in high-level language. The objective of this chapter is to explain the two popular RTOSs that ease programming complexity greatly and significantly reduce the time required to design the codes of an embedded system.

What We Will Learn

We will learn the following in this chapter:

1. Embedded systems for sophisticated real-time applications need the RTOS functions and inter-process communications (IPCs). Why is there a need for a well-tested and debugged RTOS in a multitasking and multithreading system for an application with multiple processes in it?

2. RTOS MicroC/OS-II (also known as μC/OS-II or *MUCOS* or *UCOS*) will be explained through twenty examples, Examples 10.1 to 10.20. What arguments are passed and what values are returned for each given MUCOS function will be explained. Learning the use of functions in the MUCOS is important for a reader even if another RTOS is used later. This will help greatly in understanding the advanced, sophisticated embedded RTOS later on.

3. VxWorks from Wind River® Systems is also an RTOS for sophisticated embedded systems. It has many powerful features. *VxWorks* will be learnt through seven examples, Examples 10.21 to 10.27. [Section 10.2.] Differences between the VxWorks semaphores, mailboxes, and queues with respect to that of MUCOS will become clear.

10.1   NEED OF A WELL-TESTED AND DEBUGGED REAL-TIME OPERATING SYSTEM (RTOS)

Recall Section 9.4.2. Except in the case of a small-scale embedded system, RTOS codes are also needed for the design and development of embedded systems. When designing a complex embedded system, one needs the thoroughly tested bug-free codes for the following software components.

• Multiple task functions in Embedded C or Embedded C++. [Chapter 5]

- Real-time clock-based software timers (RTCSWT codes) employing the system (hardware) clock.
- Software for a cooperative scheduler (Section 9.6.1 or 9.6.2 or 9.6.3) and its test—if it makes the sum of the CPU load around 0.7 and if worst case interrupt latencies, response times and deadlines are within manageable limits. [Section 9.6.7.]
- If a cooperative scheduler does not help and causes missing deadlines for a task then, alternatively, a preemptive scheduler should be developed, using either a Rate Monotonic Scheduling (RMS)—the task that needs the highest rate service is assigned highest priority—or an earliest deadline first (EDF) scheduling priority strategy. We can take the help of Petri tables for the software design. Each task in the preemptive scheduler has to be assigned a priority, which is included in the allocated own task control block (Section 8.1.2). The preemptive scheduler gives better response times for higher priority tasks.
- Device drivers and device manager.
- Functions for inter-process communications using the task-switching flags, semaphore-handling functions, and functions for the signals, queues, mailboxes, pipe and sockets.
- Networking functions.
- Error handling functions and exception handling functions.
- Testing and system debugging software.

Therefore, a lot of coding is needed in a complex embedded system and that happens within the reasonable amount of time that is assigned to develop a product. *A readily available RTOS package provides an advantage that the previously tested and debugged RTOS functions and error and exceptional handling functions can be ported directly as these are already well tested by thousands of users. Tools for the source-code engineering, testing, simulating, and debugging may also be available with an RTOS package.*

When designing a mission-critical real-time application, lack of appropriate error handling capability or an appropriate RTOS or a testing and debugging tool causes data loss. Even hardware loss may be caused. A readily available, well-tested and debugged RTOS thus not only simplifies the coding process greatly for a developer but also helps in building a product fast; it aids in building robust and bug-free software by thorough testing and simulation before locating the codes into the hardware.

Figure 10.1(a) shows common options available for selecting a RTOS. Figure 10.1(b) shows the basic functions expected from a kernel of a RTOS.

10.2   USE OF μC/OS-II

One popular RTOS for the above-described needs is μC/OS-II. For non-commercial use, RTOS μC/OS-II is also a freeware. Jean J. Labrosse designed it in 1992. Its name μC/OS-II is derived from <u>Micro-C</u>ontroller <u>O</u>perating <u>S</u>ystem. It is popularly also known as *MUCOS* or *UCOS*. Using this RTOS has another advantage. It is well documented in the book by its designer. [Refer to the printed book references at the end].

MUCOS codes are in C and a few CPU-specific modules are in assembly. Its code ports on many processors that are commonly used in the designing of embedded systems.

There are two types of source files. Master header file includes the '#include' preprocessor commands for all the files of both types. It is referred to as "includes .h" file. [Section 5.2.1]

1. *Processor dependent source files:* Two header files at the master are the following: (*i*) os_cpu.h is the processor definitions header file. (*ii*) The kernel building configuration file is os_cfg.h. Further, two C files are for ISRs and RTOS Timer, specifying os_tick.c and processor C codes os_cpu_c.c. Assembly codes for the task switching functions are at os_cpu_a.s12 (for 68HC12 microcontroller).

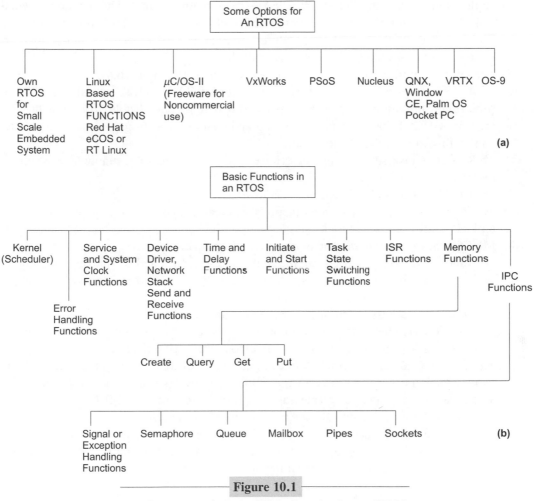

Figure 10.1

(a) Common options available for selecting an RTOS

(b) Basic functions expected from kernel of an RTOS.

2. *Processor independent source files:* Two files, MUCOS header (included in master) and C files, are ucos.ii.h and ucos.ii.c. The files for the RTOS core, timer and task are os_core.c, os_time.c and os_task.c. The memory partitioning, semaphore, queue and mailbox codes are in os_mem.c, os_sem.c, os_q.c and os_mbox.c, respectively.

The features of MUCOS are as follows:

(i) OS or OS_ (OS followed by underscore) when used as a prefix denotes that the function or variable is a MUCOS operating system function or variable. For example, OSTaskCreate ( )

is a MUCOS function that creates a task. OS_NO_ERR is a MUCOS variable that returns true in case no error is reported from a function. OS_MAX_TASKS is a constant for the maximum number of tasks in the user application. [The user in the preprocessor definitions defines this constant.]

(ii) MUCOS is scalable OS. [Only the OS functions that are necessary become part of the applications codes, thus having reduced memory requirements]. The functions needed for task servicing, inter-process communications, etc., must be predefined in a configuration file included in the user codes. [Refer to Example 10.7 Steps 1 and 2 for a clear understanding of configuration setting codes].

(iii) For multitasking, it employs a preemptive scheduler. [Section 9.6.4].

(iv) MUCOS has system-level functions. These are for RTOS initiation and start, RTC ticks (interrupts) initiation and the ISR enter and exit functions. [Table 10.1]. For the critical section, MUCOS has interrupts disabling and enabling functions that execute at entering and exiting the section, respectively.

(v) MUCOS has task service functions (for example, task creating, running, suspending and resuming). [Table 10.2].

(vi) MUCOS has task delay functions. [Table 10.3].

(vii) MUCOS has memory allocation functions for creating and partitioning into blocks, getting a block, putting into the block and querying during debugging at a block. [Table 10.4].

(viii) MUCOS has inter-process communication (IPC) functions. These are as per Tables 10.5, 10.6 and 10.7. MUCOS IPCs use the semaphores, queues and mailboxes. [Section 8.3].

(ix) MUCOS has semaphore functions, which are usable like the event flags, resource acquiring keys or counting semaphores. [Section 8.3.2]. Table 10.5 lists these.

(x) MUCOS has mailbox functions. A MUCOS mailbox has one message pointer per mailbox. [Section 8.3.4]. There can be any number of messages as MUCOS sends only the pointer (start address of the message) into the mailbox. Table 10.6 lists these.

(xi) MUCOS has queue functions. A queue permits an array of message pointers per queue from which messages retrieve in the FIFO (first-in first-out) method. [Section 8.3.3]. There can be any number of messages in a queue-element as MUCOS sends only the address of the message into the queue. Table 10.7 lists these.

The following seven subsections, 10.2.1 to 10.2.7, describe the above-mentioned MUCOS functions. For functions in each of the seven tables, 10.1 to 10.7, these sections give the details of values that are returned by the MUCOS functions and the details of parameters (arguments) that are passed by value or reference to a MUCOS function.

10.2.1 RTOS System Level Functions

Recall a function rtos_run ( ) in Example 5.3. We first use an initiate function in MUCOS, and then use a start function after the creation of the tasks by its function. These functions are OSInit and OSStart, respectively.

Recall Figure 9.7. We first initiate the system timer RTC ticks (and interrupts). MUCOS RTOS has system functions that should be executed when entering an ISR and exiting the ISR.

Recall Section 9.6.5. MUCOS RTOS has system functions that should be executed when entering a critical section of a task and exiting the critical section. [Section 8.2] Table 10.1 gives these RTOS system level functions.

Table 10.1
RTOS Initiate, Start, and ISR Functions for the Tasks

Prototype Functions	When is this OS function called?
void OSInit (void)	At the beginning prior to OSStart ( )
void OSStart (void)	After the OSInit ( ) and task-creating functions.
void OSTickInit (void)	In first task function that executes once to initialise the system timer ticks (RTC interrupts).
void OSIntEnter (void)	Just after the start of the ISR codes OSIntExit must call just before the return from the ISR.[#] [Enter and exit functions form a pair].
void OSIntExit (void)	After the OSIntEnter ( ) is called just after the start of the ISR codes and OSIntExit is called just before the return from the ISR. [Enter and exit functions form a pair].*
OS_ENTER_CRITICAL	Macro to disable interrupts. [Section 8.2.2(vi)].
OS_EXIT_CRITICAL	Macro to enable interrupts. [ENTER and EXIT functions form a pair in the critical section]. [Section 8.2.2(vi)].

Note: 1. Functions in this table pass no arguments and returns void. 2. [#] There is a global variable, OSIntNesting. It increments after the enter call. [We should not increment directly though it can be done. Let it increment automatically on enter to an ISR]. 3. * Global variable OSIntNesting decrements on exit call. [We should not decrement directly though it can be done. Let it decrement automatically on exit from an ISR.]

1. Initiating the operating system before starting the use of the RTOS functions

Function void OSInit (void) is used to initiate the operating system. Its use is compulsory before calling any OS kernel functions.

It returns no parameter.

An exemplary use of the function is as follows:

Example 10.1
1. /* Start executing the codes */*void main (void)* {
2. /* Initiate MUCOS RTOS to let us use the OS kernel functions */OSInit ( );
3. /* Create (Define Identity, stack size and other TCB parameters for the tasks using RTOS Functions in Table 10.1 */
   .
   .
   .
4. /* Create semaphore, queue and mailboxes, etc. */
   .
   .
   }.

2. Starting the use of RTOS multitasking functions and running the tasks

Function void OSStart (void) is used to start the initiated operating system and created tasks. Its use is compulsory for the multitasking OS kernel operations.

It returns no parameter.

An exemplary use as a function is as follows:

Example 10.2
1. /* Start executing the codes from Main*/*void main (void)* {
2. OSInit ( );
3. /* Create tasks and inter-process communication variables by defining their identity, stack size and other TCB parameters. */

   .
   .
   .

4. /* Start MUCOS RTOS to let us use RTOS control and run the created tasks and inter-process communication. */
OSStart ( );
      /* An infinite while-loop follows in each task. So there is no return from the RTOS. */
      }/ * End of the Main function.

---

3. Starting the use of RTOS system clock

Function void OSTickInit (void) is used to initiate the system clock ticks and interrupts at regular intervals as per OS_TICKS_PER_SEC predefined during configuring the MUCOS. Its use is compulsory for the multitasking OS kernel operations when the timer functions are to be used.

It returns or passes no parameter.

An exemplary use will be shown in Steps 2 and 8 of Example 10.7.

4. Sending a message to RTOS kernel for taking control at the start of an ISR

Function void OSIntEnter (void) is used at the start of an ISR. It is for sending a message to the RTOS kernel for taking control. Its use is compulsory to let the multitasking OS kernel control the nesting of the ISRs in case of occurrences of multiple interrupts of varying priorities.

It returns no parameter.

An exemplary use as a function is as follows:

Example 10.3
1. /* Start executing the codes of an ISR*/
   ISR_A ( ) {
2. /* sending message to RTOS kernel for taking control of ISR_N from nested ISRs loop. Increment OSIntNesting, a global variable */OSIntEnter ( );
3. /* Codes for servicing of the ISR by calling a task. */

   .
   .

---

5. Sending a message to the RTOS kernel for quitting the control at the return from an ISR

Function void OSIntExit (void) is used just before the return from the running ISR. It is for sending a message to the RTOS kernel for quitting control from the nesting loop. Its use is compulsory to let the OS kernel quit the ISR from the nested loop of the ISRs.

It returns no parameter.

An exemplary use as a function is as follows:

Example 10.4

1. to 3. Follow steps 1 to 3 from Example 10.3

.

.

4. /* Sending message to RTOS kernel for quitting the control of ISR_A from the nested loop. Decrement OSIntNesting, a global variable */
OSIntExit ( );
}/ * End of the ISR function.

6. Sending a message to the RTOS kernel for taking control at the start of a critical section

Macro-function OS_ENTER_CRITICAL is used at the start of an ISR. It is for sending a message to the RTOS kernel for disabling the interrupts. Its use is compulsory to let the OS kernel take note of and disable the interrupts of the system.

It returns no parameter.

An exemplary use as a function is as follows:

Example 10.5

1. /* Start executing the codes of an ISR*/*task_A ( ) {*
2. /* Codes for servicing of the task. */

.

.

3. /* Sending a message to RTOS kernel and disabling the interrupts.*/
OS_ENTER_CRITICAL;
4. /* Run critical section codes as follows. */

.

.

5. /* Codes for Exiting the service*/
}

7. Sending a message to the RTOS kernel for quitting the control at the return from a critical section

Macro-function OS_EXIT_CRITICAL is used just before the return from the critical section. It is for sending a message to the RTOS kernel for quitting control from the section. Its use is compulsory to let the OS kernel quit the section and enable the interrupts to the system.

It returns no parameter.

An exemplary use as a function is as follows:

Example 10.6

1. to 4. As in Example 10.5

   .

   .

5. /* Sending a message to RTOS kernel for quitting the control of critical section and enabling the interrupts. */
   OS_EXIT_CRITICAL;
   }/ * End of the ISR function.

10.2.2   Task Service Functions and Their Exemplary Uses

MUCOS service functions for the tasks are as per Table 10.2. *Service functions means the functions to task create, suspend and resume, and time setting and time retrieving (getting) functions.* These functions must be incorporated by preprocessor commands and variable assignments and prototype assignments as in step 1 and 2 of Example 10.7. The steps 1 and 2 codes are saved in a configuration file, which is included in the source code before compilation. These steps configure the MUCOS before they are used. [Example 10.7 shows these steps].

We shall see in the following examples that there is an infinite loop in every task function. This is a characteristic way of coding the tasks for preemptive scheduling. From the infinite loop, how will the CPU control return to MUCOS? In other words, how does context switching occur in the OS and how does the OS then activate the task switch to a higher priority task? The CPU control returns to MUCOS (or in any other preemptive scheduler) as soon as one of the following situations arises:

1. Any interrupt event including the timer tick interrupt occurs. Refer to Example 10.7, Step 8. [Time set at every 1 ms interrupt in step 2.]
2. On suspending the presently running task by calling OSTaskSuspend as in Example 10.8, Step 12.
3. As soon as any OS function is called, for example, a time delay function in Table 10.3 or a semaphore pending function in Table 10.5. The scheduler then switches context, preempts and thus passes the control of the highest priority assigned task by activating the task switch.
4. How does the control of the CPU return to a preempted task because of an infinite loop existing in preempting higher priority task also? It must return by an appropriate coding. For example, refer to code in Step 12 in Example 10.8. Here, the FirstTask is of priority 4 (the highest available to a user task). It suspends itself from the loop at Step 12.

Table 10.2
Service and System Clock Functions for the Tasks

Prototype of Functions	What are the parameters returned?	What are the parameters passed?	When is this OS function called?
unsigned byte OSTaskCreate [void (*task) (void *taskPointer), void *pmdata, OS_STK *task StackPointer, unsigned byte taskPriority]	*RA*	*PA*	Must call before running a task.

(Contd.)

Prototype of Functions	What are the parameters returned?	What are the parameters passed?	When is this OS function called?
unsigned byte OSTaskSuspend (unsigned byte taskPriority)	*RB*	*PB*	Called for blocking a task.
unsigned byte OSTaskResume (unsigned byte taskPriority)	*RC*	*PC*	Called for resuming a blocked task.
void OSTimeSet (unsigned int *counts*)	*None*	*PD*	When system time is to be set by the *counts* value. [Section 3.2].
unsigned int OSTimeGet (void)	*RE*	*None*	Find present *counts* when system time is read

Unsigned int means a 32-bit unsigned-integer. Abbreviations used in columns 2 and 3 are explained in text.

1. Creating a Task

Function unsigned byte OSTaskCreate [void (*task) (void *taskPointer), void *pmdata, OS_STK *taskStackPointer, unsigned byte taskPriority] is explained as follows.

A preemptive scheduler preempts a task of higher priority. Therefore, each user-task is to be assigned a *priority*, which must be set between 4 and OS_MAX_TASKS +3 [or 4 and OS_LOWEST_PRIORITY - 4]. If maximum number of user tasks OS_MAX_TASKS is 8, the priority must be set between 4 and 11. The OS_LOWEST_PRIO must be set at 15 for eight user tasks of priority between 4 and 11, because MUCOS will assign priority = 14 to a lowest priority task. The *priority* = 0 or 1 or 2 or 3 or 12 or 13 or 14 will then be for MUCOS internal uses. OS_LOWEST_PRIO and OS_MAX_TASKS are user-defined constants in preprocessor codes that are needed for configuring the MUCOS for the user application. Defining unnecessarily 20 tasks when actually 4 tasks are created by the user is to be avoided because more OS_MAX_TASKS means unnecessarily higher memory space allocation to the tasks.

Task Parameters Passing *PA*:
  (i) *taskPointer is a pointer to the codes of the task being created.
  (ii) *pmdata is pointer for an optional message data reference passed to the task. If none, we assign it as NULL.
  (iii) * TaskStackPointer is a pointer to the codes at the task being created.
  (iv) TaskPriority is the task priority and must be within 4 to 13 when OS_MAX_TASKS=10.

Returning *RA*: The lowest priority of any task OS_PRIO_LOWEST is 16. For the application program, task priority assigned must be within 4 to 13. The function OSTaskCreate ( ) returns the: (*i*) OS_NO_ERR, when creation succeeds. (*ii*) OS_PRIO_EXIST, if priority value that passed already exists. (*iii*) OS_PRIO_INVALID, if priority value that passed is more than the OS_PRIO_LOWEST. (*iv*) OS_NO_MORE_TCB returns, when no more memory block for the task control is available.

An exemplary use is in creating a task, Task1Connect, for a connecting task. OSTaskCreate (Task1_Connect, void (*) 0, (void *) *Task1_ConnectStack [100], 6)

Task Parameters passed as arguments are as follows:
  (i) Task1_Connect, a pointer to the codes of Task1_Connect for the task being created.
  (ii) The pointer for an optional message data reference passed to the task is NULL.

(iii) * Task1_ConnectStack is a pointer to the stack of the Task1_Connect and it is given the size = 100 address in the memory.

(iv) TaskPriority is task priority allotted at 6, the highest of two that can be allocated.

It will generate error parameters, OS_NO_ERR = true in case creation of Task1Connect task succeeds. OS_PRIO_EXIST, if priority 4 task is already created and exists. OS_PRIO_INVALID, if passed priority parameter is higher than OS_LOWEST_PRIO. OS_NO_MORE_TCB = false, when task control block (TCB) is available for Task1Connect. [TCB definition is in Section 8.1.1.]

Example 10.7

1. /* Preprocessor MUCOS configuring commands to define OS tasks service and timing functions as enabled and their constants*/
#define OS_MAX_TASKS 8 /* Let maximum number of tasks in user application be 8. */
#define OS_LOWEST_PRIO 15 /* Let lowest priority task in the OS be 15. */
#define OS_TASK_CREATE_EN 1/* Enable inclusion of OSTaskCreate ( ) function */
#define OS_TASK_DEL_EN 1/* Enable inclusion of OSTaskDel ( ) function */
#define OS_TASK_SUSPEND_EN 1/* Enable inclusion of OSTaskSuspend ( ) function */
#define OS_TASK_RESUME_EN 1/* Enable inclusion of OSTaskResume ( ) function */
.

.
/* End of preprocessor commands */
2. /* Specify all user prototype of the task functions to be scheduled by MUCOS */
/ * Remember: Static means permanent memory allocation */
static void FirstTask (void *taskPointer);
static void Task1_Connect (void *taskpointer);
static OS_STK FirstTaskStack [FirstTask_StackSize];
static OS_STK Task1_ConnectStack [Task1_Connect_StackSize];
/* Define public variable of the task service and timing functions */
#define OS_TASK_IDLE_STK_SIZE 100 /* Let memory allocation for an idle state task stack size be 100*/
#define OS_TICKS_PER_SEC 1000 / * Let the number of ticks be 1000 per second. An RTCSWT will interrupt and thus tick every 1 ms to update counts. */
#define FirstTask_Priority 4 /* Define first task in main priority */
#define *FirstTask_StackSize* 100 /* Define first task in main stack size */
#define Task1_Connect_Priority 6 /* Define Task1_Connect priority */
#define Task1_Connect_StackSize 100 /* Define Task1_Connect stack size */
.

.

.
3. /* The codes of the application starts from Main*/
void main (void) {
4. /* Initiate MUCOS RTOS to let us use the OS kernel functions */
OSInit ( );
5. /* Create first task that must execute once before any other. Task creates by defining its identity as FirstTask, stack size and other TCB parameters. */

*OSTaskCreate (FirstTask, void (*) 0, (void *) &FirstTaskStack [FirstTask_StackSize], FirstTask_Priority);*
/* Create other main tasks and inter-process communication variables if these must also execute at least once after the FirstTask. */
.
.
.
6. /* Start MUCOS RTOS to let us use RTOS control and run the created tasks */
OSStart ( );
/* Infinite while-loop is there in each task. So there is no return from the RTOS function OSStart ( ). */
}/ *** End of the Main function ***/
/* The codes of the application first task that creates in Main*/
7. static void FirstTask (void *taskPointer) {
8. /* Start Timer Ticks for using timer ticks later. */
OSTickInit ( ); /* Function for initiating RTCSWT that ticks at the configured time in the MUCOS configuration preprocessor commands in Step 1 */
9. /* Create a Task as per Step 2 defined by task identity, Task1_Connect, stack size and other TCB parameters. */
*OSTaskCreate (Task1_Connect, void (*) 0, (void *) &Task1_ConnectStack [Task1_Connect_StackSize], Task1_Connect_Priority)*
10. /* Create other tasks and inter-process communication variables. */
.
.
.
11. while (1) /* Infinite loop of FirstTask */
.
.
.
12.}; /* End infinite loop */
13.} /End of FirstTask Codes. */
The first main task is the only task created in Step 5 of *main*. The first task calls a function for the timer initiation (step 8). This is required when the RTOS timer functions are needed in a particular application. A task function *Task1_Connect* codes are created above in Step 9. All other tasks that are created are the private tasks using OSTaskCreate function within that first main task function].
14. /* The codes for the Task1_Connect*/
static void Task1_Connect (*taskPointer) {15. /* Initial assignments of the variables and pre-infinite loop statements that execute once only*/
.
.
.
16. /* Start an infinite while-loop. */
while (1) {
17. /* Codes for Task1_Connect*/
.
.
.
18.*}; /* End of while loop*/
19.}/ * End of the Task1_Connect function */

2. Suspending (Blocking) a Task [C]

Function unsigned byte OSTaskSuspend (unsigned byte taskPriority)

Task Parameters Passing *PB*: (*i*) taskPriority is the task priority is to suspend a task or starting task of the task being suspended. It is within 4 to 13.

Returning *RB*: The function OSTaskSuspend () returns the error parameters OS_NO_ERR when the blocking succeeds. OS_PRIO_INVALID, if priority value that passed is more than 16, the OS_PRIO_LOWEST, a constant value. OS_TASK_SUSPEND_PRIO, if priority value that passed already does not exists. OS_TASK_SUSPEND_IDLE, if attempting to suspend an idle task that is illegal.

An exemplary use is in blocking the task Task1_Connect of priority = *Task1_Connect_Priority* is as follows: OSTaskSuspend (*Task1_Connect_Priority*). Task Parameter passed as argument is 4. Recall *Task1_Connect_Priority* was assigned 4 earlier in the Example 10.7. The following error parameters will be returned by this function.

  (*i*) OS_NO_ERR = true, when the blocking succeeds.

 (*ii*) OS_PRIO_INVALID = false, as 4 is a valid priority and is not more than OS_PRIO_LOWEST.

(*iii*) OS_PRIO_LOWEST = 16.

(*iv*) OS_TASK_SUSPEND_PRIO = false, as priority value that passed already does exists.

 (*v*) OS_TASK_SUSPEND_IDLE = false, when attempting to suspend a task that was not idle.

Example 10.8

1 to 11. /* Steps Codes as in Example 10.7 */

12. /* Suspend FirstTask, as it was for initiating the timer ticks (interrupts), creating the user application tasks, and was to be run only once */

OSTaskSuspend (FirstTask_Priority); /*Suspend First Task and control of the RTOS passes to other tasks waiting execution*/

} /* End of while loop */

) /* End of FirstTask Codes */

---

3. Resuming (un-blocking) a Task

Function unsigned byte OSTaskResume (unsigned byte taskPriority) resumes a suspended task.

Task Parameters Passing *PC*: (*i*) taskPriority is the task priority of that task which is to resume and must be within 4 to 13.

Returning *RC*: The function OSTaskResume ( ) returns the OS_NO_ERR when the blocking succeeds. OS_PRIO_INVALID, if priority value that passed is more than 16, the OS_PRIO_LOWEST, a constant value. OS_TASK_RESUME_PRIO, if priority value that passed already does not exist. OS_TASK_NOT_SUSPENDED, if attempting to resume a not suspended (blocked) task.

An exemplary use is in un-blocking a task Task1_Connect of priority = *Task1_Connect_Priority* is as follows: OSTaskResume (*Task1_Connect_Priority*). Task Parameter passed as argument is 4, as *Task1_Connect_Priority* = 4. The following error parameters will be returned by the task resuming function:

  (*i*) OS_NO_ERR = true, when the un-blocking succeeds and task of priority 4 reaches running state.

   (ii) OS_PRIO_INVALID = false, as 4 is a valid priority and is not more than OS_PRIO_LOWEST.
  (iii) OS_PRIO_LOWEST = 16.
  (iv) OS_TASK_RESUME_PRIO = false, as priority value that passed already does exists.
   (v) OS_TASK_NOT_SUSPENDED = false, when attempting to resume a task that was not suspended.

Example 10.9
1. to 12. /* Steps as per Example 10.8 codes for Task1_Connect Function. */
.

.
13. /* Other codes, for example, time delay for 1 second */
.

.
14. /* Resume Task1_Connect. Control. */
OSTaskResume (Task1_Connect_Priority);
15.}; /* End of while loop*/
16.}/ * End of the Task1_Connect function.

4. Setting System Clock

Function void OSTimeSet (unsigned int *count*) returns no value. Passing parameter, PD as argument is given below.

PD: It passes a 32-bit integer for the *count* (set the number of ticks for the current time that will increment after each system RTC tick in the system).

An cxcmplary use is a function OSTimeSet (to preset the time). The function OSTimeSet (0) sets the present count = 0. It sets the present count, *presetTime* = 0. Caution: It is suggested that OSTimeSet function should never be used within a task function, as some other functions that rely on the timer will malfunction. Let the OS timer continue to be used as a free running counter. There is little need of using this because at an instant, the time can be read using a get function (Example 10.11) and at any other instant, it can be defined by adding a value to this time. Example 10.10 uses the set function in the FirstTask.

Example 10.10
void FirstTask (*taskPointer) {
1. to 8. /* The codes up to OSTickInit ( ) in Example 8.7*/
9. /* Set the timer number of ticks to 0 */
presetTime =0;
OSTimeSet (presetTime);
10. /* Other codes of FirstTask */
.

.
}/ * End of the FirstTask function.

5. Getting the System Clock

Function unsigned int OSTimeGet (void) returns current number of ticks as an unsigned integer. The passing parameter as an argument is none.

   RE: Returns 32-bit integer, current number of ticks at the system RTC

Example 10.11

1. /* The codes as for a task function */

.

.

currentTime = OSTimeGet ( );

2. /* Other codes of the task after determining current time */

.

.

}/ * End of the Task function.

10.2.3   Time Delay Functions

MUCOS time delay functions for the tasks are as per Table 10.3.

Table 10.3
Time Delay Functions for the Tasks

Prototype Function	What are the parameters returned?	What are the parameters passed?	When is this OS function called?
void OSTimeDly (unsigned short delayCount)	None	*PF*	When a task is to be delayed by count-nputs equal to delayCount -1. [RTCSWT :: delay( ). The task, which delays is the one in which this function executes.[#]
unsigned byte OSTimeDlyResume (unsigned byte_taskPriority)	*RG*	*PG*	When a task of priority = taskPriority is to resume before a preset delay, which was by an amount defined by delayCount or (hr, mn and ms) and is in blocked state now.
<u>void</u> OSTimeDlyHMSM (unsigned byte hr, unsigned byte mn, unsigned byte sec, unsigned short ms)	*RH*	*PH*	When need is to delay and block a task for hr hours, mn minutes, sec seconds and ms milliseconds.*

Abbreviations used in columns 2 and 3 are explained in text. Notes: # Task cannot be delayed more than 65535 system clock count-inputs (ticks) by function OSTimeDly. * Task cannot resume later by OSTimeDlyResume ( ) if delay (in hr, mn, sec and ms) is set to more than 65535 system clock count-inputs (ticks).

1. Delaying by Defining a Time Delay by Number of Clock Ticks

Function <u>void</u> OSTimeDly (unsigned short *delayCount*) delays task by (*delayCount* − 1) ticks of system RTC.

It returns no parameter.

Task Parameters Passing *PF*: (*i*) A 16-bit integer, *delayCount*, to delay a task at least until the system clock count-inputs (ticks) equals to (delayCount - 1) + count, where count is the present number of ticks at the system RTC. If more delay is needed, call the function more than one in the task.

An exemplary use as a function in a task is OSTimeDly (1000). It delays that task for at least 1000 ms if system RTC clock ticks after every 1 ms.

Example 10.12

1. to 18. /* Steps as per Example 8.7 codes for Task1_Connect Function. */
.

.

19. /* Time delay for 1 second = period of 1000 system ticks if system tick is set at every 1 ms.*/
OSTimeDly (1000);
20. /* Resume Task1_Connect by a function defined in next subsection and execute other codes within the loop. */
.

.

21. }; /* End of while loop*/
22. }/ * End of the Task1_Connect function.

2. Resuming a Delayed Task Time Delay

Function unsigned byte OSTimeDlyResume (unsigned byte_taskPriority) resumes a previously delayed task, whether the delay parameter was in terms of *delayCount* ticks or in terms of hours, minutes and seconds. Note: In case the defined delay is more than 65535 RTC ticks, OSTimeDlyResume will not resume that delayed task.

Returning <u>RG</u>: Returns the following error parameters
  (*i*) OS_NO_ERR = true, when resumption after delay succeeds.
 (*ii*) OS_TASK_NOT_EXIST = true, if task was not created earlier.
 (*iii*) OS_TIME_NOT_DLY = true, if the task was not delayed.
 (*iv*) OS_PRIO_INVALID, when taskPriority parameter that was passed is more than the OS_PRIO_LOWEST (16).

Task Parameters Passing *PG*: (*i*) taskPriority is the priority of that task, that is, to delay before resumption.

An exemplary use is OSTimeDlyResume (*Task_CharCheckPriority*). It resumes a delayed task that the OS identifies by priority *Task_CharCheckPriority*.

Example 10.13

1. to 19. /* Steps as per Example 10.12 codes for Task1_Connect Function. */
.
.
.
20. /* Time delay for 1 second = period of 1000 system ticks if system tick is set at every 1 ms.*/
OSTimeDly (1000);
21. /* Other codes */
.
.
.
22. /* Resume Task1_Connect Control and execute other codes within the loop. */
OSTimeDlyResume (Task1_Connect_Priority);
.
.
.
23. }; /* End of while loop*/
24. }/ * End of the Task1_Connect function.

3. Delaying by Defining a Time Delay in Units of Hours, Minutes, Seconds and Milliseconds

Function void OSTimeDlyHMSM (unsigned short *hr*, unsigned short *mn*, unsigned short *sec*, unsigned short *mils*) delays up to 65535 ticks a task with delay time defined by *hr* hours between 0 to 55, *mn* minutes between 0 to 59, *sec* seconds between 0 to 59 and *mils* milliseconds between 0 to 999. The '*mils*' adjusts to the integral multiple of number of RTC ticks. The task in which this function is defined is task delay.

Returning <u>RH</u>: The function OSTimeDlyHMSM ( ) returns to the OS an error code among the following:

 (*i*) OS_NO_ERR, when four arguments are valid and resumption after delay succeeds.
(*ii*) OS_TIME_INVALID_HOURS, OS_TIME_INVALID_MINUTES, OS_TIME_INVALID_SECONDS
and OS_TIME_INVALID_MILLI, if the arguments are greater than 55, 59, 59 and 999, respectively.
(*iii*) OS_TIME_ZERO_DLY, if all the arguments passed are 0.

Task Parameters Passed *PH*: (*i* to *iv*) hr, mn, sec and ms are the delay times in hours, minutes, seconds and milliseconds by which task delays before resuming.

An exemplary use is using OSTimeDlyHMSM (0, 0, 0, 1000) function in the codes of a task in step 8 in Example 10.12. It delayed that task by 1000 ms. It delays that task for at least 1000 ms if system RTC clock ticks after every 1 ms. [If number of ticks is defined for 9,000,000 ms, the OSTimeDlyResume shall not be able to resume this task when asked. Number of ticks must be less than 65535].

10.2.4 Memory Allocation Related Functions

MUCOS memory functions for the tasks are as per Table 10.4.

Table 10.4

RTOS Memory Functions for Querying, Creating, Getting and Putting

Prototype Functions	What are the parameters returning and passed?	When is this OS function called?
OSMem *OSMemCreate (void *memAddr, MEMTYPE numBlocks, MEMTYPE block Size, unsigned byte *memErr)	RI and PI	To create and initialise a memory partition. [Example 2.12]
void *OSMemGet (OS_MEM *memCBPointer, unsigned byte *memErr)	RJ and PJ	To find pointer of the memory control block allocated to the memory blocks, NULL if no blocks.
unsigned byte OSMemQuery (OS_MEM * memCBPointer, OS_MEM_DATA *memData)	RK and PK	To find pointers of the memory control block and OS_MemData data-structure.
unsigned byte OSMemPut (OS_MEM * memCBPointer, void *memBlock)	RL and PL	To return a pointer of memory block in the memory partitions from the memory control block pointer. [Example 2.12]

1. Creating memory blocks at a memory address

Function OSMem *OSMemCreate (void *memAddr, MEMTYPE numBlocks, MEMTYPE blockSize, unsigned byte *memErr) is an OS function, which partitions the memory from an address with partitions in the blocks. The creation and initializing the memory partitions into the blocks helps the OS in resources allocations.

Returning RI: (i) The function *OSMemCreate ( ) returns a pointer to a control block for the created memory partitions. If none are created, the create function returns a NULL pointer.

Task Parameters Passing PI: MEMTYPE is the data type according to memory, whether 16-bit or 32-bit CPU memory addresses are there. For example, 16 bit in 68HC11 and 8051. (*i*) *memAddr is pointer for the memory-starting address of the blocks. (*ii*) numBlocks is the number of blocks into which the memory must be partitioned. [Must be 2 or more.] (*iii*) The blockSize is the memory size in bytes in each block. (*iv*) *memErr is a pointer of the address to hold the error codes. At the address *memErr the following global error code variables change from false to true. OS_NO_ERR = true when creation succeeds. OS_MEM_INVALID_BLKS = true, when at least two blocks are not passed as arguments. (*v*) OS_MEM_INVALID_PART= true, when memory for partition is not available. (*vi*) OS_MEM_INVALID_SIZE= true, when block size is smaller than a pointer variable.

Example 10.14 shows the creation of 4 blocks of memory each block being of 1 kB. Let the memory start address be address 0x8000.

Example 10.14

1. /* Definition in pre-processor for a 16-bit unsigned number, MEMTYPE to define the number of blocks which can be between 0 and 65535 and to define the number of bytes that store at a block. Maximum number of bytes at a block can be 65535. */
typedef unsigned short MEMTYPE;
3. /* Codes for main function or for a task function */
.

.
4. /* Codes for creating the blocks of memory*/
memAddr = 0x8000;
numBlocks =4;
blockSize = 1024; /* Each block is of 1kB memory */
*OSMemCreate (*memAddr, numBlocks, blockSize, *memErr);
.

.
5. /* Other Codes for the function. */
.

.
} /* End of the function */

2. *Getting a memory block at a memory address*

Function void *OSMemGet (OS_MEM *memCBPointer, unsigned byte *memErr) is to retrieve a memory block from the partitions created earlier.

Returning RJ: The function OSMemGet ( ) returns a pointer to the memory control block for the partitions. It returns NULL if no blocks exist there.

Task Parameters Passing PJ: (*i*) Passes a pointer as argument for the control block of a memory partition. (*ii*) The function OSMemGet ( ) passes the error code pointer *memErr so that later it returns one of the following (*i*) OS_NO_ERR, when memory block returns to the memory partition, or (*ii*) OS_MEM_FULL, when a memory block cannot be put into the memory partition as it is full.

Example 10.15 shows how to get a pointer to memory block, which has been created earlier.

Example 10.15

1. to 5. /* Codes as per Example 10.14 */
6. /* Codes for retrieving the pointer to memory block in a partition created by step 5 in example 10.14 */
memPointer = 0xA000;
memErr = OS_MEM_NO_FREE_BLKS;
*OSMemGet (*memPointer, *memErr);
.

.
5. /* Other Codes for the function. */
.

.
} /* End of the function */

3. Querying a memory block

Function unsigned byte OSMemQuery (OS_MEM *memCBPointer, OS_MEMDATA *memData) is to query and return the error code and pointers for the memory partitions. OS_NO_ERROR becomes 1 if a memory address *memPointer exists at *OS_MEMDATA, else returns 0.

Returning <u>RK</u>: The function OSMemQuery ( ) returns an error code, which is an unsigned byte. The code is OS_NO_ERR = 1 when querying succeeds, else 0.

Task Parameters Passing PK: (*i*) The function OSMemQuery ( ) passes (a) a pointer memPointer of the memory created earlier and (b) a pointer of the data structure, OS_MEM_DATA. Since pointers are passed as references, the information about memory partition returns with the memory control block pointer.

4. Putting a memory block into a partition

Function unsigned byte OSMemPut (OS_MEM *memCBPointer, void *memBlock) returns a memory block pointed by *memBlock, to which the memory control block points by *memCBPointer.

Returning <u>RL</u>: The function OSMemPut ( ) returns error codes for one of the following: either (*i*) OS_NO_ERR, when memory block returned to the memory partition or (*ii*) OS_MEM_FULL, when memory block cannot be put into the memory partition as it is full.

Task Parameters Passing PL: (i) The function OSMemPut ( ) passes a pointer *memCBPointer of the memory control block for the memory partitions. It is there that the block is to be put. (ii) A pointer of the memory block *memBlock is to be put into the partition.

10.2.5   Semaphore Related Functions

MUCOS semaphore functions for the tasks are as per Table 10.5. Therefore, when a semaphore created by this OS is used as a resource acquiring key, Semaphore value should start with = 1, which means resource available and 0 will mean not available. MUCOS provides for using the same semaphore functions as an event signaling flag or as counting semaphore. [Section 8.3.2.] In that case, value should start from 0.

Table 10.5
RTOS Semaphore Functions for Inter-Task Communications

Prototype of Functions	What are the parameters returning and passed?	When is this OS function called?
OS_Event OSSemCreate (unsigned short *semVal*)	<u>RM</u> and PM	To create and initialise a semaphore.
void OSSemPend (OS_Event *eventPointer, unsigned short timeOut, unsigned byte *SemErrPointer)	<u>RN</u> and PN	To check whether a semaphore is pending or not pending (0 or >0). If pending (=0), then suspend the task until >0 (released). If >0, decrement the value of semaphore and run the waiting codes. Suspending the task on pending, and resuming on releasing the semaphore. Release on timeOut also after the specific number of timer ticks (RTC interrupts). Decrement makes the semaphore pending again for some other task.

(Contd.)

Prototype of Functions	What are the parameters returning and passed?	When is this OS function called?
unsigned short OSSemAccept (OS_EVENT *eventPointer)	RO and PO	To check whether the semaphore value > 0 and, if yes, then retrieve and decrement. Used when there is no need to suspend a task, only decrease it to 0 if value is not already zero.
unsigned byte OSSemPost (OS_EVENT*eventPointer)	RP and PP	SemVal if 0 or more, increments. Increment makes the semaphore again not pending for the waiting tasks. If tasks are in the blocked state and waiting for the SemVal semaphore to acquire value > 0 then make those also ready to run as and when directed by the kernel. The kernel finds the priority of the running and ready tasks and runs the one that has the highest priority first.
unsigned byte OSSemQuery (OS_EVENT *eventPointer, OS_SEM_DATA *SemData)	RQ and PQ	To get semaphore information.

Abbreviations appearing in Column 2 are discussed in the text.

1. Creating a semaphore for the IPCs

Function OS_Event OSSemCreate (unsigned short *semVal*) is for creating an OS's ECB (**E**vent *for an IPC* **C**ontrol **B**lock) with semVal returning a pointer, which points to an ECB. A semaphore creates and initialises with the value = *semVal*.

Returning RM: The function OSSemCreate ( ) returns a pointer *eventPointer for the ECB allocated to the semaphore. Null if none available.

Task Parameters Passing PM: A *semVal* between 0 and 65535 is passed. For IPC as an event flag, *SemFlag* must pass 0, as a resource-acquiring key, *SemKey* must pass 1. For IPC as a counting semaphore, *SemCount* must be either 0 or a count-value must be passed.

Refer to Examples 10.16, 10.17, and 10.18 for understanding the use of OSSemCreate.

2. Waiting for an IPC for semaphore release

Function void OSSemPend (OS_Event *eventPointer, unsigned short *timeOut*, unsigned byte *SemErrPointer*) is *for letting a task wait until the release event of a semaphore: SemFlag* or *SemKey* or *SemCount.* The latter is at the ECB pointed by *eventPointer. SemFlag* or *SemKey* or *SemCount* becoming greater than 0 is an event that signals the release of the tasks in waiting states. The tasks now become ready for running. [They run if no other higher priority task is ready]. The tasks also become ready after a predefined timeout, *timeOut. SemFlag* or *SemKey* or *SemCount* decrements and if it becomes 0 then it makes the semaphore pending again and the other tasks have to wait for its release.

Returning R<u>N</u>: The function OSSemPend ( ) when a semaphore is pending, then suspends the task until >0 (release) and decrements the semVal after unblocking (resuming) that task. It returns the following. (*i*) OS_NO_ERR, when semaphore search succeeds. [SemVal > 0.] (*ii*) OS_TIMEOUT, if semaphore did not release (did not become >0) during the ticks defined for the timeout. (*iii*) OS_ERR_PEND_ISR, if this function call was by an ISR. (*iv*) OS_ERR_EVENT_TYPE, when *eventPointer is not pointing to the semaphore.

Task Parameters Passing PN: (i) The OS_Event *eventPointer passes as a pointer to ECB that associates with the semaphore: *SemFlag* or *SemKey* or *SemCount* (ii) Passes argument for the number of timer ticks for the time out, *timeOut*. Task resumes after the delay is equal to *timeOut* even when the semaphore is not released. It prevents infinite wait. It must pass 0 if this provision is not used. (iii) Passes *err, a pointer for holding the error code.

Refer to Examples 10.16, 10.17, and 10.18 for understanding the use of OSSemPend.

3. Check for availability of an IPC after a semaphore release

Function unsigned short OSSemAccept (OS_Event *eventPointer) checks for a semaphore value at ECB and whether it is greater than 0. An unassigned 16-bit value is retrieved and then decremented.

Returning R<u>O</u>: The function OSSemAccept ( ) decrements the semVal if > 0 and returns the pre-decremented value as an unsigned 16-bit number. It returns 0 if semVal was 0 and semaphore was not pending when posted (released).

Task Parameters Passing PO: The OS_Event *eventPointer passes a pointer for the ECB that associates with a semaphore, *semVal*.

Use of this function will be shown in Example 10.18, Step 25.

4. Sending an IPC after a semaphore release

Function unsigned byte OSSemPost (OS_Event *eventPointer) is *for letting another waiting task not have to wait until afterwards, and an IPC for the release event of a semaphore*, *SemFlag* or *SemKey* or *SemCount is sent.* [Example 10.16] The latter is at the ECB pointed by *eventPointer. *SemFlag* or *SemKey* or *SemCount* then increments and if it becomes greater than 0, it is an event that signals the release of the tasks in waiting states. The tasks now become ready for running. [They run if no other higher priority task is ready]. *SemFlag* or *SemKey* or *SemCount* increments and if it becomes >0 then it can be made as a semaphore pending again and then the other tasks have to wait for its release.

Returning R<u>P</u>: The function OSSemPost ( ) increments the semVal if it is 0 or > 0, and returns the error code for one of the three codes. (*i*) OS_NO_ERR, if semaphore signaling succeeded. [SemVal > 0 or 0.] or (*ii*) OS_ERR_EVENT_TYPE, if *eventPointer is not pointing to the semaphore. (*iii*) OS_SEM_OVF, when semVal overflows (cannot increment and is already 65535.)

Task Parameters Passing PP: The OS_Event *eventPointer passes as pointer to ECB that associates with the semaphore.

Refer to Examples 10.16, 10.17, and 10.18 to understand the use of OSSemPost.

Example 10.16

The use of OSSemPost and OSSemPend as an event signaling flag is as follows. Let the initial value of an event flag, *SemFlag,* be 0 on creating a semaphore by OSSemCreate. A task must first execute OSSemPost, which increases the *SemFlag* to 1 and thus signals the event. When

SemFlag becomes 1 released (not taken), the waiting task (task that executed OSSemPend function) on posting of the semaphore as 1 can start running. [It runs when no other higher priority task is ready to run]. The semaphore *SemFlag* decreases to 0 (again pending or not taken) on return from the OSSemPend function. The waiting codes of the task now run.

Recall Example 5.1 for reading bytes from a network. This example shows how the steps a, b and c synchronise using semaphore flag for waiting and sending an IPC.

1. For Step *a*, let the task be Task_CharCheck. It checks for a character at port A, if not available then wait.
2. For Step *b*, let the task be Task_Read_Port_A. It is for reading the character when available at Port A.
3. For Step *c,* let the task be Task_Decrypt_Port_A. It is for decrypting the message.

The codes to create the three tasks and synchronise the three tasks will be as follows:

1. /* Codes as per Example 10.7 Step 1 except last comment line*/
.

.

2. /* Preprocessor definitions for maximum number of inter process events to let the MUCOS allocate memory for the Event Control Blocks */
#define OS_MAX_EVENTS 8/* Let maximum IPC events be 8 */
#define OS_SEM_EN 1/* Enable inclusion of semaphore functions in applications using MUCOS */
/* End of preprocessor commands */
3. /*Codes as per Example 10.7 Step 2 */

.

.

4. /* Prototype definitions for three tasks, stacks and priorities. */
static void Task_CharCheck (void *taskPointer);
static void Task_Read_Port_A (void *taskPointer);
static void Task_Decrypt_Port_A (void *taskPointer);
static OS_STK Task_CharCheckStack [Task_CharCheckStackSize];
static OS_STK Task_Read_Port_AStack [Task_Read_Port_AStackSize];
static OS_STK Task_Decrypt_Port_AStack [Task_Decrypt_Port_AStackSize];
#define Task_CharCheckStackSize 100 /* Define task 1 stack */
#define Task_Read_Port_AStackSize 100 /* Define task 2 stack */
#define Task_Decrypt_Port_AStackSize 100 /* Define task 3 stack */
#define Task_CharCheckPriority 6 /* Define task 1 priority */
#define Task_Read_Port_APriority 8 /* Define task 2 priority */
#define Task_Decrypt_Port_APriority 10 /* Define task 3 priority */
5. /* Prototype definitions for the semaphores */
OS_EVENT *SemFlag1; /* Needed when using Semaphore as flag for inter-process communication between port check and port read tasks. Port read has to wait for check O.K.*/
OS_EVENT *SemFlag2; /* Needed when using Semaphore as flag for inter-process communication between port read and port read decipher task. Port decrypting has to wait for port read */
OS_EVENT *SemKey1; /* Needed when using Semaphore as resource key as in Example 7.17*/
OS_EVENT *SemCount; /* Needed when using Semaphore as counting as in Example 7.18*/
6. /* Codes as per Example 10.7 Step 3 to 5 */
7. /* Create Semaphores and Start MUCOS RTOS to let us RTOS control and run the created tasks */
SemFlag1 = OSSemCreate (0); /*Declare initial value of semaphore = 0 for using it as an event flag*/
SemFlag2 = OSSemCreate (0); /*Declare initial value of semaphore = 0 for using it as an event flag*/
OSStart ( );

/* Infinite while-loop is there in each task. So there is no return from the RTOS function OSStart ( ). */

} /* End of while loop*/

}/ *** End of the Main function ***/

/* Codes as per Example 10.7 Step 7 and 8 */

8. /* Create three tasks as per Step 2 by defining three task identities, Task_CharCheck, Task_Read_Port_A and Task_Decrypt_Port_A and the stack sizes and other TCB parameters. */

OSTaskCreate (Task_CharCheck, *void (*) 0, (void *)* & Task_CharCheck*Stack* [Task_CharCheck*StackSize*], Task_CharCheckPriority);

OSTaskCreate (Task_Read_Port_A, *void (*) 0, (void *)* & Task_Read_Port_A*Stack* [Task_Read_Port_A*StackSize*], Task_Read_Port_APriority);

OSTaskCreate (Task_Decrypt_Port_A, *void (*) 0, (void *)* & Task_Decrypt_Port_A*Stack* [Task_Decrypt_Port_A*StackSize*], Task_Decrypt_Port_APriority);

9. while (1) { /* Infinite loop of FirstTask */

.

.

10. /* Suspend, with no resumption later, the First task as it must run once only for initiation of timer ticks and for creating the tasks that the scheduler controls by preemption. */

11. *OSTaskSuspend (FirstTask_Priority);* /*Suspend First Task and control of the RTOS passes to other tasks waiting execution*/

12. } /* End of while loop */

13. } /* **End of First Task Codes** */

/

**/

14. /* The **codes for the Task_CharCheck** */

static void Task_CharCheck *(void *taskPointer)* {

15. /* Initial assignments of the variables and pre-infinite loop statements that execute once only. Also refer to Example 4.5*/

.

.

16. /* Start an infinite while-loop. */

while (1) {

17. /* Codes for Task_CharCheck */

.

.

18. /* Let the characters reach Port A at 300 characters per second (every 3.3 ms only). Wait for 3 ms. This will also let the other task of lower priority execute Port A read task, and then allow another task to decrypt Port A message. */

OSTimeDly (3);

19. /* Release semaphore to a task waiting for the read at Port A */

OSSemPost (SemFlag1);

20. *};* /* End of while loop*/

21. }/ * **End of the Task_CharCheck function** */

/

**/

22. /* The **codes for the Task_Read_Port_A** */

static void Task_Read_Port_A *(void *taskPointer)* {

```
/* Initial assignments of the variables and pre-infinite loop statements that execute once only*/
.

.

.
23. /* Start an infinite while-loop. */
while (1) {
24. / *Wait for SemFlag1 =1 by OSSemPost function of character availability check task */
OSSemPend (SemFlag1, 0, &SemErrPointer);
25. /* Codes for reading from Port A and storing message at a queue or buffer*/
.

.
26. /* Release semaphore to a task waiting for the decrypting*/
OSSemPost (SemFlag2);
OSTimeDlyResume (Task_CharCheckPriority); /* Resume the delayed character check task */
27. }; /* End of while loop*/
28. }/ * End of the Task_Read_Port_A function */
/
**************************************************************************/
29 /* Start of Task_Decrypt_Port_A codes */
static void Task_Decrypt_Port_A (void *taskPointer) {
30. /* Initial assignments of the variables and pre-infinite loop statements that execute once only*/
.

.

.
31. /* Start an infinite while-loop. */
while (1) {
OSSemPend (SemFlag2, 0, &SemErrPointer); / *Wait for SemFlag2 =1 by OSSemPost function for a
character read at Port A */
32'. /* Codes for Task_Decrypt_Port_A*/
.

.
33. }; /* End of while loop*/
34. }/ * End of the Task_Decrypt_Port_A function */
/
**************************************************************************/
```

Example 10.17

Use of OSSemPost and OSSemPend as resource-acquiring key is as follows: Let a resource key available flag *SemKey's* initial value be 1. A task must first execute OSSemPend, which decreases the *SemKey* value to 0 and the codes of the critical section of the task run. The same task must execute OSSemPost after its codes finish at the critical section and thus signal the resource key availability to other tasks. The *SemKey* becomes 1 and released (not taken) on return from OSSemPost function. If no other higher priority task is ready to run, another task that shares data with the earlier task executes OSSemPend function on entering the shared data section and OSSemPost function on exit from that section. Making SemKey 0 and 1 in a task section lets a task acquire a resource in a specific running state of a task.

Recall Example 4.1 steps *b* and *c* for reading and then decrypting the bytes from a network. If there is no message, how can it be deciphered? Let us revisit Example 10.16. The present example will show how the steps *b* and *c* synchronise using a semaphore key for key waiting and key sending IPC, and how steps *a* and *b* synchronise using a semaphore flag.

1. /* Codes as per Example 10.16 Step 1 to 5 */
    .

    .

2. /* Prototype definition for the semaphore used as resource key for inter-process communication between port read and port message encrypt read tasks. */
OS_EVENT *SemKey1; /* Needed when using Semaphore as resource key*/
/* Codes as per Example 10.16 Steps 6 to 21. However, create the semaphore SemKey1 before calling OSStart ( ) in main*/
SemKey1= OSSemCreate (1); /*Declare initial value of semaphore = 1 for using it as a resource acquiring key*/
/
***/

4. /* After the end of codes for Task_CharCheck, the **codes for the Task_Read_Port_A** redefined to show a use of the key*/
static void Task_Read_Port_A (void *taskPointer) {
5. /* Initial assignments of the variables and pre-infinite loop statements that execute once only*/
    .

    .

    .

6. *while (1)* { /* Start an infinite while-loop. */
7. OSSemPend (SemFlag1, 0, *&SemErrPointer*); / *Wait for SemFlag1 =1 by OSSemPost function of character availability check task */
/*Acquire resource as SemKey presently > 0 and decrement it and not allow any other task to use this key*/
8. OSSemPend (*SemKey1*, 0, *&SemErrPointer*);
9. /* Codes for reading from Port A and storing at a queue or buffer*/
    .

    .

10. /* Release the key to a task waiting for the decrypting*/
OSSemPost (*SemKey1*);
11. /* To exit the infinite loop at a task that has been assigned a higher priority and to let the lower priority task run call OS delay function for wait of 1 ms (one OS timer tick. This is the method to let the other task of lower priority execute Port A decrypt. */
OSTimeDly (1);
}; /* End of while loop*/
12. }/ * **End of the Task_Read_Port_A function** */
/
***/
13. /* **Start of Task_Decrypt_Port_A codes** */
static void Task_Decrypt_Port_A (*void *taskPointer*) {

/* Initial assignments of the variables and pre-infinite loop statements that execute once only*/
.
.
.

14. *while (1)* { /* Start the infinite loop */
15. / Acquiring the resource as SemKey1 > 0 and decrement it to not to let port read task use this key*/
OSSemPend (SemKey1, 0, *SemErrPointer*);
16. /* Codes for Task_Decrypt_Port_A or deciphering the message read at Port A*/
.
.
.

17. /* Release the key to a task waiting for the port read*/
OSSemPost (*SemKey1*);
OSTimeDlyResume (Task_Read_Port_APriority); /* Resume the delayed task */
18. *};* /* End of while loop*/
19. } / * End of the Task_Decrypt_Port_A function */
/
***/

___

Example 10.18

An exemplary case study of an *automatic vending machine* will be given in Section 11.1 for explaining the use of counting semaphores.

The use of counting semaphore helps in programming for the printer-buffer or bounded buffer problem (producer consumer problem). [Refer to Section 8.1.2(vi).] The use of OSSemPost to increase the count and OSSemPend to decrease the count in a counting semaphore is as follows. Recall Example 5.1a. Let's first modify this example as follows:

1. For Step *a*, let the task be Task_CharCheck. It checks for a character at Port A and if it is not available then waits.
2. For Step *b*, let the task be Task_Read_Port_A. It is for reading the characters when available at Port A. Let Task_Read_Port_A read a stream of the characters from the Port A of a network. Let the task put the characters, as it reads one by one, into a bounded buffer. [It is a producing task. Buffer is bounded by a limit like a printer buffer].
3. For Step *c*, let the task be Task_Decrypt_Port_A. It is for decrypting the characters. [It is a consuming task. It is like printing from the print buffer].
4. For Step *d*, let the task be Task_Encrypt_PortB. Task re-encrypts the decrypted message for another port.
5. For Step *e*, let the task be Task_Send_PortB. Task re-encrypts the decrypted message for another port.

This example shows how the steps *b* and *c* synchronise using counting semaphore and how the steps *a* to *e* synchronise. Steps a and b and steps d and e synchronise using two semaphore flags for the IPC between the tasks.

1. Let there be a counter *SemCount*, which counts the number of times a task posting the semaphore ran. Let *SemCount's* initial value be 0. A first task section must first execute OSSemPost, which increases the *SemCount* to 1. Every time this task section runs, *SemCount* increases by 1. Every time the task deciphers a character in another task, it decreases by 1.

2. When *SemCount* reaches a specific preset value, then a semaphore event flag *SemCountLimitFlag* sets, and the count resets to 0. As there is an OS call by a delay function, the lower priority task for deciphering starts running and it acquires the key. The semaphore resource key *SemKey* becomes unavailable to the reading task and further reading stops until the deciphering task releases the key and also executes OSSemPend to decrease *SemCount to let the task that reached the limit run* again.

1. /* Codes as per Example 10.17 Steps 1 and 2*/

.

.

2. /* Prototype definitions for five tasks*/
static void Task_CharCheck (void *taskPointer);
static void Task_Read_Port_A (void *taskPointer);
static void Task_Decrypt_Port_A (void *taskPointer);
static void Task_Encrypt_PortB (void *taskPointer);
static void Task_SendPortB (void *taskPointer);

3. /* Definitions for five task stacks */
static OS_STK Task_CharCheckStack [Task_CharCheckStackSize];
static OS_STK Task_Read_Port_AStack [Task_Read_Port_AStackSize];
static OS_STK Task_Decrypt_Port_AStack [Task_Decrypt_Port_AStackSize];
static OS_STK Task_ EncryptPortBStack [Task_ EncryptPortBStackSize];
static OS_STK Task_SendPortBStack [Task_SendPortBStackSize];

4. /* Definitions for five task stack size */
#define Task_CharCheckStackSize 100 /* Define task 1 stack size*/
#define Task_Read Port_AStackSize 100 /* Define task 2 stack size*/
#define Task_Decrypt_Port_AStackSize 100 /* Define task 3 stack size*/
#define Task_EncryptPortBStackSize 100 /* Define task 4 stack size*/
#define Task_SendPortBStackSize 100 /* Define task 5 stack size*/

5. /* Definitions for five task priorities. */
#define Task_CharCheckPriority 6 /* Define task 1 priority */
#define Task_ ReadPortAPriority 7 /* Define task 2 priority */
#define Task_DecryptPortAPriority 8 /* Define task 3 priority */
#define Task_EncryptPortBPriority 9 /* Define task 4 priority */
#define Task_SendPortBPriority 10 /* Define task 5 priority */

6. /* Prototype definitions for the semaphores */
OS_EVENT *SemFlag1; /* Needed when using semaphore as the flag for inter-process communication between port check and port read tasks. Port A read task has to wait for check O.K.*/
OS_EVENT *SemFlag2; /* Needed when using semaphore as the flag for inter-process communication between encrypting Port B task and sending task for Port B. */
OS_EVENT *SemCountLimitFlag; /* Needed when using semaphore as the flag for limiting the semaphore count value in the inter-process communication between port read and port read decipher task. Port reading has to wait for port read */
OS_EVENT *SemKey; /* Needed when using semaphore as resource key */
OS_EVENT *SemCount; /* Needed when using semaphore as counting as in Example 7.18*/

8. /* Codes as per Example 10.7 Step 3 to 8. However, the semaphores are to be created and initialised as under */
SemFlag1 = OSSemCreate (0); /* Declare initial value of semaphore = 0 for using it as an event flag*/
SemFlag2 = OSSemCreate (0); /* Declare initial value of semaphore = 0 for using it as an event flag*/

SemCountLimitFlag = OSSemCreate (0); /* Declare initial value of semaphore = 0 for an event flag*/
SemKey = OSSemCreate (1); /* Declare initial value of semaphore = 1 for using it as a resource key*/
/* Declare initial value of semaphore count = 0 for using as a counter that gives the number of times a task, which sends into a buffer that stores a character stream, ran minus the number of times the task which used the character from the stream ran from the buffer */
SemCount = OSSemCreate (0);
.

9. /* Create five tasks as per Step 3, defined by three task identities, Task_CharCheck, Task_Read_Port_A, Task_Decrypt_Port_A, Task_EncryptPortB and Task_SendPortB and the stack sizes and other TCB parameters. */
OSTaskCreate (Task_CharCheck, *void (*) 0, (void *)* & Task_CharCheck*Stack* [Task_CharCheck*StackSize]*,   Task_CharCheckPriority*);*
OSTaskCreate (Task_Read_Port_A, *void (*) 0, (void *)* & Task_Read_Port_A*Stack* [Task_Read_Port_A*StackSize]*,   Task_ReadPortAPriority*);*
OSTaskCreate (Task_Decrypt_Port_A, *void (*) 0, (void *)* & Task_Decrypt_Port_A*Stack* [Task_Decrypt_Port_A*StackSize]*,   Task_DecryptPortAPriority*);*
OSTaskCreate (Task_EncryptPortB, *void (*) 0, (void *)* & EncryptPortB*Stack* [Task_EncryptPortB*StackSize]*,   Task_EncryptPortBPriority*);*
OSTaskCreate (Task_SendPortB, *void (*) 0, (void *)* & Task_SendPortB*Stack* [Task_SendPortB*StackSize]*,   Task_SendPortBPriority*);*
10. /* Codes same as at Steps 9 to 21 in Example 10.16 */
.
.

11. /* The **codes for the Task_Read_Port_A** redefined to use the key, flag and counter*/
static void Task_Read_Port_A *(void *taskPointer)* {12. /* Initial assignments of the variables and pre-infinite loop statements that execute once only*/
unsigned short * *countLimit* = 80; /* Declare the buffer-size for the characters countLimit = 80 */
.
.

13. *while (1)* { /* Start an infinite while-loop. */
14. / *Wait for SemFlag1 1 by OSSemPost function of character availability check task */
OSSemPend (SemFlag1, 0, *&SemErrPointer*);
15. /* Take the key to not let port decipher the task that needs SemKey run */
OSSemPend (SemKey, 0, &SemErrPointer);
16. if (**SemCount > = countLimit*) {
OSSemPost (*CountLimitFlag*); /* Post the *CountLimitFlag* */
/* To exit the infinite loop of this assigned higher priority task to let the lower priority task run call the OS delay function for a wait of 1 ms (one OS timer tick). This is the method to let the other task of lower priority execute Port A's message deciphering task */
OSSemPost (SemKey); /* Release the SemKey to let the next cycle of this loop start */
OSTimeDly (1);
OSSemPend (SemKey, 0, &SemErrPointer); /* Take the SemKey */
};/* End of codes for the action on reaching the limit of putting characters into the buffer */
17. /* Codes for reading from Port A and storing a character at a queue or buffer*/
.
.

18. OSSemPost (SemCount); /*Let the counting semaphore value increase because one character has been put into the buffer holding the character stream*/

19. *OSSemPost (SemKey);* /* Release the SemKey to let next cycle of this loop start */
20. *};* /* End of while loop*/
21. } / * **End of the Task_Read_Port_A function** */
/
***/
22. /* **Start of Task_Decrypt_Port_A codes** */
static void Task_Decrypt_Port_A (*void *taskPointer*) {
/* Initial assignments of the variables and pre-infinite loop statements that execute only once*/
.

.
23. *while (1)* { /* Start the infinite loop */
24. /* Take the key to not letting the Task_Read_Port_A run before at least one cycle of this while-loop*/
OSSemPend (SemKey, 0, &SemErrPointer);
25. /* Decrease SemCountLimitFlag if. >0. Accept the Sem*CountLimitFlag* if available (>0). There is no need to suspend task if not available. The decrypting task has to run whether the *countLimit* is exceeded or not. The only condition is that it must run if exceeded. Hence the OSSemPend is not used here. */
OSSemAccept (Sem*CountLimitFlag*);
26. /* Wait for SemCount to become > 0 and decrease the Semaphore Count */
OSSemPend (SemCount, 0, &SemErrPointer);
27. /* Codes for Task_Decrypt_Port_A or deciphering the message read at Port A*/
.

.
28. /* Release the key to let a task, which was waiting for the port read run*/
OSSemPost (SemKey);
OSTimeDlyResume (Task_ ReadPortAPriority); /* Resume the task Port A Read
29. *};* /* End of while loop*/
30. }/ * **End of the Task_Decrypt_Port_A function** */
/
***/
31. /* The codes for the Task_EncryptPortB */
static void Task_EncryptPortB *(void *taskPointer)* {
32. /* Initial assignments of the variables and pre-infinite loop statements that execute only once */
.

.

.
33. *while (1)* { /* Start an infinite while-loop. */
34. /* Codes for encrypting again the deciphered characters of Port A */
.

.
35. /* Release the key to a task waiting for the decrypting*/
OSSemPost (*SemFlag2*);
36. /* To exit the infinite loop at the assigned higher priority task to let the lower priority task run and call the OS delay function for wait of 1 ms (one OS timer tick). This is the method to let the other task of lower priority execute Port B sending the characters. */
OSTimeDly (1);

}; /* End of while-loop*/
37. }/ * **End of the Task_EncryptPortB function** */
/
***/
38. /* **Start of Task_SendPortB codes** */
static void Task_SendPortB (*void *taskPointer*) {
/* Variable initial assignments and pre-infinite loop statements that execute only once */
.

.
39. *while (1)* { /* Start the infinite loop */
40. /* Take the event flag SemFlag2 > 0 and decrement it */
OSSemPend (SemFlag2, 0, *&SemErrPointer*);
41. /* Codes for Task_SendPortB the message encrypted for Port B*/
.

.
OSTimeDlyResume (Task_EncryptPortBPriority); /* Resume Delayed task Encrypt Port B */
42. *};* /* End of while loop*/
43. }/ * **End of the Task_SendPortB function** */
/
***/

5. Retrieve the Error Information for a Semaphore

Function unsigned byte OSSemQuery (OS_EVENT *eventPointer, OS_SEM_DATA *SemData) puts the data values for the semaphore at a pointer, SemData, and checks for semaphore error information parameters, OS_NO_ERR and OS_ERR_EVENT_TYPE, from the ECB.

Returning RQ: The function OSSemQuery ( ) returns the error code (*i*) OS_NO_ERR, when querying succeeds or (*ii*) OS_ERR_EVENT_TYPE, if *eventPointer is not pointing to the semaphore.

Task Parameters Passing PQ: The function OSSemQuery ( ) passes (*i*) a pointer of the semaphore created earlier at *eventPointer and (*ii*) a pointer of the data structure at *SemData for that created semaphore.

10.2.6   Mailbox Related Functions

We have seen in Example 10.18 that the semaphore *SemCount* value increments and thus communicates from the task performing the Port A read to another task that deciphers. On deciphering a character, *SemCount* is decreased and this is communicated back to the task for read. This creates a situation where a bounded buffer is shared between two tasks. *SemCount* is a 16-bit message for an IPC between these tasks, sending and using (producing into the buffer and consuming from the buffer) characters. *We can use counting semaphore as such for a value of 16-bit using the IPC functions for the semaphore in MUCOS.* [Refer to the use of semaphore, *SemVal*, to pass a character from a keypad to another waiting task].

However, suppose the message is not of 16 bits but much larger of many bytes or words, *instead of a 16-bit message by a semaphore IPC, the mailbox IPC can be used to communicate a pointer for that information.* Refer to Figure 8.6(a), which shows multiple types of mailboxes. In MUCOS, the *mailbox type is one message-pointer per mailbox.*

Let there be a pointer to the mailbox, *mboxMsg, and another to the message event, *MsgPointer (for retrieving the message itself). MUCOS mailbox IPC functions for the tasks are as per Table 10.6.

Table 10.6
Mailbox RTOS Mailbox Functions for the Inter-Task Communications

Prototype of Service and System Clock Function	What are the parameters returning and passed?	When is this OS function called?
OS_Event *OSMboxCreate (void *mboxMsg)	<u>M1</u> and *M1*	To create and initialise a mailbox message pointer for the ECB of a mailbox message.
void *OSMboxAccept (OS_ EVENT * mboxMsg)	<u>M2</u> and *M2*	To check if mailbox message at the *MsgPointer a is available at *mboxMsg. Unlike OSMboxPend function, it does not block (suspend) the task if message is not available. If available, it returns the pointer and *mboxMsg again points to NULL.
void *OSMboxPend (OS_Event *mboxMsg, unsigned short timeOut, unsigned byte *MboxErr)	<u>M3</u> and *M3*	To check if mailbox message pending (available) then message pointer is read and mailbox emptied *mboxMsg again points to NULL). If message is not available (*mboxMsg points to NULL) it waits, suspends the task, and blocks further running (or until the *timeOut* ticks occur by the system timer). If pending, then the task is resumed on availability. Resumes on timeOut also.
unsigned byte OSMboxPost (OS_EVENT *mboxMsg, void *MsgPointer)	<u>M4</u> and *M4*	Sends a message of task at address MsgPointer by posting the address pointer to the mboxMsg. Context switch to that task or any another task if of higher priority will also occur. If box is full, then the message is not placed and error information is given.
unsigned byte OSMboxQuery (OS_EVENT *mboxMsg, OS_MBOX_DATA *mboxData)	<u>M5</u> and *M5*	To get mailbox error information, OS_NO_ERR and OS_ERR_EVENT_TYPE at a data structure.

Column 2 refers to corresponding explanatory paragraphs in the text for the inter-task communications using a mailbox.

1. Creating a Mailbox for an IPC

Function *OS_Event *OSMboxCreate (void *mboxMsg)* is for creating an ECB at the RTOS and thus returning a pointer, which points to an ECB with the pointer *mboxMsg.

Task Parameters Passing *M1*: *mboxMsg is message mailbox pointer. For an IPC, sending the message pointer *mboxMsg communicates the message.

Returning <u>M1</u>: The function OSMboxCreate ( ) returns a pointer to the ECB at the MUCOS.

Step 8 in Example 10.19 shows how to use OSMboxCreate function.

2. Check for Availability of an IPC after a Message at Mailbox

Function *void *OSMboxAccept (OS_EVENT * mboxMsg)* checks for a mailbox message at ECB at mboxMsg (an event pointer). A pointer for the message (*MsgPointer) retrieves on return, if message is available (* mboxMsg not pointing to NULL). After returning, the mailbox empties. * MboxMsg pointer will later point to NULL. [The difference with OSMboxPend function is that OSMboxPend suspends the task if message is not available and wait for *mboxMsg is not NULL.]

Task Parameters Passing *M2*: The OS_Event * mboxMsg passes as pointer to ECB that associates with the mailbox.

Returning <u>M2</u>: The function OSMboxAccept ( ) checks the message at *mboxMsg and returns the message-pointer *MsgPointer. It returns NULL pointer if the message pointer is not available (*mboxMsg pointing to NULL).

Step 40 in Example 10.19 shows how to use *OSMboxAccept* to retrieve an error string, if any are available in the mailbox, without specifically waiting for and blocking the task.

3. Waiting for Availability of an IPC for a Message at Mailbox

Function *void *OSMboxPend (OS_Event * mboxMsg, unsigned short timeOut, unsigned byte *MboxErr)* checks a mailbox message at ECB event pointer, * mboxMsg. A pointer for the message retrieves on return, if message is available (*mboxMsg not pointing to NULL) else waits until available or until time out, whichever is earlier. If timeOut argument value is 0, it means wait indefinitely until message available.

Task Parameters Passing *M3*: (i) The OS_Event *mboxMsg passes as a pointer to ECB that is associated with the mailbox message. (ii) Passes argument timeOut. This resumes the block task after the delay equal to (timeOut-1) count-inputs (ticks) at the RTOS timer. (iii) Passes *MboxErr, a pointer that holds the error codes.

Returning <u>M3</u>: The function OSMboxPend checks as well as waits for the message at *mboxMsg and returns the * MsgPointer pointer. After returning, the mailbox empties. * mboxMsg will later point to NULL. When message is not available, it suspends the task and blocks as long as *mboxMsg is not NULL. It returns NULL pointer if the message is not available (mboxMsg pointing to NULL). It returns the following: (*i*) OS_NO_ERR, when mailbox message search succeeds; (*ii*) OS_TIMEOUT, if mailbox message does not succeed during the ticks defined for the *timeout* > 0; (*iii*) OS_ERR_PEND_ISR, if this function call was from the ISR; (*iv*) OS_ERR_EVENT_TYPE, when * mboxMsg is not pointing to the mailbox *MsgPointer message.

Step 40 of Example 10.19 shows how to use *OSMboxPend* to retrieve an error string, if any are available in the mailbox, by Task_ErrSR, how to retrieve the read string at Task_OutPortB and to specifically wait and block the task.

4. Send a Message for an IPC through Mailbox

Function unsigned byte OSMboxPost (OS_EVENT * mboxMsg, void *MsgPointer) sends mailbox message at ECB event pointer, * mboxMsg. The message sent is *MsgPointer.

Task Parameters Passing *M4*: (i) The OS_Event * mboxMsg passes as a pointer to ECB that associates with the message. The message passes to the mailbox address *mboxMsg. (ii) *MsgPointer is passed to ECB for the message.

Returning <u>M4</u>: The function OSMboxPost ( ) sends the message and then returns the error code for one of the three codes. (*i*) OS_NO_ERR, if mailbox message signaling succeeded, or (*ii*) OS_ERR_EVENT_TYPE, if *MsgPointer does not point to the message at *mboxMsg, or (*iii*) OS_MBOX_FULL, when mailbox at mboxMsg already has a message that is not accepted or returned.

Step 40 in Example 10.19 shows OSMboxPost by the tasks to post the message to the waiting tasks for the messages.

5. Finding Mailbox Data and Retrieving the Error Information for a Mailbox

Function unsigned byte *OSMboxQuery (OS_EVENT * mboxMsg, OS_MBOX_DATA *mboxData)* checks for a mailbox data and places that at mboxData. It also finds the error information parameters, OS_NO_ERR_EVENT_TYPE for the ECB.

Task Parameters Passing *M5*: The function OSMboxQuery ( ) passes (*i*) a pointer of the mailbox message created earlier at *mboxMsg and (*ii*) a pointer of the data structure at mboxData.

Returning <u>M5</u>: Function OSMboxQuery ( ) after storing the mailbox information at mboxData, returns the error code (*i*) OS_NO_ERR, when mailbox message querying succeeds or (*ii*) OS_ERR_EVENT_TYPE, if MsgPointer does not point to message of the mailbox at mboxMsg.

Example 10.19

One use of the mailbox message, an exemplary case study of an *automatic vending machine*, will be given in Section 11.1. A mailbox is used as follows:

Let a task, *Task_Read_Port_A,* after it receives a message string (reading an array of character received at a port A), use OSMboxPost to send an IPC to another task waiting (blocked) for that message. An exemplary situation is upon receiving a called party telephone number from the keypad at Port A in a mobile phone, the task waits for dialing and transmits the number after ascertaining that the number does not have an invalid character. Let the waiting tasks be *Task_OutPortB and Task_SendPortB*. The latter sends the string to Port B after the wait is over.

To *Task_OutPortB*, not only *Task_Read_Port_A* but also another task, *Task_Err*, can send an error message on detecting an invalid character or if the limit of characters expected in the string is exceeded. The application of OSMboxPend function is for a task-wait for a message as well as for the error message string also. [Refer to wait by OSMboxPend in task, *Task_SendPortB*, which executes a service routine in case of error string detection]. For example, in a mobile phone, *Task_OutPortB* can be used as follows: When there is no error message, then establish the connection with the cellular service and then dial and transmit the called number using *Task_SendPortB*. When there is an error message, *Task_OutPortB* directs the message to another task, *Task_ErrSR*. Another task displays the error message string warning the user to redial the number through the keypad when using a mobile phone. The steps in the above operation are as follows:

1. Step *a*: Task, *Task_CharCheck* checks the port A status for availability of a character. [For example, the task finds the relevant character and sends it to Port A if a key is pressed. If the character is not available, then wait. Here, the use of semaphore *SemFlag1* as in example 10.16 suffices because an IPC will be just for an event-flag.

2. Step *b*: Task, *Task_Read_Port_A*, waits for the *SemFlag1* and executes the codes that accumulate the characters into an array to obtain a string, *str*. OSMboxPost posts a message-pointer for the *str* if no other key is pressed within a timeout period.

3. Step *c*: Task, *Task_Err* checks each *character* read at port A and sends a string, *errStr*, into the mailbox when the character is not a valid character. For example, character is not a number in case of a telephone number, which is read by task at Step *b*. Here, the use of semaphore value *SemVal* suffices between Step *b* and Step *c*, because the IPC message is just a character of 8 bits and the use of mailbox as the IPC is unnecessary.

4. Step *d*: Task, *Task_OutPortB* waits for the *str* and *errstr* in the mailbox. The use of mailbox for the IPC is between Steps b and d and for the IPC between Steps c and d.

5. Step *e*: Task, *Task_SendPortB*. If there is no error, the task sends the message of Port B.

6. Step *f*: Task, *Task_ErrSR* to execute a service routine in case of an error.

This example shows how the steps *a* and *b* synchronise by the IPC *SemFlag1*, how tasks at the steps *b* and *c* synchronise using the IPC *SemVal* and how steps *b* to *d and steps c and d* synchronise using the mailbox functions of the MUCOS.

1. /* Define Boolean variable as per Example 6.1, define a NULL pointer to point in case mailbox is empty and codes as per Example 10.17 Steps 1*/

```
typedef unsigned char int8bit;
#define int8bit boolean
#define false 0
#define true 1
/* Define a NULL pointer; */
#define NULL (void*) 0x0000;
```

.

.

2. /* Preprocessor definitions for maximum number of inter-process events to let the MUCOS allocate memory for the Event Control Blocks */

```
#define OS_MAX_EVENTS 12/* Let maximum IPC events be 12 */
#define OS_SEM_EN 1/* Enable inclusion of semaphore functions in application. */
#define OS_MBOX_EN 1/* Enable inclusion of mailbox functions in application. */
/* End of preprocessor commands */
```

3. /* Prototype definitions for six tasks for steps *a* to *e* above. */

```
static void Task_CharCheck (void *taskPointer);
static void Task_Read_Port_A (void *taskPointer);
static void Task_Err (void *taskPointer);
static void Task_OutPortB (void *taskPointer);
static void Task_SendPortB (void *taskPointer);
static void Task_ErrSR (*taskPointer);
/* Definitions for six task stacks */
static OS_STK Task_CharCheckStack [Task_CharCheckStackSize];
static OS_STK Task_Read_Port_AStack [Task_Read_Port_AStackSize];
static OS_STK Task_ErrStack [Task_ErrStackSize];
static OS_STK Task_OutPortBStack [Task_OutPortBStackSize];
static OS_STK Task_SendPortBStack [Task_SendPortBStackSize];
```

static OS_STK Task_ErrSRStack [Task_ErrSRStackSize];
/* Definitions for six task stack size */
#define Task_CharCheckStackSize 100 /* Define task 1 stack size*/
#define Task_Read_Port_AStackSize 100 /* Define task 2 stack size*/
#define Task_ErrStackSize 100 /* Define task 3 stack size*/
#define Task_OutPortBStackSize 100 /* Define task 4 stack size*/
#define Task_SendPortBStackSize 100 /* Define task 5 stack size*/
#define Task_ErrSRStackSize 100 /* Define task 3 stack size*/
4. /* Definitions for six task priorities. */
#define Task_CharCheckPriority 5 /* Define task 1 priority */
#define Task_ ReadPortAPriority 6 /* Define task 2 priority */
#define Task_ErrPriority 7 /* Define task 3 priority */
#define Task_OutPortBPriority 8 /* Define task 4 priority */
#define Task_SendPortBPriority 9 /* Define task 5 priority */
#define Task_ErrSRPriority 10 /* Define task 6 priority */
5. /* Prototype definitions for the semaphores */
OS_EVENT *SemFlag1; /* Needed when using the semaphore as a flag for inter-process communication between port status check task, Task_CharCheck and port read task, Task_Read_Port_A Port A read task has to wait for semaphore till port A interrupts. */
OS_EVENT *SemFlag2; /* Needed when using semaphore as a flag for inter-process communication between Task_OutPortB task and sending task Task_SendPortB. */
OS_EVENT *SemCountLimitFlag; /* A flag to define the limits of the semaphore count */
OS_EVENT *SemKey; /* Needed when using the semaphore as a resource key by Task_Read_Port_A and Task_Err */
OS_EVENT *SemVal; /* Needed when using the semaphore for passing the 16-bit message between steps *b* and *c*. */
OS_EVENT *SemCount; /* Needed when using the semaphore for passing the counts as the number of characters read as a 16-bit message between tasks at the steps *b* and *d*. */
6. /* Prototype definitions for the mailboxes */
OS_EVENT *MboxStrMsg; /* Needed when using the mailbox message between steps *b* and *d* and Steps *d* and *e* */
OS_EVENT *MboxErrStrMsg; /* Needed when using the mailbox message between steps *c* and *d* */
7. /* Codes as per codes in *Example 4.5* except the main function codes. These are for reading from Port A and storing a character at a queue or buffer */
.

.
8. /* *Codes as per Example 10.7 Step 3 to 8. However, before the 'OSStart ( );', the semaphore and mailbox must be created and initialised as under.* */
SemFlag1 = OSSemCreate (0) /* Declare initial value of semaphore = 0 for using it as an event flag*/
SemFlag2 = OSSemCreate (0) /* Declare initial value of semaphore = 0 for using it as an event flag*/
Sem*CountLimitFlag* = OSSemCreate (0) /* Declare initial value of semaphore = 0 as an event flag*/
SemKey = OSSemCreate (1) /* Declare initial value of semaphore = 1 for using it as a resource key*/
SemVal= OSSemCreate (0) /* Declare initial value of semaphore character be = '0' */

/ * Declare initial count as 0 as a counter that gives the number of times a task, which sends into a buffer that stores a character stream, ran minus the number of times the task, which used the character from the stream, ran from the buffer */

SemCount = OSSemCreate (0);

/* Create Mailboxes for the tasks. */

MboxStrMsg = OSMboxCreate (NULL); /* Needed when using mailbox message between steps *b* and *d* to pass a string message pointer*/

MboxErrStrMsg = OSMboxCreate (NULL); /* Needed when using mailbox message between steps *b* and *c* to pass a string message pointer*/

9.

.

.

10. /* Create six tasks as per Step 3 defining six task identities, Task_CharCheck, Task_Read_Port_A, Task_Err, Task_OutPortB and Task_SendPortB and the stack sizes, other TCB parameters. */

OSTaskCreate (Task_CharCheck, *void (*) 0, (void *) &* Task_CharCheck*Stack* [Task_CharCheck*StackSize*], Task_CharCheckPriority);

OSTaskCreate (Task_Read_Port_A, *void (*) 0, (void *) &* Task_Read_Port_A*Stack* [Task_Read_Port_A*StackSize*], Task_ReadPortAPriority);

OSTaskCreate (Task_Err, *void (*) 0, (void *) &* Task_ErrStack [Task_Err*StackSize*], Task_ErrPriority);

OSTaskCreate (Task_OutPortB, *void (*) 0, (void *) &* Task_OutPortB*Stack* [Task_OutPortB*StackSize*], Task_OutPortBPriority);

OSTaskCreate (Task_SendPortB, *void (*) 0, (void *) &* Task_SendPortB*Stack* [Task_SendPortB*StackSize*], Task_SendPortBPriority);

OSTaskCreate (Task_ErrSR, *void (*) 0, (void *) &* Task_ErrSR*Stack* [Task_ErrSR*StackSize*], Task_ErrSRPriority);

11. /* Codes same as those in Steps 9 to 17 in Example 10.16 */

.

.

12. /* Refer to Example 4.5. Wait until an interrupt occurs and sets STAF*/

while (STAF != 1) { };

/* Execute interrupt service routine for port A*/

13. if (STAF == 1) {OSSemPost (SemFlag1);} /* Post semaphore to a task waiting for the read at Port A */

enable_PortA_Intr ( ); /* Prepare for another interrupt from port A*/

OSTimeDly (3); /* Let a low priority task execute, as the next character at Port A will take time. */

}; /* End of while loop*/

}/ * **End of the Task_CharCheck function** */

/

**/

14. /* The **codes for the Task_Read_Port_A** redefined to use the key, flag and 16-bit value and mailbox*/

static void Task_Read_Port_A *(void *taskPointer)* {

/* Initial assignments of the variables and pre-infinite loop statements that execute only once*/

unsigned char *portAdata;

static void portA_ISR_Input (*portAdata);

unsigned char [ ] portAinputStr; /* Let port A input string be an array to hold the data from port A*/

unsigned char * *inputMaxSize* = 16; /* Let Maximum size of the string be 16 characters. */

/* Start an infinite while-loop and Wait for SemFlag1 1 by OSSemPost function of character availability check task */
 while (1) {
OSSemPend (SemFlag1, 0, &SemErrPointer);
15. /* Take the key to not let the Task_Err run and block this task. Remember that Task_Err also needs SemKey for running state */
OSSemPend (SemKey, 0, &SemErrPointer);
16. /* Actions on maximum size exceeding the string buffer size */
if (*SemCount > = inputMaxSize*) {
OSSemPost (Sem*CountLimitFlag*); /* Post the Sem*CountLimitFlag* */
OSSemPost (SemKey); / Release the SemKey to let the next cycle of this loop start */*
17. /* To exit the infinite loop of this higher priority assigned task to let the lower priority task error detect run by a call to the OS delay function that forces a wait of 1 ms (one OS timer tick) or until delay resume function executes. This is the method to let the other task of lower priority execute Port A message error-detecting task */
OSTimeDly (1);
OSSemPend (SemKey, 0, &SemErrPointer); / Take the SemKey */*
SemCount =0; /* Reset the Semaphore counter to 0. */
};/* End of Codes for the action on reaching the limit of putting the characters into the buffer */
18. /* Execute Port A interrupt service routine.*/
PortA_ISR_Input (&portAdata); /* Remember as soon as Port A is read, STAF will reset itself to reflect next interrupt status. */
/* If an ASCII code for start of text is found then initialize SemCount = 0. */
If (portAdata = = 0x02) {SemCount = 0;};
19. /* Write the array element that returned as Port A data into the Port A input string */
PortAinputStr [SemCount] = portAdata;
20. /*Let the counting semaphore count increase after one character has been put into the string holding the character stream*/
OSSemPost (SemCount);
21. /*Let the character pass as an IPC to Task_Err, not as a mailbox message but as a semaphore value. Note that we are not using Mailbox as an IPC of a character that has a message of 8 bits only. */
*SemVal = portAdata;
OSSemPost (SemVal);
22. OSTimeDly (1); /* Let a low priority Task_Err run in order to check the error if any in the read byte */
23. /* Prepare for another interrupt from port A*/
enable_PortA_Intr ( );
OSTimeDlyResume (Task_CharCheckPriority);/* Resume the task that was delayed to start this task*/
24. /* When the character is equal to End of Text, ASCII code at Port A (for example, the Enter key pressed) is found or last string character is received, send the message pointer string to the waiting mail box at Task_OutPortB and make initial SemCount = 0 again for next string*/
if (SemVal = = 03 || SemCount = = *inputMaxSize* - 1) {**OSMboxPost (MboxStrMsg, portAinputStr); SemCount = 0};**
25. *OSSemPost (SemKey); /* Release the SemKey to let next cycle of this while loop start */*
}; /* End of while loop*/
26. }/ * End of the Task_ReadPortA function */

/
**/
27. /* **Start of Task_Err codes** */
static void Task_Err (void *taskPointer) { /* Initial assignments of the variables and pre-infinite loop statements that execute once only*/
/* Declare an error string for using when at Step *d* an error invalid-character is found at the mailbox.
unsigned char [ ] *ErrStr1* = "Invalid Character Found";
/* Declare an error string message for task at Step *d* when the limit exceeds. */
unsigned char [ ] *ErrStr2* = "Characters in the message exceeded the Limit declared at the task for read";
boolean invalid = false; /* Declare invalid variable 'false' and will be assigned 'true' when character read is found invalid. * /
28. *while (1)* { /* Start the infinite loop */
29. /* Take the key to not to let port read task read at least one cycle of the while-loop*/
OSSemPend (SemKey, 0, &SemErrPointer);
30. /* Post Limit of Message Exceeded Message to task at step d. */
If (Sem*CountLimitFlag* = = 1) {**OSMboxPost (MboxErrStrMsg, *ErrStr2*);**};
31. /* Codes for Checking any invalid character in SemVal*/
OSSemAccept (SemVal);

.

.

32. /* Post Mailbox message to task at step *d* if an invalid character is detected */
if (Sem*CountLimitFlag* == 0 && invalid = = true) {**OSMboxPost (MboxErrStrMsg, *ErrStr1*);**};
33. /* Decrease SemCountLimitFlag (if >0) by accepting the Sem*CountLimitFlag* semaphore. Task does not suspend even if semaphore not available (not > 0). This task has to run whether countLimit is exceeded or not. The only condition is that it must run if exceeded. Hence the OSSemPend is not used here. */
OSSemAccept (Sem*CountLimitFlag*);
34. /* Release the key to let a task, which was waiting for the port read run*/
OSSemPost (SemKey);
OSTimeDlyResume (Task_ ReadPortAPriority); /* Resume the task Port A Read
}; /* End of while-loop*/
35. } / * **End of the Task_Err function** */
/
**/
36. /* The **codes for the Task_OutPortB** */
static void Task_OutPortB *(void *taskPointer)* {
37. /* Initial assignments of the variables and pre-infinite loop statements that execute once only*/
unsigned char [ ] message; /* Declare error free message string pointer*/
unsigned char [ ] errMessage; /* Declare error message pointer. */
38. *while (1)* { /* Start an infinite while-loop. */
39. /* Wait for Mailbox Message available (not NULL) */
message = **OSMboxPend (MboxStrMsg, 0, & Mboxerr);**
40. /* Check for Mailbox Error Message available (not NULL) */
errMessage = OSMboxAccept (MboxErrStrMsg);
if (errMessage ! = NULL) {**OSMboxPost (MboxErrStrMsg, errMessage);**}
else {
/*Codes for again sending the Port B string of characters to task for transmission; the message is tested to see that it has no invalid character or that it never exceeds the limits of its size.*/

OSMboxPost (MboxStrMsg, message);
/* Release the flag to a task waiting for sending the output at port B */
OSSemPost (*SemFlag2*);
};
41. OSTimeDlyResume (Task_ErrPriority); /* Let delayed higher priority task err resume. */
/* To exit the infinite loop at higher priority assigned task to let the lower priority task run, call
the OS delay function for wait of 1 ms (one OS timer tick). This is the method to let the other task
of lower priority execute Port B sending the characters. */
OSTimeDly (1);
}; / End of while-loop*/*
42. }/ * **End of the Task_OutPortB function** */
/
**/
43. /* **Start of Task_SendPortB codes** */
static void Task_SendPortB (*void *taskPointer*) {
/* Initial assignments of the variables and pre-infinite loop statements that execute only once */
.

.
44. *while (1)* { /* Start the infinite loop */
45. /* Take the event flag SemFlag2 > 0 and decrement it */
OSSemPend (SemFlag2, 0, *&SemErrPointer*);
46. /* Wait for error free message from Port B. If available, retrieve it. */
OSMboxPend (MboxStrMsg, 0, & MboxErr);
47. /* Codes for sending the valid message to a memory buffer where it is saved or to a network
for transmission */
.

.
48. OSTimeDlyResume (Task_OutPortBPriority); /* Resume Delayed Task_OutPortB */
}; /* End of while-loop*/
49. }/ * **End of the Task_SendPortB function** */
/
**/
50. /* **Start of Task_ErrSR codes** */
static void Task_ErrSR (*void *taskPointer*) {
/* Initial assignments of the variables and pre-infinite loop statements that execute only once */
unsigned char [ ] errMessage; /* Declare error message pointer. */
51. *while (1)* { /* Start an infinite while-loop. */
/* Take the event flag SemFlag2 > 0 and decrement it */
OSSemPend (SemFlag2, 0, *&SemErrPointer*);
52. /* Check for Mailbox Error Message available (not NULL) */
errMessage = **OSMboxAccept (MboxErrStrMsg**);
53. /* Codes for the action on error message. */
if (strcmp (errMessage, "Invalid Character Found") == 0) {
/* Codes for the actions that need to be taken when invalid characters are found. For example,
codes for displaying "Invalid Number Dialed. Dial Again" on an LCD. */
.

.
};

```
    if (strcmp (errMessage, " Characters in the message exceeded the limit declared at the task for read.")
    ==0) {
    /* Codes for actions needed on limit exceeded. Codes for displaying on an LCD "Message too long to
    Accept. Dial again". */
    .
    .
    .
    }
    OSTimeDlyResume (Task_ ReadPortAPriority); /* Resume the task Port A Read
54. }; /* End of while-loop*/
55. } / * End of the Task_ErrSR function */
/
    ***********************************************************************************/
```

10.2.7 Queue Related Functions

We have seen in Example 10.18 that the semaphore sends an IPC for 16 bits. We have seen in Example 10.19 that by using a mailbox we can communicate a pointer for a message that is larger. By using a queue, we can form an array of message pointers from the tasks. The message pointers can be posted into a queue by the tasks either at the back as in a queue or at the front as in a stack. A task can thus insert a given message for deleting either in the first in first out (FIFO) mode or in last in first out (LIFO) mode in case of a priority message. [Note: The IPC queue differs from a data structure queue (Section 5.4.2) with respect to the available modes for inserting an element into a queue.]

Refer to Figure 8.5. MUCOS permits a queue of an array of pointers. Let there be a pointer, **Qtop, to a queue of message pointers, and there be two pointers, * QfrontPointer and *QbackPointer, which send and retrieve, respectively, the message pointer for the message. MUCOS queue functions for the tasks' IPCs (inter-process communications) are as per Table 10.7. MUCOS permits up to 256 message pointers into a queue. [That is, the queue size can be 256]. MUCOS does not provide for the stack ECB separately like the one for a semaphore or mailbox or queue. However, there is an alternative provision. A function OSPostFront enables stacking operation on the array of message-pointers [Refer to Section 5.6]. This function enables insertion such that the waiting task does LIFO retrieval of the message pointer, hence of the priority message.

Table 10.7
Queue Functions for the Inter-Task Communications

Prototype of Service and System Clock Function	Parameters, returns and passes	When is this OS called?
OS_Event OSQCreate (void **$QTop$, unsigned byte $qSize$)	$\underline{R}$ and R	OS creates a Queue ECB. This creates and initialises an array of pointers for the queue at QTop. Queue can be of maximum size = $qSize$. QTop should point to top (0th element of an array). ECB points at the QMsgPointer.

(Contd.)

Prototype of Service and System Clock Function	Parameters, returns and passes	When is this OS called?
unsigned byte *OSQFlush void *OSQPend (OS_Event *QMsgPointer, unsigned short timeOut, unsigned byte *Qerr) (OS_EVENT *QMsgPointer)	void and *S* T̲ and *T*	Refer to Text To eliminate all the messages in the queue that have been sent. This function checks if a queue has a message pending at QMsgPointer. [The queue front pointer at the ECB is not pointing to NULL]. Function then returns all the message pointers between the queue front pointer and queue back pointer at the ECB. When it returns, an error code and QMsgPointer will point to NULL.
unsigned byte OSQPost (OS_EVENT*QMsgPointer, void *QMsg)	U̲ and *U*	Sends a pointer of the message QMsg to the QMsgPointer at the queue back. The message is at a queue back pointer in the ECB.
unsigned byte OSQPostFront (OS_EVENT *QMsgPointer, void * QMsg)	V̲ and *V*	Sends QMsg pointer to the QMsgPointer at the queue. It points to the queue front pointer in the ECB where pointer for QMsg now stores pushing other message-pointers back.[#]
unsigned byte OSQQuery (OS_EVENT *QMsg Pointer, OS_Q_DATA *QData)	X̲ and *X*	To get queue message's information and error information.

Column 2 refers to corresponding explanatory paragraph in the text.

[#] Use OSQPostFront or OSQPost functions according to the message priority. If the message has a higher priority, post it at the front; else, post as usual in a queue. We can use post-front and post functions to build a queue in which an array of message pointers are stored and ordered according to their priorities. Queue is then called a prioritised ordered queue. Use of OSQPostFront and then OSQPend enable indirectly a LIFO mode of retrieval of a message-pointer.

1. Creating a Queue for an IPC

OS_Event QMsgPointer = *OSQCreate (void **QTop, unsigned byte qSize)* is used for creating an OS's ECB for the *QTop* and queue is an array of pointers at QMsgPointer. The array size can be declared as maximum 256 (0th to 255th element). [Refer to Section 5.5.1]. Initially, the array of pointers created points to NULL.

Task Parameters Passing *R*: The **QTop* passes as pointer for an array of voids. 'qSize' is the size of this array. [Number of message-pointers that the queue can insert before a read is within 0 and 255].

Returning R: The function OSQCreate ( ) returns a pointer to the ECB allocated to the queue. It is an array of voids initially. Null if none is available.

Example 10.20 explains the use of OSQCreate function.

2. Waiting for an IPC Message at a Queue

Function *void *OSQPend (OS_Event *QMsgPointer, unsigned short timeOut, unsigned byte *Qerr)* checks if a queue has a message pending at ECB QMsgPointer. [QMsgPointer is not pointing to NULL]. The message pointer points to the queue front (head) at the ECB for the queue defined by QMsgPointer. It suspends the task if no message is pending (until either message received or wait period, passed by argument timeOut, finishes after timeout ticks of RTOS timer). The queue front pointer at the ECB will later increment to point to the next message after returning the pointer for the message.

Returning void: The function returns a queue ECB pointer. It also returns the following: (*i*) OS_NO_ERR, when the queue message search succeeds; (*ii*) OS_TIMEOUT, if queue did not get the message during the ticks defined by the timeOut; (*iii*) OS_ERR_PEND_ISR, if this function call was from the ISR; (*iv*) OS_ERR_EVENT_TYPE, when *QMsgPointer is not pointing to the queue message.

Task Parameters Passing *S*: (i) The OS_Event *QbackPointer passes as pointer to the ECB that is associated with the queue (ii) It passes 16-bit unsigned integer argument timeOut. It resumes the task after the delay equals (timeOut-1) count-inputs (ticks) at the RTOS system clock. (iii) It passes *err, a pointer for holding the error code.

Example 10.20 explains the use of OSQPend function.

3. Emptying the Queue and Eliminating All the Message Pointers

Function *unsigned byte *OSQFlush (OS_EVENT *QMsgPointer)* checks if a queue has a message pending at QMsgPointer. [The queue front pointer at the ECB does not point to NULL]. The function has to return all the message pointers between queue front pointer and queue back pointer at the ECB. It returns an error code and QMsgPointer at ECB. These will now point to NULL.

Task Parameters Passing *T*: The OS_Event *QMsgPointer passes as pointer to the ECB that is associated with the queue.

Returning T: The function OSQFlush ( ) returns one of the following error codes. (i) OS_NO_Err, if the message queue flush succeeds, or OS_ERR_EVENT_TYPE, if pointer is not pointing to the message queue.

4. Sending a Message-Pointer to the Queue

The function unsigned byte *OSQPost (OS_EVENT *QMsgPointer, void *QMsg)* sends a pointer of the message *QMsg. The message pointer *QMsg stores a queue back pointer in the ECB for *QMsgPointer*.

Task Parameters Passing *U*: The OS_Event *QMsgPointer passes as pointer to the ECB that is associated with the queue back (tail). The message pointer *QMsg* is passed for the message.

Returning U: The function OSQPost ( ) returns the error code for one of the three codes. (*i*) OS_NO_ERR, if queue signaling succeeded, (*ii*) OS_ERR_EVENT_TYPE, if *QbackPointer is not pointing to the queue and (*iii*) OS_Q_FULL, when queue message cannot post. (QSize cannot exceed, as it is already 255.)

Example 10.20 explains the use of OSQPost function.

5. Sending a Message Pointer and Inserting it at the Queue Front

The function unsigned byte *OSQPostFront (OS_EVENT*QMsgPointer, void *QMsg)* sends QMsg pointer to the QMsgPointer at the queue, but it is at the queue front pointer in the ECB where pointer for QMsg now stores, pushing other message-pointers backwards.

Task Parameters Passing *V*: The OS_Event *QMsgPointer passes as pointer to the ECB that is associated with the queue. The second argument is the message QMsg address that is the queue front address.

Returning <u>V</u>: The function OSQPostFront ( ) returns one of the following error codes; (*i*) OS_NO_ERR, if the message at the queue front is placed successfully; (*ii*) OS_ERR_EVENT_TYPE, if pointer QbackPointer is not pointing to message queue; or (*iii*) OS_Q_FULL, if qSize was declared *n* and queue had *n* messages waiting for the read.

Example 10.20 explains the use of OSQPostFront function.

6. Querying to Find the Message and Error Information for the Queue ECB

The function *unsigned byte OSQQuery (OS_EVENT *QMsgPointer, OS_Q_DATA *QData)* checks for a queue data and places that at QData. It also finds the error information parameters, OS_NO_ERR and OS_ERR_EVENT_TYPE from the ECB at *QMsgPointer*.

Task Parameters Passing *X*: The function OSQQuery passes (*i*) a pointer of the queue at *QMsgPointer ECB and (*ii*) a pointer of the data structure at *QData.

Returning <u>X</u>: The function OSQQuery returns the error code (*i*) OS_NO_ERR, when querying succeeds; or (*ii*) OS_ERR_EVENT_TYPE, if *QMsgPointer is not pointing to queue message.

Example 10.20

Let a task, *Task_ReadPortA*, receive characters put these into a queue, *QMsg*. The task uses OSQPost to send an IPC for another task waiting (blocked) for these characters. An advantage is that the messages can be used as soon as available by another task without the completion of the whole message-string as was the case in Example 10.10. The mailbox permitted only one message or one message-pointer. Queue permits any number of messages until the queue gets full. Full means that the maximum array size defined for the queue is reached. Another advantage is that Port A need not be the one sending characters or bytes, but can be an NIC (Network Interface Card) or any other input device sending a word, frame or a segment of a message that needs to be posted to another waiting task in a sequence. Further, any number of error messages sent by the Task_Err can also be posted into the same queue as priority messages. The codes get simplified. *An important difference with respect to a mailbox or semaphore is that we declare the ECB as well as an array of pointers for the queue, while in mailbox or queue we declare the ECB only.*

To *Task_MessagePortA*, not only *Task_ReadPortA* but also another task, *Task_Err*, can send an error message on detecting an invalid character or if the limit of characters expected in the string is exceeded. The use of OSQPostFront is to send the errors as the priority message. OSQPend is used to wait for an IPC for the message as well as error message string. The steps in the above operation are as follows:

1. Step *a*: Task, *Task_CharCheck* checks the Port A status for availability of a character or message. [For example, the task is activated if, at Port A, a key is pressed for sending the character. If the character is not available then wait. Here, the use of semaphore *SemFlag1* as in Example 10.16 suffices because an IPC will be only for an event-flag.

2. Step *b*: Task, *Task_ReadPortA*, waits for the *SemFlag1* and executes the codes that accumulate the characters into a queue.

3. Step *c*: Task, *Task_Err* checks each *character or message* read at port A and sends a string, *errStr*, into a general message queue when the character or message is not a valid character or message. For example, if the character or message is not a number in the case of a telephone number, which is read by task at Step *b*. Here, the use of semaphore value *SemVal* suffices between Step *b* and Step *c*, because the IPC message is just a character of 8-bits and use of mailbox for the IPC is unnecessary.

4. Step *d*: Task, *Task_MessagePortA* waits for the *characters* or messages as an array of pointers. Queue is used for the IPC between Steps b and d and for the IPC between Steps c and d.

5. *Step e:* Task, Task_ErrLogins also waits for the errors posted as front messages as priority messages in the same queue.

6. *Step f:* Task, Task_ServiceMessage waits for servicing as per the message retrieved at Task_MessagePortA.

This example shows how the steps *a* and *b* synchronise by the IPC *SemFlag1*, how tasks at the steps *b* and *c* synchronise using the IPC *SemVal* and how the steps *b* to *d and steps c, d and e synchronise* using the queue functions of the MUCOS.

1. /* Codes are the same as in Step 1 Example 10.19, except that the statements are shown in bold for the mailbox. The mailbox-related statements are replaced by the queue-related messages. */

.

.

.

2. /* Preprocessor definitions for maximum number of inter-process events to let the MUCOS allocate memory for the Event Control Blocks */
#define OS_MAX_EVENTS 12/* Let maximum IPC events be 12 */
#define OS_SEM_EN 1/* Enable inclusion of semaphore functions in applications using MUCOS */
#define OS_Q_EN 1/* Enable inclusion of queue functions in applications using MUCOS */
/* End of preprocessor commands */
3. /* Prototype definitions for six tasks for steps *a* to *e* above. */
static void Task_CharCheck (void *taskPointer);
static void Task_ReadPortA (void *taskPointer);
static void Task_Err (void *taskPointer);
static void Task_MessagePortA (void *taskPointer);
static void Task_ServiceMessage (void *taskPointer);
static void Task_ErrLogins (*taskPointer);
4. /* Definitions for six task stacks */
static OS_STK Task_CharCheckStack [Task_CharCheckStackSize];
static OS_STK Task_ReadPortAStack [Task_ReadPortAStackSize];
static OS_STK Task_ErrStack [Task_ErrStackSize];
static OS_STK Task_MessagePortAStack [Task_MessagePortAStackSize];
static OS_STK Task_ServiceMessageStack [Task_ServiceMessageStackSize];
static OS_STK Task_ErrLoginsStack [Task_ErrLoginsStackSize];
5. /* Definitions for six task stack size */
#define Task_CharCheckStackSize 100 /* Define task 1 stack size*/
#define Task_ReadPortAStackSize 100 /* Define task 2 stack size*/
#define Task_ErrStackSize 100 /* Define task 3 stack size*/
#define Task_MessagePortAStackSize 100 /* Define task 4 stack size*/
#define Task_ErrLoginsStackSize 100 /* Define task 6 stack size*/
#define Task_ServiceMessageStackSize 100 /* Define task 5 stack size*/

6. /* Definitions for six task priorities. */
#define Task_CharCheckPriority 5 /* Define task 1 priority */
#define Task_ ReadPortAPriority 6 /* Define task 2 priority */
#define Task_ErrPriority 7 /* Define task 3 priority */
#define Task_MessagePortAPriority 8 /* Define task 4 priority */
#define Task_ServiceMessagePriority 9 /* Define task 5 priority */
#define Task_ErrLoginsPriority 10 /* Define task 6 priority */
7. /* Prototype definitions for the semaphores */
OS_EVENT *SemFlag1; /* Needed when using the semaphore as a flag for inter-process communication between port status check task, Task_CharCheck and port read task, Task_ReadPortA Port A read task has to wait for semaphore till port A interrupts.*/
OS_EVENT *SemFlag2; /* Needed when using the semaphore as flag for inter-process communication between Task_MessagePortA task and sending task Task_ServiceMessage. */
OS_EVENT *SemCountLimitFlag; /* Needed when using the semaphore as flag for reaching the limits of semaphore count in the inter-process communication between port read and port read decipher task. Port reading has to wait for port read */
OS_EVENT *SemKey; /* Needed when using the semaphore as resource key by Task_ReadPortA and Task_Err */
OS_EVENT *SemVal; /* Needed when using the semaphore for passing the 16-bit message between steps *b* and *c*. */
OS_EVENT *SemCount; /* Needed when using the semaphore for passing the counts as the number of messages read as a 16-bit message between tasks at the steps *b* and *d*. */
8. /* Prototype for queue ECBs and for message-pointers array at a queue. */
OS_EVENT *QMsgPointer; /* Needed when using mailbox message between steps *b* and *d* and steps *d* and *e* */
void **QMsgPointer [QMessagesSize];** /* Let the maximum number of message-pointers at the queue be QMessagesSize. */
OS_EVENT *QErrMsgPointer; /* Needed when using a mailbox message between steps *c* and *d* */
void **QErrMsgPointer [QErrMessagesSize];** /* Let the maximum number of error message-pointers at the queue be QErrMessagesSize. */
9. /* Define both queues array sizes. */
#define **QMessagesSize** = 64; /* Define size of message-pointer queue when full */
#define QErrMessagesSize = 16; /* Define size of error message-pointer queue when full */
10. /* Codes as per codes in *Example 4.5* except the main function codes. These are for reading from Port A and storing a character or message at a queue or buffer. Alternatively, modify the code for reading from a port at NIC or any other device or peripheral. */
.

.
11. /* Codes as per Example 10.7, steps 3 to 8. However, before the 'OSStart ( );', the semaphore and mailbox must be created and initialised as under: */
SemFlag1 = OSSemCreate (0) /* Declare initial value of semaphore = 0 for using it as an event flag*/
SemFlag2 = OSSemCreate (0) /* Declare initial value of semaphore = 0 for using it as an event flag*/
Sem*CountLimitFlag* = OSSemCreate (0) /* Declare initial value of semaphore = 0 as an event flag*/

SemKey = OSSemCreate (1) /* Declare initial value of semaphore = 1 for using it as a resource key*/
SemVal= OSSemCreate (0) /* Declare initial value of semaphore character or message be = '0' */
12. / * Declare initial count as 0 as a counter that gives the number of times a task, which sends into a buffer that stores a character or message stream, ran minus the number of times the task which used the character or message from the stream ran from the buffer */
SemCount = OSSemCreate (0);
13. /* Create Two queues for the tasks, one general purpose queue and another for error logins only after selecting error messages from the general queue. */
/* Define a top of the message pointer array. QMsgPointer points to top of the Messages to start with. */

QMsgPointer = OSQCreate (&QMsg [0], QMessagesSize);
/* Define a top of the message pointer array. QMsgPointer points to top of the Messages to start with. */

QErrMsgPointer = OSQCreate (&QErrMsg [0], QErrMessagesSize); /* Needed when using mailbox message between steps *b* and c to pass a string message pointer*/
14. /* Create six Tasks as per Step 3 defining by six task identities, Task_CharCheck, Task_ReadPortA, Task_Err, Task_MessagePortA and Task_ServiceMessage and the stack sizes, other TCB parameters. */
OSTaskCreate (Task_CharCheck, *void (*) 0, (void *) &* Task_CharCheck*Stack* [Task_CharCheck*StackSize*], Task_CharCheckPriority)*;*
OSTaskCreate (Task_ReadPortA, *void (*) 0, (void *) &* Task_ReadPortA*Stack* [Task_ReadPortA*StackSize*], Task_ReadPortAPriority)*;*
OSTaskCreate (Task_Err, *void (*) 0, (void *) &* Task_ErrStack [Task_Err*StackSize*], Task_ErrPriority)*;*
OSTaskCreate (Task_MessagePortA, *void (*) 0, (void *) &* Task_MessagePortA*Stack* [Task_ MessagesPortA*StackSize*], Task_MessagePortAPriority)*;*
OSTaskCreate (Task_ServiceMessage, *void (*) 0, (void *) &* Task_ServiceMessage*Stack* [Task_ServiceMessage*StackSize*], Task_ServiceMessagePriority)*;*
OSTaskCreate (Task_ErrLogins, *void (*) 0, (void *) &* Task_ErrLogins*Stack* [Task_ErrLogins*StackSize*], Task_ErrLoginsPriority*);*
15. /* Codes same as at Steps 9 to 17 in Example 10.16 */
.
.
16. /* Refer to *Example 4.5* Wait until an interrupt occurs and sets STAF*/
while (STAF != 1) { };
/* Execute interrupt service routine for port A*/
17. if (STAF == 1) {OSSemPost (SemFlag1);} /* Post semaphore to a task waiting for the a read at Port A */
enable_PortA_Intr ( ); /* Prepare for another interrupt from Port A*/
OSTimeDly (3); /* Let a low priority task execute, as next character or message at the Port A will take time. */
}; /* End of while-loop*/
}/ * **End of the Task_CharCheck function** */
/
***/

18. /* The **codes for the Task_ReadPortA** redefined to use the key, flag and 16-bit value and mailbox*/

static void Task_ReadPortA *(void *taskPointer)* {

/* Initial assignments of the variables and pre-infinite loop statements that execute only once */

unsigned char *portAdata;

static void PortA_ISR_Input (*portAdata);

19. /* Start an infinite while-loop and wait for SemFlag1 1 by OSSemPost function of character or message availability check task */

 while (1) {

OSSemPend (SemFlag1, 0, &SemErrPointer);

/* Take the key to not let the error task detect and block the task that also needs SemKey for running state */

OSSemPend (SemKey, 0, &SemErrPointer);

20. /* Actions on maximum size exceeding the string buffer size */

if (*SemCount* > = **QMessagesSize**) {

OSSemPost (Sem*CountLimitFlag*); /* Post the Sem*CountLimitFlag* */

OSSemPost (SemKey); / Release the SemKey to let the next cycle of this loop start */*

/* To exit the infinite loop of this higher priority assigned task and to let the lower priority task error detect run by a call to the OS delay function that forces a wait of 1 ms (one OS timer tick) or until delay resume function executes. This is the method to let the other task of lower priority execute the Port A message error-detecting task */

OSTimeDly (1);

OSSemPend (SemKey, 0, &SemErrPointer); /* Take the SemKey */

SemCount =0; /* Reset the Semaphore counter to 0. */

};/* End of Codes for the action on reaching the queue array limit of putting the message pointers into the buffer */

/* Execute Port A interrupt service routine. */

PortA_ISR_Input (&portAdata); /* Remember that as soon as Port A is read STAF it will reset itself to reflect next interrupt status. */

21. /* Write the array element that returned as Port A data into the Port A input string */

OSQPost (QMsgPointer, &portAdata);

/*Let counting semaphore value increase after one character or message has been put into the String, holding the character or message stream*/

OSSemPost (SemCount);

22. /*Let character or message pass as an IPC to Task_Err not as a mailbox message but as a semaphore value. Note that we are not using mailbox as an IPC of a character or message of 8 bits only. */

*SemVal = portAdata;

OSSemPost (SemVal);

23. OSTimeDly (1); /* Let a low priority Task_Err run in order to check the error, if any, in the read byte */

24. /* Prepare for another interrupt from port A*/

enable_PortA_Intr ( );

OSTimeDlyResume (Task_CharCheckPriority);/* Resume the task that was delayed to start this task*/

25. *OSSemPost (SemKey); /* Release the SemKey to let the next cycle of this while loop start */*

}; / End of while loop*/

26. }/ * End of the Task_ReadPortA function */

/
***/

27. /* **Start of Task_Err codes** */

static void Task_Err (*void *taskPointer*) { /* Initial assignments of the variables and pre-infinite loop statements that execute only once */

/*Declare an error string for Step *d* error invalid message data found to the mailbox. */

unsigned char [ ] *ErrStr1* = "Invalid Message Data Found";

/* Declare an error string message for task at Step *d* when the limit exceeds. */

unsigned char [ ] *ErrStr2* = "Array Size exceeded the Limit. Queue Full";

boolean invalid = false; /* Declare invalid variable 'false' and will be assigned 'true' when character or message read is found invalid. * /

28. *while (1)* { /* Start the infinite loop */

29. /* Take the key to not let Task_ReadPortA wait at least one cycle of the current while loop*/

OSSemPend (SemKey, 0, &SemErrPointer);

30. /* Post message Limit of Message Exceeded to task at step d. */

If (Sem*CountLimitFlag* = = 1) {**OSQPostFront (QMsgPointer, *ErrStr2*);**};

31. /* Codes for checking any invalid character or message in SemVal*/

OSSemAccept (SemVal);

.

.

32. /* Post as priority message queue message to task at step *d* if an invalid message data is detected */

if (Sem*CountLimitFlag* == 0 && invalid = = true) {**OSQPostFront (QMsgPointer, *ErrStr1*);**};

33. /* Decrement and accept the Sem*CountLimitFlag* if available (>0). There is no requirement of task suspension. Next task must run whether countLimit is exceeded or not. The only condition is that it must run if exceeded. Hence the OSSemPend is not used here. */

OSSemAccept (Sem*CountLimitFlag*);

34. /* Release the key to let a task, which was waiting for the port read, run*/

OSSemPost (SemKey);

OSTimeDlyResume (Task_ ReadPortAPriority); /* Resume the task Port A Read

}; /* End of while-loop*/

35. } / * **End of the Task_Err function** */

/
***/

36. /* **The codes for the Task_MessagePortA** */

static void Task_MessagePortA *(void *taskPointer)* {

37. /* Initial assignments of the variables and pre-infinite loop statements that execute only once */

void * message;

38. *while (1)* { /* Start an infinite while-loop. */

39. /* Wait for Queue Message Pointer available (not NULL) */

&message = **OSQPend (QMsgPointer, 0, &QErrPointer);**

40. /* Find if the message has an invalid character or if the messages are found as Error Messages. Check for Mailbox Error Message available (not NULL) */

if (strcmp ((char *) message, "Invalid Message Data Found ")==0) {**OSQPost (QErrMsgPointer, message);**};

if (strcmp ((char *) message , "Array Size exceeded the Limit. Queue Full") == 0){**OSQPost (QErrMsgPointer, message);**};

/* To exit the infinite loop at the task that has been assigned a higher priority and to let the lower priority task run, call the OS delay function for a wait of 1 ms (one OS timer tick). This is the method to let the other task of a lower priority execute. */

OSTimeDly (1);
/* Release the flag to a task waiting for the servicing the messages received at Port A after filtering the error messages*/
OSQPost (QMsgPointer, message);
OSSemPost *(SemFlag2)*;
};
41. OSTimeDlyResume (Task_ErrPriority); /* Let delayed higher priority task err resume. */
}; /* End of while-loop*/
42. }/ * **End of the Task_MessagePortA function** */
/
***/
43. /* **Start of Task_ServiceMessage codes** */
static void Task_ServiceMessage (*void *taskPointer*) **{**
/* Initial assignments of the variables and pre-infinite loop statements that execute only once */
void *message;
.

.
44. *while (1)* { /* Start the infinite loop */
45. /* Wait for a valid message for queue from Port A. Take the event flag SemFlag2 > 0 and decrement it */
OSSemPend (SemFlag2, 0, *&SemErrPointer*);
46. /* Get the message. */
&message = **OSQPend (QMsgPointer, 0, &QErrPointer)**;
47. /* Codes for servicing as per the valid message to a memory buffer for saving or to a network or to dial and transmit */
.

.
48. OSTimeDlyResume (Task_MessagePortAPriority); /* Resume Delayed Task_OutPortB */
}; /* End of while-loop*/
49. }/ * **End of the Task_ServiceMessage function** */
/
***/
50. /* **Start of Task_ErrLogins codes** */
static void Task_ErrLogins (*void *taskPointer*) **{**
/* Initial assignments of the variables and pre-infinite loop statements that execute once only*/
void *errorLogged; /* Declare error message pointer. */
51. *while (1)* { /* Start an infinite while-loop. */
/* Take the event flag SemFlag2 > 0 and decrement it */
OSSemPend (SemFlag2, 0, *&SemErrPointer*);
52. /* Check for Mailbox Error Message available (not NULL) */
&errorLogged = **OSQPend (QErrMsgPointer, 0, &QErrPointer)**;
53.   /* Codes for the action as per the error logged in */
54.   if (errorLogged = = " Invalid Message Data Found") {
/* Codes for actions needed on invalid character or message found. For example, codes for displaying "Invalid Number Dialed. Dial Again" on an LCD. */
.

.
};

```
if (errorLogged = = " Array Size exceeded the Limit. Queue Full ") {
/* Codes for actions needed on limit exceeded. Codes for displaying on an LCD "Message too long to
Accept. Dial again". */
.
.
.
}
OSTimeDlyResume (Task_ ReadPortAPriority); /* Resume the task Port A Read
54. }; /* End of while loop*/
55. } / * End of the Task_ErrLogins function */
/
*************************************************************************/
```

10.3   USE OF VxWORKS

For sophisticated embedded systems, there is a popular RTOS, VxWorks from WindRiver (http://www.wrs.com/). VxWorks is a high-performance, scaleable RTOS that executes on the target processor. VxWorks RTOS design is hierarchical. [Refer to Section 9.10].

VxWorks has a full range of communication options. [Refer to Section 8.3]. It provides for the following:

  (*i*) Multitasking Environment
  (*ii*) Inter-process communication (IPC)
  (*iii*) Synchronisation using (a) event flag, (b) mutually exclusive access using resource key (mutex) and (c) counting mechanism using three type of semaphores in the tasks and ISRs
  (*iv*) Synchronisation using POSIX (Section 9.8.) standard semaphore and other IPCs
  (*v*) Separate contexts for the tasks and the interrupt servicing functions (ISRs) [Tasks have separate TCBs and ISRs have a common stack]
  (*vi*) Watchdog timers
  (*vii*) Virtual I/O devices using pipes and
  (*viii*) Virtual Memory Management functions.

VxWorks TCB saves the following for each task:

  (i) Control Information for the OS that includes *priority*, *stack size*, *state* and *options*.
  (ii) CPU context of the task that includes PC, SP, CPU Registers and task variables.

Further, VxWorks I/O system also provides most of the interfaces of POSIX 1003.1b standard. VxWorks also provides:

  (a) Pipe Drivers for inter-process communications as an I/O virtual device.
  (b) Network-transparent sockets.
  (c) Network drivers for shared memory and Ethernet.
  (d) RAM "disk" drivers for memory resident files.
  (e) Drivers for SCSI, Keyboard, VGA Display, Disk and Parallel port of a computer system, HDD, diskette, tapes, keyboard and displays.

The VxWorks I/O system also includes the POSIX standard asynchronous I/O and UNIX standard buffered I/O. It also provides for simulator (VxSim) [Section 12.4.1], software logic analyzer (WindView), network facilities between VxWorks and TCP/IP network systems. For many other facilities, we can refer to the VxWorks programmer's guide and the VxWorks Network Programmer's Guide provided with the product.

10.3.1 Basic Features

A summary of the important features of VxWorks essential in a sophisticated embedded system design is given here:

1. VxWorks is a *scalable OS*. [Only the necessary OS functions become part of the applications codes, thus having reduced memory requirements]. The run-time configurable feature gives a higher performance in VxWorks. The functions needed for task servicing, inter-process communication, etc., must be predefined in a configuration file included in the user codes. Preemptive latency minimisation is there since not all the functions are at the scheduler.

2. *RTOS hierarchy* includes timers, *signals*, TCP/IP *Sockets*, queuing functions library, NFS, RPCs, Berkeley Port and Sockets (Section 8.3.6), Pipes (Section 8.3.7), Unix-compatible loader, language interpreter, shell, debugging tools and linking loader for Unix. [These are similar to system tasks. The scheduler runs these, as it runs the ISRs.]

3. For multitasking, like MUCOS, VxWorks employs a preemptive scheduler [Section 9.6.4]. VxWorks is a preemptive priority base scheduler with 256 priority levels within 0 to 255 assigned during task creation. VxWorks is also configurable in time-slice round robin mode [Section 9.6.3]. A task context saves fast when the CPU access changes to a higher priority. Different tasks of the same priority execute in round robin mode. Each task in a set of tasks executing round robin runs for a given number of system clock ticks and after timeout becomes the last in a queue of the set.

4. *We can use either preemptive priority or time-slicing scheduling in VxWorks*. It is unlike POSIX functions in that using VxWorks we can schedule certain tasks in time-slicing mode and the remaining in preemptive mode. [Note: The *preemptive priority* and POSIX FIFO scheduling are identical].

5. Refer to Examples 10.19 and 10.20. The interrupt service routine interrupt flag was checked in Task_CharCheck and interrupt Service routine was called in Task_ReadPortA. *VxWorks RTOS schedules the ISRs separately and has special functions for interrupt handling.* [Refer to Section 10.3.3(iii)].

6. VxWorks has system level functions. These are for RTOS initiation and start, RTC ticks (interrupts) initiation and the ISR functions, ISR connecting to interrupt vector, and masking functions. Recall Section 8.2. For the critical section, VxWorks has the interrupts disabling and enabling functions that execute at entering and at exiting the section, respectively. [Semaphore functions as event flag, and counting should not be used since semaphore-based pending should not be invoked in the ISRs.]

7. If a task is expecting a message from another task, which is being deleted by using the task-delete function, then RTOS inhibits the deletion.

8. VxWorks has task service functions [Table 10.9]. VxWorks *task creation* (initiation) by itself does not make a task in a list of active tasks (Section 5.7.3). Active task means that it is in one of the three states, ready, running, or waiting (blocking or pending). Thus, VxWorks not only has the task creating, running, waiting (blocking or pending until a time out or until resource available), suspending (inhibiting task-execution) and resuming, but also the functions for task spawning (creating followed by activating). VxWorks also includes the task-pending cum

suspending and pending cum suspension with timeout functions. VxWorks also has the tasks, which have a state and an inherited priority. Section 10.3.4.(5) describes these.

9. VxWorks has *task delay functions and task delaying cum suspending function* [Section 10.3.2.4].

10. To improve the performance of RTOS, VxWorks provides a shared address in memory to all the tasks. This helps in fast access through the pointers. A pipe need not be allocated a separate memory space. Of course, there is an attendant risk due to a possible illegal access.

11. VxWorks has the shared memory allocation functions and bounded ring buffer allocation for sharing the memory and buffers between the tasks and ISRs.

12. VxWorks has inter-process communication (IPC) functions that are more sophisticated than MUCOS functions. Recall that MUCOS has identical semaphore functions for *event flags, resource acquiring keys and counting semaphores*. Recall the use of the semaphore *SemKey* with OSSemPend and OSSemPost functions on SemKey. SemKey was used as a resource-acquiring key by the various tasks in Examples 10.17 to 10.20. *VxWorks provides for three types of semaphores separately*. [POSIX IPCs are additional].

13. VxWorks has special features for mutual exclusiveness in a critical region. We use a *mutex semaphore* for the resource key when using VxWorks. One type of semaphore used as mutex has the following special features: (*a*) One task can be protected from being deleted by any other task. Thus, no unprotected deletion can occur when using a mutex semaphore function. Mutex creation can be done in such a way that it includes deletion protection. (*b*) Only the task that is taking the resource key through a *mutex* semaphore can release (give or post) the key. No other task can release it. This provides mutual exclusiveness. (*c*) Priority inversion can be prevented in case a task acquires the key using the mutex. The priority assignment of a high-priority task can now inherit so that in case of preemption by an intermediate priority task, the high-priority task does not get blocked. [Section 8.2.3]. *This prevents priority inversion situations*.

14. Mutex locks are available for the tasks and interrupts, for disabling interrupts or disabling preemption (task switching) and for a resource locking by mutex semaphore.

15. Let us recall Figure 8.4(a) in section 8.2.2.(iv) to understand P and V mutex semaphores used for locking the resources. *VxWorks provides for P and V semaphore functions also*.

16. The mutex lock and unlock functions in VxWorks do not cause the priority inversion problem. [Section 8.1.2(v).] The priority first inherits and then returns to the original ones.

17. Unlike MUCOS, VxWorks has no separate mailbox functions that distinguish these from a message queue. VxWorks messages can be queued. It provides for sending messages of variable length into the queues. MUCOS queue functions permit a pointer only for a message and a queue is an array of message pointers. MUCOS messages can insert such that these retrieve in FIFO (first-in first-out) method or in LIFO (if message pointer is posted in the front instead of at the back) when a message is of priority. There are other special features with the message queues in VxWorks.

18. In addition to queues, *VxWorks provides the IPC through the pipe* (Section 8.3.5) into which an ISR or task can write by invoking function *write* ( ). The task can read from a pipe by using *open* ( ) and *read* ( ) functions. *A VxWorks pipe is a queue with virtual I/O device functions.* [Refer to Section 10.3.4.(12)].

19. The scheduler design features of VxWorks are compatible with POSIX 1003.1b. The POSIX library can be included in VxWorks.
20. The timings taken by the various RTOS functions are similar to the ones given in Table 7.6.

VxWorks and kernel library functions are in the header files, "vxWorks.h" and "kernelLib.h". System and task library functions are in "taskLib.h" and "sysLib.h". For logging, the library function is "logLib.h".

The following section describes the important VxWorks functions.

10.3.2 Task Management Library at the System Library Header File

Each task divides into eight states (places). Four of these are also available in MUCOS tasks.

(a) Suspended (Idle state just after creation or state where execution is inhibited). [Refer to the use of OSTaskSuspend function for FirstTask in the Example 10.8, Step 12 and refer to the use of FirstTask codes in the Examples 10.16 to 10.20].

(b) Ready (waiting for running and CPU access in case scheduled by the scheduler but not waiting for a message through IPC).

(c) Pending (the task is blocked as it waits for a message from the IPC or from a resource; only then will the CPU be able to process further) [This is called a blocked place. Refer to OSSemPend, OSMboxPend and OSQPend functions in Examples 10.16 to 10.20].

(d) Delayed (Sent to sleep for a certain time-interval) [Refer to the use of OSTimeDly in Examples 10.16 to 10.20].

(e) Delayed + Suspended [Delayed and then suspended if it is not preempted during the delay period].

(f) Pended + Suspended [Pended and then suspended if the blocked state does not change].

(g) Pended + Delayed [Pended and then preempted after the delayed time-interval].

(h) Pended + Suspended + Delayed [Pended and then suspended after the delayed time-interval].

When we include sysLib.h, the task library gets included. Library functions are given in Table 10.8 and 10.10. Table 10.8 lists the function for task-state a transitions. Table 10.9 gives the task creation, naming, and control service functions.

1. Creating and Activating a Task by TaskSpawn Function

The function for task creating and activating is taskSpawn ( ). Prototype use is unsigned int *taskId* = taskSpawn (*name, priority, options, stacksize, main*, arg0, arg1, ..., arg8, arg9). A memory is allocated to the stack as well as to the TCB. The task identified by taskID will be assigned a stack of size *stacksize* with arg0 to arg9 passed to the stack. It is also assigned a TCB pointer, which points to the entry point of the function *main* that this task executes.

Recall that in MUCOS we were accessing a task by its priority argument, for example, OSTimeDlyResume (*Task_CharCheckPriority*). We access a task in VxWorks by its ID, *taskID*. [Refer to *taskID* argument at the functions in the following subsections]. *taskID* is a 32-bit positive number. Further, when a task of *taskID* = 0 is referred, VxWorks assumes that the calling task is being referred.

New task *taskID* gets a *name, priority*, options and stack size on spawning. If the NULL pointer is used as the argument for *name*, then conventionally the *name* is of form tN (character t followed by number N).

- Function 'unsigned int taskIdSelf ( )' returns the identity of the calling task.
- Using a function as 'unsigned int [ ] *listTasks* = taskIdListGet ( )' will return the tasked list of all existing tasks needed in array, *listTasks*.
- Function taskIDVerify (*taskId*) verifies whether task, taskId exists.

Each task has the highest priority of 100 by default. User task priorities are between 101 and 255. Lowest priority means that priority = 255. [Priority numbers below 100 are used for the system-level and scheduler-level tasks]. A task may have other calling functions. For example, there are three functions in taskLib. A calling function can find the *priority* by function 'taskPriorityGet (*taskId*, &*priority*).' We can change the task priority dynamically. Function 'taskPriorityPut (taskId, *newPriority*)' will reassign the *priority* to *newPriority* for the task that *taskId* identifies. [Note: In POSIX, the priority numbering scheme is reverse of VxWorks: (the lower the number, the lower the POSIX priority)].

The options definable on spawning are the following:

- An *option* is VX_PRIVATE_ENV. It means that the task must be executed in the private environment. The task is not public.
- An option is VX_NO_STACK_FILL. It means no stack fills with the hexadecimal bytes 0xEE.
- An option is VX_FP_TASK. It means that the task must be executed with the floating-point processor.
- An option is VX_UNBREAKABLE. It means disable the breakpoint at the task during execution. [Breakpoints are provided for help during debugging]. In place of option names, the hex values, 0x80, 0x100, 0x8 and 0x2, respectively, can be used.

Multiple options are given as an argument. For example, VX_PRIVATE_ENV | VX_NO_STACK_FILL selects both options in a task when spawning. Like the priority task, options can also be reassigned using taskOptionsSet ( ) and taskOptionsGet ( ) functions. [The | sign is used between multiple options because & sign in C refers to the address of the succeeding variable].

A sufficient stack size should be declared by *stacksize*. VxWorks reserves a part of the stack as a buffer to protect the stack from overflow, which may lead to unpredictable behavior of the system. To start with, task stack fills the bytes 0xEE. [When using simulator VxSim for the VxWorks application, VxSim adds an additional 8000 bytes to the stacksize. Thus, the stacks for interrupts that it simulates also become available]. A library function, ' unsigned int checkStack (taskId)' returns the stack usage. It first finds the unused stack area by counting the number of bytes from the end with 0xEE and then subtracts counts from the stacksize.

The argument *main* is the main routine address. MUCOS and for many RTOSs, *main* ( ) is the function, which is called by the RTOS first. Refer to Example 10.7. The *main* function is used to create a task that executes first, *FirstTask*. FirstTask, when it executes later, initiates the system timer. It then creates (activates as well) the application tasks and suspends itself to let the OS schedule and run the application tasks. In VxWorks, the main ( ) function analogue may be used. It is schedule ( ).

When using VxWorks, unlike Unix operating system or MUCOS ( ), all tasks are spawned as peers, [that is, every task is independent and no task calls or spawns another task]. This means that it is similar to FirstTask in Example 10.7; a starting task need not be spawned first. The starting task spawns daughter tasks and then suspends itself to prevent scheduling. The daughters in a parent task are spawned only when the parent is a server task that concurrently processes the daughter task.

Ten arguments from arg0 to arg9 can be passed into the main routine. These arguments give the startup parameters.

An important point to note is that MUCOS OSTaskCreate creates the task as well as activates it (puts it in the list of tasks to be scheduled). These functions are separate in VxWorks. The taskInit ( ) and taskActivate ( ) can also be used separately in place of taskSpawn function. However, unless a greater control is needed first by creation and then later by activation in an application, we should prefer to use taskSpawn function.

Example 10.23 step 4 will explain the use of taskSpawn function.

2. Task Suspending and Resuming Functions

Function taskSuspend (*taskId*) inhibits the execution of task identified by *taskId*. Function taskResume (*taskId*) resumes the execution of task identified by *taskId*. Function taskRestart (*taskId*) first terminates a task and then spawns again with its original assigned arguments. This function is used in certain situations. The priority might have been reassigned in between, and now the original priority is to be restored. Similarly, the start up parameters might have to be restored.

3. Deleting and Protecting from Deletion

Function taskDelete (*taskId*) not only inhibits permanently the execution but also cancels the allocation of the memory block for the task stack and TCB. Deletion thus frees the memory. The task deleted is one identified by argument *taskId*. Function 'exit (*code*)' deletes the task itself but stores the code at a TCB field *exitcode*. The debugger can examine the TCB using the code.

A task should not be deleted when it has a resource key because the key can then never be released. Protection is available to the calling task by using a function taskSafe ( ) before entering its critical region and using function taskUnsafe ( ) function at the end of the region. When using the mutex semaphore, we can alternatively select an option, SEM_DELETE_SAFE when creating it. [Refer to Section 10.3.4].

Why do we use the task delete or exit function? It is because often times system resources have to be reclaimed for reusing; memory may be a scarce resource for the given application. TCB and stack are the only resources that are automatically reclaimed. There is no saving of tasks spawned by other tasks by the kernel. Each task should itself execute the codes for the following.
- Memory de-allocation.
- Ensure that the waiting task gets the desired IPC.
- Close a file opened before.
- Delete daughter tasks when the parent task executes, exit ( ) function.

4. Delaying a Task to Let a Lower Priority Task Get Access

The function 'int sysClkRateGet ( )' returns the frequency (system ticks and thus RTC interrupts per second). Therefore to delay 0.25 s, the function taskDelay (sysClkRateGet ( )/4) is used. Recall the use of OSTimeDly and later OSTimeDly to resume in MUCOS [See Examples 10.16 to 10.20]. This lets a task of lower priority to run. VxWorks taskDelay (NO_Wait) will allow other tasks of the same priority or lower to run [because timeout time is zero]. No delayed task resumption is needed in the other priority task that runs after this function. Function nanosleep (1,000,000) will delay the task by 1000 μs. It is a POSIX function. Integer arguments define the number of nanoseconds for delay (sleeping).

Table 10.8
Functions for the Task State Transitions

Function	Present State	Next State	Previous function call before the call or previous states
taskResume ( )	Suspended	Delayed or Pended	taskInit ( ) [task must have been initiated for the idle state].
taskResume ( ) or taskActivate ( )	Suspended	Ready	taskInit ( )
taskSuspend ( )	Delayed	Suspended	taskSpawn ( ) or askActivate ( )
taskSuspend ( ) *After a time out*	Delayed	Ready	Suspended
taskSuspend ( ) *After a wait for a resource*	Pended	Suspended	Ready
semGive ( ) or msgQSend ( )	Pended	Ready	Suspended
semTake ( ) or msgQReceive ( )	Ready	Pended	Delayed or Suspended
taskDelay ( )	Ready	Delayed	Pended or suspended
taskSuspend ( )	Ready	Suspended	Delayed or Pended
taskInit ( )	Unknown	Suspended	
exit ( )[&]	Suspended	Terminated	
taskDelete ( )[#]	Suspended	Eliminated	

[&] Terminate the task. [#]Terminate the task and free the memory.

Table 10.9
Task Creation, Naming and Control Functions

Function	Description
taskSafe ( )	Protects the calling task from deletion.
taskUnsafe ( )	Permits deletion of the task protected earlier.
taskDelete ( )	Deletes a task.
taskRestart ( )	Restarts (create again)[#] the running task as the earlier run returned error.
taskActivate ( )	Task activates if initialised earlier.
taskSpawn ( )	Creates as well as activates.
taskName ( )	Returns the task name that associates with the taskID passed as argument.
taskNameTold ( )	Returns the taskId that associates with the taskID passed as argument.
taskIDVerify ( )	Verifies if a task of taskID in the argument is available.
taskIDSelf ( )	Returns the task ID of the task.
taskIDListGet ( )	Returns an array of all ready tasks IDs.
taskInfoGet ( )	Returns information (parameters of the task).
taskRegsGet ( )	Returns the registers of the task.
taskRegsSet ( )	Sets the registers of the task.
TaskOptionsSet ( )	Sets the task options.
taskOptionsGet ( )	Returns task options defined earlier.
taskIsSuspended ( )	Checks if the task is in suspended state.
taskIsReady ( )	Checks if the task is in ready state.
taskTcb ( )	Returns the pointer to the task control block.

(Contd.)

Function	Description
taskPriorityGet ( )	Returns the task priority.
taskLock ( )	Disables other tasks and thus rescheduling.~ No priority preemption when the task is running.
taskUnlock ( )	Enables other tasks and thus rescheduling. ^ No priority preemption when the task is running.
taskPrioritySet ( )	Sets the priority within 0 and 255.
kernelTimeSlice (int numTicks)	Defines time slice per task after enabling round robin running of the tasks. When numTicks = 50, after 50 RTC interrupts define the time slicing period.

#means allocate the memory with the stack and control block allocated at the beginning. ~means scheduler can not block when the task is running, even though a higher priority task needs to be scheduled. ^means scheduler can block when the task is running and when a higher priority task needs to be scheduled.

10.3.3   VxWorks System Functions and System Tasks

The first task that a scheduler executes is UsrRoot from the entry point of usrRoot ( ) in file install/Dir/target/config/all/usr/Config.C. It spawns the VxWorks tools and the following tasks. The root terminates after all the initialisations. Any root task can be initialised or terminated. The set of functions, tLogTask, logs the system messages without current task context I/O. The daemon (a set of large number of functions) supports the task-level network functions. The exceptions-handling functions are at tExcTask. It has the highest priority. It should not be suspended, deleted or assigned lesser priority. By using it, the system reports exceptional conditions that arise during running the scheduler and tasks.

An important set of functions that are also target specific is tWdbTask. The user creates it to service the requests from Tornado target server. It is a target agent task.

1. System Clock (RTC) and Watchdog Timer related Functions

VxWorks sysLib is the system library and is in a header file kernelLib.h. The following important functions are:

1. Function sysClkDisable ( ) disables the system clock interrupts and sysClkEnable ( ) enables the system clock interrupts.
2. Function sysClkRateSet (TICK numTicks) sets the number of ticks per second. It thus defines the number of RTC system interrupts per second. Function sysClkRateGet ( ) returns the system ticks (System RTC interrupts) per second. sysClkRateSet (1000) will set the RTC tick after every 1/1000 second (1 ms). This function should be called in the main ( ) or start up FirstTask or as a starting function. There is a 32-bit global variable, vxAbsTicks. Variable *lower* that increases after each tick and vxAbsTicks. Variable *upper* that increments after each 2^{32} ticks. 'TICK' is defined by type def as following.
   typedef struct (
   unsigned long *lower*;
   unsigned long *upper*;
) TICK
   TICK vxAbsTicks;
   /* Function' unsigned long tickGet ( )' returns vxAbsTicks.lower. */

3. Function sysClkConnect ( ) connects a 'C' function to the system clock interrupts.

4. Function sysAuxClkDisable ( ) disables the system RTC interrupts and sysAuxClkEnable ( ) enables the system auxiliary clock interrupts.

5. Function sysAuxClkRateSet (numTicks) sets the number of ticks per second for an auxiliary clock. It thus defines the number of system RTC interrupts per second. Function sysAuxClkRateGet ( ) returns the system auxiliary clock ticks (System RTC interrupts) per second.

6. Function sysAuxClkConnect ( ) connects a 'C' function to the system auxiliary RTC interrupts.

7. Function 'WDOG_ID wdCreate ( )' creates a watchdog timer. Statement 'wdtID = wdCreate ( );' creates a watchdog timer, wdtID. There is a function STATUS wdStart (*wdtID*, *delayNumTicks*, *wdtRoutine*, *wdtParameter*). The timer created starts on calling this function. The parameters that should pass as the arguments of this function are the followings: (i) *wdtID* to define the identity of the watchdog timer, (ii) *delayNumTicks* to let the timer interrupts after the number of RTC interrupts of the system equal to *delayNumTicks*, (iii) *wdtRoutine*, a function called (not task or ISR) on each interrupt and (iv) *wdtParameter*, an argument which passes to wdtRoutine. A started watchdog timer, wdtID cancels on calling STATUS wdCancel (*wdtID*). A watchdog timer, wdtID de-allocates the memory on calling STATUS wdtDelete (*wdtID*).

2. Defining the Time Slice Period for Round Robin Time Slice Scheduling

Function kernelTimeSlice (int *numTicks*) controls the round robin scheduling, and time slicing turns on and preemptive priority scheduling turns off. Suppose there is a system RTC tick every millisecond. kernelTimeSlice (50) will set the time slice period as 50 ms. This is the time each task is allowed to run before the CPU relinquishes control for another equal priority task.

Let us use the following code. #define TIMESLICE sysClkRateGet ( ) and sysClkRateSet (1000). When we use ' sysClkRateSet (1000); if kernelTimeSlice (TIMESLICE) TIMESLICE = TIMESLICE/ 60;' and time slice period is 1000/60 ms. If we do not set the clock rate, then numTicks set to 60 per s and the time slice period will be 1 s for each task.

3. Interrupt Handling Functions

Table 10.10 gives the Interrupt Service–related functions. Refer to Section 4.1.2. An internal hardware device (interrupt source or interrupt source group) autogenerates an interrupt vector address, ISR_VECTADDR, as per the device. Exceptions are defined in the user software. For exceptions, ISR_VECTADDR does not autogenerate. That has to be defined by the intVectSet ( ) function. Function intConnect ( ) connects the ISR_VECTADDR to a 'C' function for that ISR. Device driver uses this function as follows: A lock function used as 'int *lock* = intLock ( );' disables the interrupts. It returns an integer *lock*. Using the same integer as argument in the unlock function, we enable the interrupts. An unlock function used as 'intUnlock (*lock*);' enables the interrupts.

Table 10.10
The Interrupt Service Functions

Function	Description	Function	Description
intLock ( )	Disables Interrupts*	intUnlock ( )	Enables Interrupts [#]
intVectSet ( )	^Set the interrupt vector	intCount ( )	Counts number of interrupts nested together

(Contd.)

Function	Description	Function	Description
intVecGet ( )	Get interrupt vector	intVecBaseSet ( )	Sets base address of interrupt vector
intVecBaseGet ( )	Get interrupt vector base address	intLevelSet ( )	Sets the interrupt mask level of the processor
intContext ( )	Returns true when calling function is an ISR	intConnect ( )	Connects a 'C' function to the interrupt vector

The meaning of * and # has been explained in the text. Also refer to Section 8.2. It can be used in a task critical region as a last option, because it increases the interrupt latency periods of all sources. ^ For 'Exception' only. For hardware internal device interrupts, the interrupt vectors are fixed and cannot be set.

VxWorks provides for an ISR design that is different from a task design.

1. ***ISRs have the highest priorities and can preempt any running task***. A preemption arises because an ISR is needed because of internal device events (for example from on-chip timers) and because of exceptions (user-defined software interrupts on certain error conditions).
2. An ISR inhibits the execution of the tasks until return.
3. An ISR does not execute like a task and does not have regular task context. It has a special 'interrupt context.'
4. While each task has its own TCB (Task Control Block) that includes its own stack—and unless otherwise not permitted by a special architecture of a system or processor—*all ISRs use the same interrupt stack*. In case of such a special architecture, in place of interrupt servicing support functions, the VxWorks tasks can be used similar to ones used in Example 10.19 and 10.20 for the MUCOS tasks. CPUs 80x86 and R6000 are examples of permitting use of the special architectures and the use of task stacks for processing interrupts.
5. *An ISR should not wait for taking the semaphore* or other IPC. [An ISR cannot use semTake function]. An ISR should not call 'melloc ( )' for memory allocation as that function uses semaphores. ISR *should not use a mutex semaphore*. ISR can use a counting semaphore.
6. *ISR should just write the required data at the memory* (for example STAF in Example 10.19) or post (send or give) an IPC or make a non-blocking write to a message queue (Section 8.3)—so that it has short codes and most of its functions, non-critical and long time taking functions, execute at tasks.
7. ISR should not use floating-point functions as these take longer times to execute. Let these functions be passed on to the task that runs the functions for the interrupt later.

4. Signals and Interrupt Handling Functions

Function 'void sigHandler (int *sigNum*);' declares a signal servicing routine for a signal identified by *sigNum* and a signal servicing routine registers a signal as follows: signal (*sigNum*, *sigISR*). The parameters that pass are the sigNum (for identifying the signal) and signal servicing routine name, *sigISR*. Function sigHandler passes *sigNum* as well as an additional code. The *'sigHandler codes'* associates with sigHandler. A pointer *pSigCtx associates with the signal context. The signal context saves PC, SP, registers, etc., like an ISR context. The return from sigHandler restores the saved context.

Let *sigISR* be a 'C' function that services the signal interrupt. Let its address be ISR_ADDR. Let the signal be identified by *sigNum*. The function 'intConnect (I_NUM_TO_IVEC (sigNum), sigISR, sigArg)' will connect the signal interrupt service routine, sigISR, for the signal identified by sigNum to the ISR_ADDR I_NUM_TO_IVEC (sigNum), a function that uses the argument *sigNum* to find the program counter PC from the interrupt vector and uses it for ISR_ADDR. The argument *sigArg* passes for use by the 'C' function.

The sigISR may call the following functions:

1. Call 'taskRestart ( )' to restart the task, which generated the sigNum. Restarting assigns the original context on creation. Original PC, SP, arguments and options to a task restore now.
2. Call 'exit ( )' to terminate the task, which generated the sigNum.
3. Call longjump ( ). This results in starting the execution from a memory location. The location is the one that was saved when function setjump ( ) was called.

10.3.4   Inter-Process (Task) Communication Functions

Table 10.11 gives a list and description of the inter-process functions. Recall Sections 8.3.1, 8.3.5 and 8.3.6. The table does not give the functions for the signals, sockets and RPCs. Signals use the ISR for servicing and these are given in Section 10.3.3.(4) separately.

Table 10.11
The inter-process communication functions

Function	Description
semBCreate ( )	Creates a binary semaphore. #
semMcreate ( )	Creates a mutex semaphore. #
semCCreate	Creates a counting semaphore. #
semDelete ( )	Deletes a semaphore.
semTake ( )	Takes a semaphore.
semGive ( )	Releases a semaphore.
semFlush ( )	Resumes all waiting blocked tasks.
msgQCreate ( )	Allocates and initialises a queue for the messages.
msqQDelete ( )	Eliminates the message queue by freeing the memory.
msgQSend ( )	Sends into a queue.
msgQReceive ( )	Receives a message into the queue.!
pipeDevCreate ( )	Creates a pipe device. @
select ( )	A task waits for several kinds of messages, from pipes, for sockets and serial I/Os.

Refer to the text for meaning of symbols, !, # and @.

1. # means that we specify an option SEM_Q_PRIORITY for the order in which the semaphore should be taken if there are a number of waiting tasks. Specify SEM_Q_FIFO for defining taking the semaphore in FIFO mode.
2. ! means that the calling task blocks if no message is available, else the message is read by the task. According to the option parameter, insertions into a queue can be an ordered one, with priority as ordering parameter, or they can be for a FIFO read.

3. @ means that the 'statement status = pipeDevCreate ("/pipe/name", max_msgs, max_length)' will create a named pipe with maximum number of max_msgs messages in maximum pipe length max_length bytes. A task blocks if a message is not available and the pipe is empty, when a task attempts to read a pipe.

VxWorks provides three kinds of semaphores, binary (flag), mutex and counting. A mutex semaphore also takes care of the priority inversion problem. A binary semaphore does not disable interrupt. It limits the use only to associated resources. A task must take a semaphore to run. Often, the number of messages must be put in a queue for a waiting task before it can run. The VxWorks queue functions are in a library, msgQLib, which the user includes before using these. For full duplex communication between two tasks, we should create two queues, one for each task. The mqPxLib functions are compatible with POSIX 1003.1b. The symbols in the table have the following meaning. A detailed description of the three types of VxWorks semaphores and of message queues is given below.

1. Creating a Binary Semaphore for the IPCs

The function 'SEM_ID semBCreate (*options, initialState*)' creates an ECB pointed by the SEM_ID. One of the two options mentioned below must pass on calling the function.

Passing Parameters Options: (1) One option that can be selected is (i) SEM_Q_PRIORITY. (ii) The other is SEM_Q_FIFO. Let us assume that at an instant, several tasks are in the blocked state and are waiting (pending) for a binary semaphore for its posting. A waiting task can take the semaphore in one of the two ways: (a) A task higher in priority than the other waiting ones takes the semaphore first. This becomes possible by SEM_Q_PRIORITY option. (b) A task that first blocked and reached the waiting state takes the semaphore first among the waiting ones. This is possible by SEM_Q_FIFO option. The initial state of the binary semaphore passes by argument, initialState. It is zero when using the binary semaphore as an event flag. For the initial state, two options can be chosen; SEM_FULL in which the created semaphore's initial state is initialised as *taken;* and SEM_EMPTY, in which the created semaphore's initial state is initialised as *not taken.* [Recall the use of semaphores SemKey and SemFlag in MUCOS. Mostly SEM_EMPTY is used in VxWorks, while creating a binary semaphore, as there is mutex semaphore provision as a better option than the SemKey of MUCOS.] (2) Another option: *initialState to define the task initial state when it created.*

Returning Parameter: The function semBCreate ( ) returns a pointer, *SEM_ID. It returns NULL in case of an error for the ECB allocated to the *binary semaphore*. Null if none available.

Example 10.21 explains the use of semBCreate. Let us assume that Task_CharCheck checks byte availability at a port A and another task reads that byte after waiting for the semaphore availability from Task_CharCheck.

Example 10.21
1. /* Include the VxWorks header file as well as semaphore functions from a library. */
2. # include "vxWorks.h"
3. # include "semLib.h"
4. # include "taskLib.h"
5. 2. /* Task parameters declarations */
6. .
7. .

8. 3. /* Declare a binary semaphore to be used as flag. */
    SEM_ID semBCharCheckFlagID;
9. /* Create the binary semaphore and pass the options chosen selected to it. */
    semBCharCheckFlagID = semBCreate (SEM_Q_PRIORITY, SEM_EMPTY);
    .
    .
    .
10. /* Task creation codes */
11. /* Codes for Task_CharCheck*/
    .
    .
    .
12. /* At the end, make the binary semaphore full (taken) and available like SemKey set as 1 in
    MUCOS. Refer to Section 10.3.4 for an understanding of the semGive function. */
13. semGive (semBCharCheckFlagID);
14. /* Other remaining codes for the task. */
15. .
16. .
17. /* End of Task_CharCheck Codes */

2. Waiting for an IPC for Binary or Other Type of semaphore Release or Check for Availability of an IPC after Release [C]

The function 'STATUS semTake (*semId*, *timeOut*)' is for letting a task wait until release event of posting the binary or other type of semaphore. Wait until either *SemId* posts (given) by a task or until timeout occurs, whichever happens first. An exemplary use is semTake (*semBCharCheckFlagID*, WAIT_FOREVER). WAIT_FOREVER means timeout = -1 and the period is thus infinity. semTake is like OSSemPend function of MUCOS.

Passing Parameters: (1) *semID* is the semaphore for which a suspended task waits. (2) *timeOut* is the timeout period. One option that can be selected is WAIT_FOREVER if wait must be done for the posting of *semID*. The other option is NO_WAIT. Recall OSSemAccept function in MUCOS. Whenever this task is scheduled by the scheduler and this function is called, take the semaphore identified by *SemID*. Third option, NO_Wait is simply used for checking the availability of a semaphore. If available the semaphore is taken, or else the task runs other succeeding codes. [This is an equivalent of OSSemAccept ( ) of MUCOS]. For example, when an error-message occurs, there is no need to wait especially for the semaphore, only a *check* is needed.

Returning Parameter: The function semTake ( ) returns STATUS. It returns STATUS = OK in case of success in taking of the semID, else returns 'ERROR' in case of an error. Semaphore Id is invalid on timeout also. After the semTake ( ) function unblocks a task, it again becomes available (empty or not taken).

Refer to Examples 8.22 for understanding the use of semTake.

3. Sending an IPC after a binary or mutex or counting Semaphore Release (Posting)

The function 'STATUS semGive (*semId*)' is for letting a task post and release the binary or other type of semaphore. After this, a task can unblock. Which task unblocks depends on the option defined while creating the semaphore posted. Unblock can be as per SEM_Q_FIFO or as per SEM_Q_PRIORITY option. An exemplary use is semGive (*semBCharCheckFlagID*) at Step 13 in Example 10.21.

Passing Parameter: *semID, f*or which there is a wait by this or other task.

Returning Parameter: The function semGive ( ) returns STATUS. It returns STATUS = OK in case of success in taking of the semID, else returns 'ERROR' in case of an error that *semID* is invalid.

Refer to Examples 10.22 for understanding the use of semGive ( ).

4. Taking the Semaphore Multiple Times Until Unavailable Before Next Posting

Function STATUS semFlush (*semFlagID*) is for flushing. It will take the semaphore multiple times until unavailable before next posting. It lets any waiting task not wait any further. It unblocks not only the calling but also all the other tasks waiting for taking the semaphore, *semFlagID*.

Parameter Passing: The *semFlagID* passes as SEM_ID pointer at ECB that associates with the semaphore.

Returning Parameter: The function semFlush ( ) returns the STATUS and makes the semaphore state as SEM_EMPTY. It returns STATUS = OK in case of the success in flushing of the semaphore and making semFlagID state SEM_EMPTY. Else, it returns 'ERROR' in case of an error that on timeout the semaphore is unavailable or the semaphore identity is invalid. After execution of the semFlush ( ) function, all the waiting tasks waiting for this *semFlagID* unblock. [Earlier semFlagID state was SEM_FULL.]

5. Creating a Mutex Semaphore for the IPCs

A mutex semaphore is needed when there is a critical region that shares a data structure with the other tasks. [For example, the bytes in a buffer between a sending task and a receiving task.] There may also be sharing of the hardware devices or files between two tasks.

(a) An exemplary use in which we are using binary semaphore for mutual exclusion is as follows: Here, the use of semMKeyId is not as a special semaphore for resource key but as a semaphore key SemKey, as in Examples 10.17 to 10.20 for MUCOS.
   1. SEM_ID semMKeyID;
   2. semMKeyID = semBCreate (SEM_Q_PRIORITY, SEM_FULL);

(b) The function 'SEM_ID semMCreate (*options*) is for creating an ECB pointed by the SEM_ID. Section 10.3.4.(2) explained the semTake function and Section 10.3.4.(3) explained the use of semGive function. Let us assume that when entering a critical region in a task, semTake (semMReadPortAKey) executes and on leaving the critical region, semGive (semMReadPortAKey) for using the mutex semaphore, semMReadPortAKey. An exemplary use is given in Example 10.22 with two options passed in this example in Step 8 using semMCreate function. Here, we prevent the priority inversion situation (Section 8.2.3) by choosing an option. Another option is for selecting either SEM_Q_FIFO or SEM_Q_PRIORITY. However, ***when selecting SEM_INVERSION_SAFE we must select the option SEM_Q_PRIORITY***. [Reason for using | sign in place of & in case of passing multiple options by a single argument is given in Section 10.3.2.(1)].

(c) Another exemplary use with three options passing and using semMCreate function is as follows. Here, we prevent the priority inversion situation as well as protect the task from deletion by any other task until the semaphore is made empty (not taken) at the end of the critical region of a task. [Section 8.1.3]. Both options are selected as follows.
   1. SEM_ID semMReadPortAKey;
   2. semMReadPortAKey = semMCreate (SEM_Q_PRIORITY | SEM_INVERSION_SAFE | SEM_DELETE_SAFE);

Passing Parameters: (*i*) The use of option between SEM_Q_PRIORITY and SEM_Q_FIFO is identical to the binary semaphore, which section 10.3.4.(1) described earlier. (ii) The use of SEM_INVERSION_SAFE makes the critical region using the mutex safe from priority inversion situation. [Section 8.2.3]. It means the created semaphore initial state is initialised as *taken*. (iii) Recall the use of SEM_DELETE_SAFE. It protects deletion of this task when in the critical region.

Returning Parameter: The function semMCreate ( ) returns a pointer, *SEM_ID for the ECB allocated to the *mutex semaphore*. It returns NULL in case of an error if none available.

Refer to Example 10.22 for understanding the use of semMCreate.

Example 10.22
1. /* Include the VxWorks header file as well as the task and semaphore functions from a library. */
2. # include "vxWorks.h"
3. # include "semLib.h"
4. # include "taskLib.h"
5. /* Declare a semaphore key to be used as mutex. */
6. SEM_ID semMReadPortAKey;
7. /* Create the mutex and pass the options chosen selected to it. */
8. semMReadPortAKey = semMCreate (SEM_Q_PRIORITY | SEM_INVERSION_SAFE);
9. /* This makes the mutex semaphore full (taken) and available like SemKey set as 1 in MUCOS. */
10. semGive (semMReadPortAKey);
    /* Other codes for task creation */

    .
    .
11. /* End of Task Creation Codes */
12. /* Task Codes */

    .

    . .
    /* Entering Critical region. *
13. **semTake (semMReadPortAKey**, WAIT_FOREVER);

    .

    . .
14. /* Leaving the critical region. */
15. **semGive (semMReadPortAKey)**;
    /***/

For using mutex, the following must be observed:
(a) The critical region that uses mutex for the resources protection should not be unnecessarily long and should be as short as possible.
(b) Use options for SEM_DELETE_SAFE if some task uses taskDelete ( ) function.
(c) Use option SEM_INVERSION_SAFE if some priority inversion situation is likely to arise.
(d) Use semTake function in the same task at the beginning and semGive at the end of a critical region in which there are shared resources. semTake can be used recursively but the total number of times a semTake executes should be the same as the number of times semGive executes.

(e) Do not use semGive ( ) for the mutex posting.

(f) Do not use semFlush ( ). [Its use is illegal when using the mutex semaphore]. Refer to section 10.3.4.(4) for semFlush function].

6. Creating a Counting Semaphore for the IPCs

The VxWorks counting semaphore [Section 8.1.2(5)] is similar to the POSIX semaphore [Section 8.2.2.(6)]. These increment on posting (giving) and decrement on taking (wait-over) semaphore. Posting this semaphore up to 256 times is permitted before it is taken. The status becomes equal to the initial value of counting semaphore only when the number of times a semaphore is given equals the number of times it is taken. The counting semaphore helps in a bounded buffer problem, ring-buffer problem, and consumer-producer problem. [Section 8.2.2.(6).] We have seen this in Examples 10.17 to 10.20. If initial count = 0, then a task waiting for the semaphore blocks.

The function SEM_ID semCCreate (*options*, unsigned byte *initialCount*) is for creating an ECB pointed by the SEM_ID. One of the two options must be passed on calling a function. An exemplary use is as follows: 'semCCharCheckFlagID = semCCreate (SEM_Q_PRIORITY, SEM_EMPTY);'.

1. SEM_ID semCID;

2. SEM_ID = semCCreate (SEM_Q_PRIORITY, 0); /* To initial count = 0. */

Passing Parameters: (i) One option that can be selected is SEM_Q_PRIORITY and the other is SEM_Q_FIFO. For the initial state, two options can be chosen: either initialCount should pass as 0 or it should be a fixed value. It depends on whether the semaphore is to be used for (*a*) decrementing count for the tasks that are already blocked or (*b*) for incrementing counting. [Recall the use of semaphore semCount in MUCOS. (ii) We had initialised semCount to 0 in Examples 10.17 to 10.20].

Returning Parameter: The function semCCreate ( ) returns a pointer, *SEM_ID. It returns NULL in case of an error for the ECB allocated to the *counting semaphore*. Null if none is available.

Recall MUCOS Example 10.18 Steps 11 to 21. Example 10.23 shows how to use VxWorks semCCreate function. It also shows the use of the other VxWorks functions for task spawning and semaphores.

Example 10.23

1. /* Same as Steps 1 and 2 of Examples 10.21 and 10.22. */

2. /* Declare and Create Semaphores function, its identifying variables. */

/* Declare *SemFlag1ID* as the argument that passes to the task whenever called. Declare SemMKeyID and SemCCountID as the mutex and counting semaphores. */

SEM_ID *SemFlag1ID, SemMKeyID, SemCCountID;*

3. /* Create Semaphore flag and declare unblocking of the tasks priority wise. Declare initially semaphore flag unavailability. */

SemFlag1ID = semBCreate (SEM_Q_PRIORITY, SEM_EMPTY);

4. / * Create Semaphore mutex and declare unblocking of the tasks priority wise. Initially semaphore mutex is available by default. */

SemMKeyID = **semMCreate (SEM_Q_PRIORITY);** /* SEM_Q_PRIORITY | SEM_DELETE_SAFE two options can also be used. However that prolongs the execution time. We are not using safe option as taskDelete ( ) function is not used anywhere in the codes. */

5. /* Create Semaphore for counting and declare unblocking of the tasks priority wise. Initially semaphore mutex is available by default. */
unsigned byte *initialCount* = 0;
SemCCountID = semCCreate (SEM_Q_FIFO, *initialCount*);
unsigned short *COUNT_LIMIT* = 80; /* Declare limiting Count = 80 */
6. /* Declare and Create Semaphores task function, its variables and parameters. */
void Task_ReadPortA (SEM_ID *SemFlag1ID*);
int *readTaskID* = ERROR; /* Let initial ID till spawned be none */
int *Task_ReadPortAPriority* = 105; /* Let priority be 105 */
int *Task_ReadPortAOptions* = 0; /* Let there be no option. It is the only task that waits for the SemFlag1ID from the port. */
int *Task_ReadPortAStackSize* = 4096; /* Let stack size be 4 kB memory */
4. /* Create and initiate a task for reading at Port A. Task name starts with 't'. The task calling-function is Task_ReadPortA */
readTaskID = taskSpawn ("tTask_ReadPortA", *Task_ReadPortAPriority*, *Task_ReadPortAOptions*, *Task_ReadPortAStackSize*, void (* Task_ReadPortA) (SEM_ID *SemFlag1ID*), SemMKeyID, SemCCountID, &*initialCount*, *COUNT_LIMIT*, 0, 0, 0, 0, 0, 0); /* Pass SemFlag1ID as the argument of task function and pass other arguments SemMKeyID and SemCCountID as arg0 and arg1. Remaining eight arguments are 0s. */
.

.

/* Other Codes */
.

.

5. /* The codes for the Task_ReadPortA redefined to use the key, flag and counter*/
static void Task_ReadPortA (SEM_ID *SemFlag1ID*) {
6. /* Initial assignments of the variables and pre-infinite loop statements that execute only once */
; /* Declare the buffer-size for the characters countLimit = 80 */
.

.

7. *while (1)* { /* Start an infinite while-loop. We can also use FOREVER in place of while (1). */
8. / *Wait for SemFlag1ID state change to SEM_FULL by semGive function of character availability check task */
semTake (SemFlag1ID, WAIT_FOREVER);
9. /* Take the key so that another task, *port decipher* does not unblock. That task needs SemMKeyID to unblock and run */
semTake (SemMKeyID, WAIT_FOREVER); /* SemMKeyID is now not available and the critical region starts */
10. if (*initialCount* > = *COUNT_LIMIT*) {
.

.

};/* End of Codes for the action on reaching the limit of putting the characters into the buffer */
11. /* Codes for reading from Port A and storing a character at a queue or buffer*/
.

.

12. semGive (SemCCountID); initialCount ++; /*Let counting semaphore value increase because one character has been put into the buffer holding the character stream. initialCount incremented because of the need to compare later with the COUNT_LIMIT*/

13. **semGive** (SemMKeyID); /* Critical region ends. Release the mutex SemMKeyID to let next cycle of this loop start */
14. *}; /* End of while loop*/*
15. *}* / * **End of the Task_ReadPortA function** */
/
**/

7. Using POSIX Semaphores

POSIX semaphore functions can also be used for the VxWorks counting semaphores. The function 'semPxLibInit ( )' initialises the VxWorks library to permit use of these. The functions sem_open ( ), sem_close ( ) and sem_unlink ( ) initialise, close and remove a named semaphore. The functions sem_post ( ) and sem_wait ( ) unlock and lock a semaphore. sem_trywait ( ) is to lock a semaphore if not locked. The actions of these three are giving, taking and accepting the VxWorks counting semaphore or OSSemPost, OSSemPend and OSSemAccept functions of MUCOS semaphores. The function sem_getvalue ( ) retrieves the value of a POSIX semaphore. POSIX semaphore functions sem_init ( ) and sem_destroy ( ) initialise and destroy an unnamed semaphore. Destroy means de-allocate associated memory with its ECB. This effect is the same as first closing a semaphore and then unlinking it by sem_close and sem_unlink. Remember that no deletion safety like VxWorks mutex is available before using these functions. VxWorks semaphores have the additional following features: (i) Options of protection from priority inversion and task deletion. (ii) Single task may take a semaphore multiple times and recursively. (iii) Mutually exclusive ownership can be defined. (iv) Two options, FIFO and task priority, for semaphores wait by multiple tasks.

8. Creating a Message Queue for the IPCs

The function 'MSG_Q_ID msgQCreate (int maxNumMsg, int maxMsgLength, int q*Options*)' is used for creating an ECB pointed by the MSG_Q_ID. One of the two options mentioned below must pass, on calling the function. The memory allocation to the buffer that holds the bytes for the messages is according to maxNumMsg and maxMsgLength.

A point to be noted is that the message pointer passed into an array of message pointers in MUCOS. Another point to be noted is that there is no provision for setting the limit up to which a counting semaphore can be given (posted). It is fixed at 256. Therefore, we have to use COUNT_LIMIT variable, as in Example 10.23, and compare with the initialCount variable, which increments after each function 'semGive (SemCCountID)' call. Further, in a message queue, the maximum number of messages $= 2^{31} - 1$ and the maximum number of bytes in each message is $2^{31} - 1$ bytes. Instead of declaring these variables as unsigned int, declaring as int in VxWorks offers the advantage of simplicity at the expense of possibility of error in case a negative value is passed for these.

Passing Parameters: (*i*) To the function, maxNumMsg passes the maximum number of messages that can be sent to the queue. (*ii*) maxMsgLength passes the maximum number of bytes permitted to be sent as a message. (*iii*) One option that can be selected is MSG_PRI_NORMAL, when the message is sent into the queue for receiving as a FIFO. The first message sent is then read first. The other option is MSG_PRI_URGENT. When the message is sent into the queue with this option, the message is received as LIFO. Urgent messages like error logins are sent with this option selected. The last message sent is then read first. (*iv*) Another option that can be selected is MSG_Q_PRIORITY. The other is MSG_Q_FIFO. Let us assume that at an instant, several tasks are in the blocked state and are waiting (pending) for a message from the same queue for its posting (sending). A waiting task can take from the queue in one of two ways. (a) A task, higher in priority than the other waiting ones, takes the message from the queue first. This becomes possible by MSG_Q_PRIORITY option. (b) A task, which first blocked and reached the waiting state, takes the message from the queue first among the waiting ones. This is possible by MSG_Q_FIFO option.

Returning Parameter: The function msgQCreate ( ) returns a pointer *MSG_Q_ID. It returns NULL in case of an error for the ECB allocated to the *message queue*. Null if none is available or on error.

Consider the *client-server architecture of the tasks* in an application. A server task can receive client task requests by receiving from a common request queue. The above function creates a request queue. The server can send the replies to each client separately through separate reply queues. This function also creates reply queues.

Example 10.24 explains the use of MsgQCreate. Let us assume that Task_ReadPortA reads a byte from port A and sends it to another task that receives the messages from a queue after waiting for the queue message availability.

Example 10.24

1. /* Include the header files in Example 10.22 Step 1 as well as queue functions from a library. */
2. # include "msgQLib.h"
3. /* Declare message queue identity and message data type or structure. */
4. **MSG_Q_ID *portAInputID*;**
5. unsigned byte portAdata;
6. void * message; /* Pointer for the message buffer */
7. /* Create the message queue identity and pass the parameters and options chosen selected to it and let the maximum number of messages be 80 and message be of 1 byte each. */
8. *PortAInputID* = **msgQCreate (80,1, MSG_Q_FIFO | MSG_PRI_NORMAL)**
9. /* Task Creation Codes as in Example 10.23. */
10. .
11. .
12. /* Start Codes for Task_ReadPortA. */
13. void Task_ReadPortA {
14. .
15. while ( ) {
16. semTake (SemFlag1ID, WAIT_FOREVER);
17. /* Take the key to not let a port-decipher task unblock and run. Because that task also needs SemMKeyID for running. */
18. *semTake (SemMKeyID,* WAIT_FOREVER); /* SemMKeyID is now not available and the critical region starts */
19. .
20. .
21. /* At the end, the send the byte, which is read at Port A. It is sent as a message to queue, *portAInputID*. Refer to Section 10.3.4.(10) for an understanding of the msgQSend function. */
22. *message = portAData;
23. **msgQSend (portAInputID, &message, 1, NO_WAIT, MSG_PRI_NORMAL);**

24. /* Other remaining codes for the task. */
25. .
26. .
27. }
28. }/* End of Task_ReadPortA Codes */

9. Waiting for an IPC for Binary or Other Type of Semaphore Release or Check for Availability of an IPC after Release

The function 'int msgQReceive (*msgQId*, &buffer, maxBytes, timeOut)' is used *for letting a task wait until sending (posting) of a message*. Wait until either *msgQSend* function sends the message in a task or until a timeout occurs, whichever happens first.

Passing Parameters: (i) Whenever the scheduler schedules this task and this function is called, there is a wait for a message pointed by *msgQId*. (ii) *buffer* is the address of the buffer where the received messages store. (iii) maxBytes is the maximum number of bytes acceptable per message by msgQReceive function. (iv) timeOut is the timeout period. One option that can be selected is WAIT_FOREVER if wait must be done for sending the message by msgQSend function. The other option is NO_WAIT. Recall OSQFlush function in MUCOS. NO_Wait option is simply used for checking the availability of a message. The message is put into buffer if available, or the task runs other succeeding codes. In case of error message, for example, there is no waiting especially for the message. Only a check is needed.

Returning Parameter: The function msgQReceive ( ) returns an integer for number of bytes retrieved and puts it at the buffer address.

Example 10.25 explains the msgQReceive function.

Example 10.25
1. to 9. /* Codes as per Steps 1 to 9 in Example 10.24 */
10. /* **The codes for the Task_MessagePortA** */
static void Task_MessagePortA *(void *taskPointer)* {
11. /* Initial assignments of the variables and pre-infinite loop statements that execute once only*/
int maxBytes = 80;
12. *while (1)* { /* Start an infinite while-loop. */
13. /* Wait for a Queue Message sending or availability. */
msgQReceive (*msgQId*, *message*, *maxBytes*, WAIT_FOREVER); /* WAIT_FOREVER means timeout = -1 and the period is thus infinity. */
14. /* Other remaining Codes */
.

.
15. }; /* End of while-loop*/
16. }/ * **End of the Task_MessagePortA function** */
/
**/

10. Sending an IPC after a Message is Sent into the Queue

The function 'STATUS msgQSend (*msgQId*, *&buffer*, *numBytes*, *timeOut*, *msgPriority*)' is used for letting a task send into the queue.

Passing Parameters: (i) The queue identifies by msgQ*Id*. (ii) Message posts to an addressed *buffer*. The number of bytes sent into the buffer = numBytes. (iii) The timeOut is the period that the posting of the message waits in case the queue is full. (iv) *msgPriority* is specified as MSG_PRI_NORMAL or MSG_PRI_URGENT, depending upon whether the message is inserted to later on retrieve in the FIFO or LIFO mode, respectively.

Returning Parameter: The function msgQSend ( ) returns STATUS. It returns STATUS = OK in case of success in taking of the msgQId, else returns 'ERROR' in case of an error that *msgQId* is invalid.

Step 23 of Example 10.24 shows how to use of this function.

11. Using POSIX Queues

Important points in using the POSIX queues are as followings:
1. The function 'mqPxLibInit ( ) initialises' the VxWorks library to permit use of the POSIX Queues.
2. The functions mq_open ( ), mq_close ( ) and mq_unlink ( ) initialise, close and remove a named queue.
3. The function mq_setattr ( ) sets the attribute of a POSIX queue.
4. The functions mq_send ( ) and mq_receive ( ) unlock and lock a queue.
5. The function mq_notify ( ) signals to a single waiting task that the message is now available. The notice is exclusive for a single task, which has been registered for a notification. [Registered means later on takes note of the mq_notify.]. This provision is extremely useful for a server task. A Server task receives the notification from a Client task through a signal handler function (like an ISR).
6. The function mq_getattr ( ) retrieves the attribute of a POSIX queue.
7. The POSIX queue function mq_unlink ( ) does not destroy the queue immediately but prevents the other tasks from using the queue. The queue will get destroyed only if the last task closes the queue. Destroy means to de-allocate the memory associated with queue ECB.

VxWorks queues have the additional following features. (i) Timeout option can be used. (ii) Two options, FIFO and task priority for queues wait by multiple tasks. POSIX queues have the additional following feature. (i) Task notification in case a single waiting task is available. (ii) There can be 32 message priority levels in place of one priority level URGENT in VxWorks.

12. Creating a Pipe Device and Read-Write for the IPCs

A pipe in VxWorks is a queue with management by a pipe driver (like a device driver) *pipedrv*. [This is analogous to the named pipe in UNIX.] Pipes also implement the client server architecture between a set of tasks.

Function pipeDevCreate ("/pipe/*pipeName*", maxMsgs, maxMsgBytes) creates a pipe device named *pipeName* for maximum maxMsgs messages. Each message can be of maximum size = maxMsgBytes. It enters into a list of devices on creation. devs ( ) function retrieves the list of devices with the device number allotted to each device including the pipe devices.

Consider an example for creating a pipe named as *pipeUserInfo*. Assume that it can have a maximum of four messages: user name, password, telephone number and e-mail ID. Each of these can be of a maximum size of 32 bytes only. A global variable fd is an integer number for a file descriptor that identifies a device among a number of devices at the I/O system. Example 10.26 explains the codes for creating, writing and reading.

Example 10.26
1. # include "fioLib.h" /* Include the IO library functions. */
pipeDrv ( ); /* Install a pipe driver. */
2. /* Declare file descriptor. */
int *fd*;
3. /* Mode refers to the permission in an NFS (Network File Server). Mode is reset as 0 for unrestricted permission. */
int *mode;*
4. /* Create pipe named as *pipeUserInfo* for 4 messages, each 32 bytes maximum. */
pipeDevCreate ("/pipe/*pipeUserInfo*", 4, 32); mode = 0x0;
Messages can be written into a pipe by the function by first opening a pipe device and then writing into that. The function for opening is open ("/pipe/*pipeUserInfo*", *rdwrFlag*, mode). We define flag = O_RDWR, which both permits read and write. Flag O_RDONLY permits the read only option and flag O_WRONLY permits the write only option. Remember that after opening a pipe, when we finish using it, we must use the function 'STATUS close ( )'. Writing to a pipe is analogous to writing to an I/O device. To write, the coding is as following:
5. /* Open read-write device using, a pipe named as pipeUserInfo with mode = 0 for unrestricted permission. */
fd = **open** ("/pipe/*pipeUserInfo*", O_RDWR, 0);
6. /* Write a message, *info* of lBytes. */
unsigned char [ ] info; Let the message be a string of characters. */
int *lBytes*;
lBytes =12;
write (fd, info, lBytes);
The message can be read from an open pipe by the function 'int read (fd, &buffer, lBytes)' Reading from a pipe is analogous to reading from an I/O device as per file descriptor. To read, the coding is as following:
7. int numRead; /* An integer to indicate the number of bytes successfully read.* /
lBytes =12; /* Let the Message bytes to read = 12*/
numRead = **read** (fd, info, lBytes);

13. Selecting the Bits from the IPCs from the Number of Devices

A file descriptor can be used for a pipe or socket or serial device or other type of device. Let a file descriptor fd = n; There is an array of bits in which the nth bit corresponds to fd = n. There is 'C' structure, struct fd_set. This structure *fd_set* has a set of file descriptors as well as a set of functions.

Now a function of this structure, FD_SET (n, &fdSet), when executed will make n[th] bit = set. FD_SET (m, &fdSet) makes the mth bit = set. FD_CLR (n, &fdSet) will make n[th] bit = clear.

FD_ZERO (&fdSet) makes all bits of array = 0. FD_ISSET (n, &fdSet) returns true if n[th] bit in the array is set and false if reset.

Now, let us examine how a task selects and finds the number of active devices at an instance. Task finds whether a pipe, *pipeUserInfo* is active or a *pipeServerReply* is active. The function to select is 'int select (numBitWidth, pointerReadFds, pointerWriteFds, pointerExceptFds, pointerTimeOut). The arguments passed are the following: numBitWidth = Number of bits to examine in the array of bits at two pointers, pointerReadFds and pointerWritedFds. Examination is as per value at a structure that stores NULL if wait forever or a value for timeout. Timeout is the number of system RTC interrupts up to which the wait is done. The function select ( ) blocks till at least *one device* in the array of devices is ready or until timeout, whichever happens first. Select clears all the bits that correspond to the devices that are not ready. This function returns the number of active devices. It returns *ERROR* on an error.

■ SUMMARY ■

The following is summary of what we learnt in this chapter.

- It is mandatory to use a well-tested and debugged RTOS in a sophisticated multitasking embedded system. MUCOS and VxWorks are the two important RTOSs.
- MUCOS task creating, deleting, suspending and resuming functions are used for the task controlling and scheduling functions.
- There are functions for initiating the system-timer in MUCOS. Starting a multitasking system by a first task and later suspending it forever is shown as a technique in programming for a multitasking system.
- MUCOS handles and schedules the tasks and ISRs similarly.
- There are delay and delay-resume functions in MUCOS. These are shown to be useful for letting a low priority task run.
- MUCOS has the inter-process communication (IPC) functions, semaphore, mailbox and queues. The simplicity feature of MUCOS is that the same semaphore functions are used for binary semaphore, for event flag, for resource key and for counting.
- MUCOS has mailbox functions and a simple feature that a mailbox has one message pointer per mailbox. There can be any number of messages or bytes, provided the same pointer accesses them.
- MUCOS has queue functions. A queue receives from a sender task an array of message pointers. Message pointers insertion can be such that later on it can retrieve in FIFO (first-in first-out) method as well as in LIFO (last-in last-out) method from a queue. It depends on whether the post was used or post front function was used, respectively. This help in taking notice of a high priority message at the queue.
- Instead of one create function, VxWorks has three functions, task create, task activate and task spawn (create and activate).
- VxWorks also provides for system-timer functions, delay functions and delay resume functions.
- VxWorks handles and schedules the functions for the tasks and ISRs differently. It allocates highest priorities for the ISRs over the tasks.

- VxWorks has signal-servicing routines. A signal-servicing routine is a C function. It executes on occurrence of an interrupt or exception. A connect function connects the function with the interrupt vectors.
- VxWorks has an IPC called *signal*. It is used for exception handling or handling interrupts.
- Exceptions are the software interrupts. A signal setting is equivalent to a flag setting in case of hardware interrupts.
- VxWorks provides for round robin time sliced scheduling as well as preemptive scheduling.
- VxWorks provides for two ways in which a pending task among the pending tasks can un-block. One is as per task priority and another is as a FIFO when accepting (taking) an IPC.
- VxWorks has three different semaphore functions for use as IPC for event flag, for resource key and counting semaphore. VxWorks also supports POSIX semaphores.
- VxWorks, instead of queuing the message pointers only, provides for queuing of the messages. Queues can be used as LIFO for priority messages as in MUCOS.
- VxWorks also supports use of pipes and POSIX queues.
- VxWorks pipes are the queues that can be opened and closed like a file. Pipes are like virtual I/O devices that store the messages as FIFO.

▪ LIST OF KEYWORDS AND THEIR DEFINITIONS ▪

- *Well-tested and debugged RTOS*: An RTOS, which is thoroughly tested and debugged in a number of situations.
- *Sophisticated multitasking embedded system*: A system that has multitasking needs with multiple features and in which the tasks have deadlines that must be adhered to.
- *MUCOS*: An RTOS μC/OS-II from Jean J. Labrosse.
- *VxWorks*: An RTOS from Wind River® Systems.
- *Task creation*: Task is allotted a TCB (task control block) and an identity. Creation also *initiates and schedules* on creation in MUCOS.
- *Task deletion*: Task no longer has the TCB and is ignored until created again.
- *TCB*: Task control block which has the task parameters, so that on task switching, the parameters remain saved, and when RTOS re-switches it back, the task can run from the point at which it left. The task is thus an independent process.
- *Task suspension*: A task unable to run its codes further.
- *Task resumption*: Task, which was delayed or suspended, can now be scheduled when the turn comes.
- *System timer*: An RTC that can be set to interrupt at preset intervals. The time is updated regularly and system interrupts regularly. RTOS also gets control of CPU to examine if any preemption is needed. Task priority provides priority for system-timer functions, delay functions and delay resume functions.
- *Task delay:* Let a task wait for a minimum time defined by number of system ticks passed as an argument to the delay function.
- *Task spawning*: Task creation and activation.
- *Signal*: Flag-like intimation to RTOS for development of certain situations during a run that need urgent attention by executing an ISR.

- *Exception handling*: Executing a function on receiving a signal. Error is also handled by using an exception-handling function.
- *Event flag*: A flag that sets on occurrence of an event and resets on response to the event.
- *Resource key*: A semaphore that resets on the start of execution of a critical region code and sets on finishing these.
- *Counting semaphore*: It is a semaphore that increments when an IPC is given by a task or a section of the task. It decrements when a waiting task unblocks and starts running.
- *POSIX semaphores*: Semaphore functions as per IEEE POSIX standard functions.
- Message queue: A queue into which a task posts a message that is retrieved by another task.
- *POSIX queues*: IPC queue functions as per POSIX standard functions.
- *Pipe*: A queue from which one task gets the messages and other task puts the messages. VxWorks *pipe* is a queue in which the I/O device functions operate. Putting and getting messages from a pipe is similar to a file.
- *File device*: Memory block(s) in which the file read, file write, file open and file close functions operate as in case of file on a disk.

▪ REVIEW QUESTIONS ▪

1. What are the advantages of a well-tested and debugged RTOS, which is also well trusted and popular? [*Hint:* Embedded software has to be of highest quality and also there should be faster software development. Complex coding skills are required in the development team for device drivers, memory and device managers, networking tasks, exception handling, test vectors, APIs and so on].
2. How does a mailbox message differ from a queue message? Can you use message queue as a counting semaphore?
3. Explain ECB (Event Control Block).

▪ PRACTICE EXERCISES ▪

Note: Exercises 5 to 12 pertain to MicroC/OS-II and 13 to 23 to VxWorks.

4. Search the Web (for example, www.eet.com) and find the latest top RTOS products.
5. Classify and list the source files, that depend on the processor and those that are processor independent?
6. Design a table that gives MUCOS features.
7. MUCOS has one type of semaphore for using as a resource key, as a flag, as a counting semaphore, and a mutex. What is the advantage of this simplicity?
8. How do you set the system clock using function void OSTimeSet (unsigned int counts)?
9. When do you use OS_ENTER_CRITICAL ( ) and OS_EXIT_CRITICAL ( )?
10. How do you set the priorities and parameters, OS_LOWEST_PRIO and OS_MAX_TASKS, for preemptive scheduling of the tasks?
11. A starting task is first created, which creates all the tasks needed and initiates the system clock, and then that task is suspended. Why must you use this strategy?

12. List ten examples each of for: semaphore, mailbox, and message queue.
13. VxWorks kernel includes both POSIX standard interfaces and VxWorks special interfaces. What are the advantages of special interfaces for the semaphores and queues?
14. List task state transitions.
15. How do you initiate round robin time-slice scheduling? Give ten examples of the need for round robin scheduling.
16. How do you initiate preemptive scheduling and assign priorities to the tasks for scheduling? Give ten examples of the need for preemptive scheduling.
17. How do you use signals and use function 'void sigHandler (int *sigNum*), signal (*sigNum*, *sigISR*) and 'intConnect (I_NUM_TO_IVEC (sigNum), sigISR, sigArg)? Give five examples of their uses.
18. How do you create a counting semaphore?
19. OS provides that all ISRs share a single stack. What are the limitations it imposes?
20. How do you create, remove, open, close, read, write, and I/O control as a device using RTOS functions? Take an example of a pipe delivering an I/O stream from a network device.
21. Explain the use of file descriptor for I/O devices and files.
22. How do you let a lower priority task execute in a preemptive scheduler? Give four coding examples.
23. How do you spawn tasks? Why should you not delete a task unless memory constraint exists? [Hint: Refer to Table 9.11].
24. Write exemplary codes for using the POSIX functions for timer, semaphores, and queues.
25. How do you create client server architecture using sockets?

# Chapter 11

Case Studies of Programming with RTOS

What We Have Learnt

The following important aspects of embedded systems have been covered in the previous chapters:

1. *Hardware units*, processor, memories, buses, interfacing circuits, ports devices, real-time clock driven software timers, device drivers, and *interrupt handling mechanism* in an embedded system. [Chapters 1 to 4].

2. High-level language *programming* concepts, program models, software engineering approaches, *RTOS* and use of the *IPCs* semaphores, mailboxes, queues and pipes for synchronizing multiple tasks and for solving problems related to sharing of variables between processes using interrupt disabling–enabling mechanism or binary, mutex, P and V, POSIX and counting semaphores. [Chapters 8 to 10].

3. Twenty-seven exemplary uses of available functions for the system, task and IPC functions in the two popular RTOSs for the embedded systems. These are MUCOS (μC/OS-II) from Jean J. Labrosse and VxWorks from WindRiver® Systems. [Refer to Chapter 10].

What We Will Learn

In this chapter, one case study each has been given to explain and illustrate the use of RTOS programming tools in each of the four areas:

1. Consumer Electronic Systems,
2. Communication and Network Systems,
3. Control Systems in Automotive Electronic Systems and
4. Secure SoC systems.

Exemplary codes for the following case studies will be explained:

(a) Section 11.1 details the codes for a vending machine system (VS) to be designed using MUCOS as RTOS. It demonstrates (i) an interesting application of RTOS,

(ii) use of the same type of semaphore as an event flag, as a resource key and as a counter and (iii) the use of system RTC and mailboxes.

(b) Section 11.2 presents the codes and a code design for sending the TCP/IP stack (byte streams) on a *network*. The streams first pass through the multiple layers with TCP/IP protocols. A network driver finally sends the stream on a TCP/IP network. This case study will show the application of various functions of VxWorks RTOS. We shall learn the applications of VxWorks binary and mutex semaphore-types, message queues and pipes. We shall learn another VxWorks feature: We can pre-define the number of bytes for the messages sent to the queues and pipe in place of the message pointers in MUCOS. These features in VxWorks RTOS become handy when applying an RTOS to embedded networking systems.

(c) Section 11.3 has the codes and a code design for an *automotive cruise control* (ACC) system. We shall first learn embedded systems for automotive electronics, adaptive algorithm in control systems, and important standards for hardware and software including a standard RTOS, OSEK-OS. After understanding its differences with VxWorks and MUCOS, we will learn how the code design is given using VxWorks after incorporating OSEK features.

(d) The codes and the code design for a *smart card* are given in Section 11.4. This specific case is chosen in order to explain the special embedded hardware and special RTOS functions that are needed. In this case, the RTOS MUCOS and VxWorks functions do not suffice.

11.1 CASE STUDY OF CODING FOR A VENDING MACHINE USING MUCOS RTOS

Embedded systems are one of the most used systems today in the area of consumer electronics. We describe a case study for MUCOS RTOS. Assume that the codes for an automatic vending machine are to be designed.

11.1.1 Case Definition, Multiple Tasks and Their Functions

Let us first understand the system requirements. The machine has a slot for inserting the coins. The following is the case definition.

Figure 11.1(a) shows the basic system of a Vending Machine System *(VS)* machine. Figure 11.1(b) shows ports at the VS. System requirements are as follows:

1. There is a slot into which a buyer inserts the coins for buying a chocolate. The chocolate costs Rs. 8. A coin can be in one of three possible denominations: Rs. 1, 2 and 5. Whenever a coin is inserted, a mechanical system directs each coin to its appropriate port, Port_1, Port_2 and Port_5.

2. The machine should have an LCD display matrix as 'User Interface'. Let the interface port be called Port_Display. It displays the message strings in three lines.

3. It should have a bowl from where the buyer collects the chocolate through a port for delivery. Let this port be called Port_Deliver. Port_Deliver connects to a chocolate channel, which the owner of the machine gets filled, whenever the channel is left with only a few chocolates. The buyer also collects the full refund or excess amount refund also at the bowl, through the ports, Port_Refund in the case of short change and Port_ExcessRefund in the case of excess change, respectively. All ports, Port_Deliver, Port_Refund and Port_ExcessRefund communicate to Port_Collect. This port is a common mechanical interface to the bowl.

4. An RTOS has to schedule the processes (tasks) for buying from start to finish.

5. It should also be possible to reprogram the codes and relocation of the codes in the system ROM or flash or EPROM whenever the following happens: (i) the price of chocolate increases, (ii) the message lines need to be changed or (iii) machine features change.

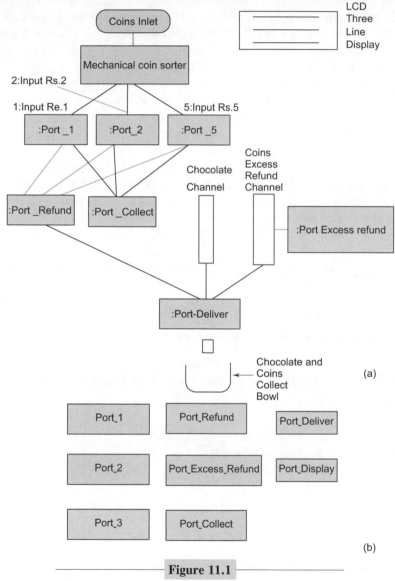

Figure 11.1

(a) Basic System (VS) of a Vending Machine (b) Ports of the VS

Let MUCOS RTOS be the choice for an embedded software development for VS. The following are the details of multiple tasks.

1. All 24 bits at three ports, Port_1, Port_2 and Port_5 are in reset state (0s) or reset on power-up.
2. Besides being attached to mechanical subsystems, each of these ports gives digital input bits to the VS being designed and receives the system output bits.

3. Mechanical subsystems also provide a facility, which helps when there are two or more coins of the same type. Thus, there can be a maximum of eight total different points at each port.

4. Use of a semaphore, *SemAmtCount* is such that within a time-out period, a buyer can thus either insert 8 coins of Rs. 1 for 8 points of Port_1, or insert a coin of Rs. 1 for Port_1 point, of Rs 2 for Port_2 point and Rs. 5 for the Port_5 point. The buyer has options for several possible combinations of coins to cover the cost of a chocolate, and the machine begins collecting coins and delivering the chocolate.

5. VS task, *Task_ReadPorts* does the following. (i) It reads the byte (8 bits) at each of the above ports. (ii) If the coins' status, as reflected by SemAmtCount, is as per the cost of the chocolate, it sends a flag to Task_Collect. The latter task initiates actions for collecting the coins into a collection unit. (iii) It also resets the bits after reading to keep the ports and all the 24 points in ready state for the next cycle of the machine. (iv) (a) It does the other actions described in Step 10, if the coins do not accumulate as per the cost within a specific timeout period or are in excess of the cost or (b) It does other actions as per Step 11 for displaying messages as per the state of the port just before switching to another task. It sends messages through three mailbox pointers.

6. VS circuit design is such that a port 0th bit, bit 0 is set (=1) when a coin is available and identified properly. Port bit 1 is also set when two are available, bit 2 also sets when three are available, and so on. There are 8 bits for 8 points at the port. The number of 1's at the port determines the number of coins at that port.

7. When a port, Port_Collect receives a direction (a signal in the form of flag, which sets) from the Task_Collect, then all 24 bits for the 24 points at the three ports release the coins using an electromechanical device. Coins collect at a collection unit. [To recover the coins, the machine owncr on a convenient day opens the unit by a lock and withdraws the money. The machine owner also fills the coin unit at Port_ExcessRefund for refunding if the buyer inserts an extra coin].

8. VS task, *Task_Collect* does the following. (i) It directs the Port_Collect to act and the unit collects all the available coins from the ports, Port_1, Port_2 and Port_5. (ii) After collection is over, it sends an IPC to another task, task_Deliver, through a Port_Deliver. This port on setting delivers the chocolate and also resets for the next cycle. (iii) It does the other actions described at Step 11 for a display.

9. When a port, Port_Refund receives a direction (a signal in the form of a flag, which sets) from a task, Task_Refund, it directs all the eight points at each of the ports to release the coins if any using an electromechanical device. The coins drop in a bowl if the amount at the three ports is found to be less than the cost. When a port, Port_ExcessRefund receives a direction (a signal in the form of a flag, which sets) from a task, Task_ExcessRefund, it directs the eight points at another port to release the excess amount using an electromechanical device and the coins drop in the bowl, provided the amount at the three ports is found to be more than the cost. [To recover the coins, the buyer reaches into a bowl to withdraw the refunded money].

10. VS task, *Task_Refund* does the following. (i) It directs the Port_Refund to act and the unit sends the coins from three ports to a bowl when the amount is short. (ii) It does the other actions described in Step 11. It sends a display message for the mailboxes, the mails in which Task_Display waits. VS task, *Task_ExcessRefund* does the following. (i) It directs the excess refunding on the unit to refund the excess amount. (ii) It does the other actions described in Step 11 and messages to the mailboxes.

11. At this step, an LCD matrix port gets a message from a Task_Display. The display is as per the state of the machine or time-date mailed to task waiting mailboxes. The displayed messages are

as follows: (i) When machine resets or is on cycle start, in the first line, a welcome message is displayed. The second line display is "Insert amount, please". The third line to display is "*time and date*" from a Task_TimeDateDisplay which sends the message every second. (ii) After a mail from Task_Collect, the first line shows a message, "Wait for a moment" and the second line displays "Vending soon." The message clears after a time out. (iii) After a mail from Task_Deliver, the first line message is "Collect the snack". The second line message is "Insert coins for more". The message clears after a time out. (iv) After an event flag from Task_Refund, a message in the first line is "Sorry"! The second line displays "Please collect the refund". The messages clear after a time out. (v) After the mails, from Task_ExcessRefund in the first line, "Collect the Chocolate and Money" is displayed. "Do not forget to collect the refund" is displayed in the second line. The message clears after a time out. (vi) After a time out or on machine reset for next cycle, repeat the entire display as in step (i).

12. OSSemPost and OSSemPend semaphore functions synchronise such that a task waits for execution of the codes until the necessary tokens, indicated by SemAmtCount, are collected within a specified timeout period. Task_Deliver sends a signal to a Port_Deliver that delivers a chocolate from VS *and assumes that it must deliver only on* collection of specific coins combinations such that the amount received is equal to or more than the chocolate cost.

The reader must have understood that there are multiple tasks to be synchronised with respect to each other, using the IPCs. The reader must have also understood the need for using an RTOS and the need for binary-semaphores, resource-key semaphores, counting semaphores and mailboxes.

11.1.2   Creating a List of Tasks, Functions and IPCs

A designer now creates a table as follows: Puts a task *name*, *priority* and *actions expected* from the task in columns 1, 2 and 3, respectively in each row. IPCs pending and IPCs posted should be listed in column 4 and 5. Mechanical or other system inputs and outputs are put in the last columns 6 and 7. Figure 11.2 shows multiple tasks at VS and their synchronisation model. Table 11.1 gives the designed table. The task synchronisation model, which is apparent from the table as well as the figure, is as follows:

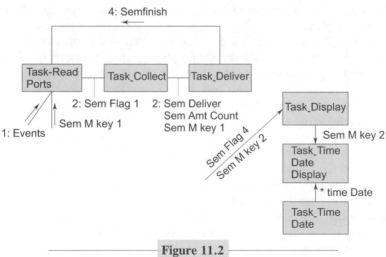

Figure 11.2

Tasks at VS tasks and their synchronisation model

1. Task_TimeDateDisplay time outs after every 1000 ticks of system clock, which are reset as 1 ms. After a timeout, it updates the time and date values at a pointer *timeDate. It posts into the mailbox *timeDate. Task_Display uses it to display in the third line, right corner of the LCD matrix.

2. Task_ReadPorts starts action only when a semaphore *SemFinish* is reset or posted by the Task_Deliver or Task_Refund or Task_ExcessRefund. It posts SemMKey1 to access the critical resource, *SemAmtCount*. The task then waits for event signals from Port_1, Port_2 or Port_5. The task then starts counting the amount by posting to the semaphore, *SemAmtCount*. Machine inputs are the coins at the points (maximum 8) at Port_1, Port_2 and Port_5. It posts an event flag, SemFlag1, to Task_Collect on collecting an amount equal to or more than the chocolate cost (= 8). It posts SemFlag2 to Task_Refund, if amount collected within a timeout is insufficient. It posts SemFlag3 to Task_ExcessRefund, if amount collected is in excess of the cost. It posts SemFlag4 to Task_Display. It posts message pointers for mailboxes waiting for the mails.

3. Task_Collect waits for taking SemFlag1 and *SemAmtCount*. It releases *SemDeliver* to let a Task_Deliver provide the chocolate by writing to the Port_Deliver. Task_Collect also releases SemDeliver when Task_ExcessRefund takes SemFlag3. This is because the buyer has already paid extra, so s/he must get a chocolate. Task_Deliver releases *SemFinish* on finishing the delivery and accepting the SemAmtCount to make it 0.

4. Task_Refund waits for taking SemFlag2 and then flushes to 0, the *SemAmtCount*, and SemMKey1. It releases *SemFinish* on finishing the refund and accepting the *SemAmtCount* to make it 0. It sends *SemFinish* to Task_ReadPorts.

5. Task_ExcessRefund waits for taking SemFlag3 and accepts *SemAmtCount* to decrease it by 8 and posts SemFlag1 and releases SemMKey1.

6. Task_Display waits for taking SemFlag4. It takes mutex, SemMKey2 before passing the bytes to a stream for Port_Display and releases it after sending. It displays the mailbox messages at the message pointers, *Collect, *delivered, *refund, and *ExcessRefund.

Table 11.1
List of Tasks, Functions and IPCs Used in Example 11.1

Task Function ID	Priority	Action	IPCs pending	IPCs posted	VS input	VS Output
Task_Read Ports	5	Waits for the coins and action as per coins collected	Event signals from Port_1, Port_2 and Port_5; Sem Finish, SemMKey1	SemFlag1, SemFlag2, SemFlag3, SemFlag4, SemMKey1, *Message Pointer *Collect*	*Coins at Port_1, Port_2 and Port_5.*	
Task_ Collect	7	Waits for coins = or > cost till timeout and act accordingly.	SemFlag1	*SemMKey1,* SemDeliver, *Message Pointer *wait*	-	*Coins at Port_ Collect*

(Contd.)

Task Function	Priority	Action	IPCs pending	IPCs posted	VS input	VS Output
Task_ Deliver	8		SemDeliver, SemAmtCount, SemMKey1	*SemFinish, SemAmtCount,* SemMKey1 Message Pointer *delivered	Chocolate from a channel for sending it.	Delivers Chocolate into the bowl.
Task_Refund	9	Waits for refund event and refunds the amount	SemFlag2, SemMKey1, SemAmtCount	*SemFinish, SemAmtCount, SemMKey1, Message Pointer *refund*	*Coins at Port_1, Port_2 and Port_5*	*Coins Flushed back to a Port_Exit*
Task_ Excess Refund	11	Refunds the excess amount	SemFlag3, *SemAmtCount, SemMKey1*	SemFlag1, *SemMKey1, SemAmtCount* Message Pointer *ExcessRefund*	Coins at the points of Port_ Excess Refund	(a) *Coins at Port_ Collect* (b) *Excess Coins from* Port_ Excess Refund
Task_Display	13	Waits for the message mails	SemFlag4, SemMKey2, *Message Pointer *delivered, *refund, *Excess Refund *timeDate*	*SemMKey2,*	Strings for line 1, 2 and 3 and time date.	Bytes for the LCD matrix display to lines 1, 2 and 3.
Task_Time DateDisplay	14	Updates time and date by counting the number of system ticks	Gets a time out message from the RTOS after every 1000 ms (1000 system clock interrupts)	Message pointer *timeDate	Interrupts from system ticks	String for time date

11.1.3 Exemplary Coding Steps

Example 11.1

1. /* <u>Define Boolean variable as per example 4.5, define a NULL pointer</u> to point in case mailbox is empty. Codes are as per Example 8.17 Step 1*/

```
typedef unsigned char int8bit;
#define int8bit boolean
#define false 0
#define true 1
/* Define a NULL pointer; */
#define NULL (void*) 0x0000;
```

/* Preprocessor commands define OS tasks service and timing functions as enabled and their constants; similar to Example 10.7. */

```
#define OS_MAX_TASKS 10
#define OS_LOWEST_PRIO 20 /* Let lowest priority task in the OS be 15. */
#define OS_TASK_CREATE_EN 1 /* Enable inclusion of OSTaskCreate ( ) function */
#define OS_TASK_DEL_EN 1 /* Enable inclusion of OSTaskDel ( ) function */
#define OS_TASK_SUSPEND_EN 1/* Enable inclusion of OSTaskSuspend ( ) function */
#define OS_TASK_RESUME_EN 1/* Enable inclusion of OSTaskResume ( ) function */
```

.

.

/* Specify all child prototype of the first task function that is called by the main function and is to be scheduled by MUCOS at the start. In step 11, we will be creating all other tasks within the first task. */
/ * Remember: Static means permanent memory allocation */

```
static void FirstTask (void *taskPointer);
static OS_STK FirstTaskStack [FirstTask_StackSize];
```

/* Define public variables of the task service and timing functions */

```
#define OS_TASK_IDLE_STK_SIZE 100 /* Let memory allocation be for an idle state task
stack size be 100*/
#define OS_TICKS_PER_SEC 1000 / * Let the number of ticks be 1000 per second. An RTCSWT
will interrupt and thus tick every 1 ms to update counts. */
#define FirstTask_Priority 4 /* Define first task in main priority */
#define FirstTask_StackSize 100 /* Define first task in main stack size */
```

.

2. /* Preprocessor definitions for maximum number of inter-process events to let the MUCOS allocate memory for the Event Control Blocks */

```
#define OS_MAX_EVENTS 24/* Let maximum IPC events be 24 */
#define OS_SEM_EN 1/* Enable inclusion of semaphore functions. */
#define OS_MBOX_EN 1/* Enable inclusion of mailbox functions for the mailing of the mes-
sage pointers to Task_Display. */
#define OS_Q_EN 1/* Enable inclusion of queue functions for sending the string pointers to
LCD matrix Port_Display * /
```

/* End of preprocessor commands */
3. /* Prototype definitions for seven tasks for steps *1* to *12* above. */

```
static void Task_ReadPorts (void *taskPointer);
static void Task_ExcessRefund (void *taskPointer);
static void Task_Deliver (void *taskPointer);
static void Task_Refund (void *taskPointer);
static void Task_Collect (*taskPointer);
static void Task_Display (void *taskPointer);
static void Task_TimeDateDisplay (void *taskPointer);
```

/* Definitions for seven task stacks. */

```
static OS_STK Task_ReadPortsStack [Task_ReadPortsStackSize];
static OS_STK Task_ExcessRefundStack [Task_ExcessRefundStackSize];
static OS_STK Task_DeliverStack [Task_DeliverStackSize];
static OS_STK Task_RefundStack [Task_RefundStackSize];
static OS_STK Task_CollectStack [Task_CollectStackSize];
static OS_STK Task_DisplayStack [Task_DisplayStackSize];
static OS_STK Task_TimeDateDisplayStack [Task_TimeDateDisplayStackSize];
```
/* <u>Definitions for seven task-stack sizes.</u> */
```
#define Task_ReadPortsStackSize 100 /* Define task 2 stack size*/
#define Task_ExcessRefundStackSize 100 /* Define task 3 stack size*/
#define Task_DeliverStackSize 100 /* Define task 4 stack size*/
#define Task_RefundStackSize 100 /* Define task 5 stack size*/
#define Task_CollectStackSize 100 /* Define task 3 stack size*/
#define Task_DisplayStackSize 100 /* Define task 1 stack size*/
#define Task_TimeDateDisplayStackSize 100 /* Define task 1 stack size*/
```
4. /* <u>Definitions for seven task-priorities.</u> */
```
#define Task_ReadPortsPriority 5 /* Define task 1 priority */
#define Task_ExcessRefundPriority 11 /* Define task 5 priority */
#define Task_DeliverPriority 8 /* Define task 3 priority */
#define Task_RefundPriority 9 /* Define task 4 priority */
#define Task_CollectPriority 7 /* Define task 2 priority */
#define Task_DisplayPriority 13 /* Define task 6 priority */
#define Task_TimeDateDisplayPriority 14 /* Define task 7 priority */
```
5. /* <u>Prototype definitions for the semaphores.</u> */

OS_EVENT *SemFlag1; /* On interrupt signals from Port_1 to Port_5, Task_ReadPorts starts running. This flag needed when using semaphore for inter-process communication between tasks reading Port_1, Port_2 and Port_5 and port, Port_Collect. Also posted by Task_ExcessRefund to deliver the chocolate also while refunding the excess. */

OS_EVENT *SemFlag2; /* Needed when using semaphore as flag for inter-process communication between Task_ReadPorts and amount sending task, Task_Refund. */

OS_EVENT *SemFlag3; /* Needed when using semaphore as flag for inter-process communication between Task_ReadPorts and Task_ExcessRefund. */

OS_EVENT *SemFlag4; /* Needed when using semaphore as flag for inter-process communication between Task_ReadPorts and displaying task Task_Display. */

OS_EVENT *SemAmtCount; /* Needed when using semaphore for passing the counts as the number of characters read as a 16-bit message between Task_ReadPorts and other tasks, collecting or excess refunding the amount. */

OS_EVENT *SemChocoCostVal; /* Needed when using semaphore as cost value for chocolate. When SemAmtCount reaches cost value or exceeds this, the inter-process communication from Task_ReadPorts initiates. */

OS_EVENT *SemMKey1; /* Needed when using semaphore as resource key by Task_ReadPorts and Task_ExcessRefund */

OS_EVENT *SemMKey2; /* Needed when using semaphore as resource key by Task_ReadPorts and Task_ExcessRefund */

OS_EVENT *SemVal; /* Needed when using semaphore for passing the 16-bit message between steps b and c. */

6. /* <u>Prototype definitions for the mailboxes and queues.</u> */

OS_EVENT *MboxStr1Msg; /* Needed when using mailbox message between steps *1* and *12*. */

OS_EVENT *MboxStr2Msg; /* Needed when using mailbox message between steps *1* and *12*. */

7. /* Codes as per codes in Example 4.5 (except the main function codes). These are for reading from Port_1, 2 and 5. */

OS_EVENT *MboxStr3Msg; /* Needed when using mailbox message between steps *1* and *12*. */

OS_EVENT *MboxTimeDateStrMsg; /* Needed when using mailbox message from Task_TimeDateDisplay */

OS_EVENT *QMsgPointer; /* Needed when using mailbox message between steps *1* and *12*. void *QMsgPointer [QMessagesSize]; /* Let the maximum number of message-pointers at the queue be QMessagesSize. */

OS_EVENT *QErrMsgPointer; /* Needed when using mailbox message between steps *1* and *12* */

void *QErrMsgPointer [QErrMessagesSize]; /* Let the maximum number of error message-pointers at the queue be QErrMessagesSize. */

8. /* Define both queues array sizes. */

#define QMessagesSize = 64; /* Define size of message-pointer queue when full */

#define QErrMessagesSize = 16; /* Define size of error message-pointer queue when full */

SemFlag1 = OSSemCreate (0) /* Declare initial value of semaphore = 0 for using it as an event flag*/

SemFlag2 = OSSemCreate (0) /* Declare initial value of semaphore = 0 for using it as an event flag*/

SemFlag3 = OSSemCreate (0) /* Declare initial value of semaphore = 0 for using it as an event flag*/

SemFlag4 = OSSemCreate (0) /* Declare initial value of semaphore = 0 for using it as an event flag*/

SemChocoCostVal = OSSemCreate (8) /* Declare initial value of semaphore = 8 as an event flag*/

SemChocoCostVal = OSSemCreate (8) /* Declare initial value of semaphore = 8 as an event flag*/

SemAmtCount = OSSemCreate (0) /* Declare initial value of semaphore = 0 on reset of the machine. */

SemMKey1 = OSSemCreate (1) /* Declare initial value of semaphore = 1 for using it as a resource key for SemAmtCount using sections in the tasks. */

SemMKey2 = OSSemCreate (1) /* Declare initial value of semaphore = 1 for using it as a Display resource key*/

SemVal= OSSemCreate (0) /* Declare initial value of semaphore character = '0' */

/* Declare initial counts as 0 as a counter that shows the number of times a task, which sends into a buffer that stores a character stream, ran minus the number of times the task which used the character from the stream ran from the buffer */

SemAmtCount = OSSemCreate (0);

/* Create Mailboxes for the tasks. */

/* The following are required for three lines of LCD matrix at Port_Display when using mailbox message from the tasks to the Task_Display. */

MboxStr1Msg = OSMboxCreate (NULL);

MboxStr2Msg = OSMboxCreate (NULL);

MboxStr3Msg = OSMboxCreate (NULL);
MboxTimeDateStrMsg = OSMboxCreate (NULL); /* For message from Task_TimeDateDisplay to Task_Display. */
9. /* Any other OS Events for the IPCs. */

.
.

10. /* The codes similar to the codes in Example 4.5, except the main function codes. These are for reading from Port A and storing a character. Here, we have three ports, Port_1, Port_2 and Port_5 for Rs. 1, 2 and 5 denomination coins. These are basically device driver codes for port_1, port_2 and port_5 and three status flags for resetting to the beginning. */
STAF _1 = 0;
STAF _2 = 0;
STAF _3 = 0;

.
.

11. /* Start of the codes of the application from **Main. Note:** Code steps are similar to Steps 9 to 17 in Example 10.16 */

.
.
.

void main (void) {
12. /* Initiate MUCOS RTOS to let us use the OS kernel functions */
OSInit ( );
13. /* Create first task, *FirstTask* that must execute once before any other. Task creates by defining its identity as FirstTask, stack size and other TCB parameters. */
*OSTaskCreate (FirstTask, void (*) 0, (void *) &FirstTaskStack [FirstTask_StackSize],* FirstTask_Priority);
14. /* Create other main tasks and inter-process communication variables if these must also execute at least once after the FirstTask. */
15. /* Start MUCOS RTOS to let us RTOS control and run the created tasks */
OSStart ( );
/* Infinite while-loop exits in each task. So there is no return from the RTOS function OSStart ( ). RTOS takes the control forever. */
16. }/ *** End of the Main function ***/
/* The codes of the application first task that *main* created. */
17. *static void FirstTask (void *taskPointer){*
18. /* Start Timer Ticks for using timer ticks later. */
OSTickInit ( ); /* Function for initiating RTCSWT that starts ticks at the configured time in the MUCOS configuration preprocessor commands in Step 1 */.
19. /* Create seven Tasks defining by seven task identities, Task_TimeDateDisplay, Task_Display, Task_ReadPorts, Task_ExcessRefund, Task_Deliver and Task_Refund and the stack sizes, other TCB parameters. */
OSTaskCreate (Task_Display, *void (*) 0, (void *)* & Task_Display*Stack*
[Task_Display*StackSize*], Task_DisplayPriority);
OSTaskCreate (Task_TimeDateDisplay, *void (*) 0, (void *)* & Task_TimeDateDisplay*Stack*
[Task_TimeDateDisplay*StackSize*], Task_TimeDateDisplayPriority);
OSTaskCreate (Task_ReadPorts, *void (*) 0, (void *)* & Task_ReadPorts*Stack*
[Task_ReadPorts*StackSize*],   Task_ReadPortsPriority);

OSTaskCreate (Task_ExcessRefund, *void (*) 0, (void *)* & Task_ExcessRefund*Stack* [Task_ExcessRefund*StackSize*], Task_ExcessRefundPriority);

OSTaskCreate (Task_Deliver, *void (*) 0, (void *)* & Task_Deliver*Stack* [Task_Deliver*StackSize*], Task_DeliverPriority);

OSTaskCreate (Task_Refund, *void (*) 0, (void *)* & Task_Refund*Stack* [Task_Refund*StackSize*], Task_RefundPriority);

OSTaskCreate (Task_Collect, *void (*) 0, (void *)* & Task_Collect*Stack* [Task_Collect*StackSize*], Task_CollectPriority);

20. while (1) { /* Start of the while loop*/

21. /* Suspend with no resumption later the First task, as it must run once only for initiation of timer ticks and for creating the tasks that the scheduler controls by preemption. */

*OSTaskSuspend (FirstTask_Priority); /*Suspend First Task and control of the RTOS passes forever to other tasks, waiting their execution*/

22. } /* End of while loop */

23. } /* **End of FirstTask Codes** */

/
***/

/*** RTOS schedules the task functions below and never returns the control to the first task. ***/

24. /* The **codes for the Task_ReadPorts** redefined to use the key, flag and 16-bit value and mailbox*/

static void Task_ReadPorts *(void *taskPointer)* {

/* Initial assignments of the variables and pre-infinite loop statements that execute once only. The variable, *i* used for counting the 1's at the 8 points of a port and j is to wait for one more cycle after first if SemAmtCount is less than the SemChocoCostVal. */

static unsigned byte *i*, j;

static unsigned char *port_1data, *port_2data, *port_5data;

static void port_1_ISR_Input (*port_1data); /* Declare function to call to retrieve port_1data. */

static void port_2_ISR_Input (*port_2data); /* Declare function to call to retrieve port_2data. */

static void port_5_ISR_Input (*port_5data); /* Declare function to call to retrieve port_5data. */

25.　　*while (1)* { /* Start an infinite while-loop */

/* Take the event flag SemFinish > 0 and if so decrement it Wait forever till 0. Take mutex for critical section. Variable SemAmtCount is used by other tasks also. */

OSSemPend (SemFinish, 0, & *SemErrPointer*);

OSSemPend (SemMKey1, 0, & *SemErrPointer*);

26. /* Post Two mails to waiting Task_Display through two Message pointers for first line and second line of LCD matrix at Port_Display. */

OSMboxPost (MboxStr1Msg, *"Welcome to the vending machine"*); OSMboxPost (MboxStr2Msg, "Insert amount, please");

/* Set initial amount counts to SemAmtCount. */

SemAmtCount = 0;

OSSemPost (SemFlag4); /* Release flag for Task_Display. */

27. /* Device driver Codes for Port_1, Port_2 and Port_5 and three status flags will set. */

/* Let us give time for the child to insert the coins in 10000 ms (10 s) and mechanical subsystem distribute the child's coins to the Port_1, Port_2 and Port_5. */
/* Wait for STAF_1 and STAF_2 and STAF_5 setting to 1 within a time limit and wait for SemAmtCount > = SemChocoCostVal*/
j = 0;
while (j < 2 | | (STAF_1 != 1 | | STAF_2 != 1 | | _STAF_5_ != 1) && SemAmtCount < SemChocoCostVal) /* Wait for 10 s in first cycle and 10 s in second cycle. */
OSTimeDly (10000); *port_1data = 0; *port_2data = 0; *port_2data = 0;
/* Refer to Example 4.5 Wait till an interrupt occurs and sets the STAF flags at Re. 1 and Rs. 2 and Rs. 5 ports. */
28. /* Execute Port_1 interrupt service routine and find the amount received at the Port_1. */
if (STAF_1 = = 1){Port_1_ISR_Input (&port_1data);}; /* Remember as soon as Port_1 is read STAF_1 will reset itself to reflect next interrupt status. */
/* Count the number of points having Rs. 1 coin. */
for (i = 0; i < 8; i++) {
If (*port_1data & 0x01) {SemAmtCount ++; *port_1data >> SemChocoCostVal; i++;};
STAF_1 = 0;}; /* Reset again Port_1 for next interrupt. */
29. /* Execute Port_2 interrupt service routine and find the amount received at the Port_2. */
if (STAF_2 = = 1){Port_2_ISR_Input (&port_2data);}; /* Remember as soon as Port_2 is read STAF_2 will reset itself to reflect next interrupt status. */
/* Count the number of points having Rs. 2 coin. */
for (i = 0; i < 8; i++) {
If (*port_2data & 0x01) {SemAmtCount = SemAmtCount +2; port_2data >> SemChocoCostVal; i++;};
STAF_2 = 0;}; /* Reset again Port_2 for next interrupt. */
30. /* Execute Port_5 interrupt service routine and find the amount received at the Port_5. */
if (STAF_5 = = 1){Port_5_ISR_Input (&port_5data);}; /* Remember as soon as Port_5 is read STAF_5 will reset itself to reflect next interrupt status. */
/* Count the number of points having Rs. 5 coin. */
for (i = 0; i < 8; i++) {
If (*port_5data & 0x01) {SemAmtCount = SemAmtCount +5; port_5data >> SemChocoCostVal; i++;};
STAF_5 = 0; };/* Reset again Port_5 for next interrupt. */
j++;
};
31. /* Posts the semaphore flags for the actions as per the amount */
if (SemAmtCount = = SemChocoCostVal) {OSSemPost (SemFlag1);}; /* Post semaphore to a task waiting Task_Collect. */
if (SemAmtCount < SemChocoCostVal) {OSSemPost (SemFlag2);}; /* Post semaphore to a task waiting Task_Refund. */
if (SemAmtCount > SemChocoCostVal) {OSSemPost (SemFlag3);}; /* Post semaphore to a task waiting Task_ExcessRefund. */
OSSemPost (SemAmtCount); /* Post counting semaphore for three waiting tasks. */
OSSemPost (SemMKey1); /* Release the resource key used for above critical section. */
enable_Port_1_Intr ( ); /* Prepare for another interrupt from port_1. */
enable_Port_2_Intr ( ); /* Prepare for another interrupt from port_2. */
enable_Port_5_Intr ( ); /* Prepare for another interrupt from port_5. */

OSTimeDly (5000); /* Let a low priority task execute as the next coin insertions will take time. The time 5 s is expected for the mechanical subsystems to deliver or refund or refund excess. */
32. }; /* End of while loop*/
}/ * **End of the Task_ReadPorts function** */
/
/***/
33. /* **Start of Task_Collect codes** */
static void Task_Collect (*void *taskPointer*) {
34. /* Initial Assignments of the variables and pre-infinite loop statements that execute once only*/
.

.
35. *while (1) {* /* Start an infinite while-loop /
/* Take the SemFlag1, SemMKey1 and accept *SemAmtCount*. */
OSSemPend (SemFlag1, 0, & *SemErrPointer*);
36. /* Post two mails to waiting Task_Display through two Message pointers for first line and second line of LCD matrix at Port_Display. */
OSMboxPost (MboxStr1Msg, "Wait for a moment"); OSMboxPost (MboxStr2Msg, "*Collect a chocolate soon*");
37. /* Codes for device drivers for Port_Collect to collect the coins from Port_1, Port_2 and Port_5. */
.

.
}
OSTimeDly (3000); /* Let task Deliver lower in priority deliver in 3 s, expected time for mechanical subsystem to deliver. */
38. }; /* End of while loop*/
39. } / * **End of the Task_Collect function** */
/
/***/
40. /* The **codes for the Task_Deliver** */
static void Task_Deliver (*void *taskPointer*) {
41. /* The initial assignments of the variables and pre-infinite loop statements that execute once only*/
.

.
42. *while (1)* { /* Start an infinite while-loop. */
43. /* Wait for flag SemDeliver from Task_Collect */
OSSemPend (SemDeliver, 0, & *SemErrPointer*);
44. /* Codes for device driver for Port_Deliver for delivering a chocolate into a bowl. */
.

.
45. /* Post two mails to waiting Task_Display through two message pointers for first line and second line of LCD matrix at Port_Display. */
OSMboxPost (MboxStr1Msg, "*Collect the chocolate*"); OSMboxPost (MboxStr2Msg, "*Insert coins for more*");
OSSemPend (SemMKey1, 0, & *SemErrPointer*);
OSSemAccept (SemAmtCount, 0, & *SemErrPointer*);

46. /* Reset SemAmtCount. Exit critical section*/
SemAmtCount =0; OSSemPost (SemMKey1);
47. /* Let delayed higher priority task err resume. */
OSTimeDlyResume (Task_CollectPriority);
OSTimeDlyResume (Task_ReadPortsPriority)
48./* Post semaphore to flag that the chocolate delivery is over. */
OSSemPost (SemFinish);
}; /* End of while loop*/
49. }/ * **End of the Task_Deliver function** */
/
**/
50. /* **Start of Task_Refund codes** */
static void Task_Refund (*void *taskPointer*) {
/* Initial assignments of the variables and pre-infinite loop statements that execute once only*/
.

.

51. *while (1)* { /* Start the infinite loop */
OSSemPend (SemFlag2, 0, & *SemErrPointer*);
52. /* Code for the device driver to let Port_Exit release the coins as refund as within two cycles of timeouts, if child fails to insert the required amount of coins. */
.

.

53. /* Post two mails to waiting Task_Display through two message pointers for first line and second line of LCD matrix at Port_Display. */
OSMboxPost (MboxStr1Msg, *"Collect the chocolate"*); OSMboxPost (MboxStr2Msg, *"Insert coins for more"*);
OSSemPend (SemMKey1, 0, & *SemErrPointer*);
OSSemAccept (SemAmtCount, 0, & *SemErrPointer*);
54. /* Reset SemAmtCount. Exit critical section*/
SemAmtCount =0; OSSemPost (SemMKey1);
OSSemPost (*SemFinish*); /* Return to Task_ReadPorts. */
55. /* Let delayed higher priority task err resume. */
OSTimeDlyResume (Task_ReadPortsPriority) /* Resume Delayed Task_ReadPorts. */
}; /* End of while loop*/
56. } / * **End of the Task_Refund function** */
/
**/
57. /* **Start of Task_ExcessRefund codes** */
static void Task_ExcessRefund (*void *taskPointer*) {
58. /* Initial assignments of the variables assignments and pre-infinite loop statements that execute once only*/
.

.

59. *while (1)* { /* Start the infinite loop */
OSSemPend (SemFlag3, 0, & *SemErrPointer*);
OSSemPend (SemMKey1, 0, & *SemErrPointer*);
OSSemAccept (SemAmtCount, 0, & *SemErrPointer*);
60. /* Reset SemAmtCount. Exit critical section*/

SemAmtCount =SemAmtCount −8; OSSemPost (SemMKey1);
61. /* Code for the device driver to let Port_ExcessRefund release the coins from a channel as per the SemAmtCount value now. AVCS thus refund the coins within two cycles of timeouts, if child fails to insert the coins of the required amount. */
.
.

62. /* Post two mails to waiting Task_Display through two message-pointers for the fist line and second line of LCD matrix at Port_Display. */
OSMboxPost (MboxStr1Msg, *"Collect the chocolate"*); OSMboxPost (MboxStr2Msg, "'*Insert coins for more*");
63. /* Also deliver the chocolate now after refunding the excess amount. */
OSSemPost (SemFlag1); /* Run Task_Collect, after Task_Deliver run. */
}; /* End of while loop*/
64. }/ * **End of the Task_ExcessRefund function** */
/
***/
65. /* The **codes for the Task_Display** */
static void Task_CharCheck *(void *taskPointer)* {
66. /* Declare string variables for the three lines and other initial assignments and pre-infinite loop statements that execute once only. */
unsigned char Str1 [ ];
unsigned char Str2 [ ];
unsigned char Str3 [ ];
unsigned char Str3Right [ ]; /* A variable to display at the right corner of line 3 of LCD Matrix */
.
.

67. /* Start an infinite while-loop. */
while (1) {
OSSemPend (SemFlag4, 0, *& SemErrPointer*); /* Wait for the flag for Display */
OSSemPend (SemMKey2, 0, *& SemErrPointer*); /* Take access to Display resources. */
/* Wait for Messages for line 1, line 2 and line 3. */
Str1= OSMboxAccept (MboxStr1Msg, 0, & MboxErrPointer);
Str2 = OSMboxAccept (MboxStr1Msg, 0, & MboxErrPointer);
Str3 = OSMboxAccept (MboxStr3Msg, 0, & MboxErrPointer);
Str3Right = OSMboxAccept (MboxTimeDateStrMsg, 0, & mboxErrPointer);
68. /* Device driver Codes for sending the four strings to a byte stream from line 0 character first to last character line through Port_Display. For using the queues for byte stream, refer to the codes in Example 8.20. */
.
.

OSSemPost (SemMKey2);
69. }/ * End of While loop
}; /* End **codes for the Task_Display** */
/
***/
70. /* The **codes for the Task_TimeDateDisplay** */
static void Task_TimeDateDisplay *(void *taskPointer)* {

71. /* Initial assignments of the variables and pre-infinite loop statements that execute once only. */
unsigned char *timeDate;
.
.
.
72. /* Start an infinite while-loop. */
while (1) {
73. /* Codes for creating a message for time and date after each 1000th interrupt from the system RTC tick. */
.
.
.
74. /* Post time and date to Task_Display.*/
OSMboxPost (MboxTimeDateStrMsg, timeDate);
75. }/* End of Codes for Task_TimeDateDisplay */
/
**/

11.2   CASE STUDY OF CODING FOR SENDING APPLICATION LAYER BYTE STREAMS ON A TCP/IP NETWORK USING RTOS VxWORKS

Embedded systems are one of the most used systems today in the area of Data Communication and Networks. RTOS is used extensively in embedded software. We describe a case study using VxWorks RTOS for such an application. When explaining the following case study, it is presumed that the reader is familiar with network basics, particularly TCP/IP and other important Internet protocols. [The reader may refer to the book *Internet and Web Technologies* by Raj Kamal from Tata McGraw-Hill. It describes bit-wise formats of TCP segment, UDP datagram, IP packet, and SLIP and Ethernet frame].

A communication may be within the same system or between remote systems connected through the network drivers. A communication may be from peer-to-peer or between client and server. Figure 8.8 shows the functions of two sockets communicating through a stack. Socket A is usually a client socket and socket B is a server socket on a network. Each network socket is identified by a distinct pair of parameters, IP address and port number. Inter-socket communication between two applications is exactly the same, whatever may be the application and OS used.

VxWorks provides the APIs (Application Interfaces) for networking. It has a library, *sockLib* for socket programming. [Section 8.3.5]. Since our objective is to learn the use of IPCs in multitasking RTOS through a case study, these APIs and *sockLib* are not used here for creating a TCP stack. [The reader may refer to VxWorks Network Programmer's Guide when using these APIs and *sockLib*. The guide's source is Wind River Systems (www.wrs.com)].

Assume a case of an embedded system in which the RTOS tasks communicate the TCP stacks from an application. An example of the application is HTTP or FTP. The latter communicates to a Web server and transfers a file, respectively. A TCP/IP stack is a stack containing the frames communicated on the network using TCP/IP suite of protocols. The application-layer byte streams are formatted at the successive layers to obtain the final stack for the network. A stream format is as per protocol specifications for in-between layers. A TCP stack typically has many frames into which a network driver writes the bytes. Example 11.2 in Section 11.2.3 shows the application of RTOS for writing into the stack. We use VxWorks functions described in Section 10.3 for the present case. At another node (end point of the network), the priorities of the tasks that receive the stack to retrieve the bytes put in by the application will obviously be in reverse order. Coding for that is left as an exercise for the reader.

The advantages of multitasking should become obvious by this example. An application may not output the strings continuously. The tasks at the other layers can therefore process concurrently during the intermediate periods. How do the semaphores that are used as a mutex guard the shared data problems when using the global variables for the I/O streams? How do the semaphores, used as event flags, provide an efficient way of synchronizing the tasks of various priorities? The use of VxWorks or any other tested RTOS is shown to simplify the coding as scheduling and IPC functions are now readily available at the RTOS.

11.2.1 Case Definition, Multiple Tasks and Their Functions

Figure 11.3(a) shows a sub-system for application, which is transmitting a TCP/IP stack. Figure 11.3(b) shows the scheduling sequence of the tasks during a TCP/IP stack transmission.

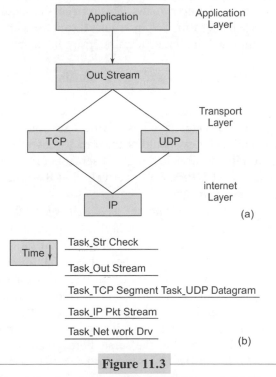

───── **Figure 11.3** ─────

(a) Subsystem for application transmitting a TCP/IP stack (b) The tasks and their scheduling sequence during a TCP/IP stack transmission.

Figure 11.4 details the model to show how the tasks can be synchronised. Top layer is the 'application' layer in a TCP/IP network.

Step **A**: Let a task, **Task_StrCheck** *check* for availability of a string at output from an application. This is the highest priority task during sending to the net. If *check* shows string availability through a status flag semaphore, *STAF*, the task gives (posts) a semaphore to a waiting task, **Task_OutStream**. Let the maximum size of the stream be *maxSizeOutStream*.

Step **B**: The waiting task unblocks on taking the semaphore. It then reads a string and sends a byte stream into a message queue. Let the SemFlag2 be the semaphore taken before sending the string from application into the queue for the buffer, *OutStream*. Let the semaphore given by Task_StrCheck and taken by Task_OutStream be *SemFlag1*. Let the second task use the message queue, *OutStream* for sending the strings from the application to the next waiting task. Let the OutStream give bytes to the message-queue identified by *QStreamInputID*.

Step **C**: Next to the application layer, there is a transmission control layer in the network. It is the equivalent of *transport layer* in the OSI model. The task is next in priority. The application task posts two other semaphore flags, besides *STAF*. One flag is to unblock a waiting task, **Task_TCPSegment**. It takes the semaphore and unblocks, if this layer protocol is TCP. The other semaphore flag is to unblock another waiting task, **Task_UDPDatagram**. It unblocks when this layer protocol is UDP.

Step **D**: Appropriate headers must be inserted at the front of the stream at the *transport layer*, and the stream formats into either a TCP segment or UDP datagram, depending on which task unblocks (undergoes to running state), Task_TCPSegment or Task_UDPDatagram. The task puts the *blocks,* each of 256 bytes, *blkSize* = 256 bytes, into a queue. This is necessary because if bytes are put into the queue, a time called context-switching time for inter-task through RTOS will be added to the over-heads. The overall execution time will increase. Thus interrupt latencies also increase. [Refer to Section 4.6].

Step **E**: A TCP segment or UDP datagram is too long compared to a *block*. The block size has to be optimised later, during a simulation run of the codes. A bigger block size lets another task wait until a block is ready for sending. A smaller block size, on the other hand, increases the task-switching overheads and the interrupt latencies. This task has a critical section. Header bytes are inserted into this section at the queue front and no bytes should be inserted by any other task at this stage. A mutex, *SemMKey1,* protects the section by resource locking. Let semaphores taken by TCP task and taken by UDP task be *SemTCPFlag* and *SemUDPFlag*, respectively. Let the task give the blocks to the message queue, identified by *QStreamInputID*.

Step **F**: Next to the transport layer is the 'internet' layer. [It is a small 'i', not a capital, in internet. We are referring to inter-networking, not the Internet.] It is the equivalent of the network layer in the OSI model. A task, **Task_IPPktStream,** waits and unblocks on the availability of a block from Task_TCPSegment or Task_UDPDatagram. Each packet has a maximum size of 2^{16} bytes. This task is next in priority. **Task_IPPktStream** is for forming an IP packet for transmission on the network through the network driver. This task inserts (writes) an IP header into a socket stream, *SocketStream,* to a network socket. The task inserts the header into the stream after the blocks received from the upper layer stack together in the packet. This task also sends a semaphore flag when the packet is ready for delivery to the network driver. Let the semaphore given by the task and taken later by Task_NetworkDrv be *SemPktFlag*. Let the task send the packets to a message queue, identified by *QPktInputID*.

Step **G**: Next to the internet layer is the 'network interface' layer. It is the equivalent of data-link layer in the OSI model. Network driver task, **Task_NetworkDrv,** waits for the packet ready flag, *SemPktFlag*. When the task is unblocking, it reads *SocketStream* and writes the frame, *frame* into a pipe, *pipeNetStream*. The pipe stores and transmits the bytes at the frame header, *frameHeader*, at the *SocketStream* data or its fragment and at the trailing bytes, *trailBytes*. The latter are usually for error control functions or frame terminal (end) functions. Let the task give the *frame,* one by one, to a pipe identified by *pipeNetStream*. Let the semaphore given by the task and taken later by the application task be *SemFinishFlag,* when no more bytes are available for transmission. A semaphore *SemFlag2* is given by the task when a block of byte is ready. This lets the next layer task unblock outStream and initiate the formation of packets whenever the RTOS schedules it next.

SLIP, PPP, Ethernet, and Token ring are examples of interface layer protocols. The *frameHeader* and *trailBytes* are as per the protocol used by the driver. Certain other protocols do not write any *frameHeader*, for example SLIP. Certain protocols do not write *trailBytes*.

Task_NetworkDrv opens a configuration file. The configuration file is a virtual file device in the embedded system. [Virtual file-device means a file not on the disk but in the memory and using similar open, read, write and close functions as for the files on the disk]. The file points to the configuration information. The function creat ( ) [Note: missing 'e' in creat function] and remove ( ) are used to create and remove a file device in VxWorks. VxWorks open, read and write functions are used as shown earlier in Example 8.26. The use of the select ( ) and close ( ) functions for a file is analogous to that for a pipe [Section 10.3.4(12)].

Exemplary configuration information in the file can be as follows: We can define, in the case of a serial link, the following parameters:
 (i) Protocol for the link (SLIP or PPP)
 (ii) Host
 (iii) Port
 (iv) Baud rate
 (v) Number of bits per character, for example 8
 (vi) Number of stop bits, for example 1.

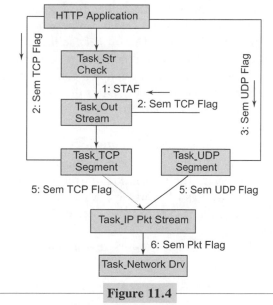

Figure 11.4

Model for how the tasks are being synchronised.

A file can describe configuration as follows, in the case of an Ethernet card used as a network driver. The first line format may begin with *net*. This specifies that the line be for specifying the network card addresses. The next three words are then '*Ethernet* 3COM 0xXYZ'. This specifies that the network is Ethernet. The card is made by 3COM. The card address at the system is hexadecimal XYZ. The next line format may begin with IP. This specifies that the IP address should be defined at the line as next word. The address is of the node that connects to the network. If another connection exists with the same network driver, there may be another line to assign another IP address. [An IP address is of 4 bytes. It is 0xFFFFFFFF for a broadcasting connection to all nodes. It can also be conventionally written as 255.255.255.255].

11.2.2   Creating a List of Tasks, Functions and IPCs

A programmer designing the codes first prepares a list of tasks, the task priorities, the layers at which the tasks function, actions by the tasks, IPCs for which each task or section waits (takes) before unblocking, IPC which it gives (posts) to let another waiting section or task unblock and the output stream, which the task sends as output. Then the coding is done. Table 11.2 lists these. The following points are to be noted from the table.

(1) Priorities have to be above 100 and in layer-wise sequence.
(2) Let us set a convention. An IPC starting with initial characters as 'Sem' means binary semaphore and 'SemM' means mutex. An IPC for flagging an event will have four characters, 'Flag'. An IPC for resource locking has three characters, 'Key'. A message queue identifier begins with character, 'Q'. A pipe identifier begins with four characters 'pipe'.
(3) Either *SemTCPFlag* or *SemUDPFlag* is given by the application task. It is as per the protocol to be employed at the transmission control layer. The IPC *SemFinishFlag* is to be taken by the application or any other task. It flags the intimation that the task of message strings put into the pipe by the network driver is finished. It is an acknowledgment. Another IPC SemFlag2 is to be taken by OutStream as an acknowledgment that the stream sent to the driver has been successfully put into the pipe and the new stream can be sent in the output.
(4) A TCP header differs from UDP header. The former facilitates the connection establishment, network flow control and management, and connection termination by proper exchanges of parameters through the network. TCP is thus called a connection-oriented protocol. The latter is a simple datagram message to the receiver. UDP is thus a connection-less protocol. The task priorities in both cases are assigned equal. Any of the flag, which specifies protocol, can be given (posted) by the application.
(5) A datagram is an independent (unconnected to previous or next) message stream of maximum size 2^{16} bytes. UDP conveys only the source and destination port numbers, stream length and checksum to the 'internet' layer. The socket stream (output of this layer) has the source and destination port numbers as well as IP addresses. A packet from a socket is a message stream of maximum size 2^{16} bytes. Each packet has an inserted IP header. Socket identifies by its IP address and port number.
(6) Network driver forms the frame as per network-driver configuration.

Table 11.2

List of Tasks, Functions and IPCs used in Example 11.2

Task Function	Priority	Layer in TCP/IP@	Action	Taken IPC	Posted IPC	Output Byte Stream
Task_Str-Check	120	Application	Get a string from the application	-	SemFlag1	None
Task_Out-Stream	121	Application	Read string and put bytes into an output stream	SemFlag1, SemFlag2	*QStream-InputID*	Out-Stream
Task_TCP-Segment	122	Transmission Control (Transport)	Insert TCP header to the stream	*SemTCP-Flag, SemM-Key1, QStream-InputID*	*QStream InputIDand SemMKey1,*	Out-Stream
Task_UDP Datagram	122	Transmission control (Transport)	Insert UDP header to the stream sent as a datagram	*SemUDP-Flag, SemMKey1 QStream-InputID*	*QStreamInput ID, SemM -Key1,*	Out-Stream
Task_IPPkt Stream	123	internet (Network)	Form the packets of 2^{16} bytes	*SemMKey1, QStream InputID,*	*SemMKey1, SemPktFlag, QPktInputID*	*socket-Stream*
Task_Net-workDrv	124	Network Interface (Data-link)	Send the packets as the frames	*SemPktFlag, QPktInputID, SemMKey1,*	SemFinishFlag SemFlag2	*pipeNet Stream*

Note @Corresponding layer name in OSI model is given in the bracket.

11.2.3 Exemplary Coding Steps

Example 11.2

1. # include "vxWorks.h" /* Include VxWorks functions. */
include "semLib.h" /* Include semaphore functions library. */
include "taskLib.h" /* Include multitasking functions library. */
include "msgQLib.h" /* Include message queue functions library. */
include "fioLib.h" /* Include file-device input-output functions library. */
include "sysLib.c" /* Include system library for system functions. */
pipeDrv ( ); /* Install a pipe driver. */
include "*netDrvConfig*.txt" /* Include network driver configuration file for frame formatting protocol (SLIP, PPP, Ethernet) description, card description/make, address at the system, IP addresss of the nodes that drive the card for transmitting or receiving from the network. */

include "***prctlHandlers**.c*" /* Include file for the codes for handling and actions as per the protocols used for driving streams to the network. */

2. sysClkRateSet (1000); /* Set system clock rate 1000 ticks per second. */

3. /* Initialise the socket parameters and other network parameters initial values*/

/* *SourcePort* means source port number of the application used. *DestnPort* means destination port number. Define a variable string, *Str*. */

unsigned short *SourcePort*; unsigned short *DestnPort*; unsigned char [ ] *Str*;

unsigned short *SourcePort* = ;

unsigned short *DestnPort* = ;

unsigned short SourceIPAddr = ;

unsigned short DestnIPAddr = ;

4. /* Declare data types of Output Byte Streams for arguments in the tasks. */

unsigned char [ ] applStr, OutStream, socketStream, pipeNetStream;

5. /* Declare data types of the maximum sizes of streams from and to the tasks. Declare data type of block size, blkSize. It is the number of bytes that must be available first before sending an IPC to a buffering stream. It avoids repeated switching from one task to the next after each byte. */

unsigned int blkSize, strSize, strSize, maxSizeOutStream, maxSizeSocketStream, maxSizepipeNetStream;

6. /* Allocate Default Values to various sizes*/

maxSizeOutStream = 1024 * 1024; /* Let application put 1 MB on the net. */

unsigned int strSize = 1; /* Let default string size from an application be 1 byte. */

blkSize = 256; /* Let default block be 256 bytes. */

maxSizeSocketStream = 64 * 1024; /* Let Socket put 64 kB packet on the net. */

maxSizepipeNetStream = 16 * maxSizeSocketStream; /* Let network driver put 16 packets = 1 MB maximum number of bytes. */

7. /* Declare all Table 11.2 ***Task function prototypes***. */

void Task_StrCheck (SemID SemFlag1); /*Task check for the string Availability. */

void Task_OutStream (SEM_ID *SemFlag1*, SEM_ID *SemFlag2*, MSG_Q_ID QStreamInputID);

void Task_TCPSegment (SEM_ID SemTCPFlag, SEM_ID SemMKey1, MSG_Q_ID QStreamInputID);

void Task_UDPDatagram (SEM_ID SemUDPFlag, SEM_ID SemMKey1, MSG_Q_ID QStreamInputID, OutStream, SourcePort, DestnPort, maxSizeOutStream, blkSize);

void Task_IPPktStream (SEM_ID SemMKey1, SEM_ID SemPktFlag, MSG_Q_ID QPktInputID);

void Task_NetworkDrv (SEM_ID SemMKey1, SEM_ID SemPktFlag, SEM_ID SemFinishFlag, SEM_ID SemFlag2, MSG_Q_ID QPktInputID socketStream, pipeNetStream, blkSize, maxSizeSocketStream, maxSizepipeNetStream, MSG_FRAME *aFrame*);

8. /* Declare all Table 11.2 ***Task IDs, Priorities, Options and Stacksize***. Let initial ID until spawned be none. No options and stacksize = 4096 for each of six tasks. */

int Task_StrCheckID = ERROR; int Task_StrCheckPriority = 120; int Task_StrCheckOptions = 0; int Task_StrCheckStackSize = 4096;

int Task_OutStreamID = ERROR; int Task_OutStreamPriority = 121; int Task_OutStreamOptions = 0; int Task_OutStreamStackSize = 4096;

int Task_TCPSegmentID = ERROR; int Task_TCPSegmentPriority = 122; int Task_TCPSegmentOptions = 0; int Task_TCPSegmentStackSize = 4096;

int Task_UDPDatagramID = ERROR; int Task_UDPDatagramPriority = 122; int Task_UDPDatagramOptions = 0; int Task_UDPDatagramStackSize = 4096;

int Task_IPPktStreamID = ERROR; int Task_IPPktStreamPriority = 123; int Task_IPPktStreamOptions = 0; int Task_IPPktStreamStackSize = 4096;

int Task_NetworkDrvID = ERROR; int Task_NetworkDrvPriority = 124; int Task_NetworkDrvOptions = 0; int Task_NetworkDrvStackSize = 4096;

9. /* ***Create and Initiate (Spawn) all the six tasks*** of Table 11.2. */

Task_StrCheckID = taskSpawn (" tTask_StrCheck", Task_StrCheckPriority, Task_StrCheckOptions,
Task_StrCheckStackSize, void (*Task_StrCheck) (SEM_ID STAF, SEM_ID SemFlag1), 0, 0, 0, 0, 0, 0, 0, 0, 0, 0);

Task_OutStreamID = taskSpawn (" tTask_OutStream", Task_OutStreamPriority, Task_OutStreamOptions,
Task_OutStreamStackSize, void (*Task_OutStream) (SEM_ID *SemFlag1*, MSG_Q_ID QStreamInputID, applStr, OutStream, maxSizeOutStream, blkSize), 0, 0, 0, 0, 0, 0, 0, 0, 0, 0);

Task_TCPSegmentID = taskSpawn (" tTask_TCPSegment", Task_TCPSegmentPriority, Task_TCPSegmentOptions, Task_TCPSegmentStackSize, void (*Task_TCPSegment) (SEM_ID SemTCPFlag, SEM_ID SemMKey1, MSG_Q_ID QStreamInputID, unsigned char [ ] OutStream, int maxSizeOutStream, int blkSize, unsigned char [16] *txtcpState*, unsigned char [16] *txtcpOpPdFormat*) unsigned short *SourcePort*, unsigned *DestnPort*, unsigned int *SequenNum*, unsigned int *AckNum*, unsigned char *TCPHdrLen*, *unsigned *TCPHdrFlags*, unsigned short *TCPChecksum16*, unsigned short *window*, unsigned short *UrgPtr*, unsigned char [optPdLen] *extras*)

Task_UDPDatagramID = taskSpawn (" tTask_UDPDatagram", Task_UDPDatagramPriority, Task_UDPDatagramOptions, Task_UDPDatagramStackSize, void (*Task_UDPDatagram) (SEM_ID SemUDPFlag, SEM_ID SemMKey1, MSG_Q_ID QStreamInputID, unsigned char [ ] OutStream, int maxSizeOutStream, int blkSize), SourcePort, DestnPort, 0, 0, 0, 0, 0, 0, 0, 0);

Task_IPPktStreamID = taskSpawn ("tTask_IPPktStream", Task_IPPktStreamPriority, Task_IPPktStreamOptions, Task_IPPktStreamStackSize, void (*Task_IPPktStream) (SEM_ID SemPktFlag, MSG_Q_ID QPktInputID, unsigned char [ ] OutStream, int maxSizeOutStream, blkSize, unsigned char [ ] SocketStream, maxSizeSocketStream), 0, 0, 0, 0, 0, 0, 0, 0, 0, 0);

Task_NetworkDrvID = taskSpawn ("tTask_NetworkDrv", Task_NetworkDrvPriority, Task_NetworkDrvOptions, Task_NetworkDrvStackSize, void (* Task_NetworkDrv) (SEM_ID SemMKey1, SEM_ID SemPktFlag, SEM_ID SemFinishFlag, SEM_ID SemFlag2, MSG_Q_ID QPktInputID socketStream, unsigned char [ ] pipeNetStream, int blkSize, int maxSizeSocketStream, int maxSizepipeNetStream, unsigned char [ ] frameHeader, unsigned char [ ] trailBytes), 0, 0, 0, 0, 0, 0, 0, 0, 0, 0);

10. /* Declare IDs and create the binary semaphore flags, keys and message queues. */
SEM_ID SemTCPFlag, SemUDPFlag; /* Declared at the application */
SemTCPFlag = **semBCreate** (SEM_Q_FIFO, SEM_EMPTY); /* Declared at the application */
SemUDPFlag = **semBCreate** (SEM_Q_FIFO, SEM_EMPTY); /* Declared at the application */
/* Note: The application posts wither SemUDPFlag or SemTCPFlag to let one of the two tasks run. [Task_TCPSegment or Task_UDPDatagram. */
SEM_ID *SemFlag1*, SemFlag2, *SemPktFlag*, SemFinishFlag, SemMKey1; /* Declared for the six tasks listed in Table 11.2. */

11. /* Create the binary semaphores, message queue for a stream of bytes from an application and pass the options selected to it. */
SemFlag1 = **semBCreate** (SEM_Q_PRIORITY, SEM_EMPTY); / *Task higher in priority takes it first. */
SemFlag2 = **semBCreate** (SEM_Q_PRIORITY, SEM_EMPTY); / *Task higher in priority takes it first. */
SemPktFlag = **semBCreate** (SEM_Q_PRIORITY, SEM_EMPTY); /*Task higher in priority takes it first.*/
SemMKey1 = **semBCreate** (SEM_Q_FIFO, SEM_EMPTY); / *Taken in FIFO */

SemFinishFlag = **semBCreate** (SEM_Q_FIFO, SEM_EMPTY); / *Taken in FIFO */
MSG_Q_ID   *QStreamInputID*;
void char [ ] msgStream; /* Pointer for the message buffer */
12. /* Create the message queue identity and pass the parameters and chosen options to it. Let maximum number of messages be 256 kB and the message be of 1 byte each. An IP packet has a maximum 2^{16} bytes. Assume that a TCP segment stream for transmitting has maximum of 256 kB. Let 64 bytes be additionally assigned for the headers. Header bytes add at the lower layers (refer column 3 in Table 11.2). */
QStreamInputID = **msgQCreate** (maxSizeOutStream, strSize, MSG_Q_FIFO I MSG_PRI_NORMAL);
QPktInputID = **msgQCreate** (maxSizeSocketStream, strSize, MSG_Q_FIFO I MSG_PRI_NORMAL);
13. /* Steps 1 to 3 as per Example 8.26 for creating a pipe. */
.

.

14. /* Create pipe named as *pipeNetStream* for including overheads and frame messages, each if 1 byte. Overheads mean header bytes as well as trailing bytes. */
STATUS   pipestatus;
pipestatus = pipeDevCreate ("/pipe/*pipeNetStream*", maxSizepipeNetStream, 1);
mode = 0 = 0x0;
15. /* Other declarations that are needed. */
.

.

16. /* Declare common functions needed in the networking tasks. */
/* Declare the functions to get 32-bit, 16-bit and 8-bit lengths from a stream or string. */
unsigned int **getLength32** (unsigned char [ ] **Str**) {
/* Codes for finding as an unsigned integer the length of a string or stream up to 2^{32} bytes * /
.

.

};
unsigned short **getLength16** (unsigned char [ ] **Str**) {
/* Codes for finding, as an unsigned short, the length of a string or stream up to 2^{16} bytes. * /
.

.

};
unsigned byte **getLength8** (unsigned char [ ] **Str**) {
/* Codes for finding, as an unsigned byte, the length of a string up to 256 bytes. * /
.

.

};
17. /* Declare Codes for adding extra padding in shorter size message or string or stream to fill 0s so that string size now equals block size, *blkSize*. */
unsigned char [ ] **StrAddPadding** (unsigned char [ ] **Str**, unsigned int *blkSize*) {
.

.

};
18. /* Exemplary codes for finding a function to get 16-bit checksum from a byte stream or string. */
unsigned short **checksum16 (unsigned char [ ] Str)** {
/* Codes for finding, as an unsigned integer, the length of a byte stream or string* /

/* StrPtr is pointer to the string or stream. For the while loop, 'i' is a 16-bit integer. Number of carries generated = *numCarry*. */
static unsigned short *strPtr, i, numCarry;
static unsigned int sum = 0;
unsigned short num = (unsigned short) getLength32 (unsigned char [ ] Str); /* num is 16-bit number for total number of bytes in the string. We are typecasting the length data type to 16-bit */
/* A 16-bit Checksum method is as follows: Sum the 16-bit words and first count the carries, which generate on successive additions. Then add these carries to get the 16-bit checksum. However, we are having the byte stream. So method is as follows: */
strPtr = (unsigned short) Str; /* Type cast address to 16-bit value pointer. */
i = num/2; /* Let us split the calculation into two parts because we are adding the bytes instead of 16-bit words to get the checksum*/
while (i − −) {
sum += *strPtr ++; /* add the byte into the sum and then pointer to next byte. */
if (i & 1){sum += *(unsigned char *) strPtr;}; /* Sum the odd byte from next address. */
while ((numCarry = (unsigned short) (sum >>16)) ! = 0) {sum = (sum & 0xFFFF) + numCarry;}
return ((unsigned short) sum); /* Type cast sum to 16- bit. */
};
/* Reader to write 32 bit checksum codes for function **checksum32** by himself (or herself). */
.

.

/* Declare a Function for Cycle Redundancy Check */
unsigned int **CRC32 (unsigned int crc, unsigned char aByte)** {
/* Codes for finding 32-bit cycle redundancy check bits as an unsigned integer for the given message frame. * /
.

.

};
19. /* On a network, while we are sending the bytes a protocol uses the different data types, such as unsigned int, unsigned short, unsigned char, etc. Hence, the definitions of the following six functions to get the lower byte and higher byte from a 16-bit short, data16 and to get four bytes byte0, byte1, byte2 and byte3 from a 32-bit int, data32 is essential. */
unsigned char **LByte** (unsigned short data16) { ... }; unsigned char **HByte** (unsigned short data16) { ... };
unsigned char **byte0** (unsigned int data32) { ... }; unsigned char **byte1** (unsigned int data32) { ... };
unsigned char **byte2** (unsigned int data32) { ... }; unsigned char **byte3** (unsigned int data32) { ... };
20. to 27./* Code for other declaration steps specific to the various networks */
.

.

/* End of the codes for creation of the tasks, semaphores, message queue, pipe tasks, and variables and all needed function declarations */
/
***/

28. /* Start **of Codes for Task_StrCheck**/

void Task_StrCheck (SEM_ID STAF, SEM_ID SemFlag1) {

.

.

29.

while (1) { /* *Start of while loop.* */

/* When character output is generated by the application, the semaphore is given. */

SemTake (STAF, WAIT_FOREVER); /* Wait for Status flag from Application. */

30. /* Codes for the task. *

.

.

semGive (*SemFlag1*);

semGive (SemMKey1); /* Release the mutex if any taken before. */

taskDelay (10); /* Delay 10 ms to let lower priority task run. */

}; /* End of while loop. */

} /* End of Task_StrCheck. */

31. /* Start of Codes for Task_OutStream. */

void **Task_OutStream** (SEM_ID   *SemFlag1*, MSG_Q_ID QStreamInputID,   *applStr*, OutStream, maxSizeOutStream, blkSize){

32. /* Initial assignments of the variables and pre-infinite loop statements that execute once only*/

int numBytes;

.

.

33. *while (1)* { /* Start an infinite while-loop. We can also use FOREVER in place of while (1). */

34. / *Wait for SemFlag1 state change to SEM_FULL by semGive function for string availability check task */

semTake (SemFlag1, WAIT_FOREVER);

35. /* Take the key to not let any application other than the present task put the message into the queue port decipher task that needs SemMKeyID run */

semTake (SemMKey1, WAIT_FOREVER); /* SemMKeyID is now not available and the critical region starts */

36. /* Codes for Encrypting applStr if any or illegal character or message check, if any */

.

.

37. /* At the end, send the byte input at OutStream as message to queue, Q*StreamInputID*. Refer to Section 10.3.4.(10) for understanding the msgQSend function. NO_Wait if string is more than the block size 256 bytes. Otherwise string is too short to deserve sending on the stream without padding it with extra 0s. Call a function to add padding bytes. */

numBytes = getLength32 (applStr);

if (numBytes < blkSize) {StrAddPadding (applStr, blkSize); numBytes = blkSize};

/* Now wait if any of the previous bytes of applStr have not been put into the socket streams. Task_IPPktStream does that. It posts SemFlag2 on successfully sending the applStr into the stream with two transport and internet layer headers. */

semTake (SemFlag2, WAIT_FOREVER);

msgQSend (Q*StreamInputID*, *applStr*, numBytes, NO_WAIT, MSG_PRI_NORMAL);}};

semGive (SemMKey1); /* Critical Region ends here. */}

38. /* Resume the delayed task, Task_StrCheck as message has been put into the message queue. */

taskDelay (20)

taskResume (*Task_StrCheckID*);
. /* Other remaining codes for the task. */
.

.
};
39. } /* End of the codes for Task_OutStream. */
40. /* Start of Codes for Task_TCPSegment. */
void **Task_TCPSegment** (SEM_ID SemTCPFlag, SEM_ID SemMKey1, SEM_ID SemFlag2, MSG_Q_ID QStreamInputID, OutStream, maxSizeOutStream, blkSize, unsigned char [16] *txtcpState*, unsigned char [16] *txtcpOpPdFormat*) {
41. /* Initial assignments of the variables and pre-infinite loop statements that execute once only*/
static int numBytes; /* Number of bytes successfully read. Equals error on timeout or message queue ID not identified. */
static int timeout = 20; /* Let after 20 system clock-ticks 20 ms*/
unsigned char [16] txtcpState;
unsigned char [16] *txtcpOpPdFormat*;
unsigned char [ ] OptionAndPds (unsigned char [16] *txtcpState*, unsigned char [16] *txtcpOpPdFormat*);
unsigned char [optPdLen] *extras*;
unsigned char [ ] *Str,* unsigned int *SequenNum,* unsigned int *AckNum,* unsigned char * *TCPHdrLen, unsigned *TCPHdrFlags,* unsigned short *window,* unsigned short *TCPChecksum16,* unsigned short *UrgPtr,* unsigned char [optPdLen] *extras*);
static unsigned char [ ] *header*;
unsigned char [ ] TCPHeader (unsigned short *SourcePort,* unsigned short *DestnPort,* unsigned char [16] txtcpState, unsigned char [16] *txtcpOpPdFormat,* unsigned char [ ] *Str,* unsigned int *SequenNum,* unsigned int *AckNum,* unsigned char * *TCPHdrLen, unsigned *TCPHdrFlags,* unsigned short *window,* unsigned short *TCPChecksum16,* unsigned short *UrgPtr,* unsigned char [optPdLen] *extras*;

.

.
42. while (1) *{* /* Start task infinite loop. */
/* Take mutex so that till header is inserted into the stream, it is not released to Task_OutStream. */
semTake (*SemTCPFlag,* WAIT_FOREVER);
semTake (SemMKey1, WAIT_FOREVER); /* SemMKeyID is now not available. Wait for entering critical region*/
43. /* Receive the message sent by Task_OutStream */
numBytes = msgQReceive (QStreamInputID, OutStream, maxSizeOutStream, 20);
if (numBytes != ERROR) {
44. /* Code for defining the txtcpState as per the connection status. */
.

.
/* Code for defining the SequenNum as per txtcpState */
.

.
/* Code for defining the AckNum as per txtcpState */
.

```
.
/* Code for defining the window as per txtcpState */
.

.
/* Code for defining the txtcpOpPdFormat as per txtcpState */
.

.
45. /* Codes for finding the additional integers as options and padding to be put as the header. */
extras = OptionAndPds (txtcpState, txtcpOpPdFormat);
46. /* Codes for retrieving the TCP header bytes */
header = TCPHeader (SourcePort, DestnPort, txtcpState, txtcpOpPdFormat, OutStream, SequenNum,
AckNum, * TCPHdrLen, *TCPHdrFlags, window, *TCPChecksum16, *UrgPtr, extras);
47. /* Send header into the front of the queue */
msgQSend (QStreamInputID, header, getLength32 (header), NO_WAIT, MSG_PRI_URGENT);
48. /* Send data to the back of the queue */
msgQSend (QStreamInputID, applStr, numBytes, NO_WAIT, MSG_PRI_NORMAL);
}; /* End of the message handling codes. */
semGive (SemMKey1); /* Critical region ends here. */}
semGive (SemTCPFlag); /* SemTCPFlag. */
taskDelay (20);
taskResume  (Task_OutStream);
}; / * End of while-loop. */
49. } /* End of codes for Task_TCPSegment */
50. /* Start of codes for Task_UDPDatagram */
void Task_UDPDatagram (SEM_ID SemUDPFlag, SEM_ID SemMKey1, SEM_ID SemFlag2,
MSG_Q_ID QStreamInputID, OutStream, maxSizeOutStream, blkSize) {
51. /* Initial assignments of the variables and pre-infinite loop statements that execute once only*/
static int numBytes; /* Number of bytes successfully read. Equals error on timeout or if message queue
ID not identified. */
static int timeout = 20; /* Let after 20 system clock-ticks 20 ms*/
static unsigned char [8] header;
unsigned char [8] UDPHeader (unsigned short SourcePort, unsigned short DestnPort, unsigned char
[ ] OutStream);
.

.
52. while (1) { /* Start task infinite loop. */
/* Take mutex so that till header is inserted into the stream; it is not released to Task_OutStream. */
semTake (SemUDPFlag, WAIT_FOREVER);
semTake (SemMKey1, WAIT_FOREVER); /* SemMKeyID is now not available. Wait for entering the
critical region*/
53. /* Receive the message sent by Task_OutStream */
numBytes = msgQReceive (QStreamInputID, OutStream, maxSizeOutStream, 20);
if (numBytes != ERROR) {
54. /* Codes for retrieving the UDP header bytes */
header = UDPHeader (SourcePort, DestnPort, OutStream);
55. /* Send Header to the front of the queue */
msgQSend (QStreamInputID, header, 8, NO_WAIT, MSG_PRI_URGENT);
56. /* Send Data to the back of the queue */
```

msgQSend (Q*StreamInputID*, *applStr*, numBytes, NO_WAIT, MSG_PRI_NORMAL);

}; /* End of the message handling codes. */

/* Critical Region ends here. */

57. semGive (SemMKey1);

semGive (*SemUDPFlag*); /* *SemUDPFlag* release for use in new application string, applStr. */

58. /* Let lower priority task, Task_IPPktStream start Higher priority one resume. */

taskDelay (20);

taskResume (*Task_OutStream*);

}; / * End of while-loop. */

59. } /* End of codes for Task_UDPDatagram */

60. /* Start of Codes for Task_IPPktStream */

void Task_IPPktStream (SEM_ID SemFlag2, SEM_ID SemPktFlag, MSG_Q_ID QPktInputID, int maxSizeOutStream, int blkSize, unsigned short *SourceAddr*, unsigned short *DestnAddr*, unsigned short IPverHdrPrioSer, unsigned char [16]   *txipState*, unsigned char [16] *txipOpPdFormat*, unsigned char *timeToLive, unsigned char *PrctlField) {

61. /* Initial assignments of the variables and pre-infinite loop statements that execute once only*/

static int numBytes; /* Number of bytes successfully read. Equals error on timeout or if message queue ID not identified. */

static int timeout = 20; /* Let after 20 system clock-ticks 20 ms*/

static unsigned char [8] *header*; unsigned char [ ] Str;

unsigned char [ ] IPHeaderSelPkt (unsigned short *SourceAddr*, unsigned short *DestnAddr*, unsigned short IPverHdrPrioSer, unsigned short *IPPktLen, char * *IPVerHdrLen,* unsigned char * *IPHdrLen, unsigned char *IPHdrFlags*, unsigned short *IPHdrFrag, unsigned short *IPChecksum16*, unsigned short UniqueID, unsigned char *timeToLive, unsigned char *PrctlField, unsigned char [ ] Str, unsigned char [ ] OutStream, char [optPdLen] *IPextras*, unsigned char [16] *txipState*, unsigned char [16] *txipOpPdFormat*);

62. while (1) { /* Start task infinite loop. */

/* Take mutex so that until header is inserted into the stream, it is not released to Task_OutStream. */

semTake (SemMKey1, WAIT_FOREVER); /* SemMKeyID is now not available. Wait for entering critical region*/

63. /* Receive the message sent by Task_OutStream */

numBytes = msgQReceive (QStreamInputID, OutStream, maxSizeOutStream, 20);

if (numBytes != ERROR) {

64. /* Codes for retrieving the IP Packet header bytes */

header = IPHeaderSelPkt (*SourceAddr*, *DestnAddr*, IPverHdrPrioSer, *IPPktLen, * *IPVerHdrLen,* * *IPHdrLen*, *IPHdrFlags*, *IPHdrFrag, *IPChecksum16*, UniqueID, *timeToLive, *PrctlField, Str, OutStream, *IPextras*, txipState, txipOpPdFormat);

65. /* Send Header into the front of the queue */

msgQSend (Q*PktInputID*, *header*, * IPHdrLen, NO_WAIT, MSG_PRI_URGENT);};

66. /* Send data to the back of the queue */

numBytes = *IPPktLen - * IPHdrLen;

msgQSend (Q*PktInputID*, Str, numBytes, NO_WAIT, MSG_PRI_NORMAL);};

/* Further send *header* and *Str* bytes into the queue if more IP packets are to be sent */

.

.

}; /* End of the message handling codes. */

/* Critical Region ends here. */}

67. semGive (SemMKey1);

semGive (*SemPktFlag*); /* *SemPktFlag* release for use in Network Driver Task. */

68. /* Let lower priority task, Task_IPPktStream start Higher priority one resume. */

taskDelay (20);

taskResume (*Task_UDPDatagramID*);

taskResume (*Task_TCPSegmentID*);

}; / * End of while-loop. */

69. } /* End of codes for Task_IPPktStream. */

70. /* Start of codes for Task_NetworkDrv. */

void **Task_NetworkDrv** (SEM_ID *SemMKey1*, SEM_ID *SemPktFlag*, SEM_ID *SemFinishFlag*, SEM_ID *SemFlag2*, MSG_Q_ID QPktInputID socketStream, unsigned char [ ] pipeNetStream, int *blkSize*, int *maxSizeSocketStream*, int *maxSizepipeNetStream*, unsigned char [ ] *frameHeader*, unsigned char [ ] *trailBytes*) {

71. /* Declare data type for specifying the headers. Let header length of frame be *length*; type of network driver be *netDrvType*; and fragment offset at the stream be *frameOffset*, which specifies the index in the byte array from which a frame starts in case the stream is sent in fragments. */

unsigned char [ ] *frame*, frameHeader, *trailBytes*, *fragment*, *trailBytes*;

unsigned short fragOffset;

*unsigned short *FRlength*, * HdrLen, * endLen, *FrameHdrFlags*, *FrameHdrFragOffset, *PrctlField;

unsigned int *FrameCRC32;

unsigned char [ ] Str, unsigned char [ ] SocketStream, unsigned char [ ] *FRextras*, unsigned char [16] *txFRState*, unsigned char [16] *txFROpPdFormat*, unsigned char [12] *netDrvType*

72. /* Declare Network Driver Function. */

void NetHdrFrTr (unsigned char [ ] *frame*, unsigned char [ ] *frameHeader*, unsigned char [ ] *trailBytes*, unsigned short *fragOffset*, unsigned char [ ] *fragment*, unsigned char [ ] *trailBytes,* unsigned short *FRlength*, unsigned short * HdrLen, unsigned short * endLen, unsigned short *FrameHdrFlags*, unsigned short *FrameHdrFragOffset unsigned short *FrameCRC32*, unsigned short *PrctlField, unsigned char [ ] Str, unsigned char [ ] SocketStream, unsigned char [ ] *FRextras*, unsigned char [16] *txFRState*, unsigned char [16] *txFROpPdFormat*, unsigned char [12] *netDrvType*);

73. /* Integers for number of bytes successfully read, message size and file device, respectively. */

int numBytes, lBytes, *fd*;

74. /* Let a pointer *netDrvConfig* define a pointer to a configuration file for a network driver. */

int fd; /* Define an integer for a file device. */

fd = open (*netDrvConfig.txt*, **O_RDWR, 0);** /* Refer Step 5 Example 8.26. */

75. /* Code for reading *netDrvType*, protocol for the link (SLIP or PPP), data link layer format (say Ethernet), host IP, port, baudrate, card specifications, etc. Use lBytes and function, read as follows. */

lBytes =12; /* Let the first message bytes read = 12*/

numBytes = **read** (fd, Str, lBytes);

.

.

close (fd);

76. /* Open the pipe for the Network Stream. Refer Step 5 Example 8.26. */

fd = **open** ("/pipe/*pipeNetStream*", O_RDWR, 0);

77. while (1) {;

/* Task reads *SocketStream* on unblocking and writes the frame, *frame* into a pipe, *pipeNetStream*. The pipe stores and transmits the bytes at the frame header, *frameHeader*, and frame *fragment* from the SocketStream in case *frame* is of smaller size than the SocketStream and at the end trailing bytes, *railBytes*. The latter are usually for error control functions or frame terminal (end) functions. Let the task give the *frame* one after another to a pipe identified by *pipeNetStream*. This lets the next layer task unblock outStream and initiate the formation of packets whenever the RTOS schedules it. */

semTake (*SemPktFlag*, WAIT_FOREVER);

semTake (*SemMKey1*, WAIT_FOREVER);

78. /* Receive the message sent by Task_IPPktStream */

numBytes = msgQReceive (QStreamInputID, SocketStream, maxSizeOutStream, 20);

if (numBytes != ERROR) {

79. /* Codes for retrieving the frame bytes */

NetHdrFrTr (*frame*, *frameHeader*, *trailBytes*, *fragOffset*, fragment, *trailBytes, *FRlength, * HdrLen, *endLen*, *FrameHdrFlags*, *FrameHdrFragOffset, *FrameCRC32*, *PrctlField, Str, SocketStream, FRextras, *txFRState*, *txFROpPdFormat*, *netDrvType*);

/* Write a message, *info* of lBytes. */

unsigned char [ ] info; /* Let the message be a string of characters. */

int *lBytes*;

lBytes =12;

write (fd, frame, *FRlength);

80. /* Further write into the pipe if more frames are to be sent */

.

.

.

; };

semGive (SemMKey1);

semGive (*SemFinishFlag*); /* Post the semaphore for next waiting task at the application layer. */

semGive (SemFlag2); /* Post the semaphore for waiting task for sending an Out Stream. */

taskResume (*Task_IPPktStreamID*); /* Resume the previously delayed higher priority task. */

}; /* End of While-loop */

81. } /* End of codes for **Task_NetworkDrv** */

/* Start of codes for Task_NetworkDrv. */

/   *

/**/

A file ***prcltHandlers.c*** has the codes for UDP, TCP, IP and network protocols for including and handling the headers functions, selecting the fragments bytes when forming IP packets and frame bytes when forming the frames. The codes are designed as follows:

1. /* Declare a Function for returning a UDP header string. *DatagramLen* means length of UDP datagram, which transmits to next layer. *Str* means the stream or string from the application layer to be sent with UDP protocol. */

unsigned char [8] **UDPHeader** (unsigned short *SourcePort*, unsigned short *DestnPort*, unsigned char [ ] *Str*) {

2. /* Codes for returning UDP Header String. Remember that UDP protocol transmits the integers with big endian. Refer to paragraph (1) in the instructions for a hardware designer in Section 2.1. Big endian means the most significant bytes transmit first. * /

unsigned char [8] UDPHStr;

unsigned short DatagramLen = getLength16 (*Str*) + 8; /* Add 8-byte header length also. */

unsigned short *UDPChecksum16* = checksum16 (*Str*);

UDPHStr [0] = HByte (*SourcePort*); UDPHStr [1] = LByte (*SourcePort*);

UDPHStr [2] = HByte (*DestnPort*); UDPHStr [3] = LByte (*DestnPort*);

UDPHStr [4] = HByte (*DatagramLen*); UDPHStr [5] =LByte (*DatagramLen*);

UDPHStr [6] = HByte (*UDPChecksum16*); UDPHStr [7] = LByte (*UDPChecksum16*);

return (UDPHStr);

};

/*   _____

_____/*

3. /* Declarations and codes for the function, TCPHeader for returning a TCP header string. *Str* means the stream from the application layer at a node, *node1* end to be sent with TCP protocol to other end, *node2*. */

/* Define the byte position (index) of the present OutStream bytes from the application. Initial value = 0*/

unsigned int SequenNum = 0;

/* Define the total number of bytes already received at an input stream from *node2* to this *node1*. Input stream is the one that was sent as TCP stacks and received at the *node1* from the receiver *node2*. This is to let an out stream simultaneously convey to the other end an acknowledgement along with a new sequence of bytes. OutStream and input stream synchronize and there is controlled flow of bytes between the two ends, *node1* and *node2*. Initial value = 0. */

unsigned int AckNum =0;

/* Declare 4-bit and 4- bit unused, reserved for future expansion. TCP headers vary between 5 to 15 unsigned integers of 32-bit each. */

unsigned char *__TCPHdrLen__*;

/* Let 16-character string txtcpState specify the state of transmission of the current TCP segment and its action required for the controlled transmission. Action is to be as desired. It may be to establish a connection, for termination or management or flow control. Refer to TCP protocol in any standard text for definitions. */

unsigned char [16] *txtcpState*;

/* Declare TCPHdrFlags for 6 bits for flags and 2 bits unused and reserved for future modifications in protocol. Bit 0 is FIN, bit 1 is SYN, bit 2 is RST, bit 3 is PUSH, bit 4 is ACK and bit 5 is URG. */

unsigned char *__TCPHdrFlags__*;

/* Let another 16-character string txtcpOpPdFormat specify format, which gives the number and meaning of optionally added integers and padded integers. For example, an optional integer may be to specify an alternative window of 32 bits in place of 16 given in the 16-bit window field at fourth integer in the TCP header. Another option integer may specify the TCP maxSizeOutStream. */

unsigned char [16] *txtcpOpPdFormat*;

4. /* Start of the codes for returning a *short*. It specifies the 4-bit TCP header length field as well as six-flag fields. These are as per transmission state and transmitting TCP options and padding format, *txtcpState* and *txtcpOpPdFormat*, respectively */

unsigned short   **TCPHdlenFlagBits**   (unsigned char [16]   *txtcpState*,   unsigned char [16] *txtcpOpPdFormat*, &TCPHdrFlags, &TCPHdrLen) {

Boolean bit 15, bit 14, bit 13, bit 12, bit 11, bit 10, bit 9, bit 8;

5. /* Codes to have four bits, bit 15, bit 14, bit 13, bit 12 in the returned integer as per the TCP header size, which specifies the total number of unsigned integers in the TCP header. The total is 5 plus the unsigned integers for options and padding. These are as per txtcpState and txtcpOpPdFormat used by TCP segment that is transmitting. The bits, bit 11, bit 10, bit 9, bit 8 are reserved. */

/* Bits 0 to 5 are as per six flags. Bits 6 to 11 are reserved. Bits 0 to 7 are at unsigned character pointer TCPHdrFlags. */

*TCPHdrFlags = *TCPFlags (txtcpState); /* Using the function find bits 0 to 7, */

.

.

6. /* Code to find TCPHdrLen from bit 15, bit 14, bit 13, bit 12. */

.

.

}; /* End of the codes for returning an integer that specifies TCP header length and flag fields. */

7. /* Declare a function TCPFlags to return TCPHdrFlags as per *txtcpState*. Start of the codes for finding the byte to represent the TCP flag fields. */

unsigned char *__TCPFlags__ (unsigned char txtcpState) {

Boolean FIN, SYN, RST, PUSH, ACK, URG, bit 6, bit 7;

/* Codes to create as per txtcpState in which TCP segment is transmitting. */

.

.

}; /* End of the codes for returning the pointer for a byte for the TCP flags field. */

8. /* Declare other TCP Header fields. */

unsigned short *window*; /* *node1* specifies by *window* as an advertisement to *node2* how many more bytes at *node1* can be buffered in its buffer beyond the ones already buffered and acknowledged by *node1* to *node2*. Buffering is an intermediate state in which bytes are still to be sent to an upper application layer by a TCP stack receiving entity. *Node2* if finds window = 0 or too small a number, then it should not send any thing to this *node1*. This 16-bit field then is a request to the other end node not to *flood* data bytes on the network unnecessarily. It is then indirectly a request to wait. */

unsigned short *TCPChecksum16*; /* Define 16-bit Checksum. */

/* Define a 16-bit Urgent pointer. */

unsigned short *Urgent*;

/ *A 16-bit value that will indicate an urgency to *node2* whenever the URG flag is set. It indicates the first byte of the start of the urgent data from *node1*. It is a request to the *node2* that it should consider the pointed bytes first in place of attending to the bytes in the buffer and the bytes after this segment header. The buffered data and beginning data may be ignored and bytes beginning from UrgPtr are requested to be given urgency. Given or not depends on *node2* program task at TCP segment that sends the bytes to its application layer. */

unsigned short *urgPtr* = *Urgent*;

unsigned short *__TCPURGENT__ (unsigned char __txtcpState,__ &TCPHdrFlags, *urgPtr) {

Boolean URG;

URG = getFlag (&TCPHdrFlags) {

/ * Codes to extract the URG bit. It is bit 5 of byte at address of TCPHdrFlags.

.

}

/* .If URG is true, then set the urgent field and return a 16-bit short. Codes to return a 16-bit short as urgent pointer as per txtcpState in which TCP segment is transmitting. */

if (URG) {

.

. };

};

unsigned char [ ] *extras*; unsigned char OptPdLen;

extras = OptionAndPds (*txtcpState, txtcpOpPdForma*t);

optPdLen = getLength8 (*extras*);

*TCPHdrLen = (optPdLen + 20) /4 ;

unsigned char [ ] OptionAndPds (unsigned char [16] *txtcpState*, unsigned char [16] *txtcpOpPdFormat*) {

9. /* Codes for returning an array of bytes for the 32-bit integers for the options and padding. These are as per the state and thus TCP header length parameters. */

.

.

};

10. /* Codes for returning TCP Header String. Remember that TCP protocol transmits the integers with big endian. Refer to Section 2.1.2. (i). It means that the most significant byte should transmit first. [Refer to Section 2.1.2. (i) also]. * /

/* Note: Instead of using many arguments, a typedef could have been used for the header data structure. However, this will consume more memory. */

unsigned char [ ] **TCPHeader** (unsigned short *SourcePort*, unsigned short *DestnPort*, unsigned char [16] txtcpState, unsigned char [16] *txtcpOpPdFormat*, unsigned char [ ] *Str*, unsigned int *SequenNum*, unsigned int *AckNum*, unsigned char * *TCPHdrLen*, *unsigned *TCPHdrFlags*, unsigned short *window*, unsigned short *TCPChecksum16*, unsigned short *UrgPtr*, unsigned char [optPdLen] *extras*) {

int i; unsigned char [ ] TCPHStr; unsigned short lenflag; unsigned char optPdLen;

unsigned short *TCPChecksum16* = checksum16 (*Str*); unsigned short *urg; unsigned char [ ] extras;

*urg = *TCPURGENT (unsigned char *txtcpState*, &TCPHdrFlags, *urgPtr)

TCPHStr [0] = HByte (*SourcePort*); TCPHStr [1] = LByte (*SourcePort*); TCPHStr [2] = HByte (*DestnPort*); TCPHStr [3] = LByte (*DestnPort*); TCPHStr [4] = Byte0 (*SequenNum*); TCPHStr [5] = Byte1 (*SequenNum*); TCPHStr [6] = Byte2 (*SequenNum*); TCPHStr [7] = Byte3 (*SequenNum*);

TCPHStr [8] = Byte0 (*AckNum*); TCPHStr [9] = Byte1 (*AckNum*); TCPHStr [10] = Byte2 (*AckNum*); TCPHStr [11] = Byte3 (*AckNum*);

*lenflag = * TCPHdlenFlagBits (unsigned char [16] *txtcpState*, unsigned char [16] *txtcpOpPdFormat* &TCPHdrFlags, &TCPHdrLen);

TCPHStr [12] = HByte (lenflag); TCPHStr [13] = LByte (lenflag);

TCPHStr [14] = HByte (window); TCPHStr [15] = Byte3 (window);

TCPHStr [16] = HByte (*TCPChecksum16*); TCPHStr [17] = LByte (*TCPChecksum16*);

TCPHStr [18] = *Urg++); TCPHStr [19] = *(unsigned char *) urg);

extras = OptionAndPds (unsigned char [16] txtcpState, unsigned char [16] txtcpOpPdFormat);

optPdLen = getLength8 (extras);

*TCPHdrLen = (optPdLen + 20) /4;

while (optPdLen > = ++i) {TCPHStr [i] = extras [i − 20];}; /* Fill the options and padding bytes. */

unsigned short TCPSegLen = getLength16 (*Str*) + (optPdLen + 20); /* Add byte of header length also. */

return (TCPHStr);

};

/* ——/*

/* Declarations and codes for the function, ***IPHeaderSelPkt*** for returning an IP header string and the packet data in socketStream from OutStream. *Str* means the stream from transmission control layer at a node, *node1,* end to be sent with header as per IP protocol to other end, *node2,* through the network driver, switches, bridges and routers. */

11. /* First let us declare IPVerHdrLen for 4 bits (bit 7, bit 6, bit 5 and bit 4) for the version number. Presently, IP version 4 is mostly used. So these bits are 0100. Another four bits (bit 3, bit 2, bit 1 and bit 0) are the numbers of unsigned integers in the header. Header length includes 5 unsigned standard integers and 0 to 10 integers for options and padding. The latter integers are used for controlling the path and flow through the routers. Some of these integers may also be used for adding network security. Options may be recording of the route of packet, time stamp on the packet, source router IP address to used before routing through a common source, any flexible source option, security option, etc. Details can be found in the classic work of D. E. Comer and D. Stevens, *Internetworking with TCP/IP Vol. 1, Principles, Protocols and Architecture* from Prentice Hall, NJ, 1995. Usually the options and padding are not present. Then the four lower bits are 0101. */

unsigned char * IPVerHdrLen;

/* Header Length as per four upper bits in IPVerHdrLen. */

unsigned char ****IPHdrLen***;

12. /* Let us assume that the options and padding are as per the format specified by a 16-characters string, txipOpPdFormat. It represents the number and meanings of optionally added integers and padded integers. */

unsigned char [16] ***txipOpPdFormat***;

13. /* Let a 16-character string txipState specify the protocol of the current IP segment and its action required for the controlled transmission. Protocol can be ICMP, IGMP, OSPF, EGP and BGP. Refer to Raj Kamal, *Internet and Web Technologies* from Tata McGraw-Hill for definitions and any standard text for details of these protocols. */

unsigned char [16] ***txipState***;

14. /* Start of the codes for returning a *short*. It specifies 4-bit version field plus 4-bit IP Header length field in * IPVerHdrLen followed by 3-bit specifying precedence of the IP packet on the network plus 5 bits for the type of service. Let the latter 8 bits be at the address pointed by IPPrioServ. Precedence bits, '000' means usual precedence on the Internet. '111' means highest precedence, the one needed for streaming the audio and video on the net service bits for quality of service to be provided in terms of security, speed, delayed or cost to be charged from the sender. The 16-bit integer returned by function, IPVerHdrPrioServBits is thus as per version and IP packet transmission state and transmitting IP options and padding format, *txipState* and *txipOpPdFormat*, respectively. */

unsigned char * IPPrioServ;

unsigned short **IPVerHdrPrioServBits** (unsigned char [16] *txipState*, unsigned char [16] *txipOpPdFormat*, &IPPrioServ, &IPHdrLen) {

15. /* Returned integer bit 15, bit 14, bit 13, bit 12, bit 11, bit 10, bit 9 and bit 8 are as per 8 bits at IPHdrLen. Three precedence bits, bit 7, bit 6 and bit 5 in the returned integer are as per the IP precedence and service bits bit 4, bit 3, bit 2, bit 1 and bit 0 are as per QOS (quality of service specified in *txipState*. */

Boolean bit15, bit14, bit13, bit12, bit11, bit10, bit9, bit8, bit7, bit6, bit5, bit4, bit3, bit2, bit1, bit0;

/* The codes to find character at IPPrioServ from *txipState*. */

.

.

.

16. /* Code to find IPHdrLen from *txipOpPdFormat* */

.

.

.

}; /* End of the codes for returning the integer that specifies IP version, Header length, precedence and service. */

17. /* Two 16-bit short integers for packet length and packet identification by receiver. */

unsigned short IPPktLen; /* Specify the length of IP Packet */

unsigned short UniqueID; /* Specify the UniqueID of the present IP Packet. This is put by the router or the transmitter, node1. It uniquely identification for the packet routed. This will help the receiver to reassemble the fragments of this IP Packet at node2 for upward transmission to transport and application layer there. Note: A fragment may be lost on the net in between routers due to some error popping in. This helps in recovering the lost fragment. */

18. /* A stream from OutStream may be bigger than 6536 bytes minus the header bytes in the IP packet. Therefore, it is to be fragmented. Fragmentation is also necessitated when the network driver and other units in between do not permit even an IP packet of 65536 bytes and need shorter frames. Function, IPFlagFragBits Codes is for returning 3 flag bits and 13 fragment-offset bits. The offset specifies fragment number. Besides precedence and QOS, *txipState* specifies what 3 flag bits and 13 fragment-field bits should be defined by this function. */

unsigned char *IPHdrFlags; unsigned short *IPHdrFrag;
/* Function for finding a stream of bytes actually been put with the packet. */
void **selectPktData** (unsigned char [ ] Str, unsigned char OutStream, unsigned char [16] *txipState*, unsigned char [16] *txipOpPdFormat*, unsigned char [ ] Str) {
.

.
};
unsigned short **IPFlagFragBits** (unsigned char [16] *txipState*, unsigned char [16] *txipOpPdFormat*, &IPHdrFlags, &IPHdrFrag), char [ ] OutStream) {
Boolean bit15, bit14, bit13, bit12, bit11, bit10, bit9, bit8, bit7, bit6, bit5, bit4, bit3, bit2, bit1, bit0;
19. /* Codes to have three bits, bit 15, bit 14 and bit 13 in the returned integer as per the flags. A flag is *mfb*. mfb = 0 means that the receiver should wait as there are more fragments to follow this fragment. *mfb* = 0 for last fragment. It refers to the IP header size, which specifies the total number of unsigned integers in the IP header. The total is 5 plus the unsigned integers for options and padding. These are as per txipState and txipOpPdFormat used by IP segment that is transmitting. The bits, bit 11, bit 10, bit 9, bit 8 are reserved. */
/* Bits, bit 12 down to bit 0 are as fragment offset bits and as per already transmitted bytes from network driver task at node1. */
/* Code to find character IPHdrFlags from bit 15, bit 14, and bit 13. The bits are as per txipState.
.

.
20. /* Code to find 16-bit short integer IPHdrFrag from bit 13 down to bit 0. These bits are as per txipState specification, OutStreamSize and the bits already transmitted by the network driver. IPHdrFrag = *i* means OutStream byte numbering $(8)^i$ is being sent with this packet. */
.

.
}; /* End of the codes for returning an integer that specifies IP Header flags and fragment-offset bits. */
21. /* Declare other header field, 16-bit Checksum, 8-bit time to live and 8-bit, protocol-filed, source *node1* IP address and destination *node2* IP address. */
unsigned short *IPChecksum16*;
unsigned char *timeToLive*; /* Note: It decrements at each router on the way to node2. */
unsigned char *PrctlField*; / Note: *PrctlField* = 17 for UDP, = 6 for TCP, = 1 for ICMP.
unsigned short SourceIPAddr;
unsigned short DestnIPAddr;
unsigned char [ ] *extras*; unsigned char OptPdLen;
IPextras = OptionAndPds (*txipState*, *txipOpPdFormat*);
optPdLen = getLength8 (*IPextras*);
*IPHdrLen = (optPdLen + 20) /4;
unsigned char [ ] **IPOptionAndPds** (unsigned char [16] *txipState*, unsigned char [16] *txipOpPdFormat*)
{
22. /* Codes for returning an array of bytes for the 32-bit integers for the options and padding. These are as per the State and thus IP header length parameters.
.

.
};
23. /* Codes for returning IP Header String. Remember that IP protocol transmits the integers with big endian. * /

/* Note: Instead of using many arguments, a typedef could have been used for the header data structure. However, this will consume more memory. */

unsigned char [ ] **IPHeaderSelPkt** (unsigned short *SourceAddr*, unsigned short *DestnAddr*, unsigned short IPverHdrPrioSer, unsigned short *IPPktLen, char * *IPVerHdrLen,* unsigned char * *IPHdrLen, unsigned char *IPHdrFlags*, unsigned short *IPHdrFrag, unsigned short *IPChecksum16*, unsigned short UniqueID, unsigned char *timeToLive, unsigned char *PrctlField, unsigned char [ ] Str, unsigned char OutStream, char [optPdLen] *IPextras*, unsigned char [16] *txipState*, unsigned char [16] *txipOpPdFormat*) {

int i; unsigned char [ ] IPHStr; unsigned short lenflag; unsigned char optPdLen; **unsigned** short new;

24. /* Code for finding a stream of bytes actually been put with the packet. */
new = IPVerHdrPrioServBits (*txipState, txipOpPdFormat*, &IPPrioServ, &IPHdrLen);
IPHStr [0] = HByte (new);
IPHStr [1] = LByte (new);
25. /* Codes for estimating assigning UniqueID to the packet. * /
.

.

IPHStr [4] = HByte (UniqueID);
IPHStr [5] = LByte (UniqueID);
26. /* Codes for assigning Time to live and Protocol field from txipState. */
.

.
IPHStr [8] = *timeToLive; IPHStr [9] = *PrctlField;
IPHStr [12] = byte0 (SourceAddr); IPHStr [13] = byte1 (SourceAddr);
IPHStr [14] = byte2 (SourceAddr); IPHStr [15] = byte3 (SourceAddr);
IPHStr [16] = byte0 (DestnAddr); IPHStr [17] = = byte1 (DestnAddr);
IPHStr [18] = = byte0 (DestnAddr); IPHStr [19] = = byte0 (DestnAddr);
IPextras = IPOptionAndPds (txipState, txipOpPdFormat); /* Find Options and Padding. */
optPdLen = getLength8 (IPextras);
*IPHdrLen = (optPdLen + 20) /4;
while (optPdLen > = ++i) {IPHStr [i] = IPextras [i − 20];}; /* Fill the options and padding bytes. */
27. /* Codes for selecting the socket stream data to be sent and then estimating IPPktLen /
selectPktData (Str, OutStream, *txipState*, txipOpPdFormat, *IPextras*);
unsigned short *IPPktLen = getLength16 (*Str*) + (optPdLen + 20); /* Add byte of header length also. */
IPHStr [2] = HByte (IPPktLen); IPHStr [3] = LByte (IPPktLen);
new = IPFlagFragBits (txipState, *txipOpPdFormat*, &IPHdrFlags, &IPHdrFrag, Str)
IPHStr [6] = HByte (new); IPHStr [7] = LByte (new);
unsigned short *IPChecksum16* = checksum16 (IPHStr); /* Find checksum of IP header part only. */
IPHStr [10] = HByte (*IPChecksum16*); IPHStr [11] = LByte (*IPChecksum16*);
return (IPHStr);
};
/* Declarations and codes for the function, NetHdrFrTr for returning a frame header string, *frameHeader* and the fragment data in *frame* [ ] array from SocketStream. *Str* means the stream from internet layer at a node, *node1,* end to be sent with header as per data link protocol and card protocol to other end, *node2,* through the pipe having byte streams from the network driver. */

```
void NetHdrFrTr (unsigned char [ ] frame, unsigned char [ ] frameHeader, unsigned char [ ]
trailBytes, unsigned short fragOffset, unsigned char [ ] fragment, unsigned char [ ] trailBytes,
unsigned short *FRlength, unsigned short * HdrLen, unsigned short * endLen, unsigned
short *FrameHdrFlags, unsigned short *FrameHdrFragOffset unsigned short *FrameCRC32,
unsigned short *PrctlField, unsigned char [ ] Str, unsigned char [ ] SocketStream, unsigned char
[ ] FRextras, unsigned char [16] txFRState, unsigned char [16 ] txFROpPdFormat, unsigned
char [12] netDrvType) {
unsigned char [ ] frame, frameHeader, trailBytes, fragment, trailBytes; unsigned short fragOffset;
unsigned short *FRlength, * HdrLen, * endLen, *FrameHdrFlags, *FrameHdrFragOffset
*FrameCRC32;
unsigned int CRC32 (unsigned int crc, unsigned char aByte);
/* Codes as per netDrvType, transmitting frame state txFRState, transmitting frame option and padding
format, txFROpPdFormat create an array of characters for the frameHeader, for the fragment and for
trailBytes. Each is of length, HdrLen, fraglen, endLen, FRlength. Other fields calculated are short
integer for fragOffset and unsigned integer for frame CRC bits, FRCRC32, etc. */
  .
  .
  .
}
/*******************************************************************/
```

11.3 CASE STUDY OF AN EMBEDDED SYSTEM FOR AN ADAPTIVE CRUISE CONTROL SYSTEM IN A CAR

Embedded systems are one of the most used systems today in automotive electronics. Since the 1990s, these have been used in automotives for the following functions as discrete separate components or as an integrated central serving system.

1. *RTC* and *watchdog timers* for the tasks. [Refer to Section 3.2]
2. *Real-time control* for all electronic, electromechanical and mechanical systems. [Section 9.4]
3. *Adaptive Cruise Control* (ACC) to maintain constant speed. Nowadays, there can be an added feature to maintain the string stability in case of multiple cars streaming on the highway. [*String stability* means maintaining the constancy of inter-car distances.]
4. *Data Acquisition System* (DAS) for the following:
    (a) Current Time and Date display, updating, recoding of instances of malfunctions and periods of normal functioning. Updating can also occur by periodically synchronizing time with time signals from radio broadcasts.
    (b) External temperature
    (c) Internal temperature
    (d) Total distance covered for odometer and for ACC
    (e) Road speed of vehicle in km/hr for the ACC, speedometer and speed warning
    (f) Engine speed in r.p.m. [revolutions per minute]
    (g) Coolant temperature, periods of its excess above 115° C and constancy within 80°C to 90°C
    (h) Illumination levels at display panels and inside the vehicle

(i) Fuel level (Empty, R1, R2, R3, 1/4, 3/8, 1/2, 5/8, 3/4, 7/8 or Full)

(j) Oil pressure for starting oil pressure warning system activation for engine speed above 5000 r.p.m. and alarm system activation above 15000 r.p.m.

(k) Presently engaged gear information

(l) Front-end car distance

5. ***Front panel switches and display controls***

6. Port for ***alarm*** signals. These are for a display panel and for a sound system for beep and buzzer. Issue of warning is by display of pictograms and by raising sound alarms and their appropriate records for diagnostic analysis. [A pictogram is a pre-recorded picture in an image data file. It displays on an LCD matrix. Pictogram displayed is as per the required message.]

7. ***Diagnostics computations*** for driving time malfunction statistics and analysis results for fast fault diagnosis by mechanic.

8. ***Multimedia Interfaces***

9. ***Control Area Network*** Interfaces

10. ***Serial Communication Interface*** (SCI), transmitter and receiver.

Cruise control is a system that takes *charge* of controlling the throttle from the driver and cruising the vehicle at a preset constant speed. Cruising control may also maintain *string stability* in case of multiple cars streaming through highways and also VIP convoys. [Refer to a paper by Chi-Ying Liang and Huei Peng, *"Optimal Adaptive Cruise Control with Guaranteed String Stability"* Journal of Vehicle System Dynamics, 31, pp. 313–330, 1999]. Generally, the driver holds the vehicle steady during the drive using the accelerator pedal. Cruise control relieves the driver from that duty and the driver lets the ACC be on when the road conditions are suitable (not wet or icy and there are no strong winds or fog) and, if the car is cruising at high speed, when there is no heavy traffic. The driver resumes being in charge in the poor conditions listed.

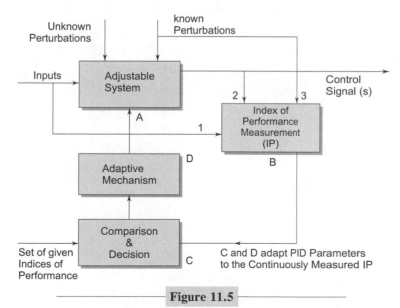

Adaptive Control Essential Components

Figure 11.5

Model for an adaptive control algorithm that adapts and functions.

An adaptive control means that the algorithm used *adapts* to the current status of the control inputs and is not using a constant set of mathematical parameters in the algorithm equations. Parameters adapt dynamically. Exemplary parameters, which are adapted continuously, are *proportionality constants*, *integration constants* and *differentiation constants*. ACC systems are commonly used in aviation electronics and defense aircrafts for cruising. Use in automotives is of recent origin. [Refer to *http://www.ee.surrey.ac.uk/Personal/R.Young/java/html/cruise.html*].

Figure 11.5 shows *how an adaptive control algorithm can adapt and function*. It calculates the output values for the control signals. [For details of the control system algorithms, the reader may refer to the standard texts in control engineering, namely *Continuous and Discrete Control Systems* by John F. Dorsey, McGraw-Hill International Edition, 2002; *Digital Control and State Variable Method* by Madan Gopal, Tata McGraw-Hill, New Delhi, 1997; *and Modern Control Systems – Analysis and Design* by Walter J. Grantham and Thomas L. Vincent, John Wiley & Sons, New York, 1993].

Figure 11.6 shows the block diagram of an ACC embedded system. A general ACC system does the following actions.

1. Get road speed from a speedometric section within the DAS unit.

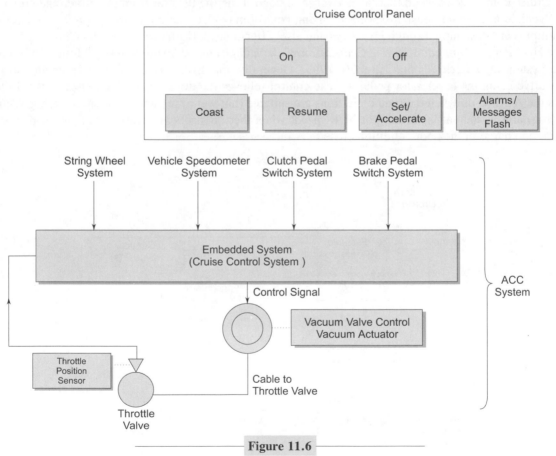

Figure 11.6

Block diagram of an ACC embedded system.

2. Get acceleration from the engine speed section within the DAS unit.
3. Get inputs from the brake switch for brake activities.
4. Send output to the pedal system for applying emergency brakes and driver non-intervention by taking charge of cruising from the ACC.
5. Run an adaptive algorithm to calculate and send the control signals to the stepper motor at the vacuum valve actuator. The orifice opening of the vacuum valve controls the throttle valve. The valve is electro-pneumatic. Vacuum creation provides the force via bellows. [A D.C. or stepper motor with a worm drive attachment to the throttle can also be directly used instead of vacuum actuator and bellows].
6. Receive throttle position by the stepper motor position sensor.
7. Control a front-end panel, which has the following: (a) Switch or Display for 'ON', 'OFF', 'COAST', RESUME', SET/ACCELERATE. The driver activates or deactivates the ACC system by pressing ON or OFF, respectively. S/he hands over or resumes the ACC system charge by pressing COAST or RESUME, respectively. S/he sets the cruise speed by SET/ ACCELERATE switch. A switch glows either green or red as per the status when the ACC activates. (b) Alarms and a message flashing unit issue appropriate alarms and message flashing pictograms.
8. Retrieve the front end-car distance information from a radar or UVHF (Ultra Very High Frequency) attachment at the front steering wheel. A stepper motor aligns the attachment so that transmitter of radar maintains the line of sight to the front-end car. The radar system maintains string stability and warns of emergency situations.

The choice of case study of an ACC is taken up to understand the RTOS needed for applications. We need the RTOS features so that we get a reliable control system in the automotive electronics system. We take here the case of an ACC system, which has the algorithm and ports such that the system maintains string stability also. A hardware system in automotive electronics has to provide functional safety. Important hardware standards and guidance at present are the following:

(a) TTP (Time Triggered Protocol)
(b) CAN (Controller Area Network) [Section 3.3.2]
(c) MOST (Media Oriented System Transport)
(d) IEE (Institute of Electrical Engineers) guidance standard exists for EMC (Electromagnetic Magnetic Control) and functional safety guidance.

Nowadays, an automotive embedded system–based control unit uses a microprocessor, microcontroller, DSP, or ASIP. Let us assume that in the present case, the following are the device-driver ports and their functions.

1. Port_Align: It is a stepper motor port. Motor steps up clockwise or counterclockwise on an interrupt signal. The motor is to align the radar or UVHF transmitting device in the line of the front-end car.
2. Port_Ranging: Time difference timeDiff is read on an interrupt signal to port device. A service routine executes and the radar emits the signal and sensor receives the reflected signal from the front-end car. The routine disables the interrupts at the beginning. Port device circuit measures the delay between the two instances. The routine enables the interrupts on exiting the critical section. Half of the delay multiplied by speed from the task, which measures the speed, gives the range distance of a front-end car.

3. Port_Speed: It is a port where, on receiving an interrupt signal, the port device sets a down counter at N_rotation and notes the time at this instance. The routine disables the interrupts at the beginning. After the down counter overflows, it again notes the time. It finds the difference in time, deltaT. The routine enables the interrupts on exiting the critical section. Port_Speed displays at the speedometer, current speed *speedNow,* on receiving a signal from Port_RangeRate

4. Port_RangeRate: It transmits on a signal to the port the rangeNow and speedNow to all other streaming cars. It sends a signal to Port_Speed to display speedNow on the speedometer.

5. Port_Brake: Port device applies the brakes or emergency brakes on an interrupt signal. The service routine disables at the beginning and enables the interrupts on exiting the critical section. It applies the brakes and signals this information to all the other streaming cars also.

Figure 11.7 shows the ACC system cycle of actions and task scheduling model. Note the marks of cycle starting and finishing positions in the figure.

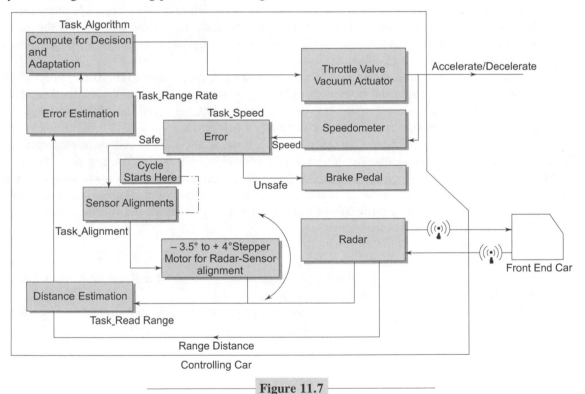

Figure 11.7

ACC-system cycle of actions and task scheduling model.

1. Cycle starts from a task, Task_Alignment. It sends the signal to a stepper motor port Port_Align. The stepper motor moves by one step clockwise or counterclockwise as directed.

2. A task Task_ReadRange is for measuring front-end car range. The task disables all interrupts as it is entering a critical section. We need real-time measurement. Port_Ranging finds timeDiff.

3. Task_Speed is the task to get the port reading at a port Port_Speed. Port sends an input, deltaT. It is the time interval between the first and the Nth rotation.

4. Task_Range_Rate sends the rangeNow for Task_Alignment so that the stepper motor aligns the sensor in line with the front-end car. It estimates the final error in maintaining the string stability from Task_ReadRange output. It estimates the error for maintaining the car speed from the Task_Speed output. It outputs both error values for use by the control system adaptive algorithm. It transmits the errors to other vehicles through Port_RangeRate. It sends the current speed, speedNow, to the Port_Speed. It is inversely proportional to the deltaT. A ratio N_rotation/ deltaT, after applying a filtering function, gives *speedNow*. Port_Speed connects to the speedometric system of the DAS, which displays speedNow after appropriate filtering function. Port_RangeRate transmits the speedNow also to other streaming cars also. Now it calculates range and rate errors. Transmits both rangeNow and speedNow through Port_RangeRate.

5. Task_Algorithm runs the main adaptive algorithm. Its inputs are as follows. It gets inputs from Task_Range_Rate. The task outputs are to Port_Throttle. Port_Throttle attaches to the vacuum actuator stepper motor. The cycle ends here. After a delay, the cycle again starts from the Task_Alignment. It receives range and rate error messages of other vehicles through Port_RangeRate. It reads the statuses of Port_Brake of this and other streaming cars.

Task_Algorithm generates output for ACC action. Output port for control signals generate for a port with a throttle valve. The signals at this port, Port_Throttle, are calculated as follows: An adjustable algorithm, *A,* gets inputs of the speed *'speed'* and front-object range *'range'* as well as the known and unknown perturbations, P_Unknown and P_Known. It adjusts the output signal to Port_Throttle. An algorithm, *B,* estimates the index of performance *IP*. An algorithm, *C,* compares *IP with a set of given IP* values. Algorithm *D* is as per the adaptive mechanism that adapts the output of *C. C* sends the new parameters that are to be adapted by *A*.

Table 11.3 lists the tasks, functions and IPCs needed for the ACC with string stability algorithm. We can note from the columns of IPCs posted and taken that the semaphores are used for defining the sequence of running of the task and cyclically running the ACC and string stability controller algorithms at Task_Algorithm.

Table 11.3

List of BCC 1# Task Functions and IPCs used in Example 11.3

BCC 1 Task Function	BCC 1 Priority	Action	IPCs pending	IPCs posted	Input	Output
Task_Alignment	101	Send signal to Port_Align	*SemReset*	*SemAlign*	deltaStep, step	*step*
Task_ReadRange	103	Disable interrupts, send signal to Port_Ranging. Port activates a radar flashing, records activation time, gets time of sensing the reflected radar signal and finds time difference, *timeDiff*. Enable interrupts.	*SemAlign*	*Sem-Range*	-	time-Diff

(Contd.)

BCC 1 Task Function	BCC 1 Priority	Action	IPCs pending	IPCs posted	Input	Output
Task_ Speed	105	Signal Port_Speed to start a timer, counter start message and wait for the 10 counts for the number of wheel rotations. It outputs deltaT.	*SemRange*	*Sem-Speed*	-	*deltaT*
Task_ Range-Rate	107	It calculates *rangeNow* from timeDiff. Get preset front car *range* and string Range from memory and compare. Get preset cruising speed, *cruise Speed*; compare it with current speed *speedNow*.	*SemSpeed*	*Sem-Reset*, *SemACC*	avgTire-*Circum*, *time-Diff*, *deltaT*, *string Range*, *CruiseSpeed*, N_rotation	range-Error, speed-Error, *range-Now*, speed-Now
Task_ Algor-ithm	109	(i) Get errors of speed and range and execute adaptive control algorithm. (ii) Get errors of other vehicles through Port_RangeRate. (iii) Get other vehicles Port_Brake status. (iv) Get present throttle position. (v) Send output, throttleAdjsut to Port _Throttle. (vi) Send signal to Port_Brake in case of emergency braking action needed. (vii) Port_Brake transmits the action needed to other vehicles also.	*SemACC*	*SemReset*	range-Error, speed-Error, All Port_ RangeRate values and Port_Brake statuses, VehicleID,	throttle-adjust, emer-gency

[#]Basic task with one task of each priority and single activation. It is called BCC 1 (Basic Conformance Class 1)

RTOSs described in Sections 10.1 to 10.3 do not suffice for this and other necessary features. Embedded software in the automotive system needs special features in its operating system over and above the MUCOS or VxWorks features and MS DOS and UNIX. *Special OS features needed are as follows:*

1. *Language* can be application-specific, need not be just C or C++ and *Data Types* should also be application specific and not RTOS specific. In VxWorks, for example, STATUS is RTOS specific. This is not permitted, as it could be the source of a bug and thus unreliable.
2. Its *OS*, every method and class and run time library *should be scalable*. This optimizes the memory needs.
3. Tasks can be classified into four types. This provides a clear-cut distinction to a programmer: which class to use for what modules in the system.

(a) Basic with one task of each priority and single activation. It is called BCC 1 (Basic Conformance Class 1). Extended task means, for example, a task, FirstTask created in the *main* function in Example 9.17.

(b) Extended with one task of each priority and single activation. It is called ECC 1 (Extended Conformance Class 1). Extended task means, for example, a task created by FirstTask in Example 9.17.

(c) Basic with multiple tasks of each priority and multiple times activation during run. It is called BCC 2.

(d) Extended with multiple tasks of each priority and multiple times activation during run. It is called ECC 2.

4. OS can schedule ISRs and tasks in distinct ways. [VxWorks scheduler does also].

5. Interrupt system disables at the beginning of the service routine and enables on return. This lets the task run in real-time environment.

6. Task can be scheduled in real-time.

7. Task can consist of three types of objects, *event* (semaphore), *resource* (statements and functions) and device. There are port devices as described above. An exemplary device is *alarm*. It displays the pictograms, messages and flashing messages and sounds the buzzer.

8. Timer, task or semaphore objects creation and deletion cannot be allowed. A run-time bug may lead to an uncalled deletion of a timer or semaphore. That is the potential source of a problem and is thus unreliable.

9. IPC, message queue posting by a task, is not allowed as a waiting task may wait indefinitely for its entire message needs. RTOS queue types, waiting infinitely or for a time out for a message, can be a potential source of trouble and thus unreliable. Similar risks may arise with semaphore as a resource key or counter. These are therefore not used.

10. Before entering a critical section and on executing a service routine, all interrupts must disable and enable on return only. [Refer to Section 6.8].

Software in automotive electronics must also be standard. Presently, the important software standards and guidance are AMI-C (Automotive Multimedia Interface Collaboration) [http://www.ami-c.org], MISRA-C (Motor Industry Reliability Association standard for C language software guidelines for automotive systems) [www.misra.org.uk] and OSEK/VDX for RTOS, communication and network management. OSEK is after the German words for Open (Offene) System (System) for Electronic (Electronik) embedded component and interface Crafts (Kraftfahrzeugen). [Refer to http://www.osek-vdx.org and also to *Programming in the OSEK/VDX Environment* by Joseph Lemieux from CMP Books, Oct. 2001.]

OSEK standard defines three standards.

1. OSEK-OS for operating system, which has greater reliability. It is because, in an OS-based-upon OSEK, the above ten points are taken care of.

2. OSEK-NM architecture for network management. As in OS, tasks are divided into four types, and the NM divides the architecture into two types. (i) Direct transfer and interchange of network messages. (ii) Indirect transfer and interchange, both between the nodes.

3. OSEK-COM architecture for IPCs between the same CPU control unit tasks and between the different CPU control unit tasks. Between different unit tasks, the data-link and physical layers exists. Different CPU physical layers connects by CAN bus architecture.

We have seen that OSEK-OS standard is reliable compared to VxWorks or MUCOS. To demonstrate ACC application, let us see the application of VxWorks and how we adapt the following in our exemplary codes (Example 11.3).

1. Use BCC 1 type of tasks, as done in VxWorks application in Example 11.2. We define each task of different priority and activate it only once in the codes.
2. Use no message queues.
3. Use no in-between creation and deletion of task.
4. Use semaphores as event flag only with no task having run-time deletion or creation of these.

Example 11.3

1. # include "vxWorks.h" /* Include VxWorks functions. */
include "semLib.h" /* Include semaphore functions Library. */
include "taskLib.h" /* Include multitasking functions Library. */
include "sysLib.c" /* Include system library for system functions. */
#include "sigLib.h" / * Initialise kernel component for signal functions. */
2. /* Set system clock rate 10000 ticks per second. Every 100 µs per tick*/
sysClkRateSet (10000);
3. /* Declare and Initialise Global parameters. */
unsigned byte VehicleID; /* Declare this car ID. */
static numCars = …; /* Numbers of cars in the string that should move with cruise speed. */
static unsigned byte *N_rotation* = ….; /* Initialise number of rotations needed when finding the speed. */
static int *avgTireCircum* = … ; /* Initialise *average tire Circumference in mm.* */
4. /* Initialise the string Range = Separation to be maintained between the two cars in the string in mm. Initialise cruise speed. */
/* All distances are in mm, speeds in km/hr and time in nanosecond unless otherwise specified. */
static int *stringRange* = …; /* In unit of mm */
static int *CruiseSpeed* = …; /* In unit of km/hr */
static float unitChange = 3600000.0; /* Km in one mm divided by hours in one nanoseconds. */
float permittedSpeedError = … ; / * In units of mm/ns error in speed permitted. It prevents oscillations in control system. */
static boolean alignment = false; /* Declare alignment = false to initiate radar transmitter alignment. */
5. /* Other Variables. */
static byte *step; /* Stepper motor step angle in degrees. */
static byte * deltaStep = 0; /* Stepper motor step angle change in degrees. */
/* Time difference between emitted signal and reflected signal from front-end car. rangeNow is present range in mm. It is timeDiff multiplied by speedNow after a filter function application. */
static unsigned long *timeDiff; static int *rangeNow;
static unsigned long *deltaT; /* *Time interval for N rotation in ns.* */
6. /* Declare pointers for variables Range error, range now, speed error, speed now. */
int rangeError, speedError, *rangeNow* = 0, *speedNow* =0; /* Speeds are in km/hr and range in mm. */
7. /* Declare arrays of size number of cars, numCars. These many cars are running as a string. Declare brakeStatus, RangeErrors, SpeedErrors, Ranges, Speeds for all the cars. */
boolean brakeStatus [numCars];
int RangeErrors [numCars], Ranges [numCars], SpeedErrors [numCars], Speeds [numCars];

boolean emergency [numCars]; /* Declare variable for emergency message sent to Port_Brake of Nth car. */

int *throttleAdjsut; /* Declare variable for throttle adjusting parameter */

8. /* Other Variables. */

.

.

9. /* Declare all Table 11.3 *Task function prototypes*. */

void Task_Alignment (SemID *SemReset*, SemID *SemAlign*, byte *step, byte *deltaStep); /*Task for aligning stepper motor in front-end car view. */

void Task_ReadRange (SEM_ID *SemAlign*, SEM_ID *SemRange*, unsigned long *timeDiff*); /*Task for receiving the timeDiff using the radar for calculating *rangeNow*. */

void Task_Speed (SEM_ID *SemRange*, SEM_ID *SemSpeed*, unsigned long *deltaT); /*Task for receiving the deltaT using the wheel counter and timer for calculating *speedNow*. */

void Task_Range_Rate (SEM_ID *SemSpeed*, SEM_ID *SemReset*, SEM_ID SemACC, int *avgTireCircum*, unsigned byte N_rotation, int *CruiseSpeed*, int *stringRange*, unsigned long * time-Diff*, unsigned long *deltaT*, int * range-Error, int *speedError, int *range-Now*, int *speed-Now*); /*Task for calculating rangeNow, speedNow, rangeError, speedError. */

void Task_Algorithm (SEM_ID *SemACC*, SEM_ID *SemReset*, boolean brakeStatus [numCars], int RangeErrors [numCars], int Ranges [numCars], int SpeedErrors [numCars], int Speeds [numCars],

boolean emergency [numCars], unsigned byte VehicleID); /* Declare array for emergency message sent to Port_Brake of Nth car. */

int *throttleAdjsut; /* Declare variable for throttle adjusting parameter */

10. /* Declare all Table 11.3 *Task IDs, Priorities, Options and Stacksize*. Let initial ID, until spawned be none. No options and Stacksize = 4096 for each of six tasks. */

int Task_AlignID – ERROR; int Task_AlignPriority = 101; int Task_AlignOptions = 0; int Task_AlignStackSize = 4096;

int Task_ReadRangeID = ERROR; int Task_ReadRangePriority = 103; int Task_ReadRangeOptions = 0; int Task_ReadRangeStackSize = 4096;

int Task_SpeedID = ERROR; int Task_SpeedPriority = 105; int Task_SpeedOptions = 0; int Task_SpeedStackSize = 4096;

int Task_RangeRateID = ERROR; int Task_RangeRatePriority = 107; int Task_RangeRateOptions = 0; int Task_RangeRateStackSize = 4096;

int Task_AlgorithmID = ERROR; int Task_AlgorithmPriority = 109; int Task_AlgorithmOptions = 0; int Task_AlgorithmStackSize = 4096;

11. /* *Create and Initiate (Spawn) all the six tasks* of Table 11.2. */

Task_AlignID = taskSpawn (" tTask_Align", Task_AlignPriority, Task_AlignOptions, Task_AlignStackSize, void (*Task_Alignment) (SemID *SemReset*, SemID *SemAlign*, byte *step, byte *deltaStep), 0, 0, 0, 0, 0, 0, 0, 0, 0, 0);

Task_ReadRangeID = taskSpawn (" tTask_ReadRange", Task_ReadRangePriority, Task_ReadRangeOptions, Task_ReadRangeStackSize, void (*Task_ReadRange) (SEM_ID *SemAlign*, SEM_ID *SemRange*, unsigned long *timeDiff*), 0, 0, 0, 0, 0, 0, 0, 0, 0, 0);

Task_SpeedID = taskSpawn (" tTask_Speed", Task_SpeedPriority, Task_SpeedOptions, Task_SpeedStackSize, void (*Task_Speed) (SEM_ID *SemRange*, SEM_ID *SemSpeed*, unsigned long *deltaT), 0, 0, 0, 0, 0, 0, 0, 0, 0, 0);

Task_RangeRateID = taskSpawn (" tTask_Range_Rate", Task_RangeRatePriority, Task_RangeRateOptions, Task_RangeRateStackSize, void (*Task_Range_Rate) (SEM_ID *SemSpeed*, SEM_ID *SemReset*, SEM_ID SemACC, int *avgTireCircum*, unsigned byte N_rotation, int *CruiseSpeed*, int *stringRange*, unsigned long * time-Diff*, unsigned long *deltaT*, int * range-Error, int *speedError, int *rangeNow*, int *speedNow*), 0, 0, 0, 0, 0, 0, 0, 0, 0, 0);

Task_AlgorithmID = taskSpawn ("tTask_Algorithm", Task_AlgorithmPriority, Task_AlgorithmOptions, Task_AlgorithmStackSize, void (*Task_Algorithm) (SEM_ID *SemACC*, SEM_ID *SemReset*, boolean brakeStatus [numCars], int RangeErrors [numCars], int Ranges [numCars], int SpeedErrors [numCars], int Speeds [numCars], boolean emergency [numCars], unsigned byte VehicleID), 0, 0, 0, 0, 0, 0, 0, 0, 0);

12. /* Declare IDs and create the binary semaphore event flags. */

SEM_ID *SemAlign*, *SemRange*, *SemSpeed, SemACC, SemReset*; /* Declared for Table 11.3 five tasks. */

13. /* Create the binary semaphores taken in FIFO and as empty to start with. */

SemAlign = semBCreate (SEM_Q_FIFO, SEM_EMPTY);

SemRange = semBCreate (SEM_Q_FIFO, SEM_EMPTY);

SemSpeed = semBCreate (SEM_Q_FIFO, SEM_EMPTY);

SemACC = semBCreate (SEM_Q_FIFO, SEM_EMPTY);

SemReset = semBCreate (SEM_Q_FIFO, SEM_EMPTY);

14. /* Declare function for starting a RTC timer at the Port_Ranging. */

void RTCtimer_Port_Ranging_Start ( ) {

.

.

};

15. /* Declare a function for starting a RTC timer at the Port_Speed. */

void RTCtimer_Port_Speed_Start ( ) {

.

.

};

16. /* Declare a function to read 64-bit time from an RTC. */

unsigned long timer_gettime (&RTC) {

.

.

};

17. /* Define a macro for calculating time between instance when control bit = true and Status flag = true. */

boolean * CB; /* Control bit */

boolean * SF; / * Status flag. */

CB =0; / Control bit = 0; */

SF = 0; / Status flag = 0; */

unsigned short * RTC; /* Pointer to a real-time clock. */

#define unsigned long calculate_TimeInterval (unsigned short * RTC, boolean * CB, boolean *SF) (

unsigned long timeInstance0, timerInstance1;

while (*CB != 1 && *SF = = 0) { }; /* Wait for read instruction to timer. */

timeInstance0 = timer_gettime (&RTC); /* Find initial time at start in the timer */

while (*SF ! = 1) { }; /* Wait for sensor status flag to be true */

timeInstance1 = timer_gettime (&RTC); /* Find initial time at start in the timer */

*SF =0; *CB =0;

return (timeInstance1 – timrInstance0);

) /* End of Macro to calculating time interval between two instances specified by CB and SF becoming true. */

18. /* Define a macro for calculating time interval deltaT between instance when control run bit = true and instance of N-th count input. */

boolean CR; /* Control Run bit. */

```
boolean countInput; / *Status flag. */
*CR =0; /* Control run bit = 0; */
*countInput = 0; /* countInput initial value. */
#define unsigned long DelT (unsigned short &RTC, boolean *CR, N_rotation, boolean *
countInput)(
unsigned long timeInstance = 0; byte N = 0;
for (byte N =0; N < N_rotation; N++) {
while (*CR != 1 && *countInput != 0) { }; /* Wait for count input true. */
timeInstance += timer_gettime (&RTC); /* Find initial time at start in the timer */
*countInput =0; /*Reset countInput. It will set on start of next rotation. */
}; *countInput =0;
return (timeInstance);
) /* End of Macro to calculate time interval for N count inputs. */
19. /* Declare Macro for sending a byte for step angle setting to Port_Align. */
unsigned short * Port_Align; /* Declare a pointer for Port_Align. */
# define Out_Alignment (&Port_Align, Step) / * Codes in Assembly Language for stepper
motor routine. * /
.
.
); /* End of Macro to sending byte for step angle to Port_Align. */
20. /* Declare Macro to find timeDiff from Port_Ranging. */
unsigned short RTC_Port_Ranging =...; /* Declare Address of RTC at Port_Ranging. * /
RTCtimer_Port_Ranging_Start ( );
#define RANGE (unsigned short * Port_Ranging, unsigned long * timeDiff) (
unsigned short * RTC_Port_Ranging = .....; /* Declare address of RTC of Port_Ranging. */
intLock ( ); /*disable interrupts. */
/ * Codes in Assembly Language for Ranging routine for Start Radar transmission by making
control bit CB = 1* /
*CB = 1;
*timeDiff = calculate_TimeInterval (&RTC_Port_Ranging, &CB, &SF);
intUnlock ( ); /* Enable Interrupts. */
) /* End of Macro to find time interval for reflected radar signals. */
21. /* Declare Macro to find deltaT from Port_Speed. */
unsigned short RTC_Port_Speed =...; /* Declare Address of RTC at Port_Speed. * /
RTCtimer_Port_Speed_Start ( );
#define SPEED (unsigned short * Port_Speed, unsigned long * deltaT) (
unsigned short * RTC_Port_Ranging = .....; /* Declare address of RTC of Port_Ranging.*/
intLock ( ); /*disable interrupts. */,
/ * Codes in Assembly Language for Speed routine for Start counting the tire rotation count
inputs by making control bit CR = 1* /
*CR = 1;
*deltaT = DelT (&RTC_Port_Speed, &CR, N_rotation, &countInput);
intUnlock ( ); /* Enable Interrupts. */
) /* End of Macro to find time interval for N count-inputs. */
23.
# define float filter_speed (float calculatedSpeed, int *speedNow, float permittedSpeedError) (/
* Codes for filtering the calculated speed. If the calculated value in mm/ns is within plus minus
limit of permittedSpeedError, in mm/ns do not change it; else modify it with new value. */
float a;
```

```
a = (float) (speedNow/ unitChange);
if (calculatedSpeed > a + permittedSpeedError || calculatedSpeed < a – permittedSpeedError) {return
(calculatedSpeed);} else return (a);
) /* End of macro for filtering the speed calculated to prevent vibrations and oscillation in controlling
vehicle. * /
24.
#define RangeRate (unsigned short * Port_RangeRate, int *speedNow, int *rangeNow, int *speedError,
int *rangeError, unsigned byte VehicleID) (
/ *Assembly codes for sending to Port_RangeRate and transmitting the range and rate parameters and
vehicleID. Port_RangeRate sends also *speedNow for display on speedometer at Port_Speed. /
.
.
); /* End of Macro to transmit range and rate parameters through Port_RangeRate. */
25 to 27. /* Code for Other Declaration Steps specific to the various functions */
.
.
/* End of the codes for creation of the tasks, semaphores, message queue, pipe tasks, and variables and
all needed function declarations */
/
*************************************************************************************/
28. /* Start of Codes for Task_Alignment. */
void Task_Alignment (SemID SemReset, SemID SemAlign, byte *step, byte *deltaStep) {
.
29.
while (1) { /* Start of while-loop. */
/* When cycle starts the semaphore is given. Wait for it. */
semTake (SemReset, WAIT_FOREVER); /* Wait for Cycle Reset Event flag. */
30. /* Codes for sending the step to the address of Port_Ranging. *
*step = *step + *deltaStep;
Out_Alignment (&Port_Align, *Step);
semGive (SemAlign);
}; /* End of while-loop. */
} /* End of Task_Alignment. */
/
*************************************************************************************/
31. /* Start of codes for Task_ReadRange. */
void Task_ReadRange (SEM_ID SemAlign, SEM_ID SemRange, unsigned long *timeDiff) {
32. /* Initial assignments of the variables and pre-infinite loop statements that execute once only*/
static unsigned short Port_Ranging = ....; /* Declare pointer to Port_Ranging. */
.
.
33. while (1) { /* Start an infinite while-loop. We can also use FOREVER in place of while (1). */
34. / *Wait for SemAlign state change to SEM_FULL by semGive function. */
semTake (SemAlign, WAIT_FOREVER);
/*Send signal to Port_Ranging. Port activates a radar flashing, records activation time, gets time of
sensing the reflected radar signal and finds time difference, timeDiff. Enable interrupts. */
RANGE (& Port_Ranging, &timeDiff);
);
```

semGive (*SemRange*);
};
35. } /* End of the codes for Task_ReadRange. */
/
***/
36. /* Start of codes for Task_Speed. */
void **Task_Speed** (SEM_ID *SemRange*, SEM_ID *SemSpeed*, unsigned long *deltaT) {
37. /* Initial assignments of the variables and pre-infinite loop statements that execute once
only*/
static unsigned short Port_Speed = ...; /* Declare pointer to Port_Speed. */

.

.
38. while (1) { /* Start task infinite loop. */
semTake (*SemRange*, WAIT_FOREVER);
/* Codes for receiving the deltaT using the wheel counter and timer for later on calculating
speedNow. */
SPEED (& Port_Speed, &deltaT);
semGive (SemSpeed);
}; / * End of while-loop. */
39. } /* End of codes for Task_Speed */
/
***/
/* Start of codes for Task_Range_Rate. */
void **Task_Range_Rate** (SEM_ID *SemSpeed*, SEM_ID *SemReset*, SEM_ID SemACC, int
avgTireCircum, unsigned byte N_rotation, int *CruiseSpeed*, int *stringRange*, unsigned long *
time-Diff, unsigned long *deltaT, int * range-Error, int *speedError, int *range-Now*, int
speed-Now) {
static unsigned short Port_RangeRate = ...; /* Declare pointer to Port_Ranging. */

.

.
40. while (1) { /* Start task infinite loop . */
semTake (*SemRange*, WAIT_FOREVER);
41. /* .Codes for calculating rangeNow, speedNow, rangeError, speedError. */
*speedNow = (int) (*unitChange* * filter_speed ((float) (*avgTireCircum* * N_rotation) / (float)
(*deltaT), int *speedNow, float permittedSpeedError));
*rangeNow = (*speedNow/unitChange) * (*timeDiff)/2.0; /* Divide by 2 because reflected
signal travels twice the distance in mm/ns. */
*speedError = cruiseSpeed - *SpeedNow;
RangeRate (& Port_RangeRate, *speedNow, *rangeNow, *speedError, *rangeError,
VehicleID); / *Send the parameters for transmission to other vehicles and Port_Speed for
displays. */
if (alignment != true) {semGive (SemReset);
42. /* Code for loop of tasks of priorities 101, 103, 105 in which the values of rangeNow are
calculated at different step values by changing deltaStep and finally at that instance alignment
is declared as true for rangeNow is minimum. Front-end vehicle is now in line of sight. */

.

.
} else semGive (SemACC); /* After alignment is perfect the control algorithm is sent the event
flag. */

```
}; / * End of while-loop. */
43. } /* End of codes for Task_Range_Rate. */
/
*************************************************************************/
/* Start of Codes for Task_Algorithm. */
void Task_Algorithm (SEM_ID SemACC, SEM_ID SemReset, boolean brakeStatus [numCars],
int RangeErrors [numCars], int Ranges [numCars], int SpeedErrors [numCars], int Speeds [numCars],
boolean emergency [numCars], unsigned byte VehicleID) {
 .

 .
44. while (1) { /* Start task infinite loop . */
semTake (SemACC, WAIT_FOREVER);
45. /* Assembly codes for getting errors of other vehicles through Port_RangeRate and other vehicles
Port_Brake statuses through Port_Brake. */
 .

 .
46. /* Assembly codes to read throttle position from Port_Throttle. */
 .

 .
47. /* Codes for cruise speed adaptive algorithm and code for string stability maintaining adaptive
algorithm to generate appropriate throttleAdjust signal to Port_Throttle. */
 .

 .
48. /* Codes for Port_Brake action, if emergency = true. Port_Brake transmits the action needed to other
vehicles also. */
if (emergency [numCars] = =1) {
 .
} else semGive (SemACC); /* After alignment is perfect the control algorithm is sent the event flag. */
49. }; / * End of while-loop. */
50. } /* End of codes for Task_Algorithm. */
/
*************************************************************************/
```

11.4   CASE STUDY OF AN EMBEDDED SYSTEM FOR A SMART CARD

Smart cards are one of the most used embedded systems today. They are used for credit-debit bankcards, ATM cards, identification cards, medical cards (for history and diagnosis details) and cards for a number of new innovative applications. [Readers may refer to a frequently updated website, http://www.sguthery@tiac.net for the answers to frequently asked questions about the card.]

Figure 1.11 showed the hardware of an exemplary card and Section 1.6.6 listed the units present in a typical card. The security aspect is of paramount importance for smart cards, when used for financial and banking related transactions. [Readers may refer to http://www.home.hkstar.com/~alanchan/papers/smartCardSecurity/ and http://www.research.ibm.com/secure_systems/scard.htm for details of the card-security requirements.]

11.4.1 Embedded Hardware

A smart card is a plastic card in ISO standard dimensions, 85.60 mm x 53.98 x 0.80 mm. It is an embedded SoC (System-On-Chip). [ISO standards are ISO7816 (1 to 4) for host-machine contact based cards and ISO14443 (Part A or B) for the contact-less cards.]

The silicon chip is just a few mm in size and is concealed in between the layers. Its very small size protects the card from bending. Let us understand the various hardware units first before pondering over the software and the OS. A few basic features of the card hardware are as follows:

1. The microcontroller used can be MC68HC11D0 or PIC16C84 or a smart card processor Philips Smart XA or an ASIP Processor. MC68HC11D0 has 8 kB internal RAM and 32 kB EEPROM and 2/3 wire protected memory. Most cards use 8-bit CPUs. The recent introduction in the cards is of a 32-bit RISC CPU. A smart card CPU should have special features, for example, a security lock. The lock is for a certain section of memory. A protection bit at the microcontroller may protect 1 kB or more data from modification and access by any external source or instructions outside that memory. Once the protection bit is placed at the maskable ROM in the microcontroller, the instructions or data within that part of the memory are accessible from instructions in that part only (internally) and not accessible from the external instructions or instructions outside that part. The CPU may disable access by blocking the write cycle placement of the data bits on the buses for the instructions and data protected part at the physical memory after certain phases of card initialisation and before issuing the card to the user. Another way of protecting is as follows: The CPU may access by using the physical addresses, which are different from the logical address used in the program. There can be a simple bus architecture like I^{2}C or extended I^{2}C for a card with free access to the memory. [Refer to Section 3.6 and Figure 3.16]. The buses for the ROM and RAM are separate in 2-Bus protocol for the protected memory accesses. There can also be 3-Bus protocol for further protection to memory and I/O system.

2. Standard ROM is used in the card. Usual size is 8 kB or 64 kB for usual or advanced cryptographic features in the card, respectively. Full or part of ROM bus activates after a security check only. Processor protects a part of the memory from access. The ROM stores the following.

   (i) *Fabrication key*, which is a unique secret key for each card. It is inserted during fabrication.

   (ii) *Personalisation key*, which is inserted after the chip is tested on a printed circuit board. Physical addresses are used in the testing phase. The key preserves the fabrication key and this key insertion preserves the card personalisation. After insertion of this key, RTOS and application use only the logical addresses.

   (iii) *RTOS codes*

   (iv) *Application* codes

   (v) A *utilisation lock* to prevent modification of two PINs and to prevent access to the OS and application instructions. It stores after the card enters the utilisation phase.

3. **EEPROM or Flash** is scalable. It means that only the required part of memory will unlock for use, which is needed for a particular operation. The authorizer will use its required part; the application will use its other part. It is protected by the access conditions stored therein. It stores the following:

   (i) PIN (Personal Identification Number), the allotment and writing of which is by the authorizer (for example, a bank) and its use is possible by the latter only by using the personalisation and fabrication keys. It is for identifying the card user in future transactions.

The card user is given this key. Alternatively, a modifiable password is given to the user and a password opens the PIN key.

(ii) An unblocking PIN for use by the authorizer (say bank). Through this key, the card circuit identifies the authorizer before unblocking. Data of the user unblocks for the authorizer and storing of information on the card is possible by the authorizer through the host.

(iii) Access conditions for various hierarchically arranged data files.

(iv) Card user data, for example, name, bank and branch identification number and account number or health insurance details.

(v) Post issuing data that the *application* generates. For example, in case of e-purse, the details of previous transactions and current balance. Medical history and diagnosis details and/or previous insurance claims and pending insurance claims record onto a medical card.

(vi) It also stores the application's non-volatile data.

(vii) Invalidation lock sent by the host after the expiration period or card misuse and user account closing request. It locks the data files of master or elementary individual files or both.

4. **RAM** stores the temporary variables and stack during the card operations by running the OS and the *application*.

5. **Chip power supply** voltage extracts by a charge pump circuit. The pump extracts the charge from the signals from the host analogous to what a mouse does in a computer and delivers the regulated voltage to the card chip, memory and I/O system. Signals can be from antenna or from clock pin. In a typical card operation using 0.18-μm technology, 1.6 V to 5.5 V is the threshold limit and for a 0.35-μm technology, 2.7 V to 5.5 V.

6. **I/O System** of chip and host interact through asynchronous serial UART (Figure 3.2a) at 9.6 k or 106 k or 115.2 k baud/s. The chip interconnects to a card hosting system (reader and writer) either through the gold contacts or through cm-size antenna coils on both the sides. The latter provides *contact-less* interconnection. I/O pins, which are used for *contact-based* interaction, the RST (Reset Signal from host) and Clock (from host). Interaction is through radiations through the antenna coils for contact-less interaction. The card and host interact through a card modem and a host modem. Modulation is with 10% index amplitude modulating carrier of 13.66 MHz. A one-sixteenth frequency sub-carrier modulates the load through BPSK (Binary Phase Shifted Keying).

11.4.2 Embedded Software

Smart cards are one of the most used systems today in the area of Secure SoC Systems. The smart card is an exemplary secure SoC. The card needs cryptographic software. RTOSs described in Chapter 10 do not suffice for this requirement and for other essential features. Embedded software in the card needs special features in its operating system over and above the MS DOS or UNIX system features. Special features needed are as follows:

1. *Protected environment*. It means it should be stored in the protected part of ROM.

2. A *restricted run-time environment*.

3. Its *OS*, every method, class and run time library *should be scalable.*

4. Code-size *generated should be optimum*. [Section 5.12] The system needs should not exceed 64 kB memory.

5. Limited use of data types; multidimensional arrays, long 64-bit integer and floating points and very limited use of the error handlers, exceptions, signals, serialisation, debugging and profiling.

6. Three-layered file system for the data. One file for the *master file* to store all file headers. A header means file status, access conditions and the file lock. The second file is a *dedicated file* to hold a file grouping and headers of the immediate successor elementary files of the group. The third file is the *elementary file* to hold the file header and its file data.

7. There is either a fixed length file management or a variable file length management with each file with a predefined offset.

8. It should have classes for the network, sockets, connections, data grams, character-input output and streams, security management, digital-certification, symmetric and asymmetric keys-based cryptography and digital signatures using DES/3, DES, RSA and SHA 1. [For explanation of terms, refer to *Internet and Web Technologies* by Raj Kamal from Tata McGraw-Hill, New Delhi, 2002].

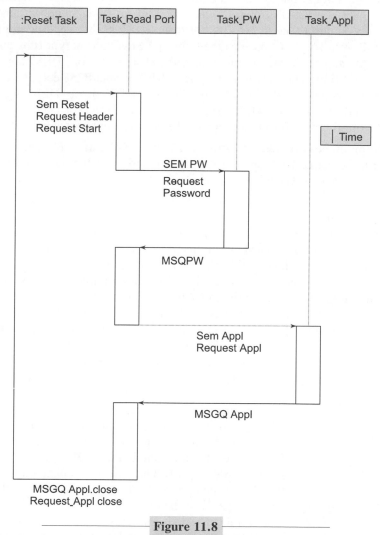

Figure 11.8

Tasks and their synchronisation model in smart card exemplary software.

Java Card[TM], EmbeddedJava or J2ME (Java 2 Micro Edition) provides one solution. [Refer to Section 5.9 and website references therein]. JVM has a thread scheduler built in. No separate multitasking OS is thus needed when using Java because all Java byte codes run in JVM environment. Java provides the features to support (i) security using class (java.lang.SecurityManager), and (ii) cryptographic needs (package java.security[*]). [*Refer to "Using Java[TM] 2 Platform" by Joseph L. Weber, Que Corporation, Reprint by Prentice Hall of India, New Delhi, May 2000*]. Java provides support to connections, datagrams, I/O streams and network sockets. Java mix is a new technology in which the native applications of the card run in C or C++ and downloadable applications run in Java or Java Card[TM]. The system has both an OS and JVM.

Assume a contact-less smart card of a bank, in which the card is not magnetic. [The earlier card used a magnetic strip to hold the non-volatile memory. Nowadays, it is EEPROM that is used to hold non-volatile application data.] SmartOS is an assumed hypothetical OS in this example, as RTOS in the card. Remember that a similar OS function name is used for understanding purposes identical to MUCOS, but actually SmartOS has to be different from MUCOS. Its file structure is different. It has two functions as follows: The function unsigned char [ ] SmartOSEncrypt (unsigned char *applStr, *EnType type*) encrypts as per encryption method, *EnType* = "RSA" or "DES" algorithm chosen and returns the encrypted string. The function unsigned char [ ] SmartOSDecrypt (unsigned char *Str, DeType *type*) encrypts as per deciphering method, DeType = "RSA" or "DES" algorithm chosen and returns the deciphered string. SmartOSEncrypt and SmartOSDecrypt execute after verifying the access conditions from the data files that store the keys, PINs and password.

Step 1: Receive from the host, on card insertion, the radiation of carrier frequency or clock signals in case of contact with the card. Extract power supply and supply to the modem, processor, memories and Port_IO (card's UART port) device.

Step 2: Execute codes for a boot up task on reset, ResetTask. Let us code in a similar way as the codes in Example 11.1 for FirstTask. The codes begin to execute from the *main* and the main creates and initiates this task and starts the SmartOS. There it is the ResetTask, which executes first.

Table 11.4 gives the tasks for the card OS in this case study. Task synchronisation model, which is apparent from the table, is as follows and is also shown in Figure 11.8.

1. **ResetTask** is the task that executes like FirstTask in Example 11.1. It suspends permanently after the following: (a) It initiates the timer for the system ticks, which are reset at 1 ms. (b) It creates three tasks, Task_ReadPort, Task_PW and Task_Appl of the system that are described below. (c) For a waiting task Task_ReadPort, it sends into a message queue MsgQStart, the request header string *requestHeader*. The latter specifies the bank-allotted PIN to the user. (d) It also sends another string *requestStart* a request for host PIN at the I/O Port, Port_IO, in order to identify the host. (e) It posts a semaphore flag, SemReset, and suspends itself so that system control does not return to it until another reset.

2. Refer to Example 4.5. A macro function ReceiveStr (&Str) uses function 'void portIO_ISR_Input (*portIOdata)' to return an input string Str. The 'portIO_ISR_Input' receives the characters one by one from the port on successive calls. Similarly, a macro function SendStr (&ApplStr) is used by the function void Port_OutStr (unsigned char [ ] *applStr) to send an output string. portIO_ISR_Output sends a character to the port.

3. **Task_ReadPort** begins only when a semaphore *SemReset* is reset or posted by the ResetTask. (a) Task_ReadPort takes the message from the queue MsgQStart and gets the pending queue messages, *requestHeader* and *requestStart*. It encrypts these two strings and sends to the Port_IO, which transmits it to the host through the modem. It receives through the modem, the host message hostStr. Host specifies, by the hostStr, the host PIN. This PIN is the one used for

the bank authorisation PIN of the card. (b) It posts the semaphore *SemPW* flag to the waiting task Task_PW, if the presently running task verifies the hostPIN. It waits for a message from the queue MsgQPW and receives *userPW* after deciphering the port input data string. (c) It posts semaphore *SemAppl,* in case the user password stored in a file at the EEPROM is verified. It posts a different semaphore *SemAppl,* if it verifies the user password. It sends in the end a close request message into a queue MsgQApplClose. Message is *requestApplClose* to the Port_IO and receives encrypted string *"Closure Permitted"*. Tasks delete on deciphering.

4. Task_PW after encryption on taking the pending *SemPW* is to send the string *requestPW*. When it takes *SemPW*, it sends the *requestPW* into the MsgQPW. Task_ReadPort will send it to the host through the I/O Port, Port_IO, in order to identify the user at the host.

5. Task_Appl runs on taking the semaphore *SemAppl* and executes the operations. The operation may (i) modify user password, (ii) print a mini-statement of bank account of the user, (iii) eject requisite cash from the host, (iv) request for accepting the envelope with cash, (v) request for a print of this transaction and (vi) request for a transfer to another party. It interacts through Task_ReadPort by sending the messages through the queue MsgQAppl.

Table 11.4

List of Tasks, Functions and IPCs Used in Example 11.4

Task Function	Prio-rity	Action	IPCs pending	IPCs posted	String or System or Host input	String or System or Host Output
Reset Task	1	Initiates system timer ticks, creates tasks, sends initial messages and suspends itself.	None	SemReset, MsgQStart	SmartOS call to the main	*request-Header*; request *Start*
Task_ Read Port	2	Wait for ResetTask Suspension, sends the queue messages and receives the messages. Starts the application and seeks closure permission for closing the application.	SemReset, Messages from MsgQStart, MsgQPW, MsgQAppl, MsgQAppl-Close	SemPW, SemAppl,	Functions SmartOS-Encrypt, SmartOS-decrypt, ApplStr, Str, close-Permitted	*request-password,* request-*Appl,* request-ApplClose
Task_PW	3	Sends request for password on verification of host when SemPW = 1.	SemPW	*MsgQPW*	*request-Password*	-
Task_ Appl	8	when SemPW = 1. runs the application program.	SemAppl	*MsgQAppl*	-	-

Example 11.4 gives the exemplary coding procedure for an application of this card.

Example 11.4

1. /* <u>Preprocessor definitions</u> for maximum number of inter-process events to let the SmartOS allocate memory for the Event Control Blocks */

#define SmartOS_MAX_EVENTS 24/* Let maximum IPC events be 24 */

#define SmartOS_SEM_EN 1/* Enable inclusion of semaphore functions in application. */

#define SmartOS_Q_EN 1/* Enable inclusion of queue functions for sending the string pointers to Task_ReadPort * /

#define SmartOS_Task_Del_En = 0 /* Disable task deletion by SmartOS at the start. */

/* End of preprocessor commands for enabling IPC functions of the SmartOS*/

2. /* Specify all user prototype of the reset task function that is called by the main function and is to be scheduled by SmartOS first at the start. In step 11, we will be creating all other tasks within the reset task. Remember: Static means permanent memory allocation. */

static void ResetTask (void *taskPointer);

static SmartOS_STK_ResetTaskStack [ResetTask_StackSize];

3. /* Define public variable of the task service and timing functions */

#define SmartOS_TASK_IDLE_STK_SIZE 100 /* Let memory allocation for an idle state task stack size be 100*/

#define SmartOS_TICKS_PER_SEC 1000 / * Let the number of ticks be 1000 per second. An RTCSWT will interrupt and thus tick every 1 ms to update counts. */

#define ResetTask_Priority 1 /* Define reset task in main priority */

#define *ResetTask_StackSize* 100 /* Define reset task in main stack size */

STAF _In = 0; /*Define flag for signaling modem interrupt for receiving a character. */

STAF_Out = 0; /*Define flag for signal from a modem interrupt after sending a character. */

/*———*/

/3. / <u>Prototype definitions for three tasks</u> for the car application codes after reset. */

static void Task_ReadPort (void *taskPointer);

static void Task_PW (void *taskPointer);

static void Task_Appl (void *taskPointer);

4. /* <u>Definitions for three task stacks.</u> */

static SmartOS_STK Task_ReadPortStack [Task_ReadPortStackSize];

static SmartOS_STK Task_PWStack [Task_PWStackSize];

static SmartOS_STK Task_ApplStack [Task_ApplStackSize];

5. /* <u>Definitions for three task stack size.</u> */

#define Task_ReadPortStackSize 100 /* Define task 2 stack size*/

#define Task_PWStackSize 100 /* Define task 3 stack size*/

#define Task_ApplStackSize 100 /* Define task 4 stack size*/

6. /* <u>Definitions for three tasks priorities.</u> */

#define Task_ReadPortPriority 2 /* Define task 2 priority */

#define Task_PWPriority 3 /* Define task 3 priority */

6. /* <u>Prototype definitions for the semaphores.</u> */

SmartOS_EVENT *SemReset; /* First task that resets the card posts it. */

SmartOS_EVENT *SemPW; /* Task_PW posts it to send request for getting user password through the host. */

SmartOS_EVENT *SemAppl; /* Needed when using Semaphore as flag for inter-process communication between Task_ReadPort and Task_PW. */

7. /* Prototype definitions for the queues. */

SmartOS_EVENT *MsgQStart; /* Needed for IPC between ResetTask and Task_ReadPort. */
void *MsgQStart [QStartMessagesSize]; /* Let the maximum number of message-pointers at the queue be QStartMessagesSize. */
SmartOS_EVENT *MsgQPW; /* Needed for IPC between Task_PW and Task_ReadPort. */
void *MsgQPW [QPWMessagesSize]; /* Let the maximum number of message-pointers at the queue be QPWMessagesSize. */
SmartOS_EVENT *MsgQAppl; /* Needed for IPC between Task_Appl and Task_ReadPort. */
void *MsgQAppl [QApplMessagesSize]; /* Let the maximum number of message-pointers at the queue be QApplMessagesSize. */
SmartOS_EVENT *MsgQApplClose; /* Needed for IPC between Task_Appl and Task_ReadPort. */
void *MsgQApplClose [QApplCloseMessagesSize]; /* Let the maximum number of message-pointers at the queue be QApplCloseMessagesSize. */

8. /* Define both queues array sizes. Assume maximum 16 strings can be sent in a queue. */
#define QStartMessagesSize 16; /* Define size of start message-pointer queue when full */
#define QPWMessagesSize 16; /* Define size of password message-pointer queue when full */
#define QApplMessagesSize 16; /* Define size of application message-pointer queue when full */
#define QApplCloseMessagesSize 16; /* Define size of application message-pointer queue when full */

9. /* Define Semaphore initial values, 0 when using as an event flag and 1 when resource key. */
SemReset = SmartOSSemCreate (0); /* Declare initial value of semaphore = 0 for using it as an event flag from ResetTask. */
SemPW = SmartOSSemCreate (0); /* Declare initial value of semaphore = 0 for using it as an event flag from Task_ReadPort */
SemAppl = SmartOSSemCreate (0); /* Declare initial value of semaphore = 0 for using it as an event flag from Task_ReadPort */
/* Define a top of the message pointer array. QMsgPointer points to top of the Messages to start with. */
MsgQStart = SmartOSQCreate (&QStart [0], QStartMessagesSize);
MsgQPW = SmartOSQCreate (&QPW [0], QPWMessagesSize);
MsgQAppl = SmartOSQCreate (&QAppl [0], QApplMessagesSize);
MsgQApplClose = SmartOSQCreate (&QApplClose [0], QApplCloseMessagesSize);

.

.

10. /* Any other SmartOS Events for the IPCs. */

.

.

11. /* Code similar to Example 4.5, except the main function codes. These were for reading from Port A and storing a character. Here, we have Port_IO and one status flag, STAF_IO. It is reset at the beginning. */
/* Prototype Declarations for modem Port_IO input and output strings. */
char [ ] Str; /* Port_IO input string to hold the data from the host through the demodulator circuit of modem. */
char [ ] ApplStr; /* Port_IO string, which the modem transfers to host after modulation. */

```
unsigned char *portIndata; unsigned char *portOutdata;
void portIO_ISR_Input (*portIndata); /* Prototype declaration for receiving an input character. */
void portIO_ISR_Output (*portOutdata); /* Prototype declaration for sending an output character. */
/* Start of Port_IO Input Interrupt Service Routine */
void portIO_ISR_Input (*portIOdata) {
disable_PortIO_InIntr ( ); /* Function for disabling another interrupt from port IO input. */
/* Insert Code for reading Port I/O bits
portIOdata = &Str;
*/
/* Start of Port_IO Output Interrupt Service Routine */
void portIO_ISR_Output (*portIOdata) {
disable_PortIO_OutIntr ( ); /* Function for disabling another interrupt from port I/O output */
/* Define a macro for sending a String */
unsigned byte i;
#define SendStr (&ApplStr) (
portIOdata = &ApplStr; i = 0; STAF_Out = 1;
while (STAF_Out = =1 && ApplStr [i] ! = NULL) {
portIOdata = ApplStr [i]; / *Pick a character from the queue message. */
portIO_ISR_Output (&portIOdata); /*Send it to the Port_IO for the modem output. */
i++; /* Be Ready for next Character */
STAF_Out =0; /*Modem interrupt when one character sent by setting STAF_Out again. */
}
ApplStr = ""; /* Clear the Queue message for the new one*/
&ApplStr = ApplStr [0];
) /* End of the macro function SendStr. */
/* Define a macro function for string comparison. Note: 'C' function strcmp is available at a C library. In
order to optimize codes, we are not using strcmp library function, but our own macro here. */
#define boolean strcmp (ApplStr, Str) (
 .
 .
 .
)
/*_____
*/
/* Define a macro for receiving a string */
#define ReceiveStr (&Str) (
while (STAF_In != 1) { }; i=0;
/* Execute interrupt service routines for each character received at modem Port_IO*/
while (STAF_In == 1 && Str [i] != NULL) {
 portIO_ISR_Input  (&portIOdata);};
STAF_In =0; /* Remember as soon as Port A is read STAF will reset itself to reflect next interrupt status.
*/
Str [i] = portIOdata;/* Write port I/O input array element from the returned data*/
i++;}
) /* End of the macro function ReceiveStr. */
12. /* Start of the codes of the application from Main. Note: Code steps are similar to Steps 9 to 17 in
Example 8.16 */
void main (void) {
/* Initiate SmartOS RTOS to let us use the OS kernel functions */
```

SmartOSInit ( );

13. /* Create Reset task, *ResetTask* that must execute once before any other. Task creates by defining its Identity as ResetTask, stack size and other TCB parameters. */

SmartOSTaskCreate (ResetTask, void (*) 0, (void *) &ResetTaskStack [ResetTask_StackSize], ResetTask_Priority*);*

14. /* Create other main tasks and inter-process communication variables if these must also execute at least once after the ResetTask. */

15. /* Start SmartOS RTOS to let us RTOS control and run the created tasks */

SmartOSStart ( );

/* Infinite while-loop exits in each task. So never there is return from the RTOS function SmartOSStart ( ). RTOS takes the control forever. */

16. } / *** **End of the Main function** ***/

/*———
*/

/* The codes of the application reset task that *main* created. */

17. *static void ResetTask (void *taskPointer){*

18. /* Start Timer Ticks for using timer ticks later. */

SmartOSTickInit ( ); /* Function for initiating RTCSWT that starts ticks at the configured time in the SmartOS configuration preprocessor commands in Step 1 */.

19. /* Create three tasks defined by three task identities, Task_ReadPort, Task_PW, Task_Appl and the stack sizes, other TCB parameters. */

SmartOSTaskCreate (Task_ReadPort, *void (*) 0, (void *)* & Task_ReadPortStack [Task_ReadPort*StackSize*], Task_ReadPortPriority*);*

SmartOSTaskCreate (Task_PW, *void (*) 0, (void *)* & Task_PW*Stack* [Task_PW*StackSize*], Task_PWPriority*);*

SmartOSTaskCreate (Task_Appl, *void (*) 0, (void *)* & Task_Appl*Stack* [Task_Appl*StackSize*], Task_ApplPriority*);*

/* Declare requestHeader */

unsigned char [ ] requestHeader;

unsigned char [ ] requestStart;

20. while (1) *{* /* Start of the while-loop*/

/* Code for retrieving two strings requestHeader for user PIN and request for host PIN string from the protected file structure. */

.

.

/* Write the array elements after encryption. */

ApplStr = SmartOSEncrypt (requestHeader, DES); /* Using an RTOS function encrypt requestHeader and post it in message queue. */

SmartOSQPost (MsgQStart, ApplStr);

ApplStr = SmartOSEncrypt (requestStart, DES); / Using an RTOS function encrypt requestStart and post it in message queue. */

SmartOSQPost (MsgQStart, ApplStr);

SmartOSSemPost (SemReset); /* Post Semaphore event flag. */

21. /* Suspend with no resumption later, the Reset task, as it must run once only for initiation of timer ticks and for creating the tasks that the scheduler controls by preemption. */

SmartOSTaskSuspend (ResetTask_Priority); /*Suspend Reset Task and control of the RTOS passes to other tasks of waiting execution*/

22. *}* /* End of while-loop */

23. } /* End of ResetTask Codes */
/**/

24. *static void Task_ReadPort (void *taskPointer) {*
while (1) {
25. /* Wait for IPC from ResetTask. */
SmartOSSemPend (SemReset, 0, & *SemErrPointer*);
26. /* Wait for a message for requestHeader from queue MsgQStart. */
&QStart = SmartOSQPend (MsgQStart, 0, & QErrPointer);
SendStr (&QStart); /* Send it to modem Port_IO. Not after sending the message in string QStart becomes null, "". */
27. /* Wait for a message for requestStart from queue MsgQStart. */
&QStart = SmartOSQPend (MsgQStart, 0, & QErrPointer);
SendStr (&QStart); /* Send it to modem Port_IO. */
/* Receive and decipher a string from the modem Port IO. */
ReceiveStr (&Str);
ApplStr = SmartOSDecrypt (Str, DES); /* Using an RTOS function, decrypt the input string from the host */
28. /* Code for saving the Host PIN and if verified, then application commands from the host is also saved. The savings are at the protected file structure. */
.
.

SmartOSSemPost (SemPW); /* Post event flag for requesting a password at MsgQPW. */
SmartOSTimeDly (100); /* Delay for 100 ms so that let lower priority task Task_PW run. */
29. /* Wait for a message for requestPassword from queue MsgQPW. If available, send request and wait for the password. */
&QPW = SmartOSQPend (MsgQPW, 0, & QErrPointer);
SendStr (&QPW); /* Send password request to modem Port_IO. */
ReceiveStr (&Str); /* Receive a String from the modem Port IO. */
ApplStr = SmartOSDecrypt (Str, DES); /* Using an RTOS function, decrypt the input string for password from the host */
30. /* Code for verifying the deciphered user password at the protected memory or file data. If verified, then application commands from the host by posing event flag SemAppl. */
.
.

SmartOSSemPost (SemAppl);
SmartOSTimeDly (100); /* Delay for 100 ms so that let lower priority task Task_PW run. */
29. /* Wait for a message for requestAppl from queue MsgQAppl. If available, send request and wait for the application command and user data. */
&QAppl = SmartOSQPend (MsgQAppl, 0, & QErrPointer);
SendStr (&QAppl); /* Send password request to modem Port_IO. */
ReceiveStr (&Str); /* Receive a String from the modem Port IO. */
ApplStr = SmartOSDecrypt (Str, DES); /* Using an RTOS function, decrypt the input string for application command and user data from the host */
30. /* Code for using the user data and executing the application command received. */
.
.

31. /* Using an RTOS function, encrypt request closing request and post it in a message queue. The closing request is from a message queue MsgQApplClose. Then retrieve it from queue in QApplClose after encryption. */

.

.

&QApplClose = SmartOSEncrypt (requestApplClose, DES);
Send (*QApplClose);
ReceiveStr (&Str);
ApplStr = SmartOSDecrypt (Str, DES); /* Using an RTOS function, decrypt the input string for password from the host. */
/* Compare deciphered string with message "Closure Permitted". If found equal, then closure permitted, then delete this Task and other low priority tasks. */
If (strcmp (ApplStr, "Closure Permitted") {
SmartOS_Task_DelEn = 1 /* Enable task deletion by SmartOS. */
OSTaskDel   (Task_ReadPortPriority);   OSTaskDel   (Task_PWPriority);   OSTaskDel
(Task_ApplPriority);};
} /* End of While loop */
/* **End of Task_ReadPorts Codes**. */
/
***/

40. ***static void Task_PW (void *taskPointer)*** {
while (1) {
41. /* Wait for IPC from Task_ReadPort. */
SmartOSSemPend (SemPW, 0, & *SemErrPointer*);
42. /* Code for retrieving one string from the protected file structure. It is for requesting the password from the user at the host of the card. */

.

.

43. /* Write the array elements after encryption. */
ApplStr = SmartOSEncrypt (*requestPassword*, DES);
SmartOSQPost (MsgQPW, ApplStr);
SmartOSTimeDlyResume (Task_ReadPortPriority); /* Resume Delayed task Task_ReadPort. */
44. } /* End of While loop */
45. /* **End of Task_PW Codes**. */
/
***/

46. ***static void Task_Appl (void *taskPointer)*** {
while (1) {
SmartOSSemPend (SemAppl, 0, & *SemErrPointer*); /* Wait for IPC from Task_ReadPort. */
47. /* Code for retrieving one string for requesting the application commands *requestAppl* from the protected file structure. It is for requesting the password from the user at the host of the card. */

.

.

48. /* Write the array elements after encryption. */
ApplStr = SmartOSEncrypt (*requestAppl*, DES);

```
SmartOSQPost (MsgQStart, ApplStr);
49. /* Resume Delayed task Task_ReadPort. */
SmartOSTimeDlyResume (Task_ReadPortPriority);
} /* End of While loop */
50. }/* End of Task_Appl Codes. */
/**************************************************************************/
```

■ SUMMARY ■

- MUCOS RTOS functions have been used in an embedded system for a *vending machine*. After initialisation by the first task, seven tasks have been shown to control the various machine functions: (a) System user coins are of three denominations Rs. 1, 2 and 5. (b) Collection of coins at a chest. (c) Refund in case of short change. (d) Excess refund from a channel in case excess amount inserted into the machine. (e) Delivering the chocolate through a port. (f) Displaying messages as per the machine status and (g) Displaying time and date. It shows how the same type of semaphores as the MUCOS can be used as event flag, resource key and counter. It also shows how to use the MUCOS mailboxes and the system RTC.

- VxWorks RTOS has been used for embedded system codes for driving a network card after generating the TCP/IP stack. The exemplary codes show a method of code designing for sending the byte streams on a *network*. A byte stream first passes through the multiple layers with TCP/IP protocols. A network driver finally sends the stream on a TCP/IP network. How VxWorks schedules the six tasks is shown. (a) A task is for checking the insertion of application strings by the application. (b) A task for creating the application stream for the transmission control layer. (c) Two task codes are given for inserting the header fields using either of two protocols, TCP or UDP, at the transport layer. (d) A task creates IP packets for the network by fragmenting the TCP segment stream. (e) A task drives the network driver and places the byte stream into a pipe (a virtual device). (f) How the tasks handle the multiple data types at the header fields was also shown. This case study gives the reader a good understanding of the application and the various functions of the system RTC in VxWorks RTOS. The reader must be able to develop solutions for an embedded networking system using the VxWorks. VxWorks is shown to give an advantage to the IPCs; binary and mutex semaphore types, message queues and pipe. We predefined, in the tasks, the number of bytes for the messages sent to the queues and pipe. One 'C' feature, typedef, is not used. This is to optimize the memory.

- There are many automobile electronics applications of embedded systems and the important ones were summarised here. A feature in modern cars, automobile cruise control system, was described. The adaptive control system was defined. Due to the greater need of reliability from the point of view of human safety, the need for special features in OS for automobiles was explained. A standard RTOS is OSEK-OS for automotive electronics. The differences with VxWorks and MUCOS were first explained so that the reader can use later the OSEK-OS for developing embedded systems for automotive applications. A case study for an *automotive cruise control* system was discussed. The code design was given using the VxWorks after retaining and incorporating the OSEK-OS features with it during the exemplary coding.

- A case study of a *smart card* was taken and special embedded hardware and special RTOS functions needed were described. The smart card case study showed that for developing codes for an embedded system, the RTOS MUCOS or VxWorks functions might not suffice. A hypothetical RTOS, SmartOS—which has MUCOS features plus the embedded system required cryptographic features and file security—accesses conditions; also, restricted access permissions is used for code designs. After an initialisation task that executes on system booting, three tasks are scheduled by SmartOS. (a) A task reads the application strings from the card data files. It sends, after encryption, the messages to UART. It receives the encrypted strings from the UART of the host. This example also shows how the multiple functions can be handled by the same task to reduce the memory needs. It is a desired feature in the smart card case. There is only 8 kB in most cases and 64 kB in extreme cases. The task for the password as well as applications interacts through the IPCs from this task. In the end, after seeking host permission, it deletes the tasks. (b) A task is for sending the password request from the user interacting with the host. (c) A task is to get the commands for executing the desired application routines and getting user data from the host.

▣ LIST OF KEYWORDS AND THEIR DEFINITIONS ▣

- *Application Layer*: A layer consisting of fields attached before placing the message on the network so that the application at the other end understands the request and service.
- *Adaptive Algorithm*: An algorithm that adjusts and adapts to the parameters and limits the changing perturbations in a control system.
- *Adaptive Control*: A control system that uses an adaptive algorithm to generate output control signals.
- *Adaptive Cruise Control*: An automobile throttle control system to maintain constant preset cruising speed.
- *Big Endian*: The most significant integer byte placed first on the network or in the memory.
- *Charge Pump*: A combination of diode and capacitor to extract and store the charge for providing a supply to the system after a regulator circuit.
- *Checksum*: A sum representing the number of carries generated by adding the 8-bit numbers or 16-bit numbers or 32-bit numbers.
- *CRC (Cyclic Redundancy Check)*: A 32-bit or 16-bit integer calculated so that if there is an error in the transmission, it can be detected by comparing the CRC of the received message. It takes a longer time to calculate but it is better than the checksum.
- *Connection-Oriented protocol*: A protocol in which inter-network communication first takes place for connection establishment, then the message flows under a flow control mechanism, and then the connection termination occurs after suitable inter-network communication. Transmission Control Protocol, TCP, is an example.
- *Connection-less protocol*: A protocol in which inter-network communication takes place without first having a connection establishment, without a flow control mechanism and without the connection termination by inter-network communication. Usually in broadcast mode, this protocol is observed. User Datagram Protocol, UDP, is its example.
- *Cryptographic Software*: Software for encrypting and deciphering a message or set of byte streams. It uses one algorithm for encrypting and another algorithm for decrypting.

- *Data Acquisition System*: A system to acquire data from multiple ports and channels.
- *Datagram*: A stream of bytes that is independent of the previous stream. The UDP datagram has a maximum size of 2^{16} bytes.
- *DES*: Data Encryption Standard.
- *DES3*: A version of DES.
- *Electromechanical Device*: A device to operate a mechanical system using electronic signals.
- *Fabrication Key*: A key embedded in ROM at the time of card fabrication so that the card gets a unique identity.
- *'internet' Layer*: A layer for inter-networking by TCP/IP protocol. The layer consists of fields attached before placing the message on the network through routers so that the routers send the message as per the source and destination IP address fields in the layer. The fields at this layer are a network-flow controlling mechanism. For example, IP header fields in TCP/IP.
- *Invalidation Lock*: A lock, that when placed in the application data files in the card, the card becomes invalid for further use.
- *Java Card*: A Java language format for smart card applications.
- *JVM* (Java Virtual Machine): The supervisory codes that execute the Java classes compiled as byte codes by the Java compiler. The codes run with the help of JVM in a computer system.
- *LCD Matrix*: A set of Liquid Crystal Displays consisting of rows and columns, like a matrix. A suitable controller shows the characters or images at the matrix.
- *Logical Address*: A memory address used for an instruction or data byte of the RTOS or application.
- *Network Driver*: A card that sends and receives the messages physically from the network and facilitates the physical connection for the stream of bytes between the different layers. A driver may use SLIP or PPP protocol for driving on the net after placing the header and trailing bytes as per Ethernet of another protocol. Each driver has a unique address.
- *OS for the Automobile Systems*: Operating system imbibed with reliability and to provide only those features that do not cause deadlock, priority inversion or unpredictable delays.
- *OSEK*-OS*: An operating system for Automobile Embedded system software.
- *Personalisation Key*: A key placed after testing the smart card circuit. The card is personalised for its own protected area of memory and its own translation scheme for conversion between physical and logical addresses during actual running of the tasks at the card. After insertion of this key, the RTOS and application use only the logical addresses, and the processor uses this key during the translation between two addresses.
- *PIN*: Personal Identification Number. Bank or hosting service allocates this PIN. The allocation unit has its own PIN, the host PIN.
- *Protection Bit*: A bit at the ROM, which the processor uses for not letting the instructions and data in the protected part on the system buses. The processor externally blocks the write cycles for these protected addresses.
- *Radar* (*Radio Detection and Ranging*): A system that uses radio waves of smaller than 1 meter to enable ranging of short distance objects by measuring time delay between transmitted signal and reflected signal.
- *RSA*: An algorithm that uses the prime numbers. RSA stands for the first letters of the last names of its three inventors: Ron Rivest, Adi Shamir, Leonard Adleman.
- *SHA*: A Security Hash Algorithm based on a hashing function.

- *SLIP*: A Serial Line Interface Protocol.
- *String Stability of Vehicles*: Stability by maintaining constant distance between multiple streaming vehicles in a convoy on highway or VIP duty.
- *TCP Header*: A header placed at the transmission control layer as per TCP protocol.
- *Throttle Valve*: A valve to control the engine thrust and hence acceleration.
- *Transmission Control Layer*: A layer for placing TCP header fields or UDP header fields on a TCP/IP network.
- *UDP*: A protocol at the transmission control layer for sending a TCP/IP network datagram.
- *UDP Header*: A header consisting of four fields—source and destination port addresses, length and checksum.
- *Unblocking PIN*: A host PIN used for unblocking a certain part of the card memory for using. For example, for permitting a modification of the user password after unblocking (permitting access by modifying the access condition fields of) the password file in the memory.

■ REVIEW QUESTIONS ■

1. Why must a designer first understand the system requirements before designing the codes? Why are the list of tasks and synchronisation model required before using RTOS functions?
2. What is meant by scalable flash?
3. Why is Java popular for smart card applications?

■ PRACTICE EXERCISES ■

Note: Questions and Exercises 4 to 7, 8 to 13, 14 to 19 and 20 to 25 pertain to case studies in Sections 11.1, 11.2, 11.3 and 11.4, respectively.

4. The while-loop of the first task contains only task suspend functions. Explain why it should be this way.
5. Explain how the task for reading ports synchronizes with the port device driver.
6. Why use a SemAmtCount counting semaphore when the count variable is an alternative available for counting the value of coins collected at the port?
7. What is the role of Task_Display? How do you code for the multiple line messages at the LCD matrix?
8. Explain the use of SemMKey in the ACVS tasks. How does the use of SemMKey in the task differ from the use of mutex?
9. We can use the number of mailbox IPC messages from a task. Explain how this has been effectively used. Why is a message queue not used in place of multiple mailboxes?
10. Model the tasks in Case Study 11.1 as CDFGs.
11. Explain the role of each semaphore used. How will the semaphore use, and consequently modify the codes in Example 11.1, in the following modification to the system? Imagine the vending machine, using a random number generator C function, delivers one chocolate free out of average eight coin insertions and refunds all the coins to the lucky ones.
12. How does the range of priority assignments differ from MUCOS assignments in VxWorks?
13. Explain how the semaphore is used to direct the use of UDP in place of TCP at the transport layer.

14. Explain the use of the statement: 'fd = open (*netDrvConfig.txt*, O_RDWR, 0);'.
15. How are streams, larger than permitted in a packet, permitted by MTUs (Maximum Transferable Units) at any time handled?
16. Explain the list of tasks for the ACC.
17. The OSEK standard for automobiles classifies tasks into four types. What are its advantages?
18. OSEK handles ISRs and tasks distinctly. What are the advantages?
19. Why does OSEK disallow message queue posting by a task?
20. Explain priority assignments.
21. List all the macros used in the codes in Example 11.3 and the uses of each.
22. How does the contact-less smart card hardware derive power?
23. Why is the use of a processor with a memory protection bit essential?
24. What are the advantages of encryption when using a fabrication key, personalisation key, utilisation lock and PIN?
25. Tabulate the features needed in the OS for a smart card.

# Chapter 12

Hardware-Software Co-design in an Embedded System

What We Have Learnt

Recall the basics from Chapter 1. We defined an embedded system as one that has computer hardware with software embedded in it as one of its most important components. An "Embedded System" has three main components.

- It has hardware.
- It has main application software. Application software may perform concurrently a series of tasks or multiple tasks.
- It has an RTOS.

We have learnt all the three main components of the embedded systems and have covered the following topics in detail:

1. Embedded system hardware consists of a processor, memory devices, I/O devices and basic hardware units—power supply, clock circuit and reset circuit.
2. I/O devices consist of I/O ports to access the peripheral and other on-chip or off-chip units. Physical device examples are UART, modem, trans-receiver, timer-counter, keypad, keyboard, LED display unit, LCD display unit, DAC and ADC and pulse dialer.
3. Real-time clock driven software timers.
4. Virtual devices.
5. Device drivers and an interrupt-handling mechanism in an embedded system.
6. Need for power dissipation management by the processor instructions during high-speed computations, organisation of processor, memory-devices and I/O devices.
7. Selection of appropriate processor and memory devices for optimum system performance.
8. Interfacing of system buses with the memory and I/O devices, and use of DMA controller to improve system performance by enabling the I/O units that have direct access to system memories.

9. The High-level language programming concepts, program models, software-engineering approaches, RTOS and use of the IPCs and the exemplary uses of RTOS MUCOS (μC/OS-II) and VxWorks functions.

10. Case studies MUCOS and VxWorks RTOS programming tools in four areas: (i) Consumer Electronic Systems, (ii) Communication and Network Systems, (iii) Control Systems in Automotive Electronic Systems and (iv) Secure SoC systems.

We have thus learnt embedded system architecture, programming and software design and development process.

There are two approaches for the embedded system design: (1) *The software life cycle ends and the life cycle for the process of integrating the software into the hardware begins at the time when a system is designed.* (2) *Both cycles concurrently proceed when co-designing a time critical sophisticated system.* [Co-design in a SoC has a different meaning.] The final design, when implemented, gives the targeted embedded system and thus the final product. Therefore, an understanding of the (a) software and hardware designs and integrating both into a system and (b) hardware–software co-designing are important aspects of designing embedded systems.

Let us refer to an interview of Jean-Louis Brelet in an article *"Exploring Hardware/ Software Co-design with Vertex-II Pro FPGAs"* [*Xcell Journal,* pp. 24–29, Summer issue, 2002]. Brelet's reply, quoted verbatim, when asked about the expertise required for successful implementation, is as follows: "Software people must understand the nature of hardware design and type of problems encountered by the hardware team. They also must understand the possibilities and capabilities of hardware. Likewise hardware teams must have a good understanding of software and how the applications operate. Both teams must have a good understanding of each other's language and a willingness to adapt."

Wayne Wolf, Burak Ozer and Tiehan Lu, Embedded Systems Group, Princeton University, reported research findings in a paper "Smart Cameras as Embedded System" by IEEE Computer, pp. 48–53, Sept. 2002. The paper demonstrates that the selection of the right hardware during hardware design and an understanding of the possibilities and capabilities of hardware during software design is critical especially for a sophisticated embedded system development.

The objective of this chapter is to understand these choices.

What We Will Learn

We will learn the following to fulfill the above objective:

1. System's project management aspects.
2. Action plan in design cycle for system hardware–software co-designing.
3. Uses of a target system circuit or its emulator.
4. Use of In-Circuit Emulators (ICE).
5. Use of simulator, use of basic modules in source code engineering tools and use of prototype development tools.
6. Use of an IDE (Integrated Development Environment).
7. Testing of hardware by simple LEDs, scope, analyser, and bit-rate meter.
8. Debugging of system monitor codes in ROM.

There are certain issues in embedded system design that need to be addressed by any development team. These include independent software–hardware design and hardware–software co-design issues in the system development process, choosing the right processor, memory, devices and bus, and porting issues of OS/RTOS. We will learn about these issues also.

12.1 EMBEDDED SYSTEM PROJECT MANAGEMENT

Recall Section 7.8 and Table 7.9, which explained software project management using Pressman's famous four Ps: people, product, process and project. The embedded systems project management is similar to this. People involved here are *a team of software development, hardware development and system integration engineers*. Table 7.9 lists these four Ps. Embedded system project management also means *organizing these four Ps*. Table 12.1 shows how these four can be organised for the development process of an embedded systems project.

Table 12.1

Embedded System Project Management's Four Components

Component	Roles	What is not advised
People *Senior Manager* *Project* *Technical manager* *or Team leader*	Same as defined in Table 7.9 (i) Selects software and hardware languages, tools and software development process life-cycle models for the software and hardware development process (ii) Tunes and reorganizes available software and hardware specifications, designs and components and existing processes and (iii) to (v) described in Table 7.9	Same as defined in Table 7.9 Lack of appreciation to implementers perceptions and uncoordinated development
Implementers	Implements software and hardware development process and integration process by using modeling, source code engineering, testing, simulating, debugging and product verification tools	Not following the agreed and accepted design and lack of coordination among fellow implementers
Customer of an embedded system	Specifies the product and its quality requirements and negotiates cost with senior manager(s)	Interference in the development process, changing the product specifications after agreeing to these
End-users	Uses the product within the suggested boundaries	Not using product as per guidelines, for example, driving with ACC control system on an icy road [Section 11.3.]
Product (embedded system)	Same as in Table 7.9	Lack of correct product specifications
Process	Using layering model (Table 9.1), partition the system into layers between the application and hardware and adapt the process similar to the one given in Table 7.9	Incorrect partitioning and adaptation of incorrect model
Project	Embedded system project management goal is to create a successful product based on the criteria listed in Table 7.9	Improper planning, incorrect effort, estimates, and lack of successful goal-oriented focus, keeping people busy in non-project activities

12.2　EMBEDDED SYSTEM DESIGN AND CO-DESIGN ISSUES IN SYSTEM DEVELOPMENT PROCESS

12.2.1　Embedded System Development Process Goal

The goal to be achieved in the last phase of an embedded system development process is to produce a thoroughly *tested and verified system*.

12.2.2　Action Plan

Recall the software development process life cycle model. Refer to Section 7.2 and Table 7.1, which describe the action plan for an embedded system development process. Even a simple small-scale embedded system needs a detailed plan.

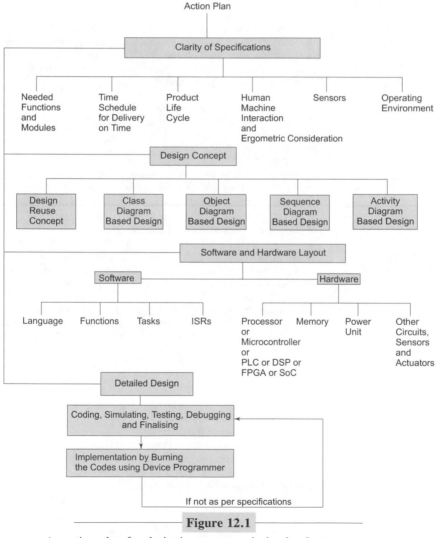

Figure 12.1

An action plan for designing a system in its development process.

Defining an action plan is the first step in any system design. Figure 12.1 shows an action plan for designing a system in its development phase.

Consider a plan for the development of a small-scale embedded system for an automatic washing machine. [A *'Clothes-In Clothes-Out'* embedded system!] We consider this example to get an easier understanding of the action plan needed in any embedded system development. For the user, it is a simple clothes-in and clothes-out system. For an embedded system designer, the things are not that easy! For the system designer, it is *'Bytes In Bytes Out'* system in place of the Clothes-In Clothes-Out machine. The designer has to design according to a full-fledged action plan. The following are the steps needed for the system development process.

12.2.3   Complete Specifications and System Requirements

The first step is to have ***complete clarity about the specifications for the needed system***. Specifications given in the first column of Table 12.2 have to be completely specified to the development team.

Table 12.2
Specifications of the System

System Specifications	Explanation
Product functions and tasks	Understanding the functions and user tasks needed for the system are essential. Consider an embedded system design for a robot. What are the functions expected from it? What are the degrees of freedom required to move of all its parts (waist, shoulder, elbows, hand and fingers)? What are the tasks it should accomplish? Is it smart?
Delivery Time Schedule	Time schedules for delivery are important. Tight delivery schedules will force high-speed rapid development model after suitably adapting linear sequential development process or force the use of a combination of the object orientation, 4th generation tools and readily available hardware design, employing general-purpose processor.
Product Life Cycle	Life cycle of the product. When a product cycle is short, it will need frequent design changes by the developer team.
Load on System	System load is an important specification. The system can fail on overloading. For example, consider a washing machine embedded system. The load can be small, medium or full. The answer will have bearing on the motor capacity, time and level of incoming water during washing and rinsing cycles. Another example is that the hardware and processor for processing an image, a video clip or a real time video will have different loads.
Human-Machine Interaction	Specifications are needed to answer these questions: What will be the human-machine interactions? This has a bearing on the plan for the keypad inputs and display outputs. What and how are the displays to be specified? The answer has a bearing on the interface circuit and the program for the displays. For example, refer to the keypad for remote-controller for a TV.
Operating Environment	Operating temperatures, humidity and environment parameters specifications are essential. A system may fail on the mountains or may fail in high temperature in the vicinity.
Sensors	Sensor specifications for sensitivity, precession, resolution and accuracy are essential for designing as per the requirement. A video-conferencing image sensor and a video image sensor will generate system input in different pixel resolutions, formats and rates.

(Contd.)

System Specifications	Explanation
Power Requirement and Environment	A system having a greater load will need greater power requirement. A battery-dependent system will need power management solutions by clever power-saving software design and hardware design. A system operating under continuous power availability condition and interrupted power availability will have different specifications. The design is simpler in case of the former as there is no memory needed to frequently save the system status.
System Cost	Maximum bearable costs must be specified to decide whether a project is acceptable to a development team and the amount of efforts to be made by the team.

System specifications for the following must be prepared before starting the design process. (i) Product functions and tasks (ii) Delivery Time Schedule (iii) Product Life Cycle (iv) Load on System (v) Human-Machine Interaction (vi) Operating Environment (vii) Sensors (viii) Power Requirement and Environment (ix) System Cost.

12.2.4 Conceptual Design

The second step is developing a *conceptual design* of the system. The question to be addressed is as follows: What will be the model of the system development process? Let us recall Section 7.10. A conceptual design model can be developed using UML approach. A conceptual design can use UML 'User Diagram', 'Object Diagram', 'Sequence Diagram', 'State Diagram', 'Class Diagram' and 'Activity Diagram'. A conceptual design helps in developing the application software and hardware structure and layout.

UML 'Class diagram', 'User Diagram', 'Object Diagram', 'Sequence Diagram', 'State Diagram' and 'Activity Diagram' help in developing a conceptual design and later on get the structure and layout of the application software and hardware.

12.2.5 Software and Hardware Layout Design

The third step is the development of a *software and hardware structure and layout* of the system. There can be two approaches.
  (1) *Independent Design Approach Followed by Integration*: Software life cycle ends and life cycle for the process of integrating into the hardware starts at the time when the system is designed.
  (2) *Concurrent Co-design approach*: Both cycles concurrently proceed when co-designing a time-critical sophisticated system.

When developing a software layout, we answer the following question. What will be the software modules needed? The exemplary three modules in an automatic washing machine system may be as follows:
  (i) A *software module*, which takes the user inputs and provides the human–machine interaction with the machine using the LED outputs. Examples of the functions *for human–machine interactions* are as follows: (a) Current default user settings on LEDs are shown at the start. Human–machine interaction can be made smart. [The default settings can be according to the user's previous preferences.] (b) A key pressing can cyclically set the clothes load (if that is

not automatic) among the three possibilities. An LED shows the selection at an instant. The user stops further pressing when the desired load is set. The processor, if it does not find any further pressing at the corresponding port, loads an input in EEPROM memory. (c) Another key pressing can cyclically set the clothes type. It can be one among the three possibilities: wool, polyester and cotton. The module also stores this user directive.

(ii) When there is a user directive to start—which signals after a key is pressed or after a remote switch signal or an Internet message arrives (in Internet compliant systems)—there has to be another *software module, which initiates the cycles and schedules each cycle-time.* It initiates the cycles (for wash, rinse and dry). (a) This has to be done after ensuring that the clothes are inserted and the door is closed. (b) The power supply output has to be according to need. (c) The power sources to the motor and solenoid valves have to be available according to need. (d) There has to be a reset of the machine status in EEPROM, if the need is to initiate the cycles from the beginning. (e) If the machine status indicates left-over cycles and functions due to some interruption during the previously run cycles in the machine, then it decides the alternative actions to be taken.

(iii) Another *module* is to *start a wash cycle.* Exemplary functions will be as follows: (a) Let the water inlet be on. (b) The water inlet must be off after the level sensors array's output shows that the water is at the level needed for the given clothes load. The water inlet should also be off after a preset time to have a watchdog action for the level sensor inputs. (c) Start the motor in slow or medium spinning mode.

Hardware and Software Implementation Tools Specifications

The foremost question before developing a layout and a detailed design is as follows: What are the elements (hardware and software requirements) of the development process? The answer can be given by selecting the elements needed to fulfill the specifications mentioned in Table 12.2. For example, the hardware requirements for an automatic washing machine are as follows: (a) Motor tank. (b) Water-level sensor-array. (c) Power sources for the motor and electronic circuit. (d) General-purpose or embedded processor or microcontroller. For example, an eight-bit *microcontroller* with adequate internal ROM, EEPROM and RAM, timers, interrupt handlers, peripheral serial or ports controller (e) An *ergonomically designed* key array and *LEDs.* Input keys are for user directives. LEDs indicate machine status and the cycles completed or remaining. (f) Interface circuits. (g) *Power source* 220V and solenoid valves for water inlet and outlet.

We may refer to different exemplary systems hardware units (Table 1.5) and refer to Sections 1.2 and 1.3 that describe hardware elements. These guide us to make a decision on the ***hardware requirements specifications***.

There are two design approaches. The first is an independent design, which follows system integration. This approach is that the software life cycle ends and the life cycle for the process of integrating into the hardware begins when a system is designed. Another approach needed for a sophisticated embedded system is a concurrent co-design approach. Both cycles concurrently proceed when co-designing a time critical sophisticated system. There are a number of software and hardware tools to implement the designed system easily with simple efforts.

12.2.6 Detailed Design

The fourth step is the ***detailed design*** of the codes and the target system by first selecting the processor and memory. [Refer to Sections 2.2 and 2.4]. Then decide about the functions that are to be implemented in the hardware and in the software. Section 12.7.1 will describe the design issues and address the issue of choice of the right platform. Software and hardware layout later helps in the detailed design for the implementation of detailed software codes and the circuit for obtaining the target system.

12.2.7 Implementation Tools

We may refer to Table 12.3, which lists hardware tools during system development process.

Table 12.3
Hardware Tools for the Detailed Design

Hardware Tools	Application
Emulator	A circuit for emulating the target system that remains independent of a particular targeted system and processor, usable during the development phase for most of the target systems that will incorporate a particular microcontroller chip. It provides great flexibility and ease for developing various applications on a single system in place of multiple targeted systems. It works independently as well as by connecting to the PC through a serial link.
In-Circuit Emulator	An emulator circuit that also emulates the target processor circuit and that must connect to the PC through a serial link and to the target system processor or microcontroller using a ribbon cable. Emulates various versions of a microcontroller family during development phase.
Logic Analysers	A power tool to collect through its multiple input lines (say, 24 or 48) from the buses, ports, etc. many bus transactions (about 128 or more). It displays these on the monitor (screen) to debug real-time triggering conditions. It helps in sequentially finding the signals as the instructions execute.
Device Programmer[@]	A programming system for a device. The device may be a PROM of EPROM chip or a unit in a microcontroller of PLA, GAL or PLC. The device is inserted into a socket (at the device programmer circuit). Programming software run at a computer using the locator output records as input. The device programs when the bytes for each address transfer to it.

[@] This is detailed in Section G.2

Hardware tools for hardware design and system integration are emulator and in-circuit emulator. Software tools are simulators, editors, compilers, assemblers, source code engineering tool, profiler (for viewing time spent at each function or set of instructions), memory scope, stethoscope-like view of code execution, memory and code coverage scope.

12.2.8 Testing

We divide the problem into small parts so that testing is easy at the initial stages. Define inputs and outputs from each stage clearly and identify and make the data flow graphs (DFGs) (Section 6.1).

Recall Sections 7.7 and 7.8. A test technique is testing by *calling the interrupt service routines*. The use of an *assert* macro is another important test technique. For example, consider a command, "assert (*pPointer* != NULL);". When the *pPointer* becomes NULL, the program will halt. We insert the codes in the program that check whether a condition or a parameter actually turns true or false. If it turns false, the program stops. We can use the assert macro at different critical places in the application program.

Testing and debugging have to occur at each stage as well as at the final stage when the modules are put together. Religiously follow the rule, *wrong until confirmed right*, for testing and debugging. Documentation in detail for each stage is also a necessity.

We may refer to Table 1.6 for a list of software tools for assembly language programming, high-level language programming, RTOS, debugging and system integration tools for making decisions on *software requirements specifications*.

12.3 DESIGN CYCLE IN THE DEVELOPMENT PHASE FOR AN EMBEDDED SYSTEM

Figure 12.2(a) shows the design cycle in the development process of any embedded system and (b) shows the edit-test-debug cycle during the implementation phase of the development process. There are cycles of editing-testing-debugging during the development phases. While the processor part once chosen remains fixed, the application software codes have to be perfected by a number of runs and tests. While the cost of the processor is quite small, the cost of developing a final targeted system is quite high and needs a larger time frame than the hardware circuit design.

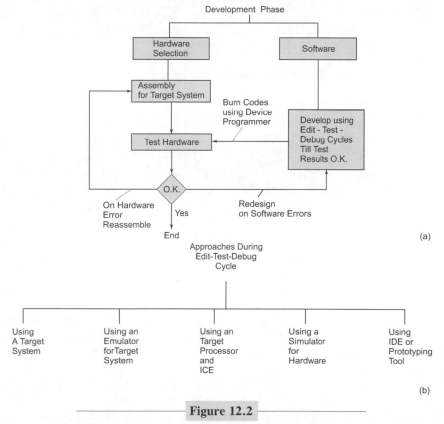

Figure 12.2

(a) Design cycle in the development process of any embedded system (b) Edit-Test-Debug Cycle during the implementation phase of the development process.

The developer uses four main approaches to the edit-test-debug cycle.

1. Uses a *target system*.
2. Uses a processor only at the target system and uses an in-between ICE (In-Circuit-Emulator). [Refer to Section 12.2.2].
3. Uses a simulator without any hardware. [Refer to Section 12.5.2].
4. Developer uses a prototype tool or IDE. [Refer to Sections 12.5.3 and 12.5.4].

12.4 USES OF A TARGET SYSTEM OR ITS EMULATOR AND IN-CIRCUIT EMULATOR (ICE)

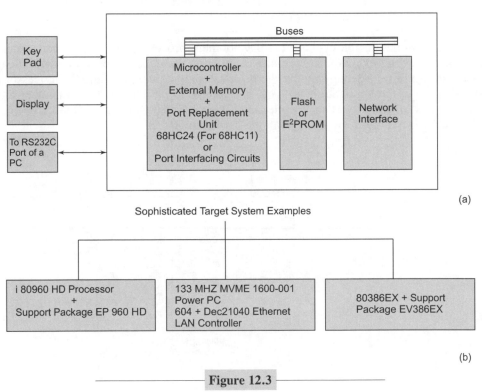

Figure 12.3

(a) Simple target system example (b) Sophisticated target system examples.

12.4.1 Using a Target System

A target system has a processor, memory, peripherals and interfaces. Some target systems have 8-MB or 16-MB flash memory and 64-MB SDRAM. [Refer to Section 2.3 for the types of memories]. A target system may possess the RS232 as well as 10/100-Base Ethernet connectivity or USB port. [Refer to Section 3.3.3 for USB].

A target system differs from a final system. It interfaces with the computer and works as a standalone system. There might be repeated downloading of the codes into it during the development phase. The target system, or its copies, simply work later as the embedded system.

Consider that a targeted system is under development. In the target system development phase, say of a *router*, the codes of the application software have to be written. These have to be embedded in the memory, flash, EEPROM or EPROM. These have to be repeatedly written or modified and tested using diagnostic, simulation and debugging tools, and embedded until a final testing in an edit-test-debug cycle shows it working according to specifications. The designer later on simply copies it into the final system or product. Also a final system may use a ROM in place of EEPROM, or EPROM or flash memory, in the target system.

An exemplary target system is a board that has an i80960HD Intel processor. Back support package is from Intel Cyclone EP960HD and it has an 8 kB data cache and 16 kB instruction cache. A support package is also from Intel for Intel 386 Ex processor. It is Intel EV 386Ex.

A WindRiver® target system board is for a 133 MHz MVME 1600-001 PowerPC 604. It has 16 kB instruction cache, 16 kB data cache and 256 kB L2 cache, a facility for on-chip nanosecond resolving PowerPC time base and a floating-point unit. It has DEC 21040 PCI Ethernet LAN controller. It sends the TCP packets through a socket of buffer size 16K. [Sixteen hundred TCP packets can queue up at the socket]. It has a TCP throughput of ~1MB/s with VxWorks RTOS.

Let us consider an exemplary sophisticated target system, VxWorks 5.4. It provides run time support by scaleable RTOS support, Internet protocols support, POSIX library support, file system and graphic supports. It has a debugging agent. It has a back-end support package for a specific processor or microcontroller. The target system parallel connects the simulator in the host computer through a target server tool with ICE (Section 12.4.2), Ethernet or serial lines from the host computer. Figures 12.3(a) and (b) show examples of simple and sophisticated target systems, respectively.

12.4.2 Emulator and ICE

Instead of the target system that is copied to obtain an embedded system, can we have a separate unit that remains independent of a particular targeted system and processor? Yes. We use an ***Emulator*** or ***ECE***. It provides a great flexibility and ease for developing various applications on a single system in place of testing multiple targeted systems. Figures 12.4(a) and (b) show an emulator and an ICE, respectively.

How does an ICE differ from an emulator? The emulator uses the circuit consisting of the microcontroller or processor itself. The emulator emulates the target system with extended memory and with codes-downloading ability during the edit-test-debug cycles. ICE uses another circuit with a card that connects to a target processor (or circuit) through a socket.

The emulator has the subunits listed and explained in Table 12.4.

Table 12.4
Emulator and ICE Subunits

Emulator subunits^	Action(s)
Interface circuit	It is for downloading ROM images into EPROM and RAM bytes from the PC into the emulator. It uses a serial (COM RS232C) port of PC. [Figures 12.4(a) and (b)]. It helps in embedding in the program memory part the large application codes directly from the PC. Codes may be developed on the PC using a high-level language. For example, in the development of the application codes, the designer finds it easier to write the large application programs instead of keying them in machine code using a small keypad at the emulator.
Socket	A multi-pin male–female socket to insert a general-purpose processor or DSP or embedded processor or microcontroller, which connects to ICE through a cable (usually a ribbon cable) and connectors. [See Figure 12.4(b), right corner socket].

(Contd.)

Emulator subunits^	Action(s)
External Memory	Additional RAM and EPROM or EEPROM, enough for use by most possible targeted systems and their applications.
Emulator-board display unit	*A single line 8- or 12- character display.* It is to show the content of memory addresses one by one. Also, it is to show the contents of registers at the various program steps.
Twenty-keys pad	It is to enter data and codes directly by the user locally at the memory addresses. These codes have to be machine codes.
Registers	Additional *system-registers* for the single step as well as full speed test runs during testing of the system.
Connectors	To plug-in this emulator to the interface circuits and other devices and peripherals that are typical to the system. A connector for the target system display module is an example. Other example is for the PC Interface circuit.
Target system keyboard	Keyboard user input board equivalent to the target system expected keyboard.
Target system driver circuit	For example, *driver* hardware for network or motor or solenoid valve or furnace or printer.
Monitor codes	These are in the emulator EPROM or EEROM.

^Emulators from Orion Instruments USA embed the logic analyser-like facility. [Section 12.6.3.] Intel provides the emulators and back support packages for its different processors and microcontrollers.

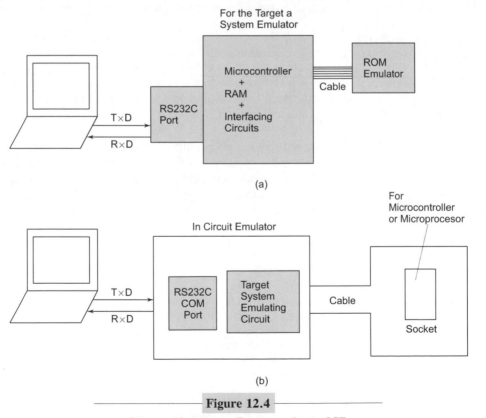

(a)

(b)

Figure 12.4

Figures 12.4 (a) An Emulator (b) An ICE.

An ICE (*In-Circuit Emulator*) consists of the following: (*i*) An *emulator pod with a ribbon cable, which extends to a processor* or microcontroller *socket* of the target system. [Figure 12.4(b)]. We can insert the processor IC in that socket and can test the target system also. [Remember this system is a one-to-one copy for the final, embedded system.] To avoid a coupling-capacity effect due to a long cable, we must use as short a cable as possible. The pod circuit emulates the target system microcontroller or processor. (*ii*) The pod links serially to COM RS232C port of a computer. Through this port, the pod gets the downloaded codes from the computer. The computer program for the emulator completely monitors the bytes at the registers and memory locations. The pod may have a card between its basic circuit and ribbon cable jumper. The replacement of this card makes it feasible to use the ICE for another version of a processor family.

What about a processor itself embedding the ICE? A feature in ARM7 and 9 processors (Section B.1) is that these processors have an ICE subunit. It helps in debugging its hardware.

The ICE or emulator disables after the development phase is over. An actual circuit forms just by copying the codes developed using the ICE. This circuit—after the inter-connection to target processor—consists of the used processor, required memory chips, keys and display units or other peripherals. This should work exactly the same and as perfectly as at the end of the development-phase that we completed using the emulator or ICE. An emulator helps in the development of the system before the final target system is ready.

Motorola provides M68HC11EVM and M68HCEVB as emulators for the 68HC11 microcontroller–based target system. These emulators have the following external connections.

When using an ICE or emulator, the software required for implementation phase are the editors, assemblers, dissembler, simulators, etc. (Section 12.5). The PC is just for downloading the codes to the emulator and for echoing-back the codes and data at the various addresses in the emulator memory. Even a highly skilled designer needs the PC to save machine-level programming time that can be too much for the sophisticated applications. We can have the additional socket connectors for the different versions of the microcontroller: for example, for emulating a 48-pin version as well as a 52-pin version of the 68HC11.

An ICE 'visionICE I' is an ICE that has the networking capabilities. The latter imbibes a 10/100 Mbps Ethernet connectivity. This lets the ICE accessible on a LAN. Remote debugging is the advantage. It also connects to the serial port of the target system.

A *ROM Emulator* (Figure 12.4(a), right side) emulates only a ROM. The target connects through a ROM socket and also connects to the computer. There is a need during the edit-test-debug cycle for downloading the codes into the target system flash or EEPROM cyclically. The ROM emulator obviates this need.

12.4.3 Use of Device Programmer for Downloading the Finalised Codes into ROM

Assume that a system design phase up to the target system is over. Refer to Figure 2.9 for a memory-map of the latter. Finalised codes must be put in nonvolatile memory at each memory address into the system. A locator output is of the final design with a booting program plus the system program (with or without a compression) plus initial data and shadow RAM data. Refer to Section 2.5. *Always remember to check before downloading that the downloaded code works when the processor resets from the processor-specific reset address.*

Refer to Table 12.3 for device programmer definition. *Burning* is a process that places the codes. Codes are the ones to be downloaded, according to ROM image (*locator* output). Section G.2 in Appendix G explains the working of a device programmer.

A burned binary image must reflect the bootstrap program and may sometimes decompress the system program before the embedded system program starts execution. The bootstrap program is the program to start up a system. [We start our days at home by strapping on our boots]. Burning is done in the laboratory using a device programmer into an erased EPROM or EEPROM or PROM. [Note: An IDE incorporates the device programmer within it].

Alternatively, we send the application software codes and data in a tabular form to a specialised manufacturer. We then get a ROM version. A ROM version is needed especially when thousands of pieces are needed. We then integrate the processor and this memory with the RAM and other hardware of the target system to obtain the final product.

12.5   USE OF SOFTWARE TOOLS FOR DEVELOPMENT OF AN EMBEDDED SYSTEM

12.5.1   Code Generation Tools (assembler, compiler, loader and linker)

Table 1.6 listed and defined, including the simulator, thirteen software modules and tools. A high-level language is machine independent. It will have an expression like $X = X + 23$, or $X = 2*Y + V*Z + 19$, etc. Each language needs a compiler. It may not be executable using an interpreter such as 'Basic'. A tool is needed for obtaining the machine codes for a target system. We may be provided a high-level language tool in C. [Refer to Section 5.1]. The designer writes the mnemonics or C program, using Editor. The mouse and keyboard combinations of the PC are for entering the program codes.

1. An *interpreter* does expression-by-expression (line-by-line) translation to the machine executable codes.

2. A *compiler* uses the complete set of the expressions. It may also include the expressions from the library routines; i.e., standard tailor-made programs. While an interpreter helps with the online execution of the codes, a compiler helps in the offline programming for obtaining the executable machine codes later. The C programs are used with an interpreter as well as with a compiler.

3. An assembly language program has the mnemonics that are machine dependent. An example of a mnemonic is SBC *A*, 0x0B. It means an instruction, which subtracts, along with the previous 'carry', the *A* register of the processor with the hexadecimal number 0x0B. An assembly mnemonic is specific to a processor or microcontroller. It is according to the instructions provided in the instruction set. The assembly mnemonics need an interpreter to translate into the machine codes that are executed on a specific processing device.

4. A *dissembler* translates the object codes into the mnemonics form of assembly language. It helps in understanding the previously made object codes.

5. An *assembler* is a program that translates the assembly mnemonics into the binary opcodes and instructions, that is, into an executable file called an object file. It also creates a list file that can be printed. The list file has address, source code (assembly language mnemonic) and object codes in hexadecimal. The object file has addresses that are to be adjusted during the actual run

of the assembly language program. A loader is a program that helps in this task by reallocating addresses before the opcode and operands load in the computer memory.

6. A *linker* links the needed object code files and library code files. This is before the *loader* reallocates the addresses and puts them at the physical addresses in the memory, and the program runs on the computer. Loader performs the analogous functions on host machine as the locator does on a target system in conjunction with a device programmer.

7. A *cross assembler* is useful to convert object codes for a microcontroller or processor to other codes for another microcontroller or processor and vice versa. Utilizing cross assemblers lets us use a processor of the PC of the development system. It later on provides the object codes. These codes would actually be needed in the finally developed system that will use another processor or microcontroller.

12.5.2 Simulator

Before flying an aircraft or fighter plane, a pilot uses the flight simulator for training. [A flight simulator may cost hundreds of millions of dollars!] Similarly, software simulates a hardware unit like emulator, peripherals, network and input-output devices on a PC. A *simulator* remains independent of a particular targeted system. It is usable during the development phase for application software for the system that is expected to employ a particular processor or processing device chip. The simulator is essentially software to simulate all functions of an embedded system circuit that includes any additional memory, peripherals devices and buses. The simulator uses a cross-compiler, linker and locator like in the actual target system.

A simulator fails in case of ASICs that may embed the IP(s) or RISC processor. [The ASIC manufacturer and RISC processor may however give an alternative debugging tool. For example, ICE in the processor ARM7 and ARM9 (Section B.1).] A simulator does not resolve timing issues and hardware-dependent problems. Recall Section 6.3.9. A simulator may fail to show a bug from shared data as it arises from an interrupt in a particular situation only. For example, a long word in four registers loaded only partly, and exactly at that moment the interrupt occurred.

A simulator helps in the development of the system before the final target system is ready with only a PC as the tool for development. Simulators are readily available for different processors and processing devices employing embedded systems, and a system designer and/or developer need not code for the simulator for application software and hardware development in the design laboratory. Figure 12.5 shows the detailed design development process using the simulator. A typical simulator is mostly run on a PC Windows environment. A typical simulator includes the following features.

1. It defines the processor or processing device family as well as its various versions for the target system.
2. It monitors the detailed information of a source code part with labels and symbolic arguments as the execution goes on for each single step.
3. It provides the detailed information of the status of RAM and ports (simulated) of the defined target system as the execution goes on for each single step.
4. It provides the detailed information of the status of peripheral devices (simulated, assumed to be attached) with the defined system.
5. It provides the detailed information of the registers as the execution goes on for each single step or for each single module. It also monitors system response and determines throughput.

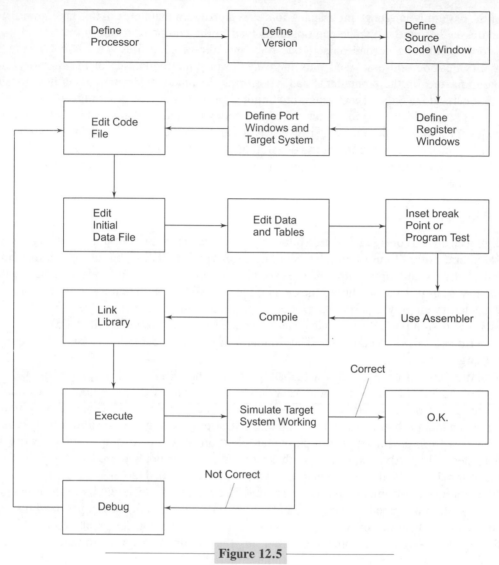

Figure 12.5

The detailed design development process using the simulator.

6. The windows on the screen provide the following:
   (i) The detailed information of the status of stack, ports (simulated) of the defined microcontroller system, and
   (ii) Program flow trace as the execution continues. A trace means the output of contents of the program counter versus the processor registers. It is an important debugging tool of an assembly language program. A trace of application software means an output of chosen variables in a function of stepping sequence. Tracescope gives a time on X-axis and a chosen parameter on Y-axis as the program continues further. [Tracescope is a Tornado® tool module to obtain a trace of the changes in the modules and tasks with time on the X-axis. An actions-list also produces with expected time scales].

7. It helps the window on the screen to provide the detailed meaning of the present command.
8. It monitors the detailed information of the simulator commands as these are entered from the keyboard or selected from the menu.
9. It incorporates the assembler, dissembler, user-defined keystroke or mouse-selected macros, and interpreters for C language expressions, as well as for assembly language mnemonics (expressions). It thus tests the assembly codes. The user-defined keystroke macro is a very useful facility. For example, we can define keystroke 1, say, for providing a particular input byte at a Port n and a particular RAM address byte.
10. It supports the conditions (up to 8 or 16 or 32 conditions) and unconditional breakpoints. There is a feature that halts a program after a definite number of times an instruction executes. Breakpoints and the trace together are the important testing and debugging tools.
11. It facilitates synchronizing the internal peripherals and delays.
12. It employs the RTOS scheduler that preempts the task.
13. It simulates the inputs from the interrupts, the timers, ports and peripherals. So it tests the codes for these.
14. It provides network driver and device driver support.
15. Recall Section 9.6.6. Consider multitasking cyclic scheduling. How do we decide the time slice after which the context switches? How do we decide the timeout? Using a simulator, we can schedule the various tasks by a *Simulated Annealing* method. It simulates the different fixed schedules. It then gradually increments the schedules of each task (timer delay settings), and continues as long as none of the tasks misses the deadline.

Section 12.5.3 describes a powerful simulating tool, VxSim. Another tool is pSOSim. It simulates RTOS pSOSystem pSOS+â kernel. pSOSim has unlimited choice of tools that exhibit the features given in Table 12.5.

Table 12.5
Features in an Exemplary Simulator, pSOSim

Supporting Features	Activities
Application Development Tools	It supports the UML and 'RougeWave'. It gives a short design cycle.
Native Development Environment	It supports several native development environments and debugging. Environment may be MS Visual C++ or GNU tools.
APIs to RTOS	Simulates use of many APIs of the RTOS for given hardware.
Debugging Capability	Powerful debugging capability enables fault finding a much easier task.
Device Simulation	Simulates devices and device driver behavior.
Network Simulation	Network simulation capabilities make it a virtual test bed, which permits modeling of complex multi-node networked systems. For example, a router or gateway. A network application can simulate internal subnet or a real network. When simulating a network, it generates stacks for various standard network protocols that include even the IP multicast and IP broadcast.
User Interface	Simulates, for example, a set-top box interface.

12.5.3 Exemplary Prototype Development, Testing and Debugger Tools for Embedded Systems

Table 1.6 describes other embedded system development tools that are not a part of the scheduler. But these tools help as power tools for using an RTOS scheduler. One concept is the use of *a prototype development tool in place of target system hardware*. An exemplary prototyping tool is Tornado **Prototyper** from WindRiver® for integrated cross-development environment with a set of tools. A set of prototyping tools in Tornado Prototyper also implements the RTOS VxWorks (Section 10.3). These tools simulate, compile, and debug with a *Browser*. The *Browser* summarises the final targeted embedded system's complete status during the development phase. Table 12.6 gives the features of a set of prototyping tools from WindRiver.

Table 12.6
Set of Prototyping Tools from WindRiver®

Tool	Features
ScopeProfile	This dynamic execution profiler lets us see, like an oscilloscope waveform, where the CPU is spending its cycles. Performance bottlenecks can then be understood. It shows how much time the processor spends in each function in the task or ISR.
MemScope	Memory usage is a critical aspect of an embedded system. Is there any wasteful use of memory? Are there any memory leak errors? A memory leak means that a pointer is incrementing into the unassigned area for a task or stack overflow or writing at the end of an array. MemScope gives the memory block usage. It detects the leak due to system call or another ported module.
StethoScope	Just as a stethoscope helps a doctor in diagnosis, it dynamically tracks the changes in any program variable. It tracks the changes in a parameter. It lets us understand the sequences of multiple threads (tasks) that execute. It records the entire time history.
TraceScope	It helps in tracing the changes with time on the X-axis and an item from the actions list. [An actions list with expected time scales was given in Table 9.11.] TraceScope lets us find the RTOS scheduler behavior during task switching and notes the times for various RTOS actions.
CodeTest Memory, Trace and Coverage	These tools help in code testing by dynamic memory allocation analysis, controlled flow view trace and code coverage under various real-world situations. The code coverage study helps in removing the extra codes and functions not needed for a specific application. It provides the scaleable system.
VxSim	A powerful simulator tool, which provides a virtual target for developing and debugging the codes. It helps in avoiding the repeated code locating in actual target board of the embedded system. Simulating the application with VxSim is of great help in the early development stage, as the RTOS task scheduling can be thoroughly simulated before implementation into the target.
VxWorks Networking Stacks	Another power tool that enhances the code development process. VxWorks RTOS prepares the stack for sending data on the Internet to test high-performance switching devices. The stack is according to protocol chosen. Protocols are the following: RFC - 1323. CIDR. IP Multicast. IP. UDP. TCP. DNS Client. DHCP Server. SMTP Server. RIPv1 support. RIPv2 support. ARP. Proxy ARP. BOOTP, RLOGIN Client and Server (for Telnet). An embedded system socket can then connect to Multi protocol LANs, ATM network or SONET or Wireless Access and Intelligent Networks.

12.5.4 Integrated Development Environment (IDE)

We can have an Integrated Development Environment (IDE) that consists of simulators with editors, compilers, assemblers, emulators, logic analysers and EPROM/EEPROM application codes burner. An IDE must have the following features.

1. It has a facility for defining a processor family as well as defining its version. It has source code engineering tools (Section 5.11) that incorporate the editor, compiler for C and embedded C++, assembler, linker, locator, logic analyser, stethoscope, and help to user.
2. It has the facility of a user-definable assembler to support a new version or type of processor. It provides a multi-user environment.
3. The design process divides into a number of sub parts. Each designer is assigned independent but linked tasks.
4. It simulates on a PC, the hardware unit like emulator, peripherals, and I/O devices. It supports conditional and unconditional breakpoints.
5. It debugs by single stepping. It has the facility for synchronizing the internal peripherals.
6. It provides Windows on the screen. These provide the detailed information of the source code part with labels and symbolic arguments, the registers as the execution continues, the detailed information of the status of peripheral devices, status of RAM and ports, the status of stack and program flow as it continues.
7. It verifies the performance of a target system that an emulator built into the development system, which remains independent of a particular targeted system, plus a logic analyser for up to 256 or 512 transactions on the address and data buses after triggering.

An IDE tool is Tornado Tool 3 from WindRiver® Systems and it employs VxWorks RTOS (Section 10.3). The Tornado Tool architectural feature is "*dynamic linking and incrementally loading the object modules*" into the target system. Exemplary target processor families that are supported are PowerPC, Intel, Motorola, Pentiums, MIPS and ARM/Strong ARM. It helps in prototype development and tests the prototype applications. There is a text editor with GNU C/C++ compilers. Debugging is done at three levels—source-code level, task level (scheduling, IPCs and interrupts study) and domain level. It includes VxSim, stethoscope and tracescope. Figures 12.6(a) and (b) show a simple and a sophisticated IDE.

An IDE for an RTOS pSOSystem is pRISM+. Target processor families that are supported are PowerPC, Motorola 68K, Intel 80x86 and Pentiums, MIPS, ARM and Mitsubishi M32/R. It has the source code engineering tools that include the pRISM+ Editor, SNIFF+ (section 5.8). It has a cross compiler, source level debugger, object browser and monitor for run-time behavior and event-to-event viewing. The object browser browses the applications behavior overtime. It graphically displays the pSOS RTOS tasks, queues, semaphores and IPC objects. A Real-Time Analysis (RTA) suite profiles the code coverage and locates runtime errors. It optimises the use of the memory.

> Code generation tools are used for creating and compiling. Then the codes are tested using simulators and a number of the latest software tools like profiler, memory scope, stethoscope, and memory and code coverage scope.

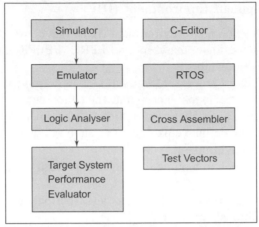

IDE for Various types and Versions of Microcontroller
with Upgradability of IDE for future Versions.

(a)

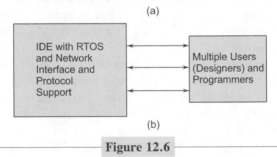

(b)

Figure 12.6

(a) Simple IDE (b) Sophisticated IDE.

12.5.5 Memory- and Processor-Sensitive Programs and Device Drivers

Standard tools for the source code engineering, emulators, simulators and prototype systems may not give optimum solutions in cases of memory- and processor-sensitive programs and for device drivers. Table 12.7 gives the examples of memory-, processor-sensitive programs and cases of device drivers.

Table 12.7
Hardware-Sensitive Programming

Program	Examples
Processor Sensitive	Recall Section 2.1. A processor has different types of structural units. It can have memory-mapped I/Os or I/O-mapped I/Os. I/O instructions are processor sensitive. A processor may be have a fixed-point ALU only. Floating-point operations when needed are handled differently than in a processor with floating-point operations. A processor may not provide for SIMD and VLIW instructions. [Section 6.3.] Programming of the modules needing these instructions is handled differently in different processors. Assembly language may sometimes facilitate an optimal use of the processor's special features and instructions. Certain processors provide compiler or optimizing compiler sub-unit to obviate need for programming in assembly.

(Contd.)

Program	Examples
Memory Sensitive	(i) Examples of memory-sensitive programs are video processing and real-time video processing. How many frames can be processed with a given number of missing frames? It will depend upon the memory available as well as latencies (Section 4.6) for the tasks of the given real-time programming model and algorithm used by a programmer. (ii) Recall Section 2.5.4. Memory addresses of I/O device registers, buffers, control-registers and vector addresses for the interrupt sources or source groups are prefixed in a microcontroller. Programming for these takes into account these addresses. These addresses can't be allotted to the tasks by an RTOS. Memory-sensitive programs must also optimise the memory use by skillful assembly language programming [Refer to Section 5.12 for memory optimisation techniques for programming].
Device Drivers	Not only the addresses of I/O device registers, buffers, control registers, vector addresses for the interrupt sources or source groups are prefixed—the control register bits and status register bits let a device's powerful features unfold. Programming, using each bit, helps in exploiting all features of a device by a programmer. Device driver codes in an assembly may become very useful. These provide a direct control of the I/O ports, registers, stack and RAM and are easy to program [Refer to Examples in Chapter 4]. Refer to Sections 4.1.4 and 9.12. There are many open source drivers available for ports, buses, and physical media attachments in Linux. Device drivers in Linux let us use each module of a class of device register, deregister and schedule like a process. Programmers can port these directly since these are open sources also.

Appropriate interface functions are needed for processor-sensitive, memory-sensitive programs and for interrupt service routines porting into the system. Appropriate drivers are needed for device-sensitive programs.

12.5.6 Use of Dynamically Linked Libraries (DLLs)

Let us recall that we mentioned that the use of standard library functions like square root ( ) saves the programmer time for coding, and a new set of library functions exists in an embedded system, specific to C or C++ compiler. Examples are for delay ( ), wait ( ) or sleep ( ). Also recall Example 5.1. We included in the preprocessor these commands: "vxWorks.h" for VxWorks functions, "semLib.h" for semaphore functions library, "taskLib.h" for multitasking management functions library, "msgQLib.h" for message queue functions library for "fioLib.h" for file-device I/O functions library and "sysLib.c" for system library for system functions (for example, a system timer for real-time software timers and "*prctlHandlers.c*" for library of protocol handler functions).

The compiler usually links the needed library functions from the source header and library files *at the compiled time*. Thus ROM image in the embedded systems has the needed library readily available. This offers an advantage of the simple and static executable file.

Recall Section 6.1.2. A CDFG graph-independent path from a decision node may be such that it is less often executed. Memory blocks are taken away by less frequently executed library functions in case of static linking at compile time. When it is feasible to dynamically link these functions, RAM space use can be optimised by using at run-time melloc ( ) and free ( ) functions. These functions can be called dynamically to link with the program and free the memory after their use. Linking program can be fetched through an I/O port or another host or server part of the system.

Assume that there are distributed embedded systems networked together. A library function may be needed only in certain specific control conditions. Needed function can be called using an inter-process communication, RPC (Section 8.3.7) and dynamically linked with the system program. A library function can also update by using dynamic linked library function.

The use of dynamic linked libraries (DLLs) saves the memory space at the risk of an out-of-memory error. [The error arises only in case of improperly tested programs]. This is because the functions of the library can load or unload. These are called in the memory only at the runtime. A DLL should however be short and have a high fraction of the needed functions compared to the functions not needed by a system.

DLLs help in reconfiguring the system when there are changes in the system configuration.

> Use of dynamic linked libraries (DLLs) is made when it is possible to link library functions from a source (port or network) and these functions save the memory space.

12.6 USE OF SCOPES AND LOGIC ANALYSERS FOR SYSTEM HARDWARE TESTS

After using the simulator, ICE and debug codes in ROM, in the last stage of debugging, we use a troubleshooting hardware diagnostic tool that records the state as a function of time and as a function of other states.

12.6.1 Logic Probe or LED tests

Let us remember that a digital signal is simply a discrete signal between two voltage ranges. In CMOS logic, '1' is the discrete range V_{DD} to 0.66 V_{DD} and '0' is 0.33V_{DD} to V_{DD}, where V_{DD} is usually 5V with respect to V_{DD} (ground potential). An analog signal varies continuously. A *logic probe* is the simplest hardware test device. It is a handheld pen like device with LEDs. Its LED glows green, when probe tip touches at the test point (a port or hardware pin) and the bit there is '1'. It glows red if it is '0'. A logic probe LED blinks fast in a probe version, when a test point is at '1' and does not blink when at '0'. The device at the other end connects by a wire to the ground potential.

A logic probe becomes an important tool when studying long delay effects (>1s) at a port. A short program for a delay and then sending the results at the port using logic probe will test the OS timer ticks.

12.6.2 An Oscilloscope

Finally, code-downloaded hardware needs testing after completing the edit-test-debug cycle, using a simulator or IDE. An oscilloscope is a scope with a screen to display two signal voltages as a function of time. It displays analog as well as digital signals as a function of time.

We must use it with DC (directly coupled) inputs. Another terminal of the input is always properly kept at ground potential. The term DC should not be confused with the direct current supply. The Oscilloscope has two selections for an input, DC and AC. DC means directly coupling the input to the scope input-amplifier. AC means input to amplifier (vertical section amplifier) after a capacitor. When using at AC, the signal shape can distort. Its use is only when viewing the signal in a form, in which the signal amplitude = 0 when averaged over time (alternating component of the given signal). [Averaging means a simple average, not a root mean square average]. If the bus signals are viewed with AC selection, a false overshoot or undershoot may show up.

A clock, if running, will show the states on the scope. The horizontal gap between the successive rising edges gives us the clock time period. For example, in an 8051 using 12 MHz crystal, there will be states, each of period 0.0825ms. A check for this and the ALE (address latch enable) simultaneously at two scope input amplifiers will test the processor activity. Another scope use is in checking the real-time clock routines and pulse width output routine. Real-time software tests and debugging are easy using the scopes. The output signal on a serial port or the output bit on a parallel port test will provide significant information. A scope is usable for testing the delay time routines. We can set three delay parameters in three registers and note the time taken for the port bit to change with each run. From these three times, we estimate the actual setting in the register for requisite delay. We then run with this delay parameter setting and test, using the scope and fine-tune to the exact delay setting.

An advantage of the scope is its use as a noise detection tool and as a voltmeter. Another use is detection of a sudden in-between transition between '0' and '1' states during a clock period. This debugs a bus malfunction. A *storage scope* is another version of an oscilloscope. It stores the signals versus time. Later we analyze the stored activity.

12.6.3 Logic Analyser

A logic analyser can easily debug a small-level embedded system. It is a more powerful tool than the scope. A scope views and checks only two signal lines. A logic analyser is a powerful software tool for checking multiple lines carrying the address data and control bits and the clock. The analyser recognizes only discrete voltage conditions, '1' and '0'.

The analyser collects, stores and tracks the multiple signals simultaneously and successively. There are multiple input lines (24 or 48). We connect the lines from the system and I/O buses, ports and peripherals. The analyser simultaneously collects for the duration of the bus transactions (128 or more). It later displays, using this tool, each transaction on each of these lines on the computer monitor (screen) and also prints. The phase differences in each input line also give important clues. It debugs the real-time triggering conditions. It helps in finding the bus signals and port signal status sequentially as the instructions are executed. A variant of the logic analyser also provides the analog measurement when needed.

A logic analyser can also show the states on the horizontal axis instead of time in its state mode of display. When displaying the logic states, a particular line displays as a function of '0' and '1' on another line.

Certain bugs that intermittently arise can also be recorded with a logic analyser by continuous and repeated runs of the system. A logic analyser does not help on a program halt due to a bug. It does not show the processor register and memory contents. If the processor uses the caches, bus examination alone may not help. We cannot modify the memory contents and input parameters as we do in a simulator. The effects of these changes are invisible.

12.6.4 Bit Rate Meter

A bit rate meter is a measuring device that finds the numbers of '1's and '0' in the pre-selected time spans. How to measure the throughput, the number of bytes per second on a network? Send 0xA55A (binary 1010010101011010) repeatedly. The number of 1s received in one second divided by 4 is the throughput in bytes. Similarly we can find another bit pattern and find the expected number of bits in the given time. We can estimate the bits, '1's and '0's in a test message and then use a bit rate meter to find whether that matches with the message.

12.6.5   System Monitor Codes for Debugging in ROM

Downloaded codes in ICE may run a 'Power On Self Test' (POST) program on bootstrapping. A ROM may also have a debug monitor. The embedded system, when coupled to the RS232C COM port or network port of a computer, uses *gdb*, a GNU debugger. It is a downloadable freeware. It may also provide the debug monitor codes for the system, which should be downloaded along with the locator binary image.

Target board manufacturers also provide the debug monitor codes and the host computer's debug tools. The following are the commands for debugging actions when using this arrangement. (*i*) Reset (kill) the program or Start (restart) program. (*ii*) A write and a read to a memory address of the system. (*iii*) A write and a read to a register of the system processor. (*iv*) Single step from current or a specified address.

12.7   ISSUES IN EMBEDDED SYSTEM DESIGN

12.7.1   Choosing the Right Platform

System design of an embedded system also involves *choosing a right platform.* A platform consists of a number of units. Table 12.8 shows a list of these units and various corresponding Sections in this book, which a designer can refer to while selecting the unit to finally obtain a right platform and right development tools.

Table 12.8

List of Units to Choose for Finally Obtaining the Right Platform and Right Development Tools

Unit to be Choose	Section describing that in detail to enable the right choice
Processor	^Sections 1.2 and 2.2
ASSP	Section 1.2.6
Multiple processors	Section 1.2.7
System-on-Chip	Section 1.7
Memory	Section 2.4
Other Hardware Units of System	Section 1.3
Buses	Sections 3.3, 3.4 and F.2
Software Language	*Sections 5.1, 5.8 and 5.9
RTOS (real-time OS)	Sections 9.4, 9.11, 10.2 and 10.3
Code generation tools	Assembler, compiler, loader, linker, emulators, simulators and debuggers and system integration tools [Section 1.4.7 and Section 12.5]
Tools for finally embedding the software into binary image	Section E.2

Note: (1) ^Processors may incorporate a compiler as a subunit. For example, Philips TriMedia TM-1300 has an optimizing C/C++ compiler subunit. [TriMedia here means data, sound and image processing capabilities]. (2) *Why should a designer/developer use C++ or Java? Refer to Sections 5.8 and 5.9 for the answer to this question. Why should a designer use C? Refer to Section 5.1.

12.7.2   Embedded System Processors Choice

A. A Processor-Less System

Recall Section 1.2.1. Can we have an alternative to a microprocessor or microcontroller or DSP? Figure 12.7(a) shows the use of a PLC in place of a processor. We can use a PLC for the clothes-in clothes-out type system. [Section 12.2.2]. A PLC fabricates by the programmable gates, PALs, GALs, PLDs and CPLDs.

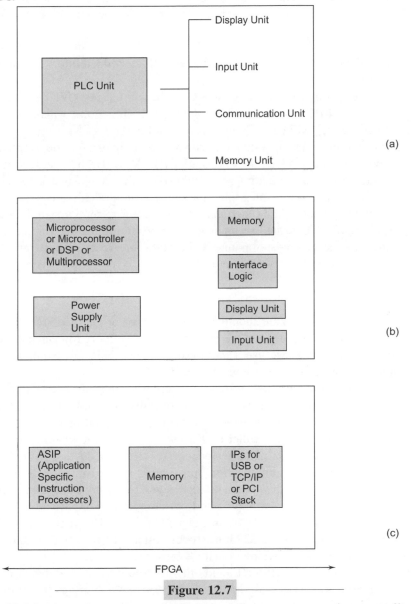

Figure 12.7

(a) Use of a PLC in place of a processor (b) Use of a microprocessor or microcontroller or a DSP (c) Processing of functions by using an IP embedded into the FPGA, instead of ALU processing.

A PLC has a very low operation speed, a very low computational ability and a very strong interfacing capability with its multiple inputs and outputs. It has system specific programmability and it is simple in application. Its design implementation is also fast. A vending machine is another exemplary application of a PLC.

B. System with Microprocessor or Microcontroller or DSP

Section 1.2 gives a detailed description of the processors described for the embedded systems in detail. Figure 12.7(b) shows the use of a microprocessor or microcontroller or a DSP.

C. System with ASIP in VLSI or FPGA

Figure 12.7(c) shows the processing of functions in using IP embedded into VLSI or FPGA instead of processing by the ALU.

A line of action in designing can: be using the IP, synthesising using a VHDL-like tool and embedding the synthesis into the FPGA. This FPGA implements the functions, which, if implemented with the ALU and designer coding, will take a long time to develop.

We implement the sophisticated operation parts on a VLSI chip using a copyrighted IP. Each IP can be synthesised at gate level using VHDL or Verilog. VHDL (VLSI High Level Description Language) and Verilog are the languages for simulating and synthesizing the gate level design. VHDL also implements concurrency and synchronisation problems and a structural hierarchy strategy. Verilog, in additional to these features, uses C-like functions. Therefore, the exception handling and timing problems are also programmable. There are two languages for programming and implementing FSM, state transitions, concurrency, synchronisation and behavioral hierarchy. These are StateCharts and SpecCharts.

12.7.3 Factors and Needed Features Taken into Consideration

Recall Section 1.2 and Table 1.6. What should be the processor or microcontroller choice? When the 32-bit system, 16kB+ on-chip memory and the need of Cache, Memory Management Unit or SIMD or MIMD or DSP instructions arise, we use a microprocessor or DSP. For example, the video game, voice recognition and image-filtering systems will need a DSP.

Microcontroller provides the advantage of on-chip memories and subsystems like the timers. Table 12.9 can help in selecting the microcontroller. A microcontroller available at a reasonable cost may not support these on-chip needs. Then decide which one will suffice, 8-bit or 16-bit or 32-bit ALU. Now make a decision about which microcontroller (and which version with what features) is needed. The first selection criterion is the on-chip memories needed for the embedded system. A second criterion is the on-chip timers and serial communication subsystems needed in each system. The third and fourth may or may not be needed in the system being designed. The third criterion is the need for input captures (interrupt and load time on an input) and out-compares (Output and interrupt when timer contents equal a comparison register). The fourth consideration is PWM and/or ADC on-chip availability. The latest versions for 8, 16 and 32 bit microcontrollers can be found from the websites of the giants: Intel, Motorola, Phillips and Microchip (for 8-bit systems).

Refer to Appendix C for 8051, 68HC11/12 and other microcontroller features. Table 12.9 gives the factors that are considered by a system designer before choosing a microprocessor or microcontroller as a processing unit.

Table 12.9
Factors and Needed Features in the Microprocessor or Microcontroller or DSP Processing Unit of the System

Factors for On-Chip Feature	Needed or which one needed	Available in chosen chip
8 bit or 16 bit or 32 bit ALU	8/16/32	8/16/32
Cache, Memory Management Unit or DSP calculations	Yes or no	Yes or no
Intensive computations at fast rate	Yes or No	Yes or No
Total external and internal Memory up to or more than 64 kB	Yes or no	Yes or no
Internal RAM	256/512 B	256/512 B
Internal ROM/EPROM/EEPROM	4 kB/8 kB/16 kB	4 kB/8 kB/16 kB
Flash	16 kB/64 kB/ 1 MB	16 kB/64 kB/ 1 MB
Timer 1, 2 or 3	1/2/3	1/2/3
Watchdog Timer	Yes or no	Yes or no
Serial Synchronous Communication Full duplex or Half	Full/Half	Full/Half
Serial UART	Yes or no	Yes or no
Input Captures and Out-compares	Yes or no	Yes or no
PWM	Yes or no	Yes or no
Single(S) or multi-channel(Mc) ADC with (W) or without(WO) programmable Voltage reference (single(S) or dual reference(D))	S/Mc W/WO V_{ref} S/D	S/M W/WO V_{ref} S/D
DMA Controller	Yes or no	Yes or no
Power Dissipation	Low or Normal	Low or Normal

12.7.4 Software Hardware Tradeoff

There is a tradeoff between the hardware and the software. It is possible that certain subsystems in hardware (microcontroller), real-time clock, system clock, pulse width modulation, timer and serial communication are also implemented by the software. The serial communication, real-time clock and timers featuring microcontroller may cost more than the microprocessor with external memory and a software implementation. Hardware implementations provide an advantage of processing speed. Hardware implementations may not only increase operation speed but also increase the power dissipation. Hardware implementation provides the following advantages. (*i*) Reduced memory for the program. (*ii*) Reduced number of chips but at an increased cost. (*iii*) Simple coding for the device drivers. (*iv*) Internally embedded codes, which are more secure than at the external ROM.

Software implementation provides the following advantages: (*i*) Easier to change when new hardware versions become available. (*ii*) Programmability for complex operations. (*iii*) Faster development time. (*iv*) Modularity and portability. (*v*) Use of standard software engineering, modeling and RTOS tools. (*vi*) Faster speed of operation of complex functions with high-speed microprocessors. (*vii*) Less cost for simple systems.

> It is possible that certain subsystems in hardware (microcontroller) are implemented by software to get an optimised performance with the lowest system cost.

12.7.5 Performance Modeling

A. System Performance Index

The performance of the finally developed embedded system is a measure of success. Its performance at each life cycle of the development process is tested for the following: each required function must show after the test that its characteristics are in conformity with the required and agreed specifications.

The system performance index can be defined as the ability to meet required functions and specifications while using the minimum amount of resources of memory, power dissipation, and devices and minimum design efforts and optimum utilisation of each resource (for example, a high CPU load).

The best embedded software and hardware is the one that achieves the balance among different performance metrics.

B. Multiprocessor System Performance

Recall Section 6.3. The multiprocessor system performance is measured by (i) an optimised partition of the program into the tasks or set of instructions between the various processors, and then (ii) an optimised scheduling of the instructions and data over the available processor times and resources. The performance cost is greater if there is more idle time left than the available time. The performance matrix (Section 6.3.6) is first obtained to calculate the total performance cost.

C. MIPs and MFLOPs as Performance Indices

One performance design metric is how long a system takes to execute the desired system functions. Processor clock frequency and MIPs (million instructions per second) and MFLOPs (million floating point instructions per second) are often quoted as design characteristics for expected system performance, which may not, however, correct metrics. A processor performance design metric is Dhrystone/s. [Refer to Appendix B for details].

D. Performance Metrics: Buffer Requirement, I/O Performance and Band Bandwidth Requirement

The buffer helps in accelerating the performance of the system. Memory or I/O buffer requirement may sometimes be a constraint. I/O performance is measured by throughput and buffer utilisation. Larger bandwidth requirement in client-server systems may be a constraint.

E. Real-time Program Performance

Recall Section 9.7. Three performance metrics were described: (i) Ratio of sum of interrupt latencies as a function of the execution times, (ii) CPU Load and (iii) Worst case execution time with respect to mean execution time.

Data communication and multimedia communication have differing performance indices. A loss of any bit requires retransmission and no frame or packet miss is tolerable. Missing frames within acceptable limits are however tolerable in video and multimedia systems.

The time of scheduling of a task can be measured by appropriate scope or analyser, instruction counts, or instruction execution-time profiler at the simulators.

The choice of an appropriate real-time programming model, with partitioning into tasks and scheduling algorithm, reflect in the following three metrics as follows:

1. System throughput. Comparative performance with respect to a previous life cycle in the development process or previous performance of the system. Relative performance = Relative increase in throughput.
2. Latency or response time of each task. [Section 4.6.] Both throughput and latency may be unrelated. Sometimes latencies may be large but throughput small.
3. Delay zitters may be a performance metric instead of response times in some cases. The delays (latencies) between retrievals of the data frames or packets or video-frames can vary. This variation is random (statistically Gaussian distributed) and is called delay zitters. The noticeable random variation in delay from the expected variation is undesired. It degrades the system performance. Image zitters may not be tolerable, but delayed retrieval within acceptable threshold is tolerable.

12.7.6 Performance Accelerators

There can be several ways to accelerate the performance. [Refer to Section B.3 also.] Examples of these are as follows:

1. Conversion of CDFGs into DFGs for example by using loop flattening (loops are converted to straight program flows) and using look-up tables instead of control condition tests to decide a program flow path.
2. Reusing the used arrays, memory, appropriate variable selection, and appropriate memory allocation and de-allocation strategy.
3. Using stacks as the data structure when feasible in-place of a queue and using a queue in place of list, whenever feasible.
4. Computing the slowest cycle first and examining the possibilities of its speed-up.
5. Code such that more words are fetched from ROM as a byte than the multi-byte words.

12.7.7 Porting Issues of OS in an Embedded Platform

The following portability issues may arise when the OS is used in an embedded platform. Table 12.10 gives the platform-dependency issues and the need for appropriate OS–Hardware interface functions for each issue.

Table 12.10

Platform-Dependency Issues and Need for Appropriate OS–Hardware Interface Functions

Platform Dependency	Need of Appropriate OS–Hardware interface functions
I/O instructions	A port instruction data type may be different on the different platforms, as follows: (i) unsigned char* (PowerPC, M68HC11/12, M68K, S390) (ii) unsigned int (ARM) (iii) unsigned long (Itanium, Alfa, SPARC) (iv) unsigned short (80x86)
Interrupt Servicing Routines	Interrupt vectors are to be defined differently. OS supports these differently on different platforms.

(Contd.)

Platform Dependency	Need of Appropriate OS–Hardware interface functions
Data types	OS should have appropriate APIs for data types. There may also be the need for Linux to declare all data types in \<asm/ types.h> and it includes in \<linux/ types.h> as the following: (i) unsigned byte (means 8-bit character also) (ii) unsigned word (means unsigned 16-bit and also unsigned short) (iii) unsigned int (means unsigned 32-bit) (iv) unsigned long (means unsigned 64-bit)
Interface specific data types	For example, a network interface card supports 32-bit unsigned integers each with a big endian.
Byte order	It may depend on the processor. Lower byte first in an integer (little endian) and upper byte first in an integer (big endian). Most-high level platforms support big endian. Some processors support both (ARM).
Data Alignment	(i) Two or three bytes stored at an address from which a processor accesses four-bytes in an access. (ii) Same data structure at 'C' source file may show differently on different platforms. ['C' takes 16-bit integer on a 16-bit processor and 32-bit integer on a 32-bit processor]. Compiler must force the alignment of data by the OS–hardware interface function.
Linked Lists[#]	An OS maintains the lists for different data structures. OS provides the standard implementation of doubly linked lists and circular linked lists. Platform-dependent device manager and drivers must include support to these.
Memory Page Size	PAGE_SIZE is 4kB in Linux. A processor may support different page sizes than this.
Time Intervals	Linux system clock ticks 10 ms each. Timeout functions need to be checked on porting an OS into a platform.

[#]Circularly linked means the last element of list linked is not to NULL pointer but to the first element of the list. Doubly linked list means that each element has two pointers, one for the next element and one for the previous element.

When porting RTOS codes into the system, the porting of I/O instructions, interrupt servicing routines, data types, interface-specific data types, byte order, data alignment, linked lists, memory page size and time intervals must be taken care of as these are platform specific. OS–hardware interface functions are needed for these.

■ SUMMARY ■

- Selection of the right hardware during hardware design and understanding of possibilities and capabilities of hardware during software design is critical especially for a sophisticated embedded system development.
- System project management is organizing software and hardware team members, people, process, product and project.
- Action plan in design cycle is needed for system hardware-software co-designing. System

requirements and specifications are a must for different parameters. System specifications for the following must be prepared before starting the design process. (i) Product functions and tasks (ii) Delivery Time Schedule (iii) Product Life Cycle (iv) Load on System (v) Human-Machine Interaction (for example by keypad and display subunits) (vi) Operating Environment (for example, temperature and humidity) (vii) Sensors (viii) Power Requirement and Environment (ix) System Cost

- Conceptual design is done later on to get the structure and layout of the application software and hardware. UML Class diagram, 'User Diagram', 'Object Diagram', 'Sequence Diagram', 'State Diagram' help in developing the conceptual design, structure and layout.

- One design approach is an independent design, which follows system integration. This approach is that the software life cycle ends and life cycle for the process of integrating into the hardware starts when designing a system. Another approach is for a sophisticated embedded system. It is a concurrent co-design approach. Both cycles concurrently proceed when co-designing a time critical sophisticated system.

- There are a number of software and hardware tools to implement the designed system easily with little effort. These are simulators, editors, compilers, assemblers, source code engineering tools, profilers (for viewing time spent at each function or set of instructions), a memory scope, a stethoscope-like view of code execution, a memory and code coverage scope, emulators, ICEs, oscilloscopes, logic probes, logic analysers, and EPROM/EEPROM application codes burner.

- System implementation and integration is done by using the target system circuit or its emulator, the In-Circuit Emulators (ICE), the device programmer for downloading the finalised codes into ROM, the simulator, and basic modules in source code engineering tool.

- Prototype development tools and IDE (Integrated Development Environment) are used to develop the fully simulated, tested and debugged sophisticated embedded systems with simpler efforts.

- Testing of hardware can be done by simple LEDs and logic probe. It is an LED-based probe, which shows a logic state as '1' or '0'. It can be used to measure port statuses and is an important tool for testing the events and responses in time scale of seconds. It is used to fine-tune the system timers.

- Logic Analysers help in collecting signals through the multiple input lines (say, 24 or 48) from the buses and ports to examine records and analyze the bus transactions (about 128 or more). It displays these on the monitor (screen) to debug real-time triggering conditions. It helps in finding sequentially the signals as the instructions execute.

- Latency times and deadline misses are measured to understand the performance of the real-time programming, scheduling models and algorithms.

- There are several ways of measuring system performance. It can be a system performance as per required and agreed specifications, power dissipation, throughputs, I/O throughputs, response time of tasks, deadline misses, response to sporadic tasks, memory buffers, bandwidth requirements and memory optimisation.

- The performance index gives the desired performance with respect to required specifications or parameters.

- Performance accelerators are used to improve the performance. Acceleration means using the same system by alternative ways such that it reduces execution times of a set of codes; reduces latencies of the tasks or increase throughput or minimise memory usage or power dissipation or reduces missing deadlines. Some ways are loop flattening, look-up tables, reusing the used arrays and memory and appropriate variable selection, appropriate memory allocation and de-allocation strategy and using stacks as data structure when feasible in-place of queue and use queue in place of list whenever feasible. We must look at the slowest computing cycle first and examining possibilities of its speed-up.

- Choosing the right processor, memory, devices and bus, and porting by OS/RTOS the processor-sensitive, memory-sensitive and device-sensitive instructions are a must. Byte order and data alignment must be according to the platform chosen.

■ LIST OF KEYWORDS AND THEIR DEFINITIONS ■

- *Action Plan*: A plan for action for the development process.
- *Target System*: A system for the targeted embedded system that is used during the development phase and the final products software and hardware are made from it.
- *ICE*: An in-circuit emulator for emulating the target system after connecting to the processor at one end and to the PC at another end.
- *In-Circuit Emulator*: Refer to ICE.
- *Emulator*: A circuit, which emulates the target system.
- *Assembler*: A tool for assembling the edited codes in mnemonics.
- *Big Endian*: An ordering in which the highest byte of a number is taken as first.
- *Little Endian*: An ordering in which the lowest byte of a number is taken as first.
- *Bit rate meter*: A meter to measure throughput in bits per second and to count the 1s and 0s during a particular interval of time.
- *Burning*: An act of placing the ROM image for code and data in uncompressed or compressed format into an EPROM or EEPROM, flash, microcontroller or some other similar device.
- *Circular Linked List*: A data structure for a list in which last element points to the first element in place of pointing to NULL in a usual list.
- *Doubly Linked List*: A data structure for a list in which each element points to the next element as well as the previous element in the list, instead of pointing to only the next element or NULL in a usual list.
- *Co-designing*: A software team designing with full knowledge of hardware capabilities and features working with a hardware team designing with full knowledge of software CDFGs and functions to be achieved. [In SoCs, co-design has different meaning. It means software implementation circuits, processor(s) circuit and devices designed as a single VLSI unit.]
- *Cross-Assembler*: An assembler that assembles code for the host machine for simulation and other purposes and later generates assembled codes for the targeted processor.
- *Data Alignment*: Data alignment means alignment in situations, for example, when (i) two or three bytes stored at an address from which processor accesses 4-bytes during an access and (ii) the same data structure at 'C' source file is showing differently on different platforms.
- *Debugging Tools*: Tools for debugging embedded system hardware and software functioning.
- *Delay Zitters*: The delay zitters mean the random variations in the delays in retrieving or arrival of successive data sets. The noticeable random variations are undesired.
- *Device Programmer*: A device for burning in the codes. [Refer to Section G.2]

- *Dissembler*: A tool for obtaining higher-level codes from the machine codes, which were assembled earlier.
- *DLL*: A library with functions that does not link at compile time but links later at run-time, when the need arises for it. This saves the storage space when imported from another system or external source.
- *Edit-Test-Debug Cycle*: A cycle in the implementation phase in which codes are edited, tested and debugged for a reported error on test.
- *Embedded System Project Management*: Organising people, processes, product and project. People involved in an embedded system development project are a team of software development, hardware development and system integration engineers.
- *Human-Machine Interactions*: Interactions of a user through tools like keypad, display unit and graphic user interfaces.
- *I/O Instructions*: Processor read, write, byte manipulation and other instructions for using a device at a port.
- *Platform Dependency*: A function, ISR, device driver, OS function, data type or data structure utilisation that is dependent on the processor or memory or devices in the system.
- *Integrated Development Environment*: Refer to IDE.
- *IDE*: An integrated tool that consists of simulators with editors, compilers, assemblers, source code engineering tool, profiler (for viewing time spent at each function or set of instructions), memory scope, stethoscope like view of code execution, memory and code coverage scope, emulators, logic analysers, and EPROM/EEPROM application codes burner.
- *Networking Stack*: A stack according to a protocol chosen. Protocol may be among the following: RFC -1323, CIDR, IP Multicast, IP, UDP, TCP, DNS Client, DHCP Server, SMTP Server, RIPv1 support, RIPv2 support, ARP, Proxy ARP, BOOTP, RLOGIN Client and Server (for Telnet). An embedded system socket can then connect to Multi-protocol LANs, ATM network, SONET or Wireless Access and Intelligent Networks.
- *Logic Probe*: It is an LED-based probe, which shows a logic state as '1' or '0'. It can be used to measure port statuses and is an important tool for testing the events and responses in a time scale of seconds. It is used to fine-tune the system timers.
- *Logic Analysers*: A powerful tool to collect the port and bus-transaction signals through its multiple input lines (say, 24 or 48) and to examine, record and analyze many bus transactions (about 128 or more). It displays these on the monitor (screen) to enable the debugging of the triggering conditions in real-time. It helps in finding sequentially the signals as the instructions execute.
- *Interpreter: Interpreter* does expression-by-expression (line by line) translation to the machine executable codes.
- *Latency:* Time taken to activate after an event or time taken in finishing certain codes before the next one starts.
- *Performance Index*: Index to measure the desired performance with respect to required specifications.
- *Performance Accelerators*: Using the same system, these are alternative ways to improve a set of codes execution time to reduce latency, increase throughput or minimise memory usage or power dissipation. A performance accelerator can also be a hardware unit interfaced to the main processor.
- *Performance Metrics*: Indices for measuring the performance using different measures.

- *Page*: A unit of memory in kB, which can be referred to as a single block from start address, and a memory address in it can be referred to as a start address plus offset.
- *Page-Size*: Size of the page taken by the memory manager.
- *Prototyping tools*: Tools for developing and co-designing a protype for the embedded system.
- *PLC*: A programmable unit to perform sequential logic control functions.
- *Right Platform*: An appropriate hardware platform with appropriate software to give best performance at minimum efforts or costs.
- *ROM Emulator*: A flash or EEPROM that is used in the implementation and testing phase and that emulates the ROM in the target system.
- *Simulator*: Simulating tool for a PC or a host machine that simulates all hardware and software functions and is of great help to the software team for testing and debugging.
- *RougeWave*: A modeling, designing and testing tool like UML.
- *Life Cycle*: Cycle of studying system requirements, specifications, conceptual design, detailed design, implementation and testing during a software or hardware development process.
- *Oscilloscope*: An oscilloscope is a scope with a screen to display two signal voltages as a function of time. It displays analog and digital signals as a function of time.
- *Storage Scope*: Also known as digital storage scope. It records the signal versus time in its memory. This enables viewing later. Time windows for recording can be preset. It is helpful in noting and measuring the delays and transition times and relative time differences.
- *Software-Hardware Tradeoff*: To appropriately plan and optimise performance at the lowest cost and choosing which set of functions and codes (for example, VLIWs) are implemented by a hardware subunit and which are implemented by a software module.
- *System Cost*: Cost for hardware and software. It includes all the costs for the development team and management efforts.
- *System Integration*: Integration of embedded software into the hardware and getting a validated product with optimised performance.
- *Throughput*: Number of processes or specified functions executed per unit of time. For I/O systems, it is the number of bytes outputted or read per unit of time.
- *VHDL and VeriLog*: Languages for designing and synthesizing the VLSI implementation of a system or part of system.

■ REVIEW QUESTIONS ■

1. What should be the goal during an embedded system development process? How does it vary from the software development process?
2. What is meant by embedded system independent design followed by system integration and by embedded system concurrent hardware–software co-design? Give five examples for each design strategy.
3. What is the action plan to follow while designing an embedded system?
4. Why are the device drivers important in the programs that are memory- and processor-sensitive?
5. What are the factors for selecting a processor during the system design phase?
6. What is a target system? How does the target system differ from the final embedded system?
7. What is meant by application software for a target system?

8. What is an emulator? What are the various components of an emulator? What are the advantages of using an ICE?
9. What is the use of a simulator in a development phase?
10. How do the readily available networking-stacks at RTOS help in faster error-free design?
11. How does a calling of an interrupt routine help in testing a design?
12. What is an assembly language program? What are the mnemonics? What are the machine codes? Give example of codes for SPI and SCI drivers.
13. What is an assembler? How is an assembler used? Write exemplary codes for ACVS input ports in assembly system. [Refer to the case study in Example 11.1].
14. What are compiler, linker, loader and interpreter?
15. What is a dissembler?
16. What is a cross-assembler?
17. What is an integrated development system?
18. What is time mode of a logic analyser? What is *state mode* of a logic analyser?
19. What is meant by a logic analyser? What is the logic analyser used for during the development phase?
20. An LED circuit is also a powerful analysis tool. How is it used?
21. What are the uses of an oscilloscope?
22. How will you use a bit rate meter to measure throughput from a real-time system?
23. How do the data align? Take the example of a 32-bit integer stored as big endian as a method for aligning bytes from an input stream.
24. How do you solve the problem of interface-specific data types?
25. When do you use DLLs in an embedded system?
26. Why is the selection of the correct platform essential during the embedded system development process?
27. Explain the software–hardware trade off. What are the advantages and disadvantages of software implementation instead of hardware implementation?

■ PRACTICE EXERCISES ■

28. Who are the *people* involved in an embedded system development project? How will you select them for the systems in Examples 11.1 to 11.4? How will the team vary when real-time video processing system is under development?
29. List the specifications with complete clarity for developing an ACC (Example 11.3).
30. Give system specifications for: (1) Product functions and tasks, (2) Delivery Time Schedule, (3) Product Life Cycle, (4) Load on System, (5) Human–Machine Interaction, (6) Operating Environment, (7) Sensors, (8) Power Requirement and Environment and (9) System Cost for a digital camera. Camera should be capable of storing 4-minute video or 500 still images. The system should include the USB port, imaging via video software, single shot timer standard as well as 10 s delay modes. Multiple resolutions 1024 x 768, 640 x 480, 320 x 240 and 160 x 120 pixels. Answer these points after conducting a web search. [Take as much time as you need to formulate a detailed answer].
31. Repeat Exercise 30 for a smart edge router. An edge router is a router that has 10/100 Mbps bandwidth, Ethernet Interfaces for LANs, Gbps Ethernet interface for connection to servers,

WAN and Internet interface of frame relay, ATM and packet over SONET/SDH. [Take as much time as you need to formulate a detailed answer].

32. Develop a software and hardware layout of an ACVS [Case study Example 11.1].

33. Refer to the case study in Example 11.4. Develop a software and hardware layout of a smart card, which stores all the medical records and history of a person.

34. Explain the use of the following hardware tools: Emulator, In-Circuit Emulator and device programmer. What are the uses of the device programmer? When is the device programmer used? [Section G.2].

35. Which are the simulators available from Motorola for 68HC11?

36. Which are the emulators available from Motorola for 68HC11? Which are the emulators available from Intel for Intel 80x86 family? Use a Web search to find a list of the ICEs available for the processor families from Intel 80x86, Motorola 68K and TI ARM families.

37. Prepare a list of emulator systems available for various microprocessors, microcontrollers and DSPs.

38. Explain with one example the use of each of the following: Application Development tools, Native Development Environment, APIs to RTOS, Debugging Capability Device Simulation, Network Simulation and User interface.

39. Explain with one example the use of each of the following software tools: Profiler scope, memory usage scope, stethoscope, scope for trace of program flow, scope for memory allocations and uses, and scope for code coverage.

40. What are the advantages and disadvantages of hardware implementation instead of software implementation?

41. Explain a hardware–software tradeoff by using the example of TPU in 68HC16 and 683xx microcontrollers. [Refer to Appendix C].

42. Why is *System performance index* defined as the ability to meet required functions and specifications while using the minimum amount of resources of memory, power dissipation, and devices and minimum design efforts and optimum utilisation of each resource (for example high CPU load)?

43. What is the performance metric for a multiprocessor-based embedded system *edge router* defined in Exercise 31?

44. How does a buffer help in improving a system performance?

45. Why is the I/O instructions platform dependent? Define throughput of an I/O system.

46. When is the minimum interrupt latency taken as an embedded system performance metric?

47. You can design an SoC by three routes: using gate arrays, using standard cells and using IPs and basic component layouts. List cases of embedded systems for each of these three routes.

48. What are the advantages of using FPSLIC (Filed Programmable System Logic IC) in an embedded system?

49. When would someone use tools like RougeWave?

50. List prototyping tools with a popular RTOS.

Appendix A

CISC and RISC Processor Architectures and an Exemplary Instruction Set

A.1 INSTRUCTIONS AND THEIR PROCESSING IN CISC AND RISC PROCESSORS

When coding in the assembly language of a processor, it is essential to understand the *formats* of instructions and data, and the *addressing modes*. They are explained below.

A.1.1 Formats of Instructions and Data

An instruction specifies an operation by bits called opcode. An instruction also specifies the operand(s). An operand specifies the bits that are employed in an operation. Consider an arithmetical instruction, *ADD z, x, y*. The operation specified is *addition*. The operand bits are from *x* and *y*. These bits operate and the resulting bits destine to *z*. The processor fetches *x* and *y* from specified registers or specified memory addresses in the instruction.

1. Consider an exemplary instruction, *ADD r4, r4, r6*. It specifies the opcode bits for *ADD*. It specifies the destination-operand register *r4 and* source-operand registers, *r4* and *r6*.

2. The *x or y* or both may be the immediate operands (not from any source register or memory) in an instruction as clarified in the following instruction-format. Consider exemplary instruction, *ADD r4, r4, # 15*. It specifies the opcode bits for *ADD* and the destination-operand register *r4*, one source-operand register *r4*, and one immediate operand 15. It adds the word in *r4* with 15 and destines the result to *r4*. The opcode bits for *ADD* differ in this instruction from the one above. This enables the ID to decode the instruction for the CU (Section 2.1) that controls its execution. The hash sign before 15 means that 15 is taken as such. It is an immediate operand.

3. Another exemplary instruction, *ADD r3, r1, [M1]*, may be supported by a particular processor. It specifies opcode bits for *ADD* and the destination-operand register *r3*, one source-operand

register *r1*, and one source-operand memory address M1. It adds the word in *r4* with the word fetched from *M1* and then destines the result to *r3*. The opcode-bits for ADD differs in this instruction from the one above. Here, square brackets mean that within them there is a memory address.

4. Consider yet another exemplary instruction, *ADD r1*, (*r7*), (*r8*), that a certain processor may support. It specifies opcode bits for *ADD* and the destination-operand register *r1*, one source-operand from a memory address pointed by register *r7* and another source-operand from a memory address pointed by the register *r8*. It adds the words from two pointers (pointed memory-addresses), and destines the result to *r4*. The opcode bits for ADD differ in instruction from the one above. Parentheses mean that within them there is a memory address pointer–register.

An instruction format of a processor consists of opcode bits followed by an operand specifying the bits. The format may provide for variable length or fixed length opcode bits. A format may provide either the source-operand(s) first and then destination-operand(s) or vice versa. An instruction format is as per the addressing modes and as per the number of addresses specified in the instruction.

When a memory address operand for a word in an instruction is defined in 68HC11, the upper and lower bytes are at the lower and upper memory addresses respectively. Upper and lower bytes of a word in Intel processors 80x86 are at the upper and lower memory addresses respectively. When a word in an instruction is defined in ARM processors, the format is configurable in the first few clock cycles after the *Reset*.

A.1.2 Addressing modes

1. The above instruction examples are for a three-address machine. In a certain processor, it may not be necessary to specify in an instruction itself all the three operands. All the three or two or one operand (source or destination) can be implied to the processor. The processor is called a zero-, one- or two-address machine (processor), respectively. This not only gives shorter instruction lengths but also gives the advantage of a less complex CU (Figure 2.1) and hence, the faster operations when processing an instruction. It may not give shorter program lengths as the number of instructions required may increase in a shorter address machine.

2. A certain processor may provide for the same *length* instructions and other the variable-lengths instructions.

3. Recall Sections 1.2.5 and 6.3. VLIW and SIMD instructions need more than three operands and a processor that permits instructions specifying up to *p* addresses is called a *p*-address machine.

An addressing mode is defined as a way in which an operand is accessed while processing an instruction. A processor supports the various addressing modes in its different types of instructions and operands of an instruction. (i) A processor may support one addressing mode for one operand and others for other operand(s). (ii) A processor circuit may support one set of addressing modes in arithmetical and logical instructions and may support a different set of addressing modes for the data transfer instruction [Non-orthogonal instruction set]. (iii) A processor may support only a few addressing modes in its instructions. [The RISC referred to in Section 2.1]. Assume a processor with 8 general-purpose registers, *r0* to *r7*. Also assume the processor as a three-address machine. The following explains the addressing modes for the operands in the exemplary instructions.

1. Consider an example *ADD r4, r4, r6*. The addressing mode is *register addressing-mode* for all three operands.
2. Consider another exemplary instruction, *ADD r4, r4, #15*. The addressing mode is an *immediate addressing-mode* for the second source-operand.
3. Consider *ADD r3, r1, [M1]*. The addressing mode is an absolute (also called extended) addressing mode for the second source operand. A 16-bit absolute memory address may be specified by a smaller number of bits in an instruction. These bits add with an implied address called the page address. For example, assume the absolute address as 0xF0B1 and the instruction specifies the short address (called a direct address) for M1 as 0xB1. The 0xF000 is the implied page address. The page address is specified in an earlier instruction (not necessarily the previous one).
4. Consider *ADD r1, (r7), (r2)*. Addressing modes are indirect (also called *index* or *base*) addressing-modes for the first as well as second source-operands.
5. Consider *ADD r1, (r6), (r7, r2)*. Addressing modes are base-index addressing-modes for the second source-operand. The *r7* is used as a base register and *r2* as the index register. The first source-operand may be referred to as the one obtained by using base addressing-mode or index addressing-mode, depending upon whether *r6* is specified as a base or index address.
6. Consider an instruction *ADD r1, (r7), (r2, #10)* that is supported by a processor. The addressing modes are as follows: *index addressing-mode* for the first source-operand and *index-relative addressing-mode* for the second source-operand. The destination is *r1*. Addition is between the word at the memory address pointed by *r7* and the word at the memory address pointed by *r2 + 10*. The hash sign before 10 means 10 is taken as such. Here '10' is the *displacement* that specifies a relative address. It is specified by the instruction. The displacement is not an unsigned integer or byte. It is a 2's complement number (signed number). Hence, the memory address can be pointed by addition or subtraction in *r2*. This mode is also called base-offset addressing-mode.
7. Consider an instruction *Load r7, (r2, #10)*. The processor supports the addressing modes as follows. For the destination-operand it is the register addressing-mode, and the *base addressing-mode* for the source-operand. The destination is *r7*. The operand is from the memory address pointed by *r2 + 10*. But after the operation of load, *r2* retains its original value. This is called *auto-index* addressing-mode.
8. Consider another exemplary instruction *ADD r1, (r7, r2, #10), (r2)*, supported by a processor. The addressing modes are as follows: *base-index relative addressing-mode* for the first source-operand and *index addressing-mode* for the second source-operand.
9. Consider an *ADD r1, (r7)+, (r2)* instruction. The addressing modes are the auto-index incrementing-register addressing-modes for the source-operand. The addition operation is between the words from the memory addresses pointed by *r7* and *r2*, after which the *r7* word increments and then points to the next memory address that is later used in a subsequent instruction.
10. Consider *LOAD r1, (r7), #10*. The addressing mode is the register addressing-mode for the first source-operand. The load operation is from a memory address pointed by *r7*. But after the operation, *r7* increments by 10. The *r7* word now points to the next memory address that is later used in a subsequent instruction. This is called *post auto-index* addressing-mode.

11. An addressing mode is a stack-addressing mode if the register for operand happens to be a stack pointer. [Refer to Section 2.5.2.(A)] The register automatically decrements on a push and increments on a pop of the bytes to or from the pointed address.
12. Recall Section 2.5.1. The CPUs with the segment registers provide for three types of addressing modes: inter-segment, intra-segment and segment-overriding modes.

A.1.3   Instruction Set

Each version of a processor has a distinct set of instructions. This is because of a distinct circuit of the structural units (Figure 2.1) that is designed to fetch and execute any instruction that the instruction set specifies. An assembly language programmer or a compiler uses instructions from the set to implement the commands of high-level languages. These instructions link the high-level language software with the structural units of the processor for implementation. An instruction set consists of the following subsets of instructions:

1. Data Transfer Instructions: There are two operands. From a source operand, the data is transferred to a destined operand. An operand may be from a register or memory (or port) address. These instructions are for transfer within: register–register, memory–register, register–memory, register–port, port–register transfer, register–stack (called Push operation) or stack–register (called Pop operation). [A port or device generally uses identical instructions for read and write.]

2. Bit transfer or manipulation instructions: There is one operand. The bits at the operand transfer within it. Examples are as follows. Lower nibble transfers (or swaps) to upper nibbles in an instruction. Lower byte may transfer or swap with the upper byte. The bits manipulation instructions can be as follows. A bit or bits at the operand may set or reset. The bits may shift left or right. The bits may rotate left or right. The shift or rotation may use the carry flag to store or to participate. The bits manipulated or swapped may be at a register or at a memory (or port or device) address.

3. Arithmetical instructions and logical instructions: There are three operands. Each operation is within two source operands and the result is at the destined operand and also in the flags. [Logical instructions like NOT need only one source operand.] These instructions use the ALU and FLPU (if present). There are up to p operands in a VLIW processing on a p-address machine. There may also be an instruction for a *comparison* of the two operands. Only the flags are affected after a hypothetical subtraction of one from another by the ALU. There may also be an instruction for a *test* between the two operands. Only the flags are affected after a hypothetical logical AND operation between the two.

4. Program-Flow Control Instructions: These instructions change the normal program flow. In normal flow, the flow is done by the program-counter or instruction pointer incrementing to the next instruction. By a program-flow-control-group instruction there is an increment and followed by a change. [Why is it a two-step process? In subroutine calls the incremented value have to be saved on the stack.] An instruction may be to implement the looping or routine-call or switch to another task. It may also be an instruction to stop, halt, wait, reset or interrupt.

A.1.4   CISC and RISC Architectures

A microprocessor may imbibe **CISC** (Complex Instruction Set Computer) features. A CISC instruction set has instructions with many addressing modes. Registers and index registers (base registers and segment registers also, in 80x86) present in a CISC provide for many of the addressing modes during the data transfer, bit-manipulation and arithmetic-logical logical group of instructions [Section A.1.2].

A microprocessor may imbibe the **RISC** (Reduced Instruction Set Computer) features. Most microprocessors are RISCs in the application areas of telecommunication, video and image processing. Table A.1.1 compares the features of CISC and RISC.

Table A.1.1
CISC and RISC Features

CISC	RISC
It provides for a *number of addressing modes*. [Refer to Section A.1.2]. More addressing modes makes it easier for the assembly language program and the machine level instructions to deal with various data structures. Implementation of the compiler design is also easier when a program is in a high-level language. A stack, too, is at a memory address space outside the processor. But at what cost are the CISC features available? (*a*) A complex instruction set, because of a number of instruction formats with many variable addressing modes for an operand. (*b*) Instructions of variable lengths. (*c*) Variable number of clock cycles for executing the different instructions. (*d*) Complex CU (instruction decoding, control and sequencing unit) [Refer to Section 2.1].	There are *very few addressing modes*. It provides a smaller instruction set for the Store, Push, Load and Pop instructions for storing in, and loading from, the memory. Most instructions and the instructions in ALU are implemented by operands from the stack or registers only. Stack also is at a register-set present within the processor. When the stack is also within the microprocessor like the registers, the accesses are fast. There may be register-windows, each with a set of registers. A window can store the values of the variable and stack in different subroutines. This results in fast switching between the processes. [Sections 4.6 and 8.1]
It has a *micro-programmed unit* with a control memory* that implements a large instruction set with a smaller hardware compared to that needed when individual instructions have separate implementation circuits.	It has *a hardwired programmed-unit without a control memory* to implement a small instruction set with a separate hardware for implementing each instruction.
An easy *compiler design*.	Needs a *complex compiler design*.
The CISC features provide ***precise and intensive calculations slower*** than a RISC because of micro-programmed control memory based implementation and the need for external memory accesses more frequently than in an RISC.	The RISC features provide ***precise and intensive calculations faster*** than a CISC, because of hardwired implementation and the need for external memory accesses much less frequently than a CISC due to availability of a number of register-sets, register-files and internal stacking registers. [Numerous sets of registers store the pushed and loaded values and also intermediate results. This greatly reduces the number of external references to the memory during processing.]

* An instruction (received at the instruction decoder) consists of a set of microinstructions and nano-instructions. A microinstruction is a set of control sequences, which generate to implement the instruction. A nano-instruction is a subset of the control sequences, which a microinstruction calls when needed. The operations during the various *micro* and *nano* instructions (control sequences) now overlap for the different instructions. This simplifies the circuit. Each set of control sequences is stored in the memory and is implemented by a common circuit. A microprogramming unit of a CISC has the control memory and the circuit. A RISC has a separate implementing circuit for each instruction.

A RISC instruction set has instructions with a register-addressing mode only for all the operands in the bit-manipulation as well as the arithmetic-logical logical group of instructions. RISC architecture is load-and-store architecture due to register addressing only in ALU and FLPU operations. Instructions are implemented faster when there is register addressing only, in the case of pipelined architecture [Section 2.1]. An RISC instruction set provides for addressing modes for data-transfer instructions–either absolute addressing (for load and store), register addressing, stack addressing (for push and pop) or by immediate addressing.

The instruction set of each processor is unique. An exemplary set of ARM7, a recent popular processor in embedded computing systems, is given in brief in Section A.2.

A microprocessor may have a ***combination of RISC and CISC features*** [examples are Intel 80960, ARM7 and ARM9]. This is seen when a processor has a RISC core for processing and there is an in-built compilation unit that first compiles the CISC instructions into RISC formats, which are then implemented by the RISC core of the processor. [There is support to a complex-addressing-modes-based instruction set, but internally the implementation for many instructions is like in a RISC (without the microprogrammed unit)].

Processors with combinations of both types of features have the following type of instruction sets. There are the additional addressing modes, indirect (index), auto-index, and index-relative addressing modes for data transfer instructions. The second operand may also be fetched by the immediate addressing mode for arithmetical and logical instructions.

The above features provide the advantage of using a CISC in terms of functionality, along with the advantage of a RISC in terms of faster program implementation and reduced code lengths. Faster implementation is due to the instant availability of the register word to the execution-unit. Reduced code lengths are due to the following. Most instructions use registers as operands. A register as operand is specified by a few bits in the instruction. [A memory address as operand and the displacement bits in the instruction are specified by 8, 16 or 32 bits].

A.2 AN EXEMPLARY INSTRUCTION SET—ARM7

The ARM7 instruction set is as per the Thumb® instruction set.

1. *Data Transfer Instructions* –Given below are the instructions for transfer between the register-memories. The memory address is as per a register used as index or index-relative or post auto-index addressing mode.
   (a) Register—load a word (*LDR*).
   (b) Register—word stores a word (*STR*).
   (c) Set a memory address into a register (*ADR*). Address is of 12 bits. [Alternative for 16 bits address setting in a register is using any register or *r15* in an arithmetic operation].
   (d) Register—load a byte (*LDRB*).
   (e) Register—byte store (*STRB*).
   (f) Register—Half Word store (*STRH*). [A word in ARM is of 32 bits].
   (g) Register—load Half Word as such or signed (*LDRH* or *LDRSH*).

The following are the instructions for a word transfer between registers:
   (a) Move (*MOV*).
   (b) Move after Negating (*MVR*).

A load, move or store instruction can be conditionally implemented. For example, *MOVLT r3*, #10. The immediate operand 10 will transfer to *r3* provided a previous instruction for comparison showed the first source as less than the second. Conditions are *LT* (signed number less than), *GT* (signed number greater than), *LE* (signed number less or equal), *EQ* (equal), *NE* (not equal), *VS* (overflow), *VC* (no overflow), *GE* (signed number greater than or equal), *HI* (unsigned number higher), *LS* (unsigned number lower), *PL* (plus, nor Negative), *MI* (minus), *CC* (carry bit reset), and *CS* (carry bit set).

2. *Bit Transfer or Manipulation Instructions*
   (a) Register—bits Logical Left Shift (*LSL*).
   (b) Register—bits Logical Left arithmetic Shift (*ASL*).
   (c) Register—bits Logical Right Shift (*LSR*).
   (d) Register—bits Logical Right arithmetic Shift (*ASR*).
   (e) Register—bits Rotate Right (*ROR*).
   (f) Register—bits Rotate Right with carry also extended for rotating (*RRX*).

3. *Arithmetical and Logical Instructions*—The following are the instructions for *arithmetical operations*: Each uses three operands from the registers. One source may, however, be by immediate operand addressing in addition and subtraction.
   (a) Add without carry two words and the result is in the third operand (*ADD*).
   (b) Add with carry two words and the result is in the third operand (*ADC*).
   (c) Subtract without carry two words and the result is in the third operand (*SUB*). [Carry bit used as borrow.]
   (d) Subtract with carry two words and the result is in the third operand (*SBC*).
   (e) Subtract reverse (second source with the first) without carry two words and the result is in the third operand (*RSB*). [Carry bit used as borrow.]
   (f) Subtract reverse with carry two words and the result is in the third operand (*RSC*).
   (g) Multiply two different registers and the result is in the destined register (*MUL*).
   (h) Multiply two source registers and add the result with the third source register and accumulate the new result in a destined register. (*MLA*). [There are four operand registers.]

The following are the instructions for *logical operations*:
   (a) Bit wise OR two words and the result is in the third operand. (*ORR*).
   (b) Bit wise AND two words and the result is in the third operand. (*AND*).
   (c) Bit wise Exclusive OR two words and the result is in the third operand. (*EOR*).
   (d) Clear a Bit (*BIC*). [There is one source for the bits; a second source for the mask and the result is at the third operand.]

An arithmetical or logical instruction can be conditionally implemented. For example, SUBGE *r1*, *r3*, *r5*. The operand from r3 is subtracted from r5 if the GE condition resulted earlier (N and V status bits equal on comparison of two signed numbers). Conditions are the same as mentioned in the above paragraph. These are the results of a *comparison* or *test*.

The following are the instructions for *compare and test operations*. The result destines to *CPSR*, which stores the four condition bits, N, V, C, and Z.
   (a) Bit wise Test two words (*TST*).
   (b) Bit wise Negated Test between two words (*TEQ*).
   (c) Compare two words and the result is at the *CPSR* condition bits (*CMP*).
   (d) Compare two negative words and the result is at the *CPSR* condition bits (*CMN*).

4. *Program-Flow Control Instructions:* The following are the instructions for branching operations. A branching instruction can be conditionally implemented. Branch to an address relative to *PC* word in *r15* (*B*) 'B #1A8' means add in *PC* 1A8 and change the program flow. 'BGE #100' means that if a *GE* condition resulted on a compare 0 test, add in *PC* 1A8. There are similar instructions for different conditions of the processor status flags.

A.3   EXEMPLARY ASSEMBLY LANGUAGE PROGRAM FOR ARM PROCESSORS

Example A.1

Recall the problem of adding three numbers, x, y and z (= 127, 29 and 40), and storing the result at a memory address, a. [$a = x + y + z$.] Using the instructions of the above instruction-set, the assembly language codes will be as follows.

1. BEGIN: MOV r2, #0x007F ; Transfer 127 into a processor register r2.
2. MOV r3, #0x001D ; Transfer 29 into a processor register r3.
3. MOV r4, #0x0028 ; Transfer 40 into a processor register r4.
4. MOV r1, #)x000 ; Transfer 0 into a processor register r1.
5. ADD r1, r1, r4 ; Add the register r4 word into the r1.
6. ADC r1, r1, r3 ; Add the register r3 word along with the carry (if any) from previous addition into the r1.
7. ADC r1, r2 ; Add the register r2 word along with the carry (if any) from previous addition into the A.
8. ADR r5, 0x800 ; Set the address into r5. Memory address set is 0x800.
9. STR [r5], r1 ; Store the r1 at the address pointed by r5.

■ LIST OF KEYWORDS AND THEIR DEFINITIONS ■

- *Absolute Addressing Mode*: Define all address bits in an instruction.
- *Auto Index*: After executing an instruction, the index register contents change automatically.
- *Instruction Set*: A definite set of executable instructions in a processor.
- *Base Addressing*: Addressing an address from where a first element of data structure starts.
- *CISC*: Complex instruction set computer (processor), which has an instruction set with many addressing modes in arithmetical, logical and other instructions.
- *Direct Address*: A directly usable address in an instruction. It is usually the address on a page in the memory.
- *Instruction Format*: Format of expressing an instruction.
- *Program Flow Instruction*: An instruction in which the program counter or instruction pointer changes in a way different from its normal changes during a program execution.
- *RISC*: Reduced instruction set computer (processor), which has an instruction set with few addressing modes—load, store, push and pop and most arithmetical, logical and other instructions are zero address instructions.
- *Zero Address Instruction*: An instruction in which an address is implied to the processor, usually to the *stack* internally within the processor, or a register set that associates with a program module.
- *RISC with CISC Functionality*: A processor with RISC implementation and user programmability similar to a CISC.
- *Thumb® instruction set*: An instruction set in which each instruction is of 16-bit on a 32-bit processor. It gives reduced code density. It is a 16-bit instruction set which enables 32-bit performance at 8/16-bit system cost. They are used by ARM processors.

Embedded System High-performance Processors

Recall Section 1.2.2 and Section 2.1. Sophisticated embedded systems for high computing performance applications needs optimised use of resources, power, caches and memory. The following are the processor performance metrics:

1. (a) High *MIPS,* (b) high *MFLOPS* and (c) high *Dhrystone benchmark* program based MIPS.
2. Optimised compiler unit performance in the processor.

The above metrics are provided by the latest innovatively designed *processors.* A high-performance processor combines capabilities with optimised use of resources, power, caches, and memory. The use of high performance processor ICs and cores in embedded systems providing billion operations per second has become feasible due to the great advances in VLSI technology.

Note: A benchmarking program is *Dhrystone,* developed in 1984 by Reinhold P. Weicker. It measures the performance of a processor for processing *integers* and strings (characters). It uses a benchmark program available in C, Pascal or Java. It benchmarks a CPU and not the performance of I/O or OS calls. Dhrystones per second is the metric used to measure the number of times the program can run in a second. 1 MIPS = 1757 Dhrystones. [Why? VAX11/780, which executed 1 MIPS, ran 1757 times. For Dhrystone benchmark program see www.webopedia.com/TERM/D/Dhrystone.html].

There is an EDN Embedded Benchmark Consortium (EEMBC) [EDN is a group that publishes the International magazine EDN, which is dedicated to Embedded System information. Refer to http://www.e-insite.net/edmag/. The EEMBC-proposed five-benchmark program suites for 5 different areas of applications of embedded systems: (a) Telecommunications, (b) Consumer Electronics, (c) Automotive and Industrial Electronics, (d) Consumer Electronics and (e) Office Automation. They are also used for measuring and comparing embedded system processor performances.

B.1 EXEMPLARY ARM PROCESSORS

Detailed information on ARM is available at http://www.arm.com. The salient points are given below.

1. ARMv4T (version 4 Thumb) microarchitecture is common to ARM7, ARM9, ARM10 and ARM11 families. Thumb is an industry standard. It is a 16-bit instruction set which enables 32-bit performance at the 8/16-bit system cost. This provides typical memory savings of up to 35%, over the equivalent 32-bit code, while retaining all the benefits of a 32-bit system (such as access to a full 32-bit address space). There are no overheads in moving between Thumb and the normal ARM state. Two states are compatible on a routine basis. The code designer has complete control over performance and code-size optimisation.

2. Enhancement of v4T is obtained with ARMv5TE architecture (1999). It has ARM DSP instruction set extensions, which improve the speed of instruction sets by up to 70% for audio DSP applications [certain applications need microcontroller data-processing features as well as DSP features in a single processor in place of multiprocessor systems].

3. An enhancement of v5TE is done with ARMv5TEJ architecture (2000). It incorporates Jazelle Java execution accelerator technology for Java. This provides significantly higher (Java execution by 8x) performance than a software-based Java Virtual Machine (JVM). There is an 80% reduction in power consumption compared to a non Java-accelerated core. This functionality gives platform developers better results. Java codes as well as OS applications can run on a single processor.

4. An enhancement of v5TEJ is obtained withARMv6 architecture (first implementation 2002), used in ARM11 microarchitecture. It has SIMD extensions (Section 6.3), optimised for applications including video and audio CODECs. SIMD execution performance is enhanced by 4x.

5. ARM9E and ARM10 families use a Vector Floating Point (VFP) ARM coprocessor, which adds full floating point operands. VFP also provides fast development in SoC design when using tools like MatLab®. Applications are in image processing (scaling), 2D and 3D transformations, font generation and digital filters.

6. ARM uses an Intelligent Energy Manager (IEM) technology. It implements advanced algorithms to optimally balance processor workload and energy consumption. It maximizes system responsiveness. IEM works with the operating system and mobile OS. An application running on a mobile phone dynamically adjusts the required CPU performance level. It uses a standard programmer's model.

7. ARM processors use the AHB (AMBA Advanced High Performance Bus) interface. AMBA is an established open source specification for on-chip interconnects. It serves as a framework for SoC designs and IP library development. AHB support is in all new ARM cores. It provides a high-performance and fully synchronous back plane. The multi-layer AHB in version ARM926EJ-S and all members of the ARM10 family represents a significant advancement. It reduces latencies and increases the bandwidth available to multi-master systems.

8. ARM codes are forward compatible with higher versions. For example, ARM7 codes are forward compatible with ARM9, ARM9E and ARM10 processors as well as Intel XScale micro-architecture.

9. ARM's debug and trace tools quickly debug real-time software, and trace instruction execution and associated program data at full core speed.

10. A wide choice of development tools and of simulation models for leading EDA (Electronic Design Automation) environments (for example, CADENCE EDA environment) and excellent debug support for SoC design are available.

Table B.1.1 gives the features and a comparison of the exemplary high-performance ARM families of processors.

Table B.1.1
ARM Family Processors

Feature	ARM7™ Thumb® Family	ARM9™ Thumb® Family	ARM11
Family members Example	(a) ARM7TDMI® (Integer Core) (b) ARM7TDMI-S™, (Synthesisable version of ARM7TDMI) (c) ARM7EJ-S™ (Synthesisable core with DSP and Jazelle technology) and ARM720T™ (cached processor macrocell[#]), 8K Cached Core with Memory Management Unit (MMU) supporting operating systems[1]	(a) ARM920T (Dual 16k caches with MMU support multiple OSs[1]. (b) ARM922T (Dual 8k caches for applications support multiple OSs[1]. (c) ARM940T™ (Dual 4k caches[#] for embedded control applications running a RTOS).	Families with ARMv6 instruction set architecture that includes the Thumb® extensions for code density, Jazelle™ technology for Java™ acceleration, ARM DSP extensions, and SIMD media processing extensions. MMU-supporting operating systems[1] and Palm OS.
Core with ARM® and Thumb® instruction sets	32-bit RISC core	32-bit RISC processor core Super scaling 5-stage integer pipeline. 8-entry write buffers. It avoids blocking the processor on external memory *writes*.	32-bit RISC processor core with 8-stage integer pipeline, static and dynamic branch prediction, and separate load-store and arithmetic pipelines to maximize instruction throughput
Application domain	Cost and power-sensitive consumer applications, for example, personal audio (MP3, WMA, AAC players), entry-level mobile phone, two-way pager, still digital camera, PDA	Set-top boxes, home gateways, games consoles, MP3 audio, MPEG4 video videophones, portable communicators, PDAs, next-generation hand-held products, digital consumer products, imaging products, desktop printers, still picture cameras, digital video cameras, Automotive Telemetric and	Battery-powered and high-density embedded applications. Embedded SoCs targeted at next generation of wireless and consumer applica-tions. Addresses the requirements of embedded application processors, advanced operating systems (OS), and multimedia,

(Contd.)

Feature	ARM7™ Thumb® Family	ARM9™ Thumb® Family	ARM11
		infotainment systems	such as audio and video CODECs. Consumer devices include 2.5G and 3G mobile phone handsets, PDAs and multimedia wireless devices, home consumer applications such as imaging and digital camera applications, home gateway and network infrastructure equipment including voice over IP and broadband modem
Performance	130 MIPS using Dhrystone 2.1 benchmark in typical 0.13µm process	Achieves 1.1 MIPS/MHz, 300 MIPS (Dhrystone 2.1) in a typical 0.13µm process	Targets a performance range of Dhrystone MIPS 400 to 1200
Code Density	High code density (comparable to 16-bit microcontroller)	High code density	High code density
Die size on silicon	Small die size portability to 0.25µm, 0.18µm and 0.13µm versions	Die Size 4.2 mm^2 in ARM940T. Portable to latest 0.18µm, 0.15µm, 0.13µm silicon processes. Frequency 185 MHz at 0.18µ in ARM 940T.	0.13µm foundry processes deliver 350 to 500+ MHz in worst case and over 1 GHz on next-generation 0.1µm processes.
Memory Coupling (Section 6.3)	No Tight coupling	Same as ARM7	Same as ARM9
Power Performance	Very low power consumption	Very low power consumption. 940T power 0.8 mW/MHz on 0.18µ silicon foundry generic process. Worst case: 1.62V, 125C, and slow silicon. Typical: 1.8V, 25C, nominal silicon	Optimum power efficiency, single-issue operation with out-of-order completion to minimise gate count, consuming less than 0.4 mW/MHz on 0.13µm foundry processes
Bus Interface	AHB	Single 32-bit AMBA bus interface	None

Note: (1) [1]Means support to Windows CE, Palm OS, Symbian OS, Linux and other OS/RTOS. There is Palm OS support in ARM920T and ARM922T processors. ARM 940T has a Memory Protection Unit (MPU) and a supporting range of Real-Time Operating Systems including VxWorks.

(2) [#]Integrated instruction and data caches.

(3) ARM architecture refers specifically to the architectural instruction sets and programmers models, such as ARMv5TE architecture, ARMv5TEJ architecture, and ARMv6 architecture in ARM11. The term ARM microarchitecture refers specifically to the implementation of architectures such as the ARM9™ family of cores and the ARM10 family of cores. For example, ARM926EJ-S™ core and the ARM1020E™ core are CPU products based on those earlier microarchitectures.

B.2   EXEMPLARY HIGH-PERFORMANCE PROCESSORS

Intel XScale and StrongARM SA-110, Motorola Power PC 860, IBM PowerPC 750X, TI OPMAP and MIPS R5000 are other examples of high-performance 32 and 32/64 bit processors. These have also been used in many applications in embedded systems. Table B.2.1 gives the features of an exemplary high-performance IBM PowerPC 750X. [http://www-3.ibm.com/chips/techlib/techlib.nsf/techdocs/852569B20050FF77852569930058A78D.htm]

Some processors are specially dedicated to a particular performance. For example, a very recent X10 family network processor delivers 10 Gbps port performance for IPv6 (broadband Internet).

Table B.2.1
IBM PowerPC 750x Family Processors

Feature	IBM PowerPC 750™ Thumb® Family
Family members Example	(a) IBM PowerPC750™ PID 8t (b) IBM PowerPC750™ PID 8p
Core	32-bit RISC core, 64-bit data bus, 32-bit address bus, CMOS 7S processor technology with copper-based levels, Dynamic power management, Integrated thermal management assisting unit, High Speed L2 cache, 60x Bus Interface, Supports up to 233 MHz SRAMs
Cache	1 MB 2-way associative L2 Cache Controller, 128 byte sector size for 1MB L2 (64 byte when 256kB)
Application domain Example	Entry level mobile devices
Performance	33.9 SPECint95 and 14.6 SPECtp95 at 500 MHz operation with 1 MB L2 Cache
ALU and FLPU	One cycle hardwired multiply and divide fixed-point unit, 64-bit Floating-point unit with an optimised single precession multiply/add
Die size on silicon	Small die size portability to 0.26μm and 5 metal level version (PID 8t) and 0.20μm and 6 metal level version (PID 8p) versions
Power Performance	2.5 V to 2.75V Core and 3.3 V IO for PID 8t and 2.0 V to 2.1V Core and 1.8 V, 2.5 and 3.3 V IO for PID 8p. Very low power consumption 0.6 W at 500 MHz in PID 8p
Bus Interface	Parity Checking, Fast reset, Powerful diagnostic test interface using a CCP (Common Chip Processor) and IEEE 11491 JTAG interface

B.3   ACCELERATORS

An accelerator accelerates the code execution. It may be an ASIC, IP core or FPGA. An accelerator may include the bus interface unit, DMA, read and write units, registers and accelerator cores. An accelerator uses a programming model to accelerate, unlike the coprocessor, which has instruction sets for specific tasks. A processor controls the registers, and the registers interact and connect to an accelerator through the buses so that the processor controls the accelerator for obtaining a higher performance.

For example, a very recent JA108 from Nazonin Communications is a Java accelerator, which accelerates by 15 to 60 times the JAVA code run. Another example is a video accelerator, which accelerates video-processing tasks.

■ LIST OF KEYWORDS AND THEIR DEFINITIONS ■

- *Accelerator*: ASIC, IP core or FPGA, which accelerates the code execution and which may also include the bus interface unit, DMA, read and write units and registers with their cores.
- *Java Accelerator*: An accelerator that helps in the execution of Java codes faster than a JVM.
- *JVM*: Machine codes that use the compiled byte codes of a Java program and run the program on a given system.
- *Video Accelerator*: An accelerator for the video output.
- *AMBA*: An established open-source specification for on-chip interconnects which serves as a framework for SoC designs and IP library development.
- *AHB*: A high-performance version of the AMBA used in ARM processors.
- *ARM*: A family of high-performance, reduced code density ARM7, ARM9, ARM10 and ARM 11 processors, which are used in embedded systems as a chip, or as a core in an ASIC or SoC.
- *Dhrystone*: A benchmarking program, which measures the performance of a processor for processing *integers* and strings (characters). It uses a benchmark program available in C, Pascal or Java. It benchmarks a CPU not the performance of I/O or OS calls. 1 MIPS = 1757 Dhrystone/s.
- *EDA*: A powerful tool for Electronic Design Automation.
- *EEMBC*: EDN Embedded Benchmark Consortium.
- *Performance Benchmarking*: A metrics set for evaluating the performance of a system.
- *Microarchitecture*: When a processor architecture refers specifically to the architectural instruction sets and programmers models, the term microarchitecture refers specifically to the implementation of those architectures. A processor may have CISC architecture with an RISC microarchitecture implementation.

Appendix

C

|||||||||||||||||||||||||||||||||||||||||||||||||||||||||||||||||||||||||||||||||||||||||||||||||||||||||||||||||||||

Embedded System 8/16/32-Bit Microcontrollers and an Overview of their Architecture

Recall Section 1.2.3. Embedded systems for a large number of applications such as automotive electronics and computer networks should include the following single chip or core features:

1. Processors with required 8-bit, 16-bit or 32-bit *MIPS*.
2. Devices, RAM and ROM, timers, interrupt controllers, PWMs, ADCs, serial and parallel I/Os.

They are provided in microcontrollers. A microcontroller should possess the capabilities of a microcomputer to enable systems to function with a single or few ICs.

C.1   OVERVIEW OF THE ARCHITECTURE OF EXEMPLARY INTEL, MOTOROLA AND PIC FAMILY CISC-BASED MICROCONTROLLERS

Recall the functional circuits present in a microcontroller, shown in Figure 1.2. The application-specific units in a specific version of a given microcontroller family is also shown. Table 2.2 gives a comparison of processor specific features in exemplary microcontrollers from Intel 8051 and Motorola 68HC11. Table C.1.1 gives a comparison of processor-specific features in two exemplary microcontrollers from Motorola 68HC12A4 and PIC family 16F84. Remember, there are variations of the peripherals in different members of same family.

Table C.1.1
Processor-Specific Features in Microcontrollers

Capability	Motorola 68HC12A4	PIC 16F84
Processor instruction cycle in μs *(typical)*	0.125	0.2
Internal Bus Width in Bits	16	14
CISC or RISC Architecture	CISC	CISC
Program Counter bits with reset value	22 [(0xFFFE)]	13
Stack Pointer bits with initial reset value in case they are defined by a processor	16	13
Super- scalar Architecture	No	No
On-Chip RAM and/or Register file Bytes	1024	68 x 8b + 64 x 8b EEPROM
Instruction Cache	No	No
Data Cache	No	No
Internal Program memory EPROM/EEPROM	4 kB	1 K x 14b
External Program memory capacity in Bytes (if Separate)	Not separate	-
Data Memory Capacity	4MB + 128kB	-
External Interrupts	24	1
Bit Manipulation Instructions	Yes	Yes
Floating Point Processor	No	No
Interrupt Controller	Yes	No
DMA Controller Channels	No	No
On-Chip MMU	No	No

The Motorola microcontroller chips 68HC11 and 68HC12 possess an interesting compatibility. Not only are the assembly codes compatible, but the executable file of HC11 also runs on 68HC12 and the assembly codes for 68HC11 also run on 68HC12. 68HC12 has greater memory address space with in-chip memory mapping to give 4MB + 128kB address space compared with 64 kB in 68HC11. 68HC12 operates at a higher clock speed (internal clock 8MHz vs. 2 MHz in HC11). It has more RAM (1024B) and more EEPROM (4 kB against 2 kB in HC11), and more ports and 8-channel timers (with 8 input-capture/out-compare registers). 68HC12 has a 16-bit pulse accumulator against the 8-bit in 68HC11. It has a changed index-addressing method to make the codes generated in assembly shorter. A part of the address can be encoded in the instruction byte. It helps in having shorter codes in the program and reduced hardware. The basic framework for these chips is the same. The code is about 30% shorter by the C-compiler compared to that in HC11. More 8-bit ports are available—totaling 12, (including one ADC, three for 24 key wake-up, one for interrupts and one for serial I/Os) compared to a total of 5 in 68HC11. More timers help to reduce the need for real-time clock–based software timers to control robot motors. The 24 wake-up lines in 68HC11 help greatly in a system with keypad input. The larger number of ports in 68HC12 than HC11 gives additional advantage in robotic applications. A robot has four or more degrees of freedom, needing four or more motors for movements of waist, neck, shoulders, elbows, and palm. More ports or multiplexing of the ports are necessary.

The PICF84 microcontroller has no external address and data buses. It is suitable for small single chip systems, for example, keypad or display controller systems. [For PIC microcontrollers details, refer to Tim Wilmshurst, *An Introduction to the Design of Small-Scale Embedded Systems, with Examples from PIC, 8051, and 68HC05/08 Microcontrollers*, Palgrave, Great Britain, 2001].

Table C.1.2 gives a comparison of the commonly available capabilities of a few select popular microcontrollers in their specific versions. Recent microcontrollers have advanced devices like incorporating the serial EEPROM, CAN bus interface, I^2C bus interface, I/O expander and USB 1.1 or 2.0 Interface and PCI-X interface.

Table C.1.2
Specific Features of Microcontrollers

Capability	Intel 8051 and Intel 8751	Intel 80196	Motorola M68HC12A4	Motorola M68HC11E2	PIC 16C76
Input-output Ports of 8 pins each	4	4	12	5	2 + 6 bit
Timers	2	2	8	1	2
Serial input-output SYN and ASYN	Yes	Yes	Two SPI and SCI	SPI and SCI	SSP + USART
Real time detection of an event or signal (Capture and Compare time on an event)	No	Yes	Yes	Yes	Yes
Pulse Width Modulation for DAC	No	Yes	Yes (16)	Yes (8)	Yes
Analog to Digital conversion (bits)	No	Yes	Yes (8) 8 channels	Yes (8) 4 channels	Yes (8)
Modulation Demodulation	No	No	No	No	No
Digital Signal Processing Instructions	No	No	No	No	No
Nonlinear controller Instructions	No	No	No	No	No
Power down mode	No	Yes	Yes	Yes	No
Idle Mode	No	Yes	Yes	Yes	No
Watchdog Timer	No	Yes	Yes	Yes	No

Note: 1. There are 3 timers in 8052 versions.
2. The main timer is of the 68HC11 Family. Others are additional.
3. Real time event detection, ADC, PWM, watchdog timer and input capture facilities also exist in advanced versions of 8051. A new version is 80251.
4. The Data Manual of an IC version must be consulted for the exact specifications of a microcontroller version in the family. This is because a manufacturer may continuously revise or add certain features and offer other versions in the family.
5. PIC 16F876 is flash memory version of PICF76.
6. 24 key wake-up lines exist in 68HC12 at three additional ports with interrupt capability compared to 68HC11.

C.2 EXEMPLARY MOTOROLA FAMILY CISC- AND RISC-BASED NEW GENERATION MICROCONTROLLERS

Table C.2.1 gives a comparison of exemplary microcontrollers from the Motorola family.

Table C.2.1

Specific Features of Recent Higher Performance Microcontrollers of the Motorola Family

Features	M68HC16 Family	M683XX family	MCORE MMC2001
CPU features	16-bit Enhanced 68HC11	32-bit M68000 Opcode Compatibility	32-bit load/store RISC architecture, ultra minimal power consumption
4-stage Instruction pipe line, Fast interrupt support with 16-entry dedicated alternate register file, fast, Context switching Vectored and auto-vectored interrupt support, On-chip emulation support, Reset unit,	No	No	Yes
Enhancements for Virtual Memory, Vector Base Register, Loop Mode, Addressing Modes, Instructions	No	Selectable M68010 and 20 enhancements	No, Each instruction is 16-bit and Best code density^
Table Look-Up and Interpolate (TABL) instruction	No	Yes	No
Data + Program Memory	1MB + 1 MB	Program 16 MB	No$
DSP Functions	Yes	No	No$
General Purpose Timer - Two free-running 16-bit counters with programmable prescalars, 16 channels, each associated with an I/O pin, a 16-bit capture register, a 16-bit compare register, and a 16-bit greater than or equal to comparator	Yes, 9-stage pre-scalar, 16 bit FRC*	Yes	No
Configurable Timer Module	Yes	Yes	Time of Day timer, Periodic Interrupt timer
Timer Processing Unit with Micro-programmed control unit, control memory and a library of more than 20 different timing functions	TPU or TPU2	Yes, TPU/TPU2	No
CAN	Yes, CAN version 2.0 A/B	Yes	No

(Contd.)

Features	M68HC16 Family	M683XX family	M*CORE MMC2001
Multi-channel Communication Interface	1 SPI + 2 SCI	Yes	No
Queued Serial Module	Yes	Yes	No
System Integrator	12 programmable chip select		Yes No
Flash or EEPROM or Flash EEPROM	Flash EEPROM	Yes	256 kB Maskable ROM n$^\$$
SRAM	With 1 to kB block programmability		Yes 32 kB$^\$$
Dynamic Bus-Resizing	8 bit or 16 bit	8 bit or 16 bit	No$^\$$
High Level Language Support	Yes	Yes	No$^\$$
Clock Rate	16 MHz, 20 MHz, 25 MHz	16 MHz, 20 MHz, 25 MHz	33 MHz for 31 MIPS performance$^\$$
Operation Voltage	5V or 2.7 V to 3.6 V	5V or 2.7 V to 3.6 V	1.8 V and IO 3.6 V$^\$$
Serial input-output SYN and ASYN	Yes	Yes	2 UARTs$^\$$, Interval Mode SPI,
Real time detection of an event or signal (Capture and Compare time on an event)	Yes	Yes	No
Pulse Width Modulation	Yes		6 Channels$^\$$
for DAC ADC	8 or 10 bit programmable with alternate voltage reference and programmable sample hold	Yes	No
Queued ADC	Yes, 16 channel	Yes	No
Modulation Demodulation	No	No	No
Nonlinear controller instructions	No	No	No
Power down mode	Yes	Also Low Power Stop (LPSTOP) instruction	Yes
Idle Mode	Yes	Yes	Yes
Watchdog Timer	Yes	Yes	Yes

Note: (1) * means 16-bit free running counter (2) $^\$$ Data for first MCORE family member MMC 2001 (3) ^ Like ARM Thumb instruction set.

◾ LIST OF KEYWORDS AND THEIR DEFINITIONS ◾

- *High-level language support*: A supporting unit with a processor structure, which facilitates program coding in 'C' or other high-level languages and enables running them like machine codes by an internal compilation.
- *Queued ADC*: An ADC which performs ADC operations for the channels placed in a queue on a FIFO basis.
- *Modulation and Demodulation Unit*: A modem.

- *Free Running Counter (FRC)*: A counter which runs nonstop and which cannot be reset. It finds many timer applications. It obtains input after the prescaling of the processor clock frequency. [If prescaling factor is configured as 64, the FRC inputs are with 0.125 ms (1/8 MHz) for a processor clock of 8 MHz.]
- *Timer Processing Unit (TPU)*: A timer may have a number of interrupts and corressponding ISRs and functions. Timer interrupts may be, for example: input capture, outcompare, real time clock, counter, timer overflows and software timers. TPU helps creating by processing the ISRs independently.
- *Input Capture*: To capture the FRC reading when there is an input event and generate an interrupt for a call to the corrseponding ISR.
- *Out Compares*: To compare the FRC reading with a prefixed value in an outcompare register. It is used for alarm-like functions. When both values are equal, an interrupt may be generated and an output event may also be intiated.
- *Watchdog Timer*: A timer which is set in advance. Its overflow indicates that a process is stuck somewhere and therefore the processor resets and restarts.

Appendix D

Embedded Digital Signal Processors

Recall Section 1.2.5. Embedded systems for applications like the video-recorder and mobile phone should give the following processor performances with optimised use of resources, power and memory:

1. High MIPS and/or high MFLOPS
2. High MACs per sec performance

A digital signal processor (DSP) gives these performances. A digital signal processor should possess capabilities for speech and video processing. The DSP-ICs and DSP cores in embedded systems possess the real time speech and video processing capabilities.

D.1 ARCHITECTURE OF DIGITAL SIGNAL PROCESSORS

The architecture of a DSP can be understood by considering an exemplary DSP of the TMSC64x™ DSP generation.

The main structural units in the TMSC64x™ DSP generation and their functions are as given in Table D.1.1. Figure D.1.1 shows the interconnections between twenty-five structural units by a block diagram for a processor structure. Table D.1.2 gives the additional structural units and the functions in the processors, TMS320C64x64™ VelociTI™ VLIW architecture Extension.

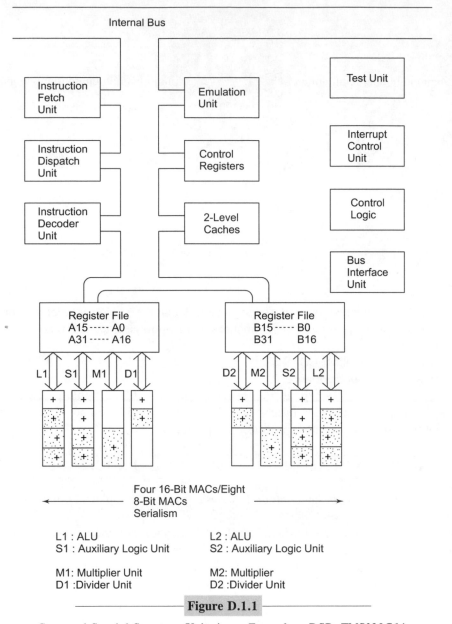

Figure D.1.1

Core and Special Structure Units in an Exemplary DSP, TMS320C64x

Table D.1.1

Structural Units and Functions of the Processor in the DSP Core

Structural Unit in Core	Functions
Basic Units	MDR, Internal Bus, Data bus, Address bus, control bus, Bus Interface Unit, Instruction fetch register, Instruction decoder, Control unit, Instruction Cache, Data Cache, multistage pipeline processing, multiline superscalar processing for obtaining processing speed higher than one instruction per clock cycle, Program counter similar to Table 2.1
Instruction Dispatch Register	For dispatch of an instruction to the appropriate unit
Control Registers	Control Registers associated with the control unit of the processor
Emulation Unit	Emulation
Register File A	Set of on-chip registers used during processing instructions in Data path 1. These are named A0… A 15 and A16 …A31. Recall that a register file is a file that associates with a unit like ALU or FLPU.
Register File B	Set of on-chip registers used during processing instructions in Data path 2. These are named A0… A 15 and A16 …A31.
Prefetch unit	The CPU fetches eight 32-bit RISC-like instructions at each cycle.
Processing unit	Two multipliers and six arithmetical units, highly orthogonal, compiler and assembly optimiser, execution resources.
Arithmetic Logical sub unit	It is a sub-unit to execute arithmetical or logical instructions according to current instructions in IR.
Auxiliary Logic sub Unit	It is a sub-unit used during subtraction. [Finds 2's complement before addition and then adds in order to subtract.]
Multiplier sub Unit	Multiply
Floating Point Processing sub unit	This sub-unit in C67x ^{TM} is separate from ALU for floating point processing. [Refer to Section D.3]
Assembly Optimiser	Optimises assembled codes
C compiler	Highly Efficient

Note: Orthogonal with frequently used instructions is available in more functional units

Table D.1.2

Additional Structural Units and Functions of Processors in TMS320C64x64^{TM} VelociTI ^{TM} VLIW Architecture Extension

Structural Unit in Core	Functions
Packed Data processing	8-bit or 16-bit data packed and processed as 32-bit data
Parallel Execution MAC units	Quad 16-bit MAC/Octal 8-bit MAC [Refer to Section 2.1 and D.2]
Special Instructions	Broadband infrastructure and image processing VLIWs
Level 2 Cache	Enhances performance of each fetch cycle
Instruction Packing unit	Instructions packed as VLIW execute in parallel

D.2 DSP PROCESSOR VERSUS CONVENTIONAL PROCESSORS

The simplest signal is sinusoidal. A distorted sinusoidal signal which repeats itself in a period T, is the sum of a series having sinusoidal signals of fundamental frequency f (= 1/T) and sinusoidal signals of frequencies that are integral multiples of *f*. Digital filtering is a process by which the selected components are separated. A signal distorted by noise can be retrieved by noise elimination using digital filtering. A telephonic tone can be detected by digital filtering. An echo in a signal is eliminated by filtering. When the ultrasonic scan signals from ultrasound sensors are filtered, they show to the physician the region of interest only. Digital filtering is one of the important basic processing operations done by a DSP.

Any analog signal has to be converted to digital signals by sampling. Let a signal sample be between intervals of T. Then the sampled signal can be multiplied by a constant. Each sample, x_n can be encoded to get a signal, y_n. Consider the following equations.

$y_n = ax_n + \Sigma (b_{-j}. y_{-j}) + \Sigma (a_{-k}. x_{-k})$, where j varies from 1 to *m* and k varies from 1 to *n* D.1
and $y_n = \Sigma (a_i.x_{n-i}) + \Sigma (b_l.y_{n-l})$, where i varies from 0 to N and *l* varies from 1 to M .... D.2.

Equation D.1 can be understood as follows. Let us assume that *y* is the 10th component of the output. The first term is the 10th input multiplied by a constant. The second term is the sum of the last but one, last but two until the last but *m* products. The product is the result of the multiplication of an output 'y' with a coefficient 'b'. The third term is the sum of the last, and last but one, and last but two up to the last but *n* products. The product is the result of the multiplication of an input 'x' with a coefficient '*a*'. Equation D.2 can be similarly understood.

Equation D.2 can be written in a standard form (called difference equation) used in DSP operations as follows:

$$y_n = \Sigma a_i.x. (n-i) + \Sigma b_l.y. (n-l),$$ D.3

where the subscripts in equation D.2 has been replaced by the coefficients. It means a *p*-th coefficient is a simple *p*.

A special case of this term is $y_n = \Sigma (a_i.x_{n-i})$ referred to in Section 2.1, and is used in FIR (Finite Impulse Response Filter). FIR filters have coefficients $b_l = 0$. It means that previous outputs of samples are not required. The signal phase is linear. Another filter is called an IIR (Infinite Impulse Response) filter and used when the signal phase is non-linear.

The MAC (Multiply and Accumulate) unit(s) in a DSP provides fast multiplication of two operands and the accumulation of results at a single address. It quickly computes Sum = Sum + [a_i. x. (n-i)], an expression such as the second or third term of Equation D.3. A DSP may have dual MAC units to calculate both first and second summation terms simultaneously in Equation D.2 or D.3.

The DSP therefore should do summations of many terms faster than a conventional GPP (general purpose processor). It should be capable of processing VLIWs [Section 6.3]. The *number of MACs rather than MIPS in a conventional processor measures a DSP performance.*

D.3 FIXED-POINT ARITHMETIC VERSUS FLOATING-POINT ARITHMETIC

Let a coefficient be 0.35967 and the variable = 1.4824931.

(a) Fixed-Point Arithmetic

Firstly, 35967 and 14824931 can be multiplied during fixed-point arithmetic processing and later the intermediate result is divided at the final stage by 10^{12} to obtain the final result.

(b) Floating-Point Arithmetic

0.35967 and 1.4824931 are multiplied by first representing them by a floating-point format (usually IEEE 754) to save the values as bits. The processor follows a multistage process during floating-point arithmetic processing.

(c) Loss of precession in Fixed-Point Arithmetic

Fixed-point arithmetic must take care of any overflows, which may occur. Overflow means an operation in which the result exceeds the capacity of the processor register(s) for storing that number as a single word. Overflows are taken care of only by sacrificing the precision in decimal numbers. If necessary, two decimal numbers being multiplied are first divided by a suitable number such that the intermediate result does not exceed the maximum storable number in the register. In the 32-bit and 64-bit processors, this maximum number is 2147483648 (= 2^{31}) for the 32-bit integer operations and the maximum number is 9223372036854775808 (= 2^{63}) for the 64-bit long integer operations.

Let us take a simple example to understand the loss of precision. Suppose 0.29 and 0.15 are to be multiplied and an 8-bit processor, which is capable of storing final result as maximum 8-bit number in a register is used. Thus maximum and minimum 8-bit integer numbers for storing a result are 127 and –128, respectively. Consider a processor having only the fixed-point arithmetic unit. Then, when 29 and 15 are multiplied, there is an overflow. It is because intermediate result (= 435) cannot be stored in the register. Therefore, let us choose a divisor of 4, the multiplication of intermediate result by which will give us the final stage result. Therefore, 0.29/2 = 0.145 and 0.15/2 = 0.075 are chosen to start with the fixed-point arithmetic. Now, the processor multiplies the integers 14 and 7, sacrificing the last digits in each number. The intermediate result is 98. Actually, the intermediate result should be 108.75 (from 0.29x0.15/4 = 0.435/4 = 0.010875).

Fixed-point arithmetic is faster in a processor than floating-point arithmetic. This is because more steps are needed in a floating-point operation. A typical processor having 4800 MIPS performance shows 1350 MFLOPS performance [remember MIPS means million instructions per second and MFLOPS means million floating-point operations per second].

A 64-bit fixed-point operation is used and the precession is sacrificed a little during a DSP operation, provided the result of processing is within an acceptable accuracy. A 32-bit fixed-point operation is used and the precision is sacrificed a little more during a DSP operation, provided the result of processing is within an acceptable accuracy. A DSP with a 32-bit fixed-point operation processes roughly two times faster than the 64-bit operation. Fixed-point operations are approximately four times faster than floating-point operations with the same 32-bit numbers stored in a computer.

Occasionally the precision needed is such that it may require floating-point operations. An example where precision of the coefficients b_1 becomes important is the IIR filter.

D.4 DSP FOR EMBEDDED SYSTEMS

Texas Instruments Inc. has DSP platforms called TMS320C2000™, TMS320C5000™ and TMS320C6000™. There is complete code-compatibility across all devices that allows reuse of existing codes during future system enhancement [refer to http://www.dspvillage.ti.com for details].

D.4.1   TMS320C2000™ Platform

The TMS320C2000 family consists of high-precision control 32-bit DSP controllers, and performance is up to 150 MIPS. It integrates peripherals and offers a unique combination of on-chip peripherals by incorporating up to 2.5K of RAM and 32k words of Flash with code security in memory, ultrafast down to 375 ns conversion time A/D converters, timer, watchdog timer, SPI, SCI, and robust CAN modules (Section 11.3). Examples of the applications are as following: (a) Industrial application, (b) Automotive control applications, (c) Optical networks, (d) Handheld Power tools and (e) Intelligent Sensors. It provides more flexibility, control-optimised peripherals and powerful processors than the earlier 8- or 16-bit microcontrollers.

D.4.2   TMS320C5000™ Platform

Power efficiency, high performance systems and optional peripherals in the devices give a system designer an edge for applications in portable Internet systems and wireless communication systems. This DSP platform provides optimised performance embedded systems for applications in portable media and communication products like digital music players, GPS (Geographical Positioning System) receivers, portable medical equipment, feature phones, modems, 3G cell phones, and portable imaging.

When developing an application, the following tools are needed.

1. The IDE are the Code Composer Studio and evaluation modules (EVMs) and free evaluation tools.
2. DSK starter kit.
3. XDS560 Emulator.
4. When developing a client-side telephone, a client-side developers kit.
5. *Chip support library.* It includes the peripheral drivers.
6. DSP and Image libraries. There are free platform-specific downloadable libraries (called *TI foundation software*). Use of these functions reduces the system development time. It has modules for high-level and optimised DSP functions.
7. The Code Composer Studio IDE is a DSP specific code composer. It provides an environment similar to MS Visual C++. An IDE environment consists of the following. (a) Multi-level (C and DSP assembly) Debugger, (b) C and DSP assembly specific editor, (c) Probe points, (d) File I/O, (e) Comprehensive data visualisation displays and (f) GEL scripting language based on C.
8. The *C5000C* is a compiler, assembler and linker for TMS320C5000™ platform DSK. A compiler has three goals: (a) General purpose C code with performance approaching that of hand-coded assembled codes. (b) Programming interface to C run time environment. This enables implementation of critical DSP algorithms in assembly languages, giving optimised extreme performance. (c) Easy to use tool for high performance applications in C.

The TMS320C5000™ platform has generations of TMS320C54x™ DSPs TMS320C54x™ (DSPs + RISC) and TMS320C55x™ DSPs.

D.4.3   TMS320C6000™ Platform

The TMS320C6000™ platform uses new-technology devices, which have performance with RISC like coding and optimum efficiency. Performances are much higher than 5000 series with lower power dissipation per MHz operations. These devices have 2-level caches. The memory, peripherals and co-processor together give applications in broadband services, high-performance audio and imaging applications.

The TMS320C6000^{TM} platform has generations of fixed-point operations TMS320C62x^{TM} DSPs, TMS320C64x^{TM} and floating-point operations TMS320C67x^{TM} DSPs. All generations have code compatibility.

When developing an application, the following tools are needed.

1. The IDE are Code Composer Studio and evaluation modules (EVMs) and free evaluation tools.
2. Network Video Developer kit
3. Imaging Developer kit
4. Networking Developer kit
5. DSK starter kit that connects to the PC through a parallel port, a 16-bit data converter TLC320AD535, and a power management device.
6. JTAG controller for emulation and debugging
7. XD560 Emulator
8. Chip support library. It includes the peripheral drivers, and platform-specific libraries are free downloadable, having TI foundation software. They have high-level optimised DSP functions.
9. DSP and Image libraries. There are free platform-specific downloadable libraries (called *TI foundation software*). Use of these functions reduces the system development time. It has modules for high-level and optimised DSP functions.
10. A Code Composer Studio IDE is a DSP specific code composer. It provides an environment similar to MS Visual C++. The IDE environment consists of the following. (a) Multi-level (C and DSP assembly) Debugger, (b) C6000C compiler and DSP assembly optimiser, (c) Enables RISC-like assembly codes and scheduling for optimum performance and efficiency Probe points, (d) File I/O, (e) Comprehensive data visualisation displays and (f) GEL scripting language based on C.
11. High-performance C6000C engine with a compiler that leverages the architecture to sustain maximum performance. It speeds up the design development time for high-performance applications with a balance code size with free downloadable optimisation tools.
12. Reference Frameworks.

D.4.4 TMS320C24x and C28X generations of DSPs

The TMSC24x core gives a 16-bit data Fixed-Point DSP core for digital control applications. It offers SCI, SPI, CAN, A/D, event manager, watchdog timers and on-chip Flash memory. C24x core offers 20-40 MIPS of computational bandwidth. It runs numerous sophisticated control algorithms in real-time. It provides the followings: (a) sensor-less speed control, (b) Random PWM, (c) Power factor correction, (d) Code compatibility with other C2000 family devices and (e) High code efficiency.

TMSC28x core gives highest performance DSP core for digital control applications. It offers 32-bit data Fixed-Point DSP and SCI, SPI, CAN, 12-bit A/D, McBSP, watchdog timers and on-chip Flash memory. C28x core offers up to 400 MIPS of computational bandwidth. It runs numerous sophisticated control algorithms in real-time and provides the following: (a) sensor-less speed control, (b) Random PWM, (c) Power factor correction, (d) Code compatibility with other C2000 family devices and (e) High code efficiency. [McBSP means high-speed communication Multi-channel Buffered Serial Port. Refer to http://www.ti.com/sc/docs/psheets/ abstract/apps/spra638a. htm.]

The TMS320C24x has 15 devices, with options in terms of ADC number of channels, CAN modules, serial ports and flash. Table D.4.1 gives the features of devices. An embedded system designer can accurately select one of these as per the requirements of the system.

Table D.4.1

Features in TMS320C2000^{TM} Platform and TMS320C24x^{TM} DSPs with an Example of TMS320FC2402A

Feature in C2000^{TM} series Platform	Feature in C24x^{TM} series DSPs	Example from 15 devices in TMS320C24x ^{TM} generation DSPs, TMS320FC2402A
Clock Rate	20 MHz to 40 MHz	40 MHz
DSP Core Processor	Fixed Point DSP	Fixed Point DSP
RAM in words	544 words or 1k	1k
ROM in words	0, 4 k, 6 k, 8k, 16 k or 32 k	0k
Flash in words	0, 4 or 16 k	8k
ADC 10 bit channels	2, 5, 8 or 16	8
ADC conversion time in ns	6100, 900, 500, 425 or 375	500
CAN module	None or yes	No
PWM Channels	*7 or 8 or 12*	8
Timers	2, 3 or 4	2
Serial Ports	1 or 2	1
Boot Loader	None or ROM or flash	ROM
Performance in 20 to 400 MIPS	20 or 40 MIPS	40 MIPS
External Memory Interface	None or yes	None

D.4.5   TMS320C54x and TMS320C55x generations of DSPs

The TMS320C54x generation has more than 17 code-compatible devices, which give options for a broad range of performance and peripherals options, low-power operation and innovative architecture and instruction sets. It gives a system designer effective ways of achieving high-performance, low-power operations at a low system cost. C5470 and C5471 system-level DSPs integrate a DSP, RISC, ready-to-use operating system and full development support enabling a new generation of designs.

The TMS320C55x generation gives the most power-efficient DSPs. Therefore they have applications in portable Internet appliances to high-speed wireless communications. They have ultra-low power performance, which is achieved through advanced power management techniques. Power is automatically down in inactive peripherals, memory and core functional units. The C55x DSP core is an OMAP 5910 processor [core means a section which integrates with other devices on a chip].

When an OMAP integrates with a C55x DSP core having a TI-enhanced ARM925, on a single chip, it gives an optimal combination of high performance with low power consumption. This is a unique architecture. Embedded system DSP and ARM developers obtain low power real-time signal processing capabilities of a DSP coupled with the command and control functionality of an ARM.

Table D.4.2 gives the features of over 20 generations of devices in TMS320C5000^{TM} platforms, each including the processor OMAP5910. An embedded system designer can accurately target one of these as per the requirements of the system.

Table D.4.2

Features in TMS320C5000^{TM} Platform and TMS320C54x^{TM} DSPs with an Example of TMS320LC54V90

Feature in C5000^{TM} series Platform	Feature in C54x^{TM} series DSPs	Example from 17 devices (16 distinct) in TMS320C54x^{TM} generation DSPs, TMS320LC54V90
Clock Rate	40 MHz to 160 MHz	117MHz/58.98 MHz
DSP Core Processor	OMAP5910	OMAP5910
Cycle time	8.33 ns to 25 ns	8.5.ns/17 ns
Power Dissipation per unit MHz operation	0.33 mA/MHz	0.33 mA/MHz
Performance in MIPS	up to 900 MIPS	117.96/58.98 MIPS
Memory Interface	Advanced Multibus Architecture with 3 separate 16-bit data memory buses and 1 program memory bus	Advanced Multibus Architecture with 3 separate 16-bit data memory buses and 1 program memory bus
Memory in words	64K data memory, 1M to 8 M program memory, 5 k to 40 k RAM 2 k to 128 k ROM	64K data memory and 8M program memory 40 k RAM and 128 k ROM
Timers (16-bit)	0 or 1 or 2	2
Communication I/F	HPI (Host Port Interface)/HPI 8/16	HPI 8/16
External DMA Channels Support	0 or 6	6
Standard Serial Ports	0 to 2	0
Total Serial Ports	0 to 2	2
TDM Serial Ports	0 to 1	0
Buffered Serial Ports	0 to 3 (option McBSPs[#])	2
Miscellaneous Serial Ports	UART, DAA*	1 UART + 1 DAA
Boot Loader	Optional	0

Note: (1) * DAA means Direct Access Arrangement (http://e-insite.net/edmag/). The DAA serial port has analog input and output and has up to 1 master and 7 slave CODECs. CODEC means a unit for digital coding by ADC and other operations and decoding of analog signals by DAC and other operations at the output and input, respectively. For example, a Video CODEC unit does: (*i*) The encoding, which preprocesses for noise reduction then controls the rate of transmission after estimating the motion picture rates. It compresses, synchronizes audio and finally bit streams are sent to output and a streaming network. (*ii*) Decoding, which receives the bit streams, decompresses and separates audio and video and eliminates noise by preprocessing. (2) [#]Refer to Section D.4.4.

D.4.6 TMS320C62x, 64x and C67x generations of DSPs

The TMS320C62x generation has fixed-point DSPs represented by with a new technology, which enables the use of new equipment and energizes existing implementations. These give multi-channel and multi-function applications. Examples of the applications are: (a) Wireless base stations, (b) Digital subscriber loop (xDSL) systems, (c) Remote access servers (RAS), (d) advanced imaging/biometrics, (e) Industrial scanners and security systems and (f) Multi-channel telephony.

The TMS320C64x generation also has fixed-point DSPs. Devices of this generation can operate up to a range of 600 MHz and 4800 MIPS performance. New instructions accelerate the performance in key application areas. Examples of new applications are: (a) Digital communications infrastructure and (b) Video and image processing.

The TMS320C67x generation has a floating-point DSP generation. It enables new innovations in cost-sensitive applications. Examples of applications are: (a) Voice and Speech recognition. (b) High end graphics and imaging. (c) Industrial automation.

Table D.4.3 gives the features of over 14 devices in the TMS320C67x^{TM} generation. An embedded system designer can accurately select one of these as per the requirements of the system.

Table D.4.3
Features in theTMS320C67x^{TM} generation with an example of TMS320C6711-100

Feature in C6000^{TM} series Platform	Feature in C67x^{TM} series DSPs	Examples from over 14 devices in TMS320C67x ^{TM} generation DSPs, TMS320C6711-100
DSP Core Processor	Floating Point Processor	Floating Point Processor
Cycle time	4. 4 ns to 10 ns	10 ns
Performance in 1200 to 4800 *MIPS*	up to 4800 MIPS	1200 MIPS
Performance in 600 to 1350 MFLOPS (67x Series^)	up to 1350 MFLOPS	600 MFLOPS
DMA Channels	4 DMA or 16 EDMA	16 EDMA
Memory in words	512 k program and 512 k data bits, or 32 k bits L1P program cache and 32 or 64 L1D data cache, and 512 k bits L2 cache in 6711 and higher version	32 k bits L1P program cache and 32 L1D data cache and 512 k bits L2 cache
Host Port or Expansion Bus or PCI	None or 16-bit HPI	16-bit HPI
General Purpose I/O	0 or 1 or 1 (16 pins)	0
McBSP	2	2
MsASP	0 or 2	0
Timers (32-bit)	2	2
External Memory Interface	32-bit/PYP 16-bit/GDP 32-bit	32-bit
I^{2}C	0 or 2	2
Core Supply Voltage	1.2 V to 1.8 V	1.8 V
I/O supply Voltage	3.3 V	3.3 V

D.4.7 OMAP5910 Embedded Processor DSP in RISC environment

A recent DSP with an RISC environment is the OMAP5910 dual core. It integrates TMS320C TMS320C55x with ARM925 and has 192 kB RAM with USB 1.1. It has 1 host and client and MMC/ SP card interface.

D.4.8 SoC Based Solution Texas DSP TMS320DM310 for Digital Media

A recently announced innovation of a DSP with an RISC environment for digital media applications is the Texas DSP TMS320DM310 for SoCs. It integrates TMS320C55x with ARM925 and has 192 kB

RAM with USB 1.1. It has 1 host and client and MMC/SP card interface. See Table D.4.4 for a list of features in this DSP for digital media.

Table D.4.4
Features in TMS320DM310

Feature	Feature in TMS320DM310 DSP Digital Media for SoC
DSP Core Processor	TMS320C55x
RISC Core	ARM9 with Cache
Co-processor	Image co-processor for real-time decoding 640 x 480 VGA MPEG2/ MPEG4 video and real-time encoding 352 x 288 CIF for MPEG1, MPEG 2 and MPEG4
Clock	125 MHz DSP core and 160 MHz ARM RISC
Performance MIPS	Up to 2125, Programmable
USB host for direct I/O	Yes
Applications	6M pixel still digital camera 1 s capture, digital imaging and audio, MPEG4 and JPG portable devices, web-pads

◾ LIST OF KEYWORDS AND THEIR DEFINITIONS ◾

- *MAC unit*: A unit used in DSP operations for fast calcuation of $[ax_n + \Sigma (b_{-j} \cdot y_{-j})]$ like terms.
- *Code Composer Studio*: An IDE for TI DSP specific code composing which provides an environment similar to MS Visual C++. It consists of the following: Multi-level (C as well as DSP assembly) Debugger, compiler, assembly optimiser, RISC-like assembly codes and RISC-like scheduling for optimum performance and efficiency probe points, file I/O functions, comprehensive data visualisation displays and GEL scripting language based on C.
- *CODEC*: A unit for digital *coding* after ADC and other operations and *decoding* of analog signals by DAC and other operations at the output and input, respectively. It is used in processing audio or video signals.
- *Code Compatibility*: Usability of codes by various generations in a family.
- *Upward code Compatibility*: Usability of codes by various enhanced generations in a family.
- *Backward Code compatibility*: Usability of codes by various previous generations in a family.
- *DAA*: Direct Access Arrangement, for example, a typical DAA serial in and out port directly transferring the analog input and output using up to 1 master and 7 slave CODECs.
- *McBSP*: A high-speed communicating Multi-channel Buffered Serial Port.
- *DCT*: Discrete Cosine Tranformation function used in a number of DSP functions, for example, the MPEG2/MPEG4 compression.
- *Digital Filtering*: A filter for the signals that use DSP functions.
- *Echo*: A signal received after a delay and which superimposes over the original signal. For example, in a hall or at the hills we hear the original sound as well as the echoed sound.
- *Echo Cancellation*: A process of eliminating echoes.
- *Fixed-Point Arithmetic*: Arithmetic using signed or unsigned integers employing processor registers or memory.

- *Floating-Point Arithmetic*: Arithmetic using processor registers or memory, where the decimal numbers and fractional numbers are stored in a standard floating-point representation.
- *Noise Elimination*: A process, which eliminates unrelated randomly introduced signal components.
- *OMAP 5910 processor*: A TI processor of unique architectutre in DSP chips of high performance with low power consumption.
- *Real-time video processing*: Processing of video signals such that all or most incoming frames are processed in a time frame such that each processed frame maintains constant phase differeces and the expected intervals between them.

New Innovative Processors for Embedded Systems

Recall Sections 1.2.5 and 1.2.6. Sophisticated embedded systems for certain applications like a mobile phone with streaming images and wireless Internet should give each of the following *processor performances* with optimised use of resources, power and memory:

1. High *MIPS* and high *MFPS*
2. High *MACs/sec* DSP performance
3. High *Mbps* I/O transreceiver bandwidth.

The latest innovatively designed *media processors* provide this. A media processor combines the capabilities of the *processor*, *video processor* and *I/O processor. The use of media processor ICs and cores in embedded systems is a recent innovation.* Mobile processors, for example Intel centrino, introduced in March 2003 is another new innovation. Mobile embedded systems with 802.11 LAN wireless data links are even easier to design.

E.1 MEDIA PROCESSORS FOR EMBEDDED SYSTEMS

A *video-processor* should provide for the following processing functions.

1. A VLIW (Section 6.3), fixed- as well as floating-point arithmetic
2. Discrete Cosine Transformation (DCT) processing unit
3. Quantiser unit [Quantisation means analog signals from camera or micrcophone interface with signals being converted to digital quantised (bit-streamed) output after encoding]
4. Processing of library functions for graphics, two dimensional text (text with fixed dimensions), MPEG and motion JPG (jpg frames from a moving system)
5. Image colour and hue correction, image rotation, image scaling, shadow enhancement, detecting image edges and sharpening the image
6. Video encoding, which preprocesses for noise reduction and then controls the rate of transmission after estimating the motion picture rates; compresses, synchronises audio and finally bit streams are sent to a streaming network

7. Video decoding, which receives the bit streams, decompresses and separates audio and video and eliminates noise by preprocessing
8. Noise reduction and echo-cancellation

Streaming Networks are defined as networks in which the packet flow is such that clients retrieve all packets or data frames continuously from a network through a server, just as in real-time audio and video communication. Steaming networks, digital smart cameras and entertainment electronic products need media processors.

Here are examples of the applications of C programmable real-time *media processors*:

1. Stream networks for audio web-casting (radio through Web)
2. Streaming audio and video web-servers and Internet connectivity devices
3. Video-Conferencing
4. Consumer Electronics—Entertainment products like DVD
5. Automotive Communications Electronics
6. Lightweight cordless phones and faxes
7. Mobile phones with Internet and video images accessibility
8. Medical Electronics in tomography, cardiography, and cardiovascular and X-ray display storages
9. Security monitors and observation systems
10. Speech processing
11. Noise reduction and echo cancellation

The image-frames received using DSP or GPP need to be stored first in real-time in memory, then decompressed, decrypted and processed. A camera image frame format may be of horizontal and vertical pixels, 1024x768 or 640x480 or 320x240 or 160x120. Frames arrive at incoming rates in either of two standards—25 frames per second or 30 frames per second, in the case of video signals and 10 to 15/s frames in video conferencing. A frame may be in either of the standard formats defined for videos.

Similarly when transmitting the images using a DSP or GPP, the image frames need to be buffered first in the memory, then decompressed, decrypted and processed; and then transmitted in real-time. *A real-time video processor has much faster processing power than a DSP processor or a GPP (general purpose processor). It requires much smaller program and data memory because of a lesser need for multiple-frame storing and buffering before processing.*

Media processors facilitate a seamless fusion of mobile telephony with broadband Internet. They facilitates voice-based Web accesses, speech recognition, text to speech conversion, VoIP (Voice over Internet Protocol) and a voice-based version of mobile Net-standard XML.

Philips Semiconductors has recently introduced C programmable real-time *media processors Trimedia 1x00 series* [the three media are data, audio and video]. They have processing capability for video, audio and streaming network processing functions. Real-time multimedia data-streams are facilitated. The Tri-Media TM-1300 media processor for these functions is a members of this series. Refer to http://www.semiconductors.philips.com/trimedia/ for details of the TriMedia series. For H.261 and H.263 video CODEC libraries for the Philips TriMedia power real-time performance with software flexibility, refer to http://www.4i2i.com/h_263_philips_trimedia.htm. [Note: (a) TriMedia is a media processing technology wholly owned by TriMedia Technologies Inc. TriMedia is a trademark of TriMedia Technologies Inc. (b) CODEC mean coding and decoding of incoming and outgoing signals, respectively].

Philips Semiconductors has very recently introduced the Nexperia PNX1300 Series and PNX1500 series. Figure E.1.1 gives a block diagram of the PNX1300 media processor. Table E.1.1 gives the features of PNX1300 and the earlier TM-1300 processors.

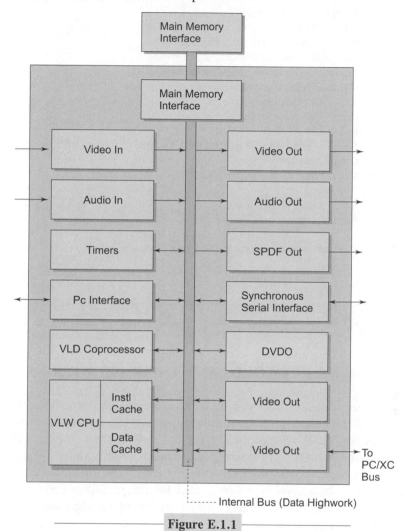

Figure E.1.1

Nexperia PNX1300 Series Media Processor Block Diagram (Reproduced with permission from Philips Semiconductors).

Table E.2.1

Features of the Nexperia PNX1300 Series Compared to the Earlier TM-1300 Processors

Feature	TM-1300 processors	Nexperia PNX1300 Series
Clock Rate	100 MHz	Up to 200MHz of processing power to a variety of multimedia applications
Memory Interface	Slower than PNX-1300	Faster Compared to TM-1300 Series

(Contd.)

Feature	TM-1300 processors	Nexperia PNX1300 Series
Pin Compatibility	Compatible with PNX 1300	100% pin compatibility with their TM-1300 predecessors
Structural units	Main Memory Interfaces on a common data-highway 32-bit internal bus to SDRAM, Video input unit, Video input unit, audio input unit, audio output unit, Timers, I^2C interface, synchronous serial interface, VLD coprocessor, DVDO, VLIW CPU with instruction and data caches, dedicated image coprocessor and PCI-X I/O Interface.	
C/C++ program-ability	Built-in optimised code generating compiler unit	Built-in optimised code generating compiler unit
Billion Operations per second	~ B/s. Much Slower than PNX 1300	Seven billion
Media Processing	Video, audio, graphics, and communications data streams processing	Video, audio, graphics and communications data streams processing faster than before
Operation Voltage series in power-constrained multimedia designs	Optional low-voltage media processor	Option of low-voltage operation

■ LIST OF KEYWORDS AND THEIR DEFINITIONS ■

- *Media Processor:* Media processor is a new innovative processor with high system performance for real-time video-performance, audio processing and data streaming. Its structural units are VLIW (Section 6.3) fixed- as well as floating-point arithmetic, Discrete Cosine Transformation (DCT) processing unit, Quantiser unit, units for processing of library functions for graphics, two dimensional text (text with fixed dimensions), MPEG and motion JPG (jpg frames from a moving system), units for image colour and hue correction, image rotation, image scaling, shadow enhancement, detecting image edges and sharpening the image, and Video encoding and decoding.

- *Media Processor with Respect to GPP and DSP*: A media processor is a new innovative processor that is superior to several DSPs together for real-time video-performance, audio processing and data streaming, with respect to system performance, memory buffer requirement and frame storage requirement. Conventional processors also provide for fixed-point as well as floating-point arithmetic operations (Section 2.1) and are also fast but their ALU processes add, multiply, subtract and divide like instructions only. Their performance becomes sluggish when processing data from video, audio or streaming networks.

- *Real-Time Video Processing*: Processing of video signals for encryption, decryption, compression, decompression, encoding, decoding and other operations such that there is maintenance of synchronisation between input rates and outgoing rates. Suppose 30 frames per second are the inputs in a video input. There should be 30 frames processed per second without internal buffer build up and without delay zitters at the output. However, frame misses up to an acceptable limit can be permitted.

- *Media Processor Clock Rate*: The clock rate is the processor clock rate, which controls the number of maximum operations feasible per second. It is 100 MHz for the TM-1300 series and 200 MHz for the Nexperia PNX 1300 series [operations at higher clock rates also need higher power dissipation].
- *Mobile Processor*: Process with advanced processing capabilities for wireless short distance (100 m) 802.11 LAN and for mobile technology applications.
- *Multimedia Processor Bus*: A bus of 32-bit high performance that connects to the SDRAM of the motherboard. It also fetches codes for the instruction and data caches. A high performance bus is a bus in which data fetch rates are several times faster than the clock speed.
- *Structural Units of PNX Media Processors*: They are as follows. Main Memory Interface, Common data-highway 32-bit internal bus to SDRAM, Video input unit, Video output unit, audio input unit, audio output unit, Timers, I^2C interface, synchronous serial interface, VLD coprocessor, DVDO, VLIW CPU with instruction and data caches, dedicated image coprocessor and PCI-X IO Interface.
- *Using Media Processors in an Embedded System*: Media processors are increasingly being used in place of DSPs in sophisticated embedded systems for stream networks for lighter weight, low-power dissipation and optimised memory performance real-time systems. They are particularly appropriate for web-casting (radio through Web), streaming audio and video web-servers and Internet connectivity devices, video-conferencing, Consumer Electronics and Entertainment products like DVDs, Automotive Communications Electronics, Lightweight cordless phones and faxes, Mobile phones, Medical Electronics in tomography, cardiography, and cardiovascular and X-ray display storages, Security monitors and observation systems, and Speech processing. The number of applications of media processor embedded systems continues to grow.
- *Web-casting*: Distribution (broadcasting) of information to many Web-registered addresses.
- *Streaming Network:* Networks in which the packet flow is such that clients retrieve all packets or data frames from a network continuously from a server just as in real-time audio and video communication.
- *Streaming Audio*: Audio message received through a network like a stream, in real time.
- *Streaming Video*: Video message received through a network (usually broadband Internet) like a stream, in real time.

Appendix
F

Serial and Parallel Buses

F.1 EMERGING SERIAL BUS STANDARDS (USB 2.0, IEEE1394)

Section 3.3 describes serial communication buses, 'I^{2}C', 'CAN', 'USB' and advanced serial buses. Table F.1.1 gives the salient features of serial buses including the emerging bus standards, USB 2.0 and IEEE1394.

Table F.1.1
Serial Buses to the Devices

Feature	Details
Serial Bus Options in order of Increasing bit rate	UART (512 Baud/s), 1-wire CAN (33 kbps), Industrial I^{2}C (100kbps), SM I^{2}C Bus (100 kbps), SPI (100 kbps), Fault tolerant CAN (110 kbps), Serial Port (230 kbps), MicroWire (300 kbps), I^{2}C (400 kbps, 2 meter), High Speed CAN (1 Mbps), IEEE 1284 (2.4 Mbps), High Speed I^{2}C (3.4 Mbps, 0.5 meter), USB 1.1 (Low Speed Channel 1.5 Mbps and 3 meter, High Speed Channel 12 Mbps and 25 meter), SCSI parallel (40 Mbps), Fast SCSI (8M to 80 Mbps), Ultra SCSI-3 (8M to 160 Mbps), FireWire/IEEE 1394 (400 Mbps, 72 meter), High Speed USB 2.0 (480 Mbps, 25 meter)
UART	Simple, no plug and play, non-secure, low speed, no master-slave mode
USB 2.0	Speed close to IEEE 1394, facilitates UHF transmission, plug and play, bus or external powered, interfaces to serial Port, IrDA and Ethernet, master-slave mode
IEEE 1394	Up to 63 devices to a host, 4.5 m Cable hops with a maximum of 16 hops, Constant transfer rate, Ensures fixed bandwidth, two shielded pairs of buses for data and strobe, power pair may be included in a single pair, IEE1394 specified connectors, bus Standard for DVD I/Os and MPEG2, High data throughput, data rates standards 100 Mbps, 200 Mbps, 400 Mbps, data rate standard 3.2 Gbps, plug and play, device auto-reconfiguration, master-slave mode, hot plugging allowed
CAN	Secure, slow compared to USB and I^{2}C, no plug and play, complex addressing, master-slave mode
I^{2}C	Simple plug and play, cost effective, fast with 3.4 Mbps only

F.2 EMERGING PARALLEL BUS STANDARDS (COMPACT PCI, PCI-X)

Section 3.4 describes host or computer system parallel communication interfaces, ISA, PCI, PCI-X and advanced buses. Table F.2.1 gives the salient features of parallel PCI buses including emerging bus standards, PCI-X and Compact PCI (cPCI).

Table F.2.1
Parallel PCI Buses

Standard	Features
PCI 2.2	64-bit bus, 66 MHz option, 32-bit 33 MHz throughput = 133 MBps, full component level, connector (94-pin connector with 50 signals) and board specifications, multiplexed AD0: 31 bus, dual address 64-bit support. An un-terminated bus, and the signal relay reflected on signal to attain the final value [refer to Section 3.4 for details].
PCI-X (PCI extended)	Backward compatible with existing PCI cards, improves upon the speed of PCI from 133 MBps to as much as 1 GBps, used in high bandwidth devices (Fiber Channel, and processors that are part of a cluster and Gigabit Ethernet).
Compact PCI (cPCI)	Electrically identical to the PCI specification but uses the Euro (VME) card 3U/6U format with 2mm connectors and its bus uses 8, 16, 32, or 64 bit transfers. Maximum 264MBps throughput, 6U cards contain additional pins for user-defined I/Os, live insertion support (Hot-Swap), supports two independent buses on the back plane (on different connectors), supports Ethernet, Infiniband, and Star Fabric support (Switched fabric based systems).
PXI	cPCI additional enhancement to Instrumentation, additional signals and chassis requirements identical to the VXI additions to VME [unlike the VXI, PXI does not require that the card be enclosed in a metal shield].

■ LIST OF KEYWORDS AND THEIR DEFINITIONS ■

- *USB 2.0:* A new 480 Mbps USB bus standard for facilitating UHF data transfers between the host and plug and play devices up to 25 meters and which interfaces with Serial Port, IrDA and Ethernet.
- *IEEE1394 Bus*: A new bus standard for DVD I/Os and MPEG2 and high data throughput. Refer to text in Table F.1.1 for details.
- *Plug and Play*: A device which can be directly interfaced to the bus and used without executing device initialisation and configuring codes.
- *Device Auto-reconfiguration*: A host controller on a bus may automatically sense a new device that attaches or detaches. On sensing the newly attached device, it runs the device initilisation and configuration codes automatically.
- *Hot plugging*: Attaching a device on a running system, or attaching without a system reset or reconfiguration.
- *Master-Slave Mode*: A mode in which there is a master (host) controller to control other devices called slave devices (devices other than host devices).

Devices in Embedded Systems

The following sections give the memories, RAM/ROM/flash devices, device programmer, SPI, SCI, UART, TIMERS, PWM and Watchdog Timer used in embedded systems.

G.1 VARIOUS FORMS OF ROM DEVICES

Table G.1.1 gives the various forms of ROM devices used in embedded systems. It compares their features. Erasing a ROM or a sector of memory in a flash means placing 1s in all the cells or in a sector of cells, respectively. Programming involves changing the 1s to 0s with a device programmer. [Section G.2]. This table explains where a ROM is programmed, how is it erased, how is it burnt (writing files is called burning) and how many times each type of ROM can be programmed.

Table G.1.1
Various forms of ROM and Comparative Features

Form of a ROM	Where programmed	How the bits are programmed (changed to 0s)	How the bits are erased (changed to 1s)	Bits erased at a time	Number of times it is programmable
Masked ROM	At the mask-manufacturer of the system ROM	Appropriately masking the circuit on silicon	Not erasable	Not erasable	None
EPROM	By a device-programmer at the system designer or system manufacture site	Bits are program-mable by using the device programmer in the laboratory	It erases (writes all bits as 1s) by exposing its quartz window to ultraviolet light*	All at a time	100 plus

(Contd.)

Form of a ROM	Where programmed	How the bits are programmed (changed to 0s)	How the bits are erased (changed to 1s)	Bits erased at a time	Number of times it is programmable
PROM (OTP ROM)	At system design or system manufacturer, site once only	Bits are programmable by using the device programmer in the laboratory	Does not erase as it is for programming once only	Not erasable	Once only
EEPROM	In the initially made system by a device programmer and later on in-circuit as per the embedded program	Programs byte by byte either in-circuit at the system or at device programmer by writing twice– firstly all 8 bits as 1s and then the bits as per needed byte	Erases either in-circuit at the system or at the device programming circuit by writing all 1s.	A byte at a time	1 million plus
Flash	In the system made initially by a device programmer and later on in-circuit as per the embedded program	Programs either in-circuit or at the device programmer by writing the needed byte at a time if its sector has been previously erased	Erases in a flash either in-circuit or at device programming circuit by writing all 1s to a selected sector	A sector of bytes in a flash	10 K plus (in the latest flashes)

Note: (1) A device programmer uses an appropriate software tool that has a program specific to that device for its programming. (2) * Ultraviolet exposure is by ~12 mW/cm^2 2.5 cm above the chip's quartz window for about 1000s. (3) ROM/EPROM/EEPROM can be verified in different versions of a microcontroller family after programming [verification means reading a programmed EPROM or EEPROM check whether the programmed bits are there or not].

G.2 ROM DEVICE PROGRAMMER

Consider a device that has a 512 kB programmable memory. It means that it has eight signals (D_0 to D_7) and 19 [= $\log_2$ (512* 1024)] address A_0 to A_{18} signals. There are a total of 524,288 (=512* 1024) arrays of cells with each array having 8 cells. Each cell has for output the D bit as logic 1 in the unprogrammed (fresh) state. Programming the device (locating the desired bytes) means replacing 1s with 0s according to each byte needed at each cell array address. Bytes saved are as generated by the locator program of the embedded system after the programming [after programming, the device memory part will hold the final bytes according to the need after the software development, testing and debugging cycle of design finishes.]

A ROM Device Programmer is a programming system for the PROM or EPROM chip or unit in a microcontroller or device. A device inserted into a socket (at the device programmer circuit) is programmed on transferring the bytes for each address using a software tool at the computer and interconnecting the computer with this circuit. The device program needs the locator output records in the input. This output must reflect the final design, only then can the device program put the final inputs into the ROM. Records that are input to a device programmer as per the device being programmed, are in three formats. The formats are as follows.

G.2.1 Binary Image

Binary Bit mapped (binary Image) means bytes are sent in a sequence as per starting address to the end-address.

G.2.2 Motorola S-Record Format

Motorola S-Record format is an industry standard for storing the locator file, before its use by the device programmer or ROM-mask designer. [It is called S-record because it has first character as 'S' in each line. A line is as follows: first character is S, second character is 2 (for specifying the record type), third and fourth characters are, say 14 (to specify that there are the 20 bytes in that line), the remaining 40 characters (nibbles) divide as the address (3 bytes), and data (16 bytes) and checksum (1 byte). Table G.2.1 shows a typical S-record as a locator output and device programmer input. It is left as an exercise to the reader to show that *Addr* for line 6 in the record of Table G.2.1 will be 0x000037.

Table G.2.1
An Exemplary Motorola S-Record Format

Line Number[+]	First Character	Second Character [#]	Third and fourth Characters for N[&]		Address, Addr[*]	N_d [\$] Bytes for storage in ROM from Addr [Maximum value of N_d can be 253 decimal]	Checksum[~]
0	S	2	1	0	000000	aa bb cc dd ee ff xx yy zz bb cc dd	cs0
1	S	2	0	C	00000C	cc aa cc dd ee ff xx yy	cs1
2	S	2	1	2	000014	dd bb cc dd ee ff xx yy zz bb cc dd aa xx	cs2
3	S	2	0	5	000022	0A	cs3
4	S	2	0	8	000023	dd bb cc dd	cs4
5	S	2	1	4	000027	dd bb cc dd ee ff xx yy zz bb cc dd aa ff 01 c0	cs5

[+]Line number is not in the record

[#]2 means the availability of data record in this line. A byte from the data sequentially burns at the ROM.

[&] N = 10 means that there are 16 hexadecimal bytes in this line including the 3 bytes for the address and 1 byte for checksum at the end of the line. Number of data bytes for storing specified in this line, N_d = 12 decimal. 0C means N = 12 and N_d = 8 decimal.

[*] Starting address of 3 bytes, 000000 means the next 12 bytes store between address 0x0000 to 0x000B. Therefore in the next line the starting address is Addr = 0x000C.

[\$] Bytes for burning in ROM are in this line and numbering = N_d. Each character in this column represents a nibble.

[~] cs0, cs1, ... are the checksums of 1 byte each of all the bits in line number 0, 1, ..., respectively.

G.2.3 Intel Hex-File Format

The Intel Hex-File format is another industry standard for storing the locator file, before its use by the device programmer or ROM-mask designer. A line is as follows: first character ':' (colon), second and third characters for data counts (assume = 10 in hexadecimal in case N_d = 16) in the line (address bytes, checksum byte and data type bye excluded, only actual data bytes at the line, which are to be burned in ROM are counted), fourth to seventh address (2 bytes), sixth and seventh as 0 and 0 to specify data as ROM data, and the remaining 32 characters as the data (16 bytes) and 2 characters for the checksum (1 byte). Table G.2.2 shows an Intel hex-file, which corresponds to same data as at the Motorola S-record in Table G.2.1 as a locator output and device programmer input. It is left as an exercise to the reader to show that *Addr* for line 6 in the record of table G.2.2 will be 0x0037.

Table G.2.2
An Exemplary Intel Hex-File Format

Line Number[+]	First Character	Second and Third Characters for C_d		Address, Addr[*]	Sixth and seventh Characters [#]		N_d [$] Bytes for storage in ROM from Addr [Maximum value of N_d can be 253 decimal.	Checksum[~]
0	:	0	C	0000	0	0	aa bb cc dd ee ff xx yy zz bb cc dd	cs0
1	:	0	8	000C	0	0	cc aa cc dd ee ff xx yy	cs1
2	:	0	F	0014	0	0	dd bb cc dd ee ff xx yy zz bb cc dd aa xx	cs2
3	:	0	1	0022	0	0	0A	cs3
4	:	0	4	0023	0	0	dd bb cc dd	cs4
5	:	1	0.	0027	0	0	dd bb cc dd ee ff xx yy zz bb cc dd aa ff 01 c0	cs5

[+] Line number is not in the record.
[&] Number of data bytes for storing specified in this line, C_d = 12 decimal. 0C means C_d = 12.
[*] Starting address of 2 bytes, 0000 means the next 12 bytes store between address 0x0000 to 0x000B. Therefore in the next line starting address Addr = 0x000C.
[#] 0 and 0 - means the availability of data record in this line is for the ROM. A byte from the data sequentially burns at the ROM.
[$] Bytes for burning in ROM are in this line and numbering = N_d. Each character in this column represents a nibble.
[~] cs0, cs1, ... are the checksums of 1 byte each of all the bits in line number 0, 1, ..., respectively.

G.2.4 Programming Method of the Device Programmer

A 512 kB device cell array (at the address defined by A_0 to A_{18} signals) stores the '0's as per 0s at D_0 to D_7 when a strobe pulse of a few microseconds duration is applied in the presence of a high voltage of 12V by the device programmer circuit. A device programmer that programs its memory unit performs the following eight steps in a sequence using the software tool, computer and the device programmer circuit. (*i*) Applies the A_0 to A_{18} bits as needed at a selected address input of the array of cells. (*ii*) Applies as inputs, the D_0 to D_7 bits that are meant for that address. (*iii*) Applies a high voltage to make programming feasible for the needed duration in microseconds. (*iv*) Applies a programming pulse for a sufficient duration to cause fusing of the desired links in the array, to convert a '1' to '0'. (*v*) Switches off the high voltage. (*vi*) Applies the next higher address than the previous one. (*vii*) Repeats the above steps (*ii*) to (*iv*) for writing (converting) the logic states of the D_0 to D_7 bits at the current instance at the new address, and (*viii*) Continues until a cell array at the last desired address is programmed.

The process of writing into the device is by converting logic 1 to logic 0 and it is done by fusing the *links*, applying high voltage and programming a pulse for a short duration.

Figure G.2.1 shows the method for burning-in the EPROM or in the EEPROM the S records or hex records generated by a locator. An EPROM part of the processing device (for example, the microcontroller) is first erased in UV light. Erasing makes all of the 8 bits at each of the addresses into '1's. An erase-facility provides reusability of the memory whenever the application software changes for another version of the system. The software executed in the computer programs the EPROM as well as *verifies* the bytes burned into the EPROM with the help of an interfacing circuit between the

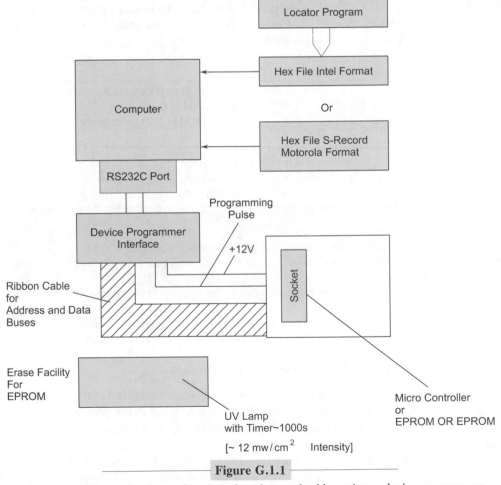

Figure G.1.1

Burning-in of the application software codes, data and tables using a device programmer.

EPROM and the computer's RS232C serial port. The EPROM interface circuit receives a byte serially from the computer through the TxD line and later sends, along with its address, this byte for burn in the processing device's EPROM. Burn-in of codes is done as follows. During the period when the appropriate address and data is available from this circuit, it also switches ON a high Voltage ~V_pVolt (Programming Voltage, V_p= +12.75V in some devices or 19V in 68HC11) and applies a program pulse for a needed period. This circuit is sequentially programmed at each address by increasing the address after every program pulse.

An EPROM interface circuit also receives another byte serially from the computer through a TxD line and sends this byte again for burning-in the processing device at the appropriate address. The bytes at the successive addresses are received by the computer in a verify mode through the interface and RxD line. Some processing devices have an auto-program mode for its EPROM in which it can automatically copy the codes and data from an IC.

The working of a device programmer is according to the processing and memory device. An explanation for the working of a device programmer for programming the internal EEPROM of 68HC11 is as follows. The 68HC11 has a control register CONFIG. It is for system configuration control. It keeps the data bits like an internal *EEPROM address*. It is also called an EEPROM register whenever the programming unit is used within 68HC11. There is another register, the EEPROM register, at the address 0x003B. This register is kept to program both the CONFIG and the *EEPROM addresses* in 68HC11 [the on-chip *EEPROM addresses* in 68HC11 are from 0xB600 to 0B7FF]. An *EEPROM* burning needs 19V in 68HC11. If CONFIG.0 bit (EPROM) is '0', then the erase or write (burn-in) does not become feasible by any instruction. To burn-in, first, the 0th bit of CONFIG (EEPROM) register is made '1'. Only then is the EEPROM programming voltage ON. If the 1st bit is also made '1' then the *EEPROM addresses* and their data are latched with the help of programming units within 68HC11. If the register EEPROM.3 and .4 bits become 00 (0 and 0, respectively), a bulk erase takes place at the EEPROM addresses [bulk erase refers to all the *EEPROM addresses*]. If 01 is written a byte is erased [byte erase means erase at one *EEPROM address* only]. If 10, then a row of 16 bytes is erased. If the 2nd bit in EEPROM is made '1' only then is the erase function enabled. An erase means all bits at an EEPROM address are made '1's. The erase time is a **total** of 10 ms in all these three modes. When the erase function is disabled (due to 2nd bit = '0' but programming voltage becoming enabled because 0th bit = '1'), the burn-in of bytes takes place by the execution of the write instruction for the appropriate address.

A computer's RS232C UART port that sends and receives at 1200 baud connects to RxD and TxD pins in 68HC11 through a line receiver and a line driver, respectively. VDD, VRH, IRQ, XIRQ pins are at +5V. VSS and VRL are at '0'. The reset circuit and 8 MHz crystal circuit connect as usual. Once 68HC11 is configured in the bootstrap mode, the bits at its *EEPROM addresses* are programmable using the software in the computer that is a part of the development system. An external computer transfers at 1200 baud in bootstrap mode when using an 8-MHz crystal with 68HC11. Further, the transfer is first of a byte with all '1's (0xFF). The computer then transfers 256 bytes to 68HC11. These bytes load between 0x0000 to 0x00FF internal RAM addresses. 68HC11 automatically sets its program counter at the end of transmission after reading its vector from 0xFFFE. It, therefore, starts executing a program called a bootstrap program. This can be a test program that includes the write to CONFIG or EEPROM register and to *EEPROM addresses*.

G.3   RAM DEVICES

RAMs are made from cells consisting of MOSFETs. There is one cell each for a bit in the memory. It is, however, volatile and looses data immediately after the power-off. Two forms of RAM are SRAM (Static RAM) and DRAM (Dynamic RAM). A battery-backed RAM is called NVRAM (nonvolatile RAM). A RAM can be loaded with a very large number of bytes for a program and data. A RAM can be written an infinite number of times. How does a write to an EEPROM differ from that to an NVRAM? A write to an NVRAM is like that to a RAM. The difference is that the NVRAM is nonvolatile. A write to an EEPROM is done by writing the byte 0xFF first (for a byte erase) and then writing the needed byte. Further, an EEPROM writes only when it receives a high voltage.

G.3.1   Static and Dynamic RAMs

There are a total of 4 MOSFETs per cell in a static RAM. The static means that once a processor cycle writes a bit in a cell, it remains unchanged until it is modified in the processor cycle or until the power switches off. An advantage of the SRAM is that a write to it is static as long as the power is on. Static memories do not need any refreshing circuits. It has two disadvantages. First, it has about four times less memory density per chip for a given level of MOSFET channel length. This is obviously due to 4 MOSFETs per cell. Second, the speed of operation in a CMOS pair is less when compared to an n-MOSFET of the same channel-length-based circuit.

A cell can also be made by an n-channel MOSFET gate as in the dynamic RAM (DRAM). When a channel does not conduct in logic state '1', it has the gate and channel capacitance through which currents leak. There is no additional circuit to maintain a constant current. Each cell of a DRAM should be read and written again [a process called *refresh*] within ~ 4 ms or less, to retain a bit in any of its cells. This provides two advantages. One, the n-MOS operations are quicker than the CMOS pair. The other advantage is that there are four times fewer MOSFETs than in the SRAM. Memory density is greatly enhanced. There has been some apprehension regarding the refresh process, which can be dismissed by the following explanation. A DRAM refresh controller IC refreshes each cell within a programmed time, performs a parity check, and issues a RAM parity-error interrupt to the processor in case of a DRAM byte check failure. This chip reads and writes repeatedly in a DRAM without hindering the other activities of the processor bus. Consider the exemplary RAM IC MCM32257A-20. It is a 256 kB Static RAM. There are 18 address input signals A_0 to A_{17}. The 51100-12 dynamic RAM has 10 lower and 10 higher bits multiplexed [D_0 to D_9 multiplexes with D_{10} to D_{19}] and each cell must be refreshed within 4 ms. Certain bits in the data bus are for accessing the parity check bits. Using these bits the dynamic refresh controller checks for loss of bits.

As an embedded system finds use mostly in application specific circuits, its memory need is not massive like that of a computer. Also, SRAM speeds are fast when the HCMOS below ~0.3 μm channel length cells are used. Lately, 0.07 μm/0.09 μm technology DRAMs have been designed and made available.

G.3.2   EDO RAM

Can there be a *zero-wait state* RAM? A zero-wait state RAM is essential in high-speed processors. Zero-wait state means that between a demand for the bits by the processor and the placements of the bits on the bus, there is no wait state. [A clock cycle has one or more states. At the states on the rising and falling edges, during high and low, there is some bus activity. A state defines a minimum period between two sequential actions]. There is a DRAM version called EDO RAM (Extended Data Out RAM). While the processor reads the first bit(s) for the fetching into the cache, the DRAM internally works out the next bit(s) so that just after the first bit(s) is transferred to the cache, the second is immediately available. However, even the EDO RAM does not act fast when the processor speed increases above 100 MHz.

G.3.3 SDRAM

There is an improved version of the EDO RAM called SDRAM (Synchronous DRAM). Each cell is organised in rows and columns. Instead of working on the next bit when the first bit is being read, it works on the next row while the previous rows and columns are being read. Now the next row becomes readily available. The speed multiplies by the number of columns in a cell organisation. The SDRAM does not act fast when the processor speed increases to 1 GHz plus.

G.3.4 RDRAM

RDRAM stands for Rambus DRAM. Rambus is the name of the developer company. Currently, the SDRAM is a low-cost option and therefore a chip set may still work with it even when the processor works at 1 GHz plus. This is because it does have internal caches providing instructions and data in advance. A chip set may also offer two configurable options, SDRAM and RDRAM. [Its bus technology is for using a 16-bit bus in place of 32 or 64 bits to improve the bus bandwidth]. A wider bus of 32 or 64 bits reduces the bus bandwidth, which is according to the rate that bits are accessed from a SDRAM. The improvement in bus bandwidth in RDRAM (in spite of narrower 16-bit bus) occurs due to compensation provided by the four successive fetches in the case of Rambus compared to a single fetch in the case of SDRAM.

G.3.5 Parameterised Distributed RAM

The slices of RAM in a system may be distributed and allotted statically or dynamically into the different units or subunits. A distributed RAM in a typical system can hold 8 Kbits to 256 Kbits. There can be 1 K to 16 K slices in a typical system. The size, rows and columns in a distributed RAM can be scaled and configured. For example, for use as a video or display sub-system RAM (buffer RAM to hold the pixels for screen display), a distributed RAM can be allotted statically and shared between the display processor and system processor. A row at the RAM can be scaled to the number of pixels in the row. A distributed RAM for use as an I/O buffer can be shared between the I/O processor and system processor, and it can be dynamically allotted to get bit or byte streams of varying sizes. Suitable scaling is done to distribute the RAM to various processes and I/O devices. Also, a part may be located close to the system, I/O, or other logic units in case high-speed buses are used.

G.3.6 Parameterised Block RAM

A block of the memory may associate with a specific hardware block of the embedded system. [An example of a block is a memory block for a 16 x 32 multiplier unit for MAC operations. MAC is a term used in DSP operations for multiply and accumulate]. There can be 4 to 168 blocks in typical systems. A block can hold 4 kB to 32 kB in typical systems. The block size may also be configurable in certain systems.

G.4 PARALLEL PORTS IN MICROCONTROLLERS

Many ports can be made available as the memory addresses present in memory-mapped I/O processors. A few parallel ports may however exist on-chip internally with the processor. Table G.4.1 gives the features of processors with on-chip parallel ports in three microcontroller families.

Table G.4.1

Processor with On-Chip Parallel Ports in Microcontrollers

Features	Intel 8051 and Intel 8751	Motorola M68HC11E2	Intel 80196
Memory-Mapping for External I/O Ports supported	Yes	Yes	Yes
Internal I/O Ports in single chip Mode[@]	P0, P1, P2, and P3	PA, PB, PC, PD, and PE	P0, P1, P2, P3, and P4
On-chip Ports in Expanded Mode[@]	None	None	P1
Port Driving Capability	open-drain (quasi-bi-directional in expanded mode)	1 mA input, 10 mA Output	P0 input port, max 1 mA, P1-P2 quasi-bi-directional, P3-P4 open-drain (quasi-bi-directional in expanded mode)
Handshaked Ports	None	PC with two separate handshaking signals, STRA and STRB	None

[@] Expanded mode means that the processor uses the external port pins as buses. The buses are to the external memories and for the external interrupts, serial I/Os, timer events and multi-channel analog inputs.

G.5 SERIAL COMMUNICATION DEVICES

A system (or a pair of systems) may posses a master-slave serial device. It is extremely suitable for inter-processor and intra-processor communication in a system. The master processor, device, or system synchronously or asynchronously controls the output to several different processors, devices or systems, called slaves. The master can choose to send output to an addressed slave if a slave has a distinct address. The *master* can choose to receive input from any slave it selects at an instant. The *slave* processor, device or system, controls and receives input from the master processor, device or system. This *slave* is the one having a distinct address and is chosen by the master. Consider two cases, one of synchronous master-slave communication and another of asynchronous master-slave communication.

G.5.1 SPI and SCI in Motorola 68HC11

The Motorola 68HC11 has separate hardware devices for synchronous and asynchronous communication, SPI (Serial Peripheral Interface) and SCI (Serial Communication Interface), respectively.

(*a*) The SPI gives full duplex synchronous communication. Its serial-in and -out are at a programmable rate for the clock bits. Serial-out of the data bits is down to the interval of 0.5ms for an 8 MHz crystal at 68HC11. The instances of the occurrence of -ve edges and +ve edges within an interval of a bit of the serial-out and -in data are also programmable in a SPI. The open-drain or totem pole output from a master to a slave and the device selection as master or slave by the hardware and software are programmable as well. If a slave select pin connects to '1', the SPI is featured as master, and if to '0', as slave.

(*b*) The SCI gives a full duplex UART asynchronous communication. Baud rates are the same (not individually programmable) and communication is in a full duplex mode. A baud rate can be selected from 32 possibilities by three rate bits and two prescalar bits. The receiver *wake up* feature is programmable. It is enabled if RWU (1st bit of SCC2) is set, and is disabled if RWU is reset. If RWU is set, the receiver of a slave is not interrupted by the succeeding frames. T8 and R8 provide the inter-processor (Master-Slave) UART communication in the 11-bit format.

Remember that the master and slave in a system need not be with the same processor family. Also, a slave can transmit to a master by executing the appropriate driver software.

G.5.2 Serial Communication Devices in Microcontrollers

Table G.5.1 gives the features of processors with on-chip serial devices in select microcontrollers.

Table G.5.1
Processor with On-chip Serial Ports (Devices) in Microcontrollers

Features	Intel 8051 and Intel 8751	Motorola M68HC11E2	Intel 80196
Synchronous Serial Port (Half or Full Duplex)	Half	Full	Half
Asynchronous UART Port (Half or Full Duplex)	Full	Full	Full
Programmability for 10 as well as 11 bits per byte	Yes	Yes	Yes
Separate un-multiplexed Port Pins for Synchronous and UART Serial ports	No	Yes (Separate 4 Pins)	No
Synchronous Serial Port Master and Slave defined by a Software or its Hardware	Software	Hardware and Software	Software
UART Serial Port Master and Slave Definition by Software programming for P bit	Yes	Yes	Yes

(Contd.)

Features	Intel 8051 and Intel 8751	Motorola M68HC11E2	Intel 80196
Synchronous Serial Port Registers	SCON, SBUF and TL-TH 0-1	SPCR, SPSR and SPDR	SPCON SPSTAT BAUD_RATE and SBUF
UART Serial Port Registers	SCON, SBUF and TL-TH 0-1 (Timer 2 in 8052)	BAUD, SCC1, SCC2, SCSR, SCIRDR and SCITDR	SPCON SPSTAT BAUD_RATE and SBUF
Uses Internal Timer or Uses Separate program- mable BAUD rate generator	Timer	Separate	Separate as well as the Timer

Note: Intel 80960 and PowerPC 604 do not possess internal serial ports.

The 8051 family processor in a system has an on-chip common hardware device, USART (Universal Synchronous and Asynchronous Receiver and Transmitter) called SI (Serial Interface). Its features are as follows. The SCON is a special function register for serial control as well as serial status of the SI, and sets the mode of communication. The SFR SBUF is a serial buffer which is either a *read* or a *write* in an instruction in both synchronous and UART communications. There are no programmable instances of occurrence of -ve edges and +ve edges within an interval of a bit of the serial-out data. The serial receiver in the 8051 SI is double buffered. There is an intermediate register that holds the received data bits of the last frame while the SBUF is still to be read. When the SBUF is *read* the intermediate register transfers the values to the SBUF. An overrun error occurs only when the SBUF is not *read* until the intermediate register gets a new value of the last frame. The SI operates as one of the following.

(*i*) A half duplex synchronous mode of operation, called mode 0. When a 12 MHz crystal is at 8051, attached to the processor, the clock bits are at the intervals of 1 μs.

(*ii*) A full duplex asynchronous serial communication, called mode 1 or 2 or 3. In mode 1 and 3, using a timer, the baud rate is variable according to the programmed timer bits. In mode 2, using an SMOD bit at an SFR called PCON, the baud rate is programmable for two rates only. It is 1/64 or 1/32 of the oscillator frequency at 8051. T8 and R8 provide the inter-processor (Master-Slave) communication in the 11-bit format.

Intel 80196 has an on-chip common hardware device called SI. It is like an USART. Its features are as follows. Programmable rates for loading bytes at the BAUD_RATE are registered twice. There are no programmable instances of occurrence of -ve edges and +ve edges within an interval of bits of the serial-out data. The serial receiver in the 80196 SI also double buffers. This SI operates as one of the following.

1. A half duplex synchronous serial communication. Bit rate is according to timer T2 clock or baud rate register. Minimum duration of a bit is 1.33 ms. [The register should never program for all bits 0s].

2. A full duplex UART mode of the USART circuit (an asynchronous serial communication. interface). Baud rate is programmable per timer T2 clock or BAUD_RATE register. [The register should never program for all bits 0s.] There is a PEN bit (the 3rd bit of SPCON). This

lets the P bit of a 11-bit format in UART be usable as an odd parity check bit, or even parity bit, or T8 and R8 for inter-processor communication.

G.6 TIMERS IN MICROCONTROLLERS

Two timers are the main timers, timer0 and timer1, in 8051 families. These count down in the timer mode. The 8052 family has an additional timer, timer2. A watchdog timer, timer3 is also available in late versions of 8051. The registers for programming these are the 8-bit special function registers (SFRs) between the addresses 0x80 to 0xFF.

The 68HC11 has the following timers:

1. **Main Timer TCNT**

   Timers for the various timer functions in Motorola 68HC11 has the addresses, names and bits of the control, compare, capture, flags and interrupt mask registers.

   (a) The TCNT is a 16-bit timer in an FRC mode. Its value at any instance is in the *read only* registers TCNT (H) and TCNT (LO). All other functions, multiple out compares, multiple input captures, and real-time clock ticks take place using the TCNT.

   (b) The FRC count-input pre-scales using two bits, PR1-PR0. If by default, resetting on a *reset* of the microcontroller or by a user resetting, these bits are 00, respectively, the prescaling factor, p, is 1. For an 8 MHz crystal, $T = 0.5\mu s$. If the PR1- PR0 bits are 01, 10 and 11, then the $p = 4$ or 8 or 16, respectively. Input to the FRC is thus as per PR1-PR0 bits, and is after successive time intervals, $p.T = dT = 0.5\mu s$, $2\mu s$, $4\mu s$ or $8\mu s$. PR1 and PR0 are the lowest two bits (1st and 0th) of the timer interrupt mask2 register, TMSK2. It must be written within 64 clock cycles after a processor *reset*.

   (c) The FRC *count*, x, overflows. When PR1-PR0 is 00, 01, 10 or 11, the overflow occurs after each $(65,536)*0.5 \mu s$, $(65,536)*2 \mu s$, $(65,536)*4 \mu s$, and $(65,536)*8 \mu s$, respectively. On overflow, a flag (bit) TOF is set. A set TOF indicates an overflow, so it must be set before the next overflow by the interrupt service routine codes. The codes are executed if the primary level I bit is reset.

   (d) An input capture register captures a value of x at an instant. A capture occurs in one of the 16-bit input capture registers, TIC1, TIC2, or TIC3 in 68HC11. There are three input pins to define a capture instance. The capture happens when an edge occurs at one of the 3 PORTA bits, PA2, PA1 and PA0 that are the inputs from the external sources, devices or circuits. Whether a +ve edge input at one of the 3 PA bits or a –ve edge causes the capture of x in a TIC register, is decided as per control bits in the register, timer control register2, TCTL2. It is a *write*-only register. The 6 bits of it, TCTL.5, TCTL.4, TCTL.3, TCTL.2, TCTL.1 and TCTL.0, are the edge1B, edge1A, edge2B, edge2A, edge3B, edge3A, respectively, corresponding to the inputs at the 3 PA bits. If these control bits, edgeA and edgeB, are 00, an input capture is disabled; if 01, the capture is enabled for a +ve edge at the input; if 10, it is enabled for a -ve edge; and if 11, it is enabled for any +ve as well as -ve edge. A user software can capture an instance of edge occurrence by reading the input capture register, adding a 16 bit's value in it and storing it in an output compare register to trigger an event or execution of a service routine after a predetermined delay.

   (e) The x, at each of its increments, compares with 5 output-compare registers of 16 bits each. These 5 registers are TOC1, TOC2, TOC3, TOC4, and TOC5, respectively. Each compare register is equivalent to an OCR. Each byte of the TOC registers has a separate address.

2. Real Time Clock Timer for Interrupts at Regular Intervals

There is a register called Pulse Accumulator Control Register, PACTL and two lowest significance bits, RT1-RT0 (1st and 0th). PACTL is *write* only. If the RT1-RT0 pair is 00, an interrupt can occur after 2^{13} pulses of the E clock. If the E clock pulses are of 2 MHz and T is 0.5 μs, the interrupt from a real-time clock occurs after every 4.096 ms. If the RT1-RT0 pair is 01, an interrupt can occur after 2^{14} pulses of the E clock, that is, after 8.192 ms. If the RT1-RT0 pair is 10, the interrupt can occur after 2^{15} pulses of the E clock, that is, after 16.384 ms. If the RT1-RT0 pair is 11, an interrupt can occur after 2^{16} pulses of the E clock, that is, after 32.768 ms. The real-time clock is a free running counter. RT1-RT0 bits control its prescaling factor.

The interrupts from a real-time clock are disabled or enabled by the **I** bit in the CC register of the microcontroller. The interrupts from real-time clocks are also locally masked by the 6th bit, RTI in timer interrupt mask register2, TMASK2. This bit is set to unmask and reset to mask the real-time interrupt locally. If **RTI** and **I** bits permit the interrupt request for real-time, the microcontroller fetches the lower and higher bytes of the interrupt servicing routine address from the addresses 0xFFF0 (higher byte) and 0xFFF1 (lower byte). This is the vector address for real-time interrupts. The interrupt service routine must clear ('0') the RTIF, which is an interrupt flag for the real-time interrupts. The RTIF is a bit in the timer interrupt flag register 2, TFLG2. The TFLAG2 is at the address 0025H. It is set by an interrupt from the real-time clock interrupt. It must be cleared in order to enable a future interrupt before returning from the corresponding service routine and before a next real-time clock interrupt occurs.

3. Pulse Accumulator Counter (PACT)

The timer is essentially a counter, which is necessarily given inputs at regular intervals. A counter may, however, be given inputs at variable intervals. The PACTL is a pulse accumulator control 8-bit register. It is a write-only register for the software. The use of PACNT helps in counting the pulses reaching the PAI input from an external circuit. A servicing routine instruction *reads* the current time shown by the FRC. Let PACNT load a value *q1*. If it is then compared with the overflow time, the time taken by the (256 - *q1*) input pulses at PAI input. An accurate method is the use of an input capture-register in conjunction.

4. Watchdog Timer

There are two registers, CONFIG and COPRST. They are for programming the watchdog timer interrupts. CONFIG is a system configuration control register in which there is a bit, NOCOP. It configures by writing at the address 0x003F. NOCOP is the 2nd bit of CONFIG. If this bit is '0' the COP facility is enabled. [COP means computer (68HC11) operating properly]. COP facility provides for keeping a watch on the user program execution time. When the user program takes a longer time in routine than planned or expected by the user software the user provides for storing at desired intervals; firstly, the 0x55 and then the 0xAA at the computer-reset control register COPRST. By keeping a watch it is meant that as soon as the watchdog timer overflows (time outs), the program counter is reset to the 16 bits as per lower byte and higher byte from the addresses 0xFFFA and 0xFFB, respectively. If these 16 bits are the same as the bits in 0xFFFE and 0xFFFF, then the microcontroller executes instructions as when it resets on power up. The 0th and 1st bit of option register, OPTION, at the address 0.0039 are the CR1 and CR0 bits. IF NOCOP resets ('0') and CR1-CR0 =0-0, watchdog timer time out occurs after every 2^{16} pulses. As $T = 0.5 \mu$s for the processor E clock output at 2 MHz, the

WDT time-out will occur every 16.384 ms ($2^{16} * 0.5\mu s$) unless the user software stores at desired intervals before a time out, first the 0x55 and then the 0xAA at the computer reset control register COPRST. [After 2^{15} pulses if CR1-CR0 =0-1, 2^{14} pulses for 1-0, 2^{13} pulses for 1-1].

G.7 INTERRUPT SOURCES AND THEIR CONTROLS IN VARIOUS FAMILIES OF PROCESSORS

Table G.7.1 summarizes the internal maskable interrupts for the internal devices in various families of processors. It also gives their priorities and groupings. Table G.7.2 summarizes the interrupt controller system features in the various families of processors used in embedded systems. It also gives the priorities and groupings from the internal non-maskable sources, external maskable and non-maskable sources.

Table G.7.1
Internal Device Interrupts

Processor	Maskable Interrupting Internal Devices
8051/52 and 8751	{TI, RI} (0), TF1 (1), TF0 (3), SI (4), TF2 (5), [WDTI (7) in certain versions.]
M68HC11E2	SCI {TI, TCI, OR, RDR, ILI, FEI, NI} (1), SPI (2), PAI (3), PAOV (4), TOF (5), OC5I (6) to OC1I (10), IC3I (11) to IC1I (13), RTI (14)
Intel 80196KC	T1OVF (0), ADConv. over (1), HSI_Received data (2), HSO_Data-Outs (3), HSI.0I (4), {SWT0 to SWT3} (5), SI_INT (6), TI (8), RI (9), HSI_FIFO 4th entry (10), T2CAP (11), T2OVF (12), HSI_FIFO full (14), WDTI (16)

p_{hw} is the number in parentheses. Same source group is in curly brackets. In 68HC11 a source can be assigned highest priority during first 64 clock cycles after reset.

Table G.7.2
Internal Interrupt Specific Features

Capability	8051/52	Motorola M68HC11E2	Intel 80196KC	Intel 80960CA	80x86
From an ISR during Diversion to another higher Priority Maskable Interrupt Source	Yes	No	Yes	Yes	Yes
A specific maskable source declarable on initialisation as non-maskable	No	Yes	No	No	No
All Interrupt-Sources of Vectored Priorities	Yes	Yes	Yes	Yes	Yes
Primary Level Masking Bit for all nonmaskable interrupt-sources.	Yes (IE.7)	Yes (CCR.4)	Yes (PSW.9)	Yes	Yes (Flag IF)
Number of Interrupt Sources in the Basic Version of Family	10	26	28	248	242

(Contd.)

Capability	8051/52	Motorola M68HC11E2	Intel 80196KC	Intel 80960CA	80x86
Number of Vector Addresses for the Source-Groups in basic Version	8	21	18	248 (Int 8 to 255)	242
Software overriding of priority level p_{sw} by psw for a source	0 - 1	None	None	$1^{\#}$ - 31	None
Other than the Program Counter Context Saving	None	Yes, CPU registers	Context Switching *	Context Switching	EFLAGS
Use of DMA Channel for Devices Interrupt Servicing	No	No	Yes using PTS	four Channels	No
Non-Maskable External Interrupt Pins	None	XIRQ (NM*) (16)	NMI (15)	NMI (31)	NMI $n_{type} = 2^{\%}$
Maskable External Interrupt Pins	INT0 (6), INT1 (2) {EXF2 in 8052} (5)	{IRQ, STRA} INTR	EXINT1 (15)	EXINT (7) (13)	XINT0- XINT7@
Non-maskable Internal Interrupts	None	CME (19), NOCOP (18), Illegal Op-code (17), SWI instruction (0)	Trap Instruction, Unimplemented Opcode	-	$n_{type} = 0$, 2, 4, 6 and 8 &

Note: See meanings and explanations of symbols in the text. In 80x86, $p_{hw} = 256 - n_{type}$.

In both tables, p_{hw} is a number in brackets with a higher value, meaning higher priority for vectoring to the corresponding ISR_VECTADDR and fetching ISR. A source group is within curly brackets. In table G.7.2, the meanings of the symbols are as follows. For 68HC11, NM* means declarable as non-maskable only in the initial 64 clock cycles after reset. For 80196KC, PTS means there is a Peripheral Transactions Server providing a DMA like feature. The sign @ means 80960 has three pins with dedicated vectors. Five pins are configurable for encoded interrupt inputs. The $^{\#}$ for 80960 means p_{sw} will be 0 and the interrupt priority not defined. Assignments are from 1 to 31. For 80x86, & means division by zero and debug interrupts after each instruction if the TF flag is set and over-flow, array bound check, illegal opcode, and device not available, and double fault with n_{type} from 0, 1, 4 to 8, respectively. The n_{type} is an 8 bit interrupt vector type (level). $^{\%}$ means NMI external pin internally generating and executing the two-byte instruction *Int* 2. There is an instruction, type3 that emulates *Int* 3 instruction. Context Switching * means that the switching of windows for the registers causes fast context reference to the new ISR. [Full context switches, the stack frame pointer, program counter and local register set]. Context Switching ** means reference to local register and stack frame reallocation and de-allocation for a fast context reference to the new ISR. Interrupt latency is reduced by the reallocation of sets and fast context switching.

G.8 INTERRUPTS IN THE 80x86 PROCESSORS

The IBM PC uses the 80x86 family processors. On a *call* for the ISR, the EFLAGS (Extended 32- Flag bits Register) and 32-bit address are to be pushed on to a stack. [Refer to Section 2.2].

There is a type assigned to all interrupts, whether hardware or software. Let n_{type} denote an 8-bit interrupt vector number for an ISR_VECTADDR. The higher the n_{type} the lower the p_{hw}. ISR_VECTADDR = 0x00000004 multiplied by n_{type}. There are 242 ISR_VECTADDRs from 244 sources that are available for programming.

(a) There is a two-byte instruction, Int n_{type}. Therefore, each hardware interrupt source has a software instruction that emulates the instruction.

(b) There is a one-byte instruction *type3* that emulates the *Int* 3 instruction.

(c) There is a one-byte instruction *INTO* that emulates the *Int* 4 two-byte instruction and overflow exception.

(d) When an NMI external pin interrupt occurs, it emulates the two-byte instruction *Int* 2.

(e) The n = 255 down to 32 are the software interrupts by a two-byte instruction, Int n. n_{type} = 31 down to 18, 15 and 9 are not available.

(f) The n_{type} is 0, 1, 4 to 8, respectively, for the interrupts, division by zero, if the TF flag is set debug interrupts after each instruction, over-flow, array bound check, illegal opcode, device not available, and double fault.

(g) The n_{type} is 10 for the exception generated for an invalid call using Jump, Call, IRET or INT instruction.

(h) The n_{type} is 11, 12, 13, 14, 16 and 17 for the interrupts, unavailable segment access, stack fault, protected memory access, page fault, floating-point error and unaligned memory access, respectively. The software instructions also emulate these thirteen.

(i) The n_{type} is 16, 24 and 28 for the interrupts of timers, keyboard interrupts and real-time clock ticks in the IBM PC ROM BIOS (Basic Input Output System), 'if unavailable segment access', 'stack fault', 'protected memory access', 'page fault', 'floating-point error' and 'unaligned memory access', respectively. The software instructions also emulate these thirteen.

G.9   INTERRUPTS IN 68HC11

G.9.1   Servicing of Interrupts

When an interrupt occurs, (*i*) 68HC11 first completes the current instruction. (*ii*) If *I* bit is not '0', it does not do not recognize any maskable interrupt, and continues with the existing set of instructions. Otherwise, the processor (a) pushes the CPU registers on to the memory stack and (b) internally masks *I* bit '0' temporarily until there is *return* from an ISR. This disables any further interrupt from any low or high priority interrupt. [There is one special case]. (*iii*) Fetches the corresponding ISR_VECTADDR among the 21 vector addresses. The processor takes the steps at the end of each instruction of the program. The 27 permissible sources can interrupt the program.

G.9.2   Sources of Interrupts

In 68HC11, there are 27 sources with 21 vector addresses from which the processor fetches an ISR_VECTADDR when the service is provided to a source. The interrupt controller system pre-assigns the 20 priority levels, for example, from p_{hw} = 19 down to 0 with p_{hw} = 0 the lowest and p_{hw} = 19 the highest priority. The 27 sources have following characteristics. The nonmaskable Interrupts are from two sources and each has a vector address. These sources are illegal opcode trap (highest but 2) and SWI (lowest priority). They have no flags.

In 68HC11, there is a special kind of non-maskable interrupt, from a pin XIRQ, with a vector address. To make it nonmaskable, it is enabled once after initialisation within 64 clock cycles. The X bit at the CCR.6 should be made '0' to use this interrupt source later (after the *reset* after power up or otherwise) as a nonmaskable interrupt source. There is no flag associated with this interrupt source.

■ LIST OF KEYWORDS AND THEIR DEFINITIONS ■

- *Asynchronous Serial Communication*: Data bytes or frames do not maintain uniform phase differences in a serial communication.
- *Synchronous Communication*: Data bytes or frames do maintain uniform phase differences in a serial communication.
- *Baud Rate*: Rate at which serial bits are rceived at the line during an UART communication.
- *Bulk Erase*: Bits becoming '1' throughout a memory device.
- *Device Programming*: Programming of bits by burning-in a memory, microcontroller, PLA, PAL, CPLD or any other device.
- *Device Programmer*: A system or unit for programming a device by burning-in.
- *Burning-in*: A process in which bits are modified from '1's to the '0s' in a device.
- *Distributed RAM*: Slices of RAM in a system may be distributed and allotted statically or dynamically into the different units or subunits.
- *DRAM*: Dynamic RAM, which has to be refreshed continuously by a device called *DRAM refresh controller*. Once programmed, it auto reads and writes the same set of bits repeatedly by scanning throughout the DRAM.
- *Erase Time*: Time taken for device erasing.
- *Flash:* A memory in which a or set of sectors erase simultaneously.
- *Master*: A processor, device or system which synchronously or asynchronous controls the output to several different processors, devices, or systems called slaves. The master can choose to send output to an addressed slave if a slave has a distinct address. The *master* can choose to receive an input from any slave selected by it at an instant.
- *Slave*: A processor or device or system, which receives the input from the master processor or device or system. This *slave* is the one having a distinct address and is chosen by the master.
- *On-chip Parallel Port*: The port on a chip which receives or sends 8 or 16 bits at an instnace.
- *On-chip Serial Port*: The port on a chip which receives or sends a bit serially at an instnace with a definite rate in kbps [Baud rate in UART].
- *Pulse Accumulator Counter*: A counter that counts the input pulses during a select interval. When used as a timer with a repeatedly loadable value, it functions as a pulse width modulator.

Appendix H

Important Topics in Embedded Systems Architecture, Programming and Design

Given below are the important topics to be learnt in the architecture, programming and design of Embedded Systems. They are given here to enable professors and course designers to frame an appropriate syllabus.

H.1 SUGGESTED SYLLABUS UNITS

Level I

Unit 1: **INTRODUCTION TO EMBEDDED SYSTEMS**: Definition and classification, Overview of Processors and hardware units in an embedded system, Software embedded into the system, Exemplary embedded systems, Embedded systems on a chip (SoC) and the use of VLSI designed circuits

Unit 2: **PROCESSOR AND MEMORY ORGANISATION**: Structural units in a processor, Memory devices, Processor and memory selection, Memory map and allocation, Memory blocks for different data structures, DMA, interfacing of Memory, Processor and I/O devices and use of glue logic.

Unit 3: **DEVICES AND BUSES FOR DEVICES NETWORK**: I/O Devices, Device I/O Types and Examples, Synchronous, Iso-synchronous and Asynchronous Communications from Serial Devices, Examples of Internal Serial-Communication Devices, UART and HDLC, Parallel Port Devices, Sophisticated Interfacing Features in Devices/Ports, Timer and Counting Devices, 'I^2', 'USB', 'CAN' and advanced I/O Serial high speed buses, ISA, PCI, PCI-X, cPCI and advanced buses.

Unit 4: **DEVICE DRIVERS AND INTERRUPTS SERVICING MECHANISMS**: Device drivers, Device servicing by interrupts and interrupt service routines, Linux Internals as Device Drivers and Network Functions, Writing Physical Device Driving ISRs in a System, Concept of Virtual Devices, Exemplary device driver programs, Interrupt service (handling) routine, Hardware and Software related Interrupt sources, Software Error Related Hardware interrupts, Software Instruction Related Interrupts Sources, Context Switching, and deadline-latency Priorities.

Unit 5: **PROGRAMMING CONCEPTS AND EMBEDDED PROGRAMMING IN C AND C++**: Programming in assembly language (ALP) vs. High-level language, C Program Elements, Macros and functions, Use of Pointers, NULL Pointers, Use of Function Calls, Multiple function calls in a Cyclic Order in the Main, Function Pointers, Function Queues and Interrupt Service Routines Queue Pointers, Data types, Use of queues for implementing the protocol for a Network, Queuing of Functions on Interrupts, Use of the FIPO (First-In Provisionally-Out) Queues for Flow Control on a Network, Data Structures, Arrays, stacks, queues, trees, List and ordered List, Uses of a List of Active Device Drivers (Software Timers), Uses of a List of Tasks in a Ready List, Concepts of Embedded Programming in C++, Objected Oriented Programming, Embedded Programming in C++, 'C' Program compilers, Cross compiler and Optimisation of memory Codes.

Unit 6: **PROGRAM MODELING CONCEPT IN SINGLE AND MULTIPROCESSOR SYSTEM SOFTWARE DEVELOPMENT PROCESS**: Use of Data Flow graphs for Program Analysis, Use of Control Data Flow graphs for Program Analysis, Finite States Machine Model, Petri Net Model, FSM as a Special Case of Petri Net, Use of Petri Table for Real Time Programming, *MODELING OF MULTIPROCESSOR SYSTEMS:* Issues in Multiprocessor Systems, The Synchronous Data Flow Graph (SDFG) Model, The Homogeneous Synchronous Data Flow Graph (SDFG) Model, The Acrylic Precedence Expansion Graph (APEG) Model, Timed Petri Nets and Extended Predicate/Transition Net Models, Multi Thread Graph (MTG) System Model, Applications of the graphs and Petri Nets to Multiprocessor Systems.

Level II

Unit 1: **SOFTWARE ENGINEERING PRACTICES IN EMBEDDED SOFTWARE DEVELOPMENT PROCESS**: Software algorithm Complexity, Development Process Life Cycle (Waterfall) model, Uses of the Linear Sequential Model (Waterfall Model or Lifecycle Model) for Software Development Process, RAD (Rapid Development Phase) Model, Incremental Model, Concurrent Model, Uses of the Component-based (Object-Oriented) Software Development Process Model, Use of the Fourth Generation Tools Based Software Development Process Model, Use of the Object-Oriented-Based and Fourth Generation Tools–Based Approach, Software Analysis, design and implementation phases, Testing, Verifying and Validating, Debugging, Real-time Program Development issues, Software Project Management, Project Metrics, Software maintenance and Concept of UML.

Unit 2: **INTER PROCESS COMMUNICATION AND SYNCHRONISATION**: Definitions of process, tasks and threads, Clear-cut distinction between Functions, ISRs and Tasks by their Characteristics, Shared data problem, Use of Semaphore(s), Priority Inversion Problem and Deadlock Situations, Inter-Process Communications using Signals, Semaphore Flag or mutex as Resource Key, Message Queues, Mailboxes, Pipes, Virtual (Logical) Sockets, and Remote Procedure Calls (RPCs).

Unit 3: **OPERATING SYSTEMS**: Operating System Services, Goals, Structures, Kernel, Process Management, Memory management, Device Management, File System Organisation and Implementation, I/O Subsystems, Network OSs, Mobile OSs, RTOSs, Schedule Management for Multiple Tasks by an RTOS, Interrupt Routines Handling in RTOS.

Unit 4: **REAL TIME OPERATING SYSTEMS**: RTOS Task scheduling Models, Handling of task scheduling and latency and deadlines as Performance Metrics, Cooperative Round Robin Scheduling, Cyclic Scheduling with Time Slicing (Rate Monotonic Cooperative Scheduling), Preemptive Scheduling Model strategy by a Scheduler, Critical Section Service by a Preemptive Scheduler, Fixed (Static) Real Time Scheduling of Tasks, Precedence Assignment, Performance Metrics for real-time program, Sporadic Task Model, IEEE Standard POSIX 1003.1b functions, Basic Actions at a preemptive Scheduler, A strategy for synchronisation of multiple processes, Embedded Linux Internals for Device Drivers and embedded Systems, Security issues in an OS.

Unit 5: **REAL TIME OPERATING SYSTEM PROGRAMMING TOOLS AND EXEMPLARY APPLICATIONS**: Need for a well-tested and debugged RTOS, exemplary Applications, Study of MicroC/OS-II or VxWorks or Any other popular RTOS, RTOS System Level Functions, Task Service Functions, Time Delay Functions, Memory Allocation Related Functions, Semaphore Related Functions, Mailbox Related Functions, Queue Related Functions, Case Studies of Programming with RTOS, Understanding Case Definition, Multiple Tasks and Their Functions, Creating a List of Tasks, Functions and IPCs, Exemplary Coding Steps.

Unit 6: **HARDWARE-SOFTWARE CO-DESIGN AND INTEGRATION IN EMBEDDED SYSTEMS**: Embedded System Project management, Embedded System Design and Co-design issues during the development process, Embedded System Development Process Goal, Action Plan, Complete Specifications and System Requirements, Implementation Tools, Testing, Design Cycle in the development Phase of an Embedded System, Target Systems, Emulators, ICE, Use of Device Programmer for Downloading the finalised codes into ROM, Uses of Code generation tools (Assembler, Compiler, loader and linker), Simulator, Exemplary Prototype Development, Testing and Debugger Tools for Embedded Systems, Integrated Development Environment (IDE), Memory and Processor Sensitive Programs and Device Drivers, Use of Dynamically Linked Libraries (DLLs) Issues in System design, Choosing the Right Platform, Embedded System Processor(s) Choice, Factors and Needed Features to be taken into Consideration, Software–Hardware Tradeoff, Performance Modeling, Porting Issues of the OS in an Embedded Platform, Uses of Scopes, logic analyzers, Probes, LED Tests, Oscilloscope, Logic Analyzer, Bit Rate Meter, System Monitor Codes for Debugging in ROM.

H.2   LIST OF TOPICS THAT COVERS THE CDAC EMBEDDED SYSTEM COURSE SYLLABUS

The CDAC (Centre for Development of Advanced Computing) has designed a popular course on Embedded Systems. A list of the topics included in the CDAC syllabus modules is given below. The chapters, which may be referred to, are given in brackets.

Programming Concept Module: Programming Concepts, Introduction to Data Structures, Complexity of an algorithm, Arrays, stacks, queues, trees, Review of C-Programming [Chapter 5].

Object-Oriented Programming Concepts Module: Embedded programming in C / C++, Embedded Systems programming [Chapter 5].

Software Engineering Approach Module: Software development phases, Software lifecycle models, Design, Implementation and testing, Software Project Management, Software maintenance, Unified modeling language (UML), Real-Time Programming Issues [Chapter 7].

Real-time Operating Systems Module: OS services, goals and structures, Process management, Memory management, File System Organisation and Implementation, I/O subsystem, Network operating systems, Real Time/Embedded operating systems, Real-Time task models and Performance metrics, Scheduling and Resource Management, Inter-process communication, Interrupt routines in RTOS environment, OS security Issues, Embedded LINUX Internals and Mobile OS [Chapters 8 and 9].

Embedded-Systems programming Module: Programming using the RTOS [Chapters 10 and 11].

Embedded-System Design Module: Project Management in Embedded Systems, Embedded System Design Issues, Co-design issues, Integrated development environment (IDE), Code generation tools (Assembler, Compiler, loader, linker), Emulators, Simulators and Debuggers, Memory and processor sensitive programs, Device drivers, DLLs, Choosing the right platform, Performance modeling, Porting issues of OS in Embedded platform [Chapter 12].

The 8/16/32-bit Microcontrollers and Interfacing Module: Microcontroller architecture overview, 8-bit Microcontroller and their architecture [Appendix C].

Devices Module: Timers, UART, SPI, PWM, WDT, ADC and FLASH [Appendix G].

Emerging Bus Standards (Compact PCI, PCI-X, USB 2.0, etc.) Module: [Chapter 3; Appendix F].

Digital Signal Processing Module: Architecture of Digital Signal Processors, DSP processor vs. Conventional processor, Fixed-point arithmetic vs. floating-point arithmetic, DSP for embedded systems [Appendix D].

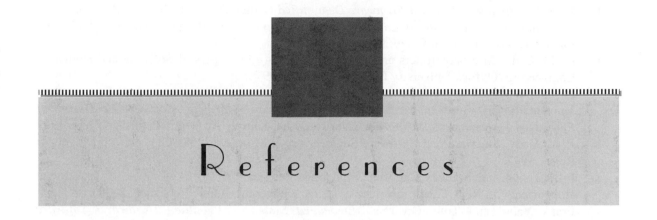

References

Following are the references of the printed books, websites and journal papers for further in-depth study of the topics related to the Embedded Systems, which has been covered in the previous Chapters.

A. Printed Books References

1. Al Williams, *Embedded Internet Design*, McGraw Hill, July 2002
2. Alessandro Rubini, *Linux Device Drivers*, O'Reilly, USA, June 1999
3. Arnold S. Berger, *Embedded Systems Design- An Introduction to Processes, Tools and Techniques*, CMP Books, Nov. 2001
4. B. Demuth and D. Eisenreich, *Designing Embedded Internet Devices*, Butterworth Heinemann, July 2002
5. Barry Kauler, Flow Design for Embedded Systems – A simple unified Object Oriented Methodology, CMP Books, Feb. 1999
6. Bob Zeidman, Designing *with FPGAs and CPLDs*, CMP Books, Sept. 2002
7. Bruce Powel Douglass, *Real-Time UML – Developing Efficient Objects for the Embedded Systems*, Addison Wesley Object Technology Series, 1998
8. Craig Hollabaugh, *Embedded Linux Hardware and Software*, Addison Wesley March 2002
9. D. Lewis, Fundamentals of Embedded Software: Where C and Assembly meet, Prentice Hall, Feb. 2002
10. Daniel Tabak, *Advanced Microprocessors*, McGraw-Hill, USA 1995.
11. David E. Simon, *An Embedded Software Primer*, Addison Wesley Longman, Inc., USA, (Pearson Education Asia) Singapore, USA 1999 (India Reprint 2000)
12. Dreamtech Software Team, *Programming for Embedded Systems- Cracking the Code*, Hungry Minds, April 2002

13. Ed Sutter, *Embedded System Firmware Demystified* (with CD), CMP Books, Feb. 2002

14. Eric Giguere, *Java 2 Micro Edition- The ultimate Guide to Programming Handheld and Embedded Devices*, John Wiley, USA, Canada 2000.

15. F. .M. Cady, Microcontrollers and Microcomputers — Principles of Software and Hardware Engineering, Oxford University Press, New York, 1997

16. F. Balarin, M. Cliodo, A. Jurecska, H. Hsieh, A. L. Lavagno, C. Paasserone, A. E. Sangiovanni-Vincentelli, E. Sentovich, K. Suzuki, and B. Tabbara, *Hardware-Software Co-Design of Embedded Systems: A Polis Approach* Norwell, MA, Kluwer Academic Publishers, June 1997

17. F. M. Cady, Software and Hardware Engineering — Motorola M68HC11, Oxford University Press, 1997.

18. Filip Thoen and Francky Catthoor, Modeling, Verification and Exploration of Task-Level Concurrency in Real-Time Embedded Systems, Kluwer Academic Publishers 2000

19. Frank Vahid and Tony Givargis, *Embedded System - A unified Hardware/ Software Introduction*, John Wiley and Sons, Inc. 2002

20. Franz J. Rammig (Ed.), *Distributed and Parallel Embedded Systems*, Kluwer Academic Publishers, Netherlands, 1999

21. Fred Halsall, *Data Communication, Computer Networks and Open Systems*, 4th Edition, Pearson Education, 1996 (Fourth Indian Reprint, 2001)

22. G. D. Greenfield, *The 68HC11 Microcontroller,* Saunders College Publishing, 1991.

23. G. F. Franklin, J.D. Powell and A. Emami-Naeini, *Feedback Control of Dynamic Systems*, 3rd Ed., Addison Wesley, Reading, MA, USA, 1994

24. Gajski, Daniel D., Frank Vahid, Sanjiv Narayan and Jie Gong, *Specification and Design of Embedded Systems*, Englewood Cliffs, NJ, Prentice Hall, 1994

25. Gary Nutt, Operating *Systems - A Modern Perspective*, Addison Wesley Longman, Inc., USA, 2000 (Pearson Education Asia Singapore, India Reprint 2000).

26. George Pajari, *Unix Device Drivers*, Pearson Education, Indian Reprint, 2002

27. J. B. Peatman, *Design with Microcontrollers and Microcomputers*, McGraw-Hill, 1988.

28. J. W. Stewart, *The 8051 Microcontroller — Hardware, Software and Interfacing*, Prentice Hall, 1993

29. Jack G. Ganssle, *Art of Programming Embedded Systems* Academic USA, 1992.

30. Jack G. Ganssle, *Art of Programming Embedded Systems*, Butter-worth Heinemann, Newton, Mass., USA, 1999.

31. Jack W. Crenshaw, *Math Toolkit for Real-Time Programming*, CMP Books, Aug. 2000.

32. Jane W.S. Liu, *Real Time Systems*, Pearson Education, 2000 (First Indian Reprint 2001)

33. Jean J Labrosse, Embedded Systems Building Blocks, 2nd Edition, CMP Books, Dec. 1999

34. Jean J. Labrosse, *MicroC/OS-II The Real Time Kernel*, R&D Books, an Imprint of Miller Freeman, Inc. Lawrence, KS 66046, USA, 1999. (Also 2nd Edition in 2002 from CMP Books)

35. Jeremy Bentham, *TCP/IP Lean Web Servers for Embedded Systems*, CMP Books, USA 2000. (Also 2nd Edition, 2002)

36. Jim Ledin, *Simulation Engineering- Build Better Embedded Systems faster,* CMP Books, Aug. 2001.

37. John A. Stankovic, Marco Spuri, Krithi Ramamritham and Giorgio C Buttazzo, *Deadline Scheduling for Real-Time Systems – EDF and Related Algorithms*, Kluwer Academic Publishers, Netherlands, Oct. 1998

38. John Forrest Brown, *Embedded System Programming in C and Assembly*, Van Nostrand, Reinhold, New York, USA, 1996.

39. John Hyde, *USB Design by Example*, John Wiley & Sons, Inc., New York, 1999

40. John Uffenbeck, *The 80x86 Family*, 3rd Ed., Pearson Education India, 2002

41. Joseph L. Weber, *Using Java^{TM} 2 Platform*, Que Corporation, Reprint by Prentice Hall of India, New Delhi, May 2000

42. Joseph Lemieux, *Programming in the OSEK/VDX Environment*, CMP Books, Oct. 2001

43. K. J. Hintz and Daniel Tabak, *Microcontrollers — Architecture, Implementation and Programming*, McGraw-Hill, 1992.

44. Kirk Zurell, *C Programming for Embedded Systems*, CMP Books, Feb. 2002

45. Luis Miguel Silveira, Srinivas Devadas, Ricardo A. Reis, *VLSI: Systems on a Chip*, Kluwer Academic Publishers, Dec. 1999

46. M. Ali Mazidi and J.G. Mazidi, *The 8051 Microcontroller and Embedded Systems*, Pearson Education, 2000, First Indian Reprint, 2002

47. M. Tim Jones, *TCP/IP Applications Layer Protocols for Embedded Systems*, Charles River Media, June 2002

48. M.C. Calcutt, F.J.Cowan, and G.H.Parchizadeh, *8051 Microcontrollers — Hardware, Software and Applications*, Arnold (and also by John Wiley), 1998.

49. M.Costanzo, *Programmable Logic Controllers — The Industrial Computers*, Arnold (and also John Wiley) 1997.

50. Macii, Benini and Poncino, *Modern Design Technologies for Low Energy Embedded Systems*, Kluwer Academic Publishers, March 2002

51. Michael Barr, *Programming Embedded Systems in C and C++*, O'Reilly, USA Aug. 1999 Reprinted Shroff Pubs. India Reprint August, 1999.

52. Michael J. Pont, *Embedded C*, Addison Wesley, April 2002

53. Miro Samek, *Practical StateCharts in C/C++ - Quantum Programming for Embedded Systems*, CMP Books, July, 2002

54. Myke Predko, *Programming and Customizing the 8051 Microcontroller*, McGraw-Hill, 1999, Third Reprint, Tata McGraw-Hill, 2002

55. Niall Murphy, *Front Panel – Designing Software for Embedded User Interface*, CMP Books, June 1998

56. Peter Marwedel, and Gerl Gossens, *Code Generation for Embedded Processors*, Kluwer Academic Publishers, June, 1995

57. Peter Spasov, *Microcontroller Technology- The 68HC11*, 2nd Edition, Prentice Hall, Englewood Cliffs, NJ, 1996

58. Phillip A. Laplante, *Real-Time Systems Design and Analysis* – An Engineer's Handbook, 2nd Edition, IEE Press, USA, 1997 (Prentice Hall of India, Third Indian Reprint, April, 2002)

59. Rainer Laeupers, *Code Optimization Techniques for Embedded Processors: Methods, Algorithms and Tools*, Kluwer Academic Publishers, Oct. 2000.

60. Raj Kamal, *Internet and Web Technologies*, Tata McGraw-Hill, 2002

61. Raj Kamal, *The Concepts and Features of Microcontrollers (68HC11, 8051 and 8096) -Includes Programmable Logic Controllers*, S. Chand & Co. (Originally Wheeler Pubs.), New Delhi, 2000.

62. Randall S. Janka, *Specification and Design Methodology for Real-Time Embedded Systems*, CMP Books, Nov. 2001.

63. Rick Grehan, Robert Moote and Ingo Cyliax, Real-Time Programming – A guide to 32-bit Embedded Development, Addison Wesley, 1998

64. Rogers S. Pressman, Software Engineering, 20th Edition, McGraw-Hill, 2001

65. S. L. Pfleeger, *Software Engineering Theory and Practices*, Pearson Education, USA Singapore, India Reprint 2001.

66. Scott Rixner, *Stream Processor Architecture* Kluwer Academic Publishers, Nov. 2001

67. Silberschatz and P.B.Galvin, *Operating Systems*, Addison Wesley, Reading, MA, USA, 1996

68. Sommerville, *Software Engineering*, Addison Wesley, Reading, MA, USA, 2000

69. Steve B. Farber, *ARM System-on-Chip Architecture,* 2nd Edition, Addison Wesley & Benjamin Cummings, 2002

70. Steve Heath, *Embedded System Design: Real World Design*, Butter-worth Heinemann, Newton, Mass. USA, May 2002

71. Steve White, Digital Signal Processing, Thomson Learning – Delmar, 2000 (First Indian Reprint, Vikas Publishing House, 2002)

72. Stuart R. Ball, *Debugging Embedded Microprocessor Systems*, Butter-worth Heinemann, Newton, Mass. USA, 1998

73. Stuart R. Ball, *Embedded Microprocessor Systems: Real World Design*, Butter-worth Heinemann, Newton, Mass. USA, 1996. (2nd Edition, May 2002)

74. Sundrajan Sriram, and Survra S. Bhattacharya, *Embedded Multiprocessors- Scheduling and Synchronization*, Marcel Dekker, Inc., NewYork, USA 2000.

75. Thomas D Burd and Robert W Brodersen, *Energy Efficient Microprocessor Design* Kluwer Academic Publishers, Oct. 2001

76. Tim Wilmshurst, *An Introduction to the Design of Small Scale Embedded Systems - with examples from PIC, 8051, and 68HC05/08 Microcontrollers*, Palgrave, Great Britain, 2001.

77. Todd D. Morton, *Embedded Microcontrollers*, Prentice Hall, New Jersey USA 2001.

78. Walter J. Grantham and Thomas L. Vincent, *Modern Control Systems- Analysis and Design*, John Wiley, 1993

79. Wayne Wolf, *Computers as Components - Principles of Embedded Computing System Design*, Academic Press (A Harcourt Science and Technology Company), USA, 2001.

80. Wayne Wolf, *Modern VLSI: System on Chip Design* Pearson, Jan. 2002

81. William A. Shay, *Understanding Data Communications and Networks*, 2nd Edition, Thomson Learning – Brooks/Cole, 1999 (First Indian Reprint, Vikas Publishing House, 2001)

B. Website References

1. http://www-3.ibm.com/chips/techlib/techlib.nsf/techdocs/ 852569B20050FF77852569930058 A78D [For PowerPC 750 Microprocessor]

2. http://www.4i2i.com/h 263 philips trimedia.htm [For Philips Trimedia Processor]

3. http://www.arm.com [For ARM Processors]

4. http://www.ami-c.org

5. http://www.cs.ucr.edu/esd [For Computer Sciences Embedded System Design website of University of California, Riverside]

6. http://www.dspvillage.ti.com [For Texas Instruments DSP Processor]

7. http://www.e-insite.net/edmag/ [For subscribing to a popular embedded system magazine]

8. http://www.EETAsia.com [For subscribing to a popular embedded system magazine]

9. *http://www.ee.surrey.ac.uk/Personal/R.Young/java/html/cruise.html* {For Section 11.3]

10. http://www.eembc.org [For benchmarking of performances of embedded Systems]

11. http://www.eet.com/embedsub [For subscribing to a popular embedded system magazine]

12. http:// www.embedded-computing.com/eletter [For subscribing to a popular embedded computing system magazine]

13. http://www.eg3.com(For articles and tutorials]

14. http://www.goembedded.com [A popular embedded system new site]

15. http://www.home.hkstar.com/~alanchan/papers/smartCardSecurity/ [For Section 11.4]

16. http://www.i2Chip.com [For an ASIP chip]

17. http://www.instantweb.com/~foldoc/contents.html

18. http:// www.java.sun.com/ products/ javacard [For Section 11.4]

19. http://www.linuxdoc.org [For Section 9.11]

20. http://www.mentorg.com/seamless

21. http://www.misra.org.uk [For Section 11.3]

22. http://www.osek-vdx.org [For Section 11.3]

23. http://www.research.ibm.com/securesystems/scard.htm [For Section 11.4]

24. http://www.semiconductors.philips.com/trimedia/ [For Section E.1]

25. http:// www.sguthery@tiac.net

26. http://www.ti.com/sc/docs/asic/modules/arm7.htm and arm9.htm [For Section D.4]

27. http://www.ti.com/sc/docs/psheets/abstract/apps/spra638a.htm [For Section D.4]

28. http://www.vsi.org

29. http://www.webopedia.com/TERM/N/operating system.htm [For Section 9.13]

30. http://www.wrs.com [For Section 10.3]

C. Printed Journal Paper References

1. Chi-Ying Liang and Huei Peng, *"Optimal Adaptive Cruise Control with Guaranteed String Stability"* Journal Vehicle System Dynamics, 31, pp. 313–330, 1999

2. Jean-Louis Brelet, *"Exploring Hardware/ Software Co-design with Vertex-II Pro FPGAs" [Xcell Journal, pp. 24–29, Summer issue, 2002*

3. Lars-Berno, Fredriksson and Kvaser, *"CAN for Critical Embedded Automotive Networks"* IEEE Micro, 22(4), 28–35, 2002

4. Tadao Murata, *"Petri Nets, Properties, Analysis and Applications"*, Proc. IEEE, 77(4), 541–580), 1989

5. Wayne Wolf, Burak Ozer and Tiehan Lu, *"Smart Cameras as Embedded Systems"*, IEEE computer, 22(5), 48–55, 2002.

Index